International Series in Operations Research & Management Science

Volume 156

Series Editor
Frederick S. Hillier,
Stanford University, CA, USA

Special Editorial Consultant
Camille C. Price
Stephen F. Austin State University, TX, USA

For further volumes:
http://www.springer.com/series/6161

Peter Kall • János Mayer

Stochastic Linear Programming

Models, Theory, and Computation

2nd Edition

Springer

Peter Kall
Professor emeritus
University of Zürich

Böniweg 10
CH-8932 Mettmenstetten, Switzerland
kall@ior.uzh.ch

János Mayer
Institute of Operations Research
University of Zürich
Moussonstrasse 15
CH-8044 Zürich, Switzerland
mayer@ior.uzh.ch

ISSN 0884-8289
ISBN 978-1-4614-2745-2 ISBN 978-1-4419-7729-8 (eBook)
DOI 10.1007/978-1-4419-7729-8
Springer New York Dordrecht Heidelberg London

Printed on acid-free paper

Springer is part of Springer Science+Business Media (www.springer.com)

To Helene and Ilona

Preface

The beginning of stochastic programming, and in particular stochastic linear programming (SLP), dates back to the 50's and early 60's of the last century. Pioneers who—at that time—contributed to the field, either by identifying SLP problems in particular applications, or by formulating various model types and solution approaches for dealing adequately with linear programs containing random variables in their right–hand–side, their technology matrix, and/or their objective's gradient, have been among others (in alphabetical order):

E.M.L. Beale [12], proposing a quadratic programming approach to solve special simple recourse stochastic programs;
A. Charnes and W.W. Cooper [41], introducing a particular stochastic program with chance constraints;
G.B. Dantzig [49], formulating the general problem of linear programming with uncertain data and
G.B. Dantzig and A. Madansky [53], discussing at an early stage the possibility to solve particular two-stage stochastic linear programs;
G. Tintner [326], considering stochastic linear programming as an appropriate approach to model particular agricultural applications; and
C. van de Panne and W. Popp [333], considering a cattle feed problem modeled with probabilistic constraints.

In addition we should mention just a few results and methods achieved before 1963, which were not developed in connection with stochastic programming, but nevertheless turned out to play an essential role in various areas of our field. One instance is the Brunn-Minkowski inequality based on the investigations of H. Brunn [36] in 1887 and H. Minkowski [235] in 1897, which comes up in connection with convexity statements for probabilistic constraints, as mentioned e.g. in A. Prékopa [266]. Furthermore, this applies in particular to the discussion about bounds on distribution functions, based on inequalities published by G. Boole in 1854 and by C.E. Bonferroni in 1937 (for the references see A. Prékopa [266]), and on the other hand, about bounds on the expectation of a convex function of a random variable, leading to a lower bound by the inequality of J.L. Jensen [148], and to the Edmundson–Madansky upper bound due to H.P. Edmundson [83] and A. Madansky [210].

Among the concepts of solution approaches, developed until 1963 for linear or nonlinear programming problems, the following ones, in part after appropriate modifications, still serve as basic tools for dealing with SLP problems:

Besides Dantzig's simplex method and the Dantzig–Wolfe decomposition, described in detail in G.B. Dantzig [50], the dual decomposition proposed by J.F. Benders [14], cutting plane methods as introduced by J.E. Kelley [180], and feasible direction methods proposed and discussed in detail by G. Zoutendijk [355], may be recognized even within today's solution methods for various SLP problems. Of course, these methods and in particular their implementations have been revised and improved meanwhile, and in addition we know of many new solution approaches, some of which will be dealt with in this book.

The aim of this volume is to draw a bow from solution methods of (deterministic) mathematical programming, being of use in SLP as well, through theoretical properties of various SLP problems which suggest in many cases the design of particular solution approaches, to solvers, understood as implemented algorithms for the solution of the corresponding SLP problems.

Obviously we are far from giving a complete picture on the present knowledge and computational possibilities in SLP. First we had to omit the area of stochastic integer programming (SILP), since following the above concept would have implied to give first a survey on those integer programming methods used in SILP; this would go beyond the limits of this volume. However the reader may get a first flavour of SILP by having a look for instance into the articles of W.K. Klein Haneveld, L. Stougie, and M.H. van der Vlerk [189], W. Römisch and R. Schultz [288], M.H. van der Vlerk [335], and the recent survey of S. Sen [301].

And, as the second restriction, in presenting detailed descriptions we have essentially confined ourselves to those computational methods for solving SLP problems belonging to one of the following categories:

Either information on the numerical efficiency of a corresponding solver is reported in the literature based on reasonable test sets (not just three examples or less!) and the solver is publicly available;

or else, corresponding solvers have been attached to our model management system SLP-IOR, either implemented by ourselves or else provided by their authors, such that we were able to gain computational experience on the methods presented, based on running the corresponding solvers on randomly generated test batteries of SLP's with various characteristics like problem size, matrix entries density, probability distribution, range and sign of problem data, and some others.

Finally, we owe thanks to many colleagues for either providing us with their solvers to link them to SLP-IOR, or for their support in implementing their methods by ourselves. Further, we gratefully acknowledge the critical comments of Simon Siegrist at our Institute. Obviously, the remaining errors are the sole responsibility of the authors. Last but not least we are indebted to the publisher for an excellent co-operation. This applies in particular to the publisher's representative, Gary Folven, to whom we are also greatly obliged for his patience.

Zürich,
September 2004

Peter Kall
János Mayer

Comments on the 2nd edition

Since fall 2004, when we finished the 1st edition of this volume, the scope of features for the field of stochastic optimization has broadened substantially, extending the variety of model types considered and the corresponding solvers designed, as well as spreading the areas of application and the related models.

Just to mention a few of these recent activities, we list the following topics:

– *Risk measures and dominance concepts.*
There is an ongoing discussion on using various kinds of risk measures as well as stochastic dominance concepts within stochastic optimization models. Particular concepts of dealing with risk are for instance the ICC, the *Integrated Chance Constraints* (joint as well as individual ICC) due to W.K. Klein Haneveld and M.H. van der Vlerk [191]. Furthermore, CVaR, the *Conditional Value at Risk*, as analyzed e.g. in T.R. Rockafellar and S.P. Uryasev [283], is increasingly included into stochastic optimization problems, either within constraints or else (additively) in the objective. On the other side, stochastic dominance of first or higher order as discussed for instance in D. Dentcheva and A. Ruszczyński [66] receives more attention in modeling risky situations. More generally, the class of polyhedral risk measures, favourable for stochastic programs with risk measures in the objective, is analyzed in A. Eichhorn and W. Römisch [84].

— *Increasing consideration of risk in applications.*
The above mentioned risk measures and dominance concepts got a wider impact in modeling stochastic programs for real situations in various areas, thus aiming towards more realistic results for the respective problems.
As examples for using risk measures in stochastic programming models within finance we just mention the investigations of Klein Haneveld – Streutker – Van der Vlerk [190] dealing with ICC in ALM (asset liability management), A. Künzi-Bay [197] aiming for CVaR minimization in multistage ALM models for Swiss pension funds, Mansini – Ogryczak – Speranza [215] involving the CVaR in portfolio optimization, or Dentcheva and

Ruszczyński [67] considering portfolio optimization with dominance constraints.
To give just one example of risk considerations in energy problems, we mention the thesis of M. Densing [64] discussing a coherent multi-period risk measure, as generalization of the one-period CVaR, to be used for a hydroelectric power plant dispatch problem incorporated into a multi-stage stochastic program.
Another broad area of applications is concerned with comparing efficiency among finitely many similar ventures called DMU (*decision making unit*) and modelled in the frame of DEA, i.e. *Data Envelopment Analysis*. Originally stated as models combining inputs and outputs by deterministic linear constraints for any particular DMU to check whether it is efficient (*non dominated*), this setup became questionable for various applications, and the first DEA models incorporating—at least partly—chance conctraints were discussed (where obviously efficiency with respect to chance constraints had to be defined appropriately). For this setup there are many references; a recent one is e.g. Talluri – Narasimhan – Nair [323]. A more general view was taken in the thesis of S. von Bergen [339], considering to model the case of stochastic outputs via constraints on special *risk functions*, which encompass chance constraints, ICC constraints and CVaR constraints, at least. Efficiency was redefined, taking into account the risk function formulation of the model constraints, yielding conditions being verifiable by solving two-phase nonlinear programs.

— *Growing need for stochastic optimization in engineering*
Formerly the attention of engineers regarding randomness was focused mainly on analyzing reliability of systems or structures. Meanwhile, an increasing interest in stochastic optimization models and methods can be observed in various fields of engineering, like for instance in structural optimization or in robot control. The reader may get an impression of the particular approaches in this area of applications in the recent volume of K. Marti [225].

— *Tools for modelling SLP problems and links to solvers*
In principle there are two kinds of modelling tools:

- Systems, controlled by programming languages, containing declarations of data structures and model types, typically used in SLP, including the syntax to specify the particular problem features (like various single-stage versions, multi-stage recourse including the recourse type, etc.) and to manipulate probability distributions used in the current model on the one hand, and on the other hand
- menu-driven systems, for instance with pull down menus, allowing to declare model types and data structures, to edit data (arrays), to specify the random model data and edit the corresponding distributions, etc.

As to the first variant, following earlier work of H.J. Gassmann, K. Fourer, et al. on languages usable in addition to AMPL (A Mathematical Programming

Language of Fourer – Gay – Kernighan [97]), there have been recently some further attempts as e.g. the paper of Colombo – Grothey – Hogg – Wooksend – Gondzio [45] and the language StAMPL (A filtration-oriented modeling tool for multistage stochastic recourse problems) of Fourer – Lopes [99]. Concerning the second variant, we have continued to work on our system SLP-IOR, extending its features. In this context, in his thesis M.T. Bielser [19] designed a programming language SEAL (Stochastic Extensions for Algebraic Languages) which is aimed to create (as its output) the information necessary to supply to SLP-IOR as input for starting up the system.

Fortunately it is not necessary to deal in detail with all these results (and others) —although containing very interesting considerations—in this edition, since the material presented should be sufficient to follow most of the recent developments.

Following a suggestion of F.S. Hillier, we have added exercises at the end of several sections, where we thought it could be helpful. At the end of the volume the reader finds hints for dealing with them in the Chapter *Exercises: Hints for answers.*

In this context we considered it as meaningful to provide access for the reader to the features of our model management system; we therefore prepared an executable version "SLP-IOR" (student version) for open access (download), for which on the COSP web page `http://stoprog.org` (see section Software & TestSets) the corresponding link can be found. Questions or comments concerning this software are welcome to `mayer@ior.uzh.ch`.

Finally, we are greatly indebted

- to many colleagues for pointing out various mistakes and/or inaccuracies in the system SLP-IOR as well as in the 1st edition of this book,
- to Fred S. Hillier, the editor of this series of books, for his encouragement to prepare the 2nd edition of this volume,
- to Neil Levine and his colleagues for their support on behalf of the publisher,
- and not least to Silvia von Bergen, formerly assistant at our Institute, for the careful reading of parts of the manuscript and for several helpful comments.

Nevertheless, the sole responsibility for any inconsistencies lies with the authors.

Zürich, July 2010

P. K.
J. M.

Contents

Notations

One–stage models: Joint chance constraints

$A, B, C, \cdots$	:	arrays (usually given real matrices)
$a, b, c, \cdots$	:	arrays (usually given real vectors)
$x, y, z, \cdots$	:	arrays (usually real or integer variable vectors)
$(\Omega, \mathscr{F}, P)$	:	probability space
$\mathbb{N}$	:	set of natural numbers
$(\mathbb{R}^r, \mathbb{B}^r)$	:	$\mathbb{R}^r$ endowed with the Borel σ-algebra $\mathbb{B}^r$
$\xi : \Omega \to \mathbb{R}^r$	:	random vector, i.e. a Borel measurable mapping, such that $\xi^{-1}[M] \in \mathscr{F}\ \forall M \in \mathbb{B}^r$, inducing the probability measure $\mathbb{P}_\xi$ on $\mathbb{B}^r$ according to $\mathbb{P}_\xi(M) = P(\xi^{-1}[M])\ \forall M \in \mathbb{B}^r$
$T(\xi),\ h(\xi)$	:	random array and random vector, respectively, defined as:
$T(\cdot) : \mathbb{R}^r \to \mathbb{R}^{m_2 \times n_1}$	:	$T(\xi) = T + \sum_{j=1}^{r} T^j \xi_j\,;\ T,\ T^j \in \mathbb{R}^{m_2 \times n_1}$ fix
$h(\cdot) : \mathbb{R}^r \to \mathbb{R}^{m_2}$	:	$h(\xi) = h + \sum_{j=1}^{r} h^j \xi_j\,;\ h,\ h^j \in \mathbb{R}^{m_2}$ fix
$\overline{\xi}$	:	expectation $\mathbb{E}_\xi[\xi] = \int_{\mathbb{R}^r} \xi\, \mathbb{P}_\xi(d\xi) = \int_\Omega \xi(\omega)\, dP$
$\overline{T}, \overline{h}$	:	expectations $\mathbb{E}_\xi[T(\xi)] = T(\overline{\xi})$ and $\mathbb{E}_\xi[h(\xi)] = h(\overline{\xi})$, respectively
$\widehat{\xi}$	:	realization of random ξ

$\widehat{T}, \widehat{h}$	:	realizations $T(\widehat{\xi}), h(\widehat{\xi})$, respectively

One–stage models: Separate chance constraints

$t_i(\cdot)$	:	i-th row of $T(\cdot)$
$h_i(\cdot)$	:	i-th component of $h(\cdot)$

Two–stage recourse models

$W(\xi), q(\xi)$	:	random array and random vector, respectively, defined as:
$W(\cdot) : \mathbb{R}^r \to \mathbb{R}^{m_2 \times n_2}$	:	$W(\xi) = W + \sum_{j=1}^{r} W^j \xi_j$; $W, W^j \in \mathbb{R}^{m_2 \times n_2}$
$q(\cdot) : \mathbb{R}^r \to \mathbb{R}^{n_2}$	:	$q(\xi) = q + \sum_{j=1}^{r} q^j \xi_j$; $q, q^j \in \mathbb{R}^{n_2}$
$\overline{W}, \overline{q}$	:	expectations $\mathbb{E}_\xi[W(\xi)] = W(\overline{\xi})$ and $\mathbb{E}_\xi[q(\xi)] = q(\overline{\xi})$, respectively
$\Lambda, \widehat{m}_\Lambda(\xi)$	:	$\widehat{m}_\Lambda(\xi) := \prod_{k \in \Lambda} \xi_k$ for $\Lambda \subset \{1, \cdots, r\}$, $(\widehat{m}_\emptyset(\xi) \equiv 1)$
μ_Λ	:	$\mu_\Lambda := \int_\Xi \widehat{m}_\Lambda(\xi) \mathbb{P}_\xi(d\xi)$, $\forall \Lambda$, joint mixed moments

Multi–stage recourse models

$\xi : \Omega \to \mathbb{R}^R$	:	random vector $\xi = (\xi_2, \cdots, \xi_T)$ with $\xi_t : \Omega \to \mathbb{R}^{r_t}$, $t = 2, \cdots, T$ and $\sum_{t=2}^{T} r_t = R$
$\zeta_t : \Omega \to \mathbb{R}^{R_t}$	:	the **state** of the process at stage t, defined as random vector $\zeta_t = (\xi_2, \cdots, \xi_t)$, $t \geq 2$, or else $\zeta_t = (\eta_1, \cdots, \eta_{R_t})$ with $R_t = \sum_{\tau=2}^{t} r_\tau$, with the corresponding marginal distribution of ξ
$A_{t\tau}(\cdot) : \mathbb{R}^{R_t} \to \mathbb{R}^{m_t \times n_\tau}$	:	$A_{t\tau}(\zeta_t) = A_{t\tau} + \sum_{\kappa=2}^{t} \sum_{\nu=R_{\kappa-1}+1}^{R_\kappa} A_{t\tau\nu} \eta_\nu$, where $A_{t\tau}, A_{t\tau\nu} \in \mathbb{R}^{m_t \times n_\tau}$ and $R_1 = 0$, with $1 \leq \tau \leq t$ and $2 \leq t \leq T$
$b_t(\cdot) : \mathbb{R}^{R_t} \to \mathbb{R}^{m_t}$	:	$b_t(\zeta_t) = b_t + \sum_{\kappa=2}^{t} \sum_{\nu=R_{\kappa-1}+1}^{R_\kappa} b_{t\nu} \eta_\nu$, where $b_t, b_{t\nu} \in \mathbb{R}^{m_t}$ and $2 \leq t \leq T$
$c_t(\cdot) : \mathbb{R}^{R_t} \to \mathbb{R}^{n_t}$	:	$c_t(\zeta_t) = c_t + \sum_{\kappa=2}^{t} \sum_{\nu=R_{\kappa-1}+1}^{R_\kappa} c_{t\nu} \eta_\nu$,

where $c_t, c_{tv} \in \mathbb{R}^{n_t}$ and $2 \le t \le T$

Multi–stage recourse models: Discrete distribution

$\xi : \Omega \to \mathbb{R}^R$: random vector with discrete distribution $\{(\widehat{\xi}^s, q_s); s = 1, \cdots, S\}$, i.e. **scenarios** $\widehat{\xi}^s = (\widehat{\xi}_2^s, \cdots, \widehat{\xi}_T^s) = (\widehat{\eta}_1^s, \cdots, \widehat{\eta}_R^s)$ with $\mathbb{P}_\xi(\xi = \widehat{\xi}^s) = q_s, s \in \mathcal{S} := \{1, \cdots, S\}$

$\zeta_t : \Omega \to \mathbb{R}^{R_t}$: discrete set $\{\widehat{\zeta}_t^s = (\widehat{\xi}_2^s, \cdots, \widehat{\xi}_t^s); s \in \mathcal{S}\}$ of states defining $k_t \ge 1$ different equivalence classes $U_t^\nu \subseteq \mathcal{S}$, with $s_i, s_j \in U_t^\nu \Leftrightarrow \widehat{\zeta}_t^{s_i} = \widehat{\zeta}_t^{s_j}$ and an associated set of different states at stage t which may be defined by $\mathcal{S}_t := \{\rho \mid \rho \text{ minimal in one of the } U_t^\nu\}$ as $\{\widehat{\zeta}_t^\rho \mid \rho \in \mathcal{S}_t\}$ with the distribution $\mathbb{P}_\xi(\zeta_t = \widehat{\zeta}_t^\rho) = \pi_{t\rho} = \sum_{s \in \mathcal{S}} \{q_s \mid \widehat{\zeta}_t^s = \widehat{\zeta}_t^\rho\}$ (see Fig. 1 with e.g. $\mathcal{S}_2 = \{1, 6, 11\}$)

Multi–stage recourse models: The scenario tree

$(\mathcal{N}, \mathcal{A})$: tree with nodes $\mathcal{N} \subset \mathbb{N}$, where $n = 1$ is the (unique) root and $|\mathcal{N}| = \sum_{t=2}^{T} |\mathcal{S}_t| + 1$

t_n : the stage to which $n \in \mathcal{N}$ belongs; there is a bijection

$$(t_{\{\cdot\}}, \rho(\cdot)) : \{\mathcal{N} - \{1\}\} \to \bigcup_{t=2}^{T} \{(t, \mathcal{S}_t)\}$$

such that $n \leftrightarrow (t_n, \rho(n)),\ n \ge 2$; hence we assign with any node $n \ge 2$

$\widehat{\zeta}^n$: $\widehat{\zeta}^n = \widehat{\zeta}_{t_n}^{\rho(n)}$ with $\{\widehat{\zeta}_{t_n}^{\rho(n)},\ \rho(n) \in \mathcal{S}_{t_n}\}$ uniquely determined by $n \in \mathcal{N}$ (state in node n)

$\mathcal{D}(t) \subset \mathcal{N}$: set of nodes in stage t, $1 \le t \le T$

h_n : the parent node of node $n \in \mathcal{N}, n \ge 2$ (immediate predecessor)

$\mathcal{H}(n) \subset \mathcal{N}$: set of nodes in the path from $n \in \mathcal{N}$ to the root, ordered by stages, including n (history of n)

$\mathcal{S}(n) \subset \{1, \cdots, S\}$: $\mathcal{S}(n) = \{s \in \mathcal{S} \mid \widehat{\zeta}_{t_n}^s = \widehat{\zeta}^n\}$, i.e. the index set of those scenarios, for which the scenario path contains $n \in \mathcal{N}$. $\mathcal{S}(n)$ and the related set of scenarios are called the **scenario bundle** of

		the corresponding node n
p_n	:	probability of $\mathscr{S}(n)$: $p_n = \mathbb{P}_\xi(\zeta_{t_n} = \widehat{\zeta}^n) = \pi_{t_n \rho(n)}$
$\mathscr{C}(n) \subset \mathscr{N}$	:	set of children (immediate successors) of n
$\mathscr{G}_s(n) \subseteq \mathscr{N}$	:	future of node n along scenario $s \in \mathscr{S}(n)$, including n (and hence $\mathscr{G}_s(n) = \emptyset$ if $s \notin \mathscr{S}(n)$)
$\mathscr{G}(n) \subseteq \mathscr{N}$	:	$\mathscr{G}(n) = \bigcup_{s \in \mathscr{S}(n)} \mathscr{G}_s(n)$, the future of $n \in \mathscr{N}$

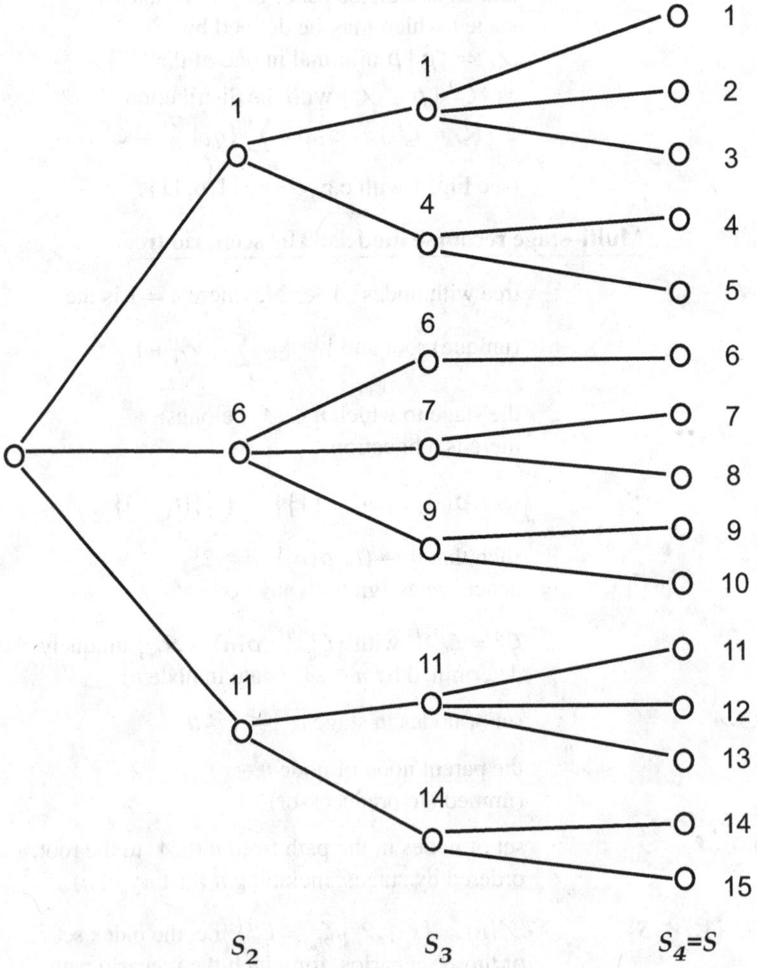

Fig. 1 Scenario tree: Assigning states to nodes.

Chapter 1
Basics

1.1 Introduction

Linear programs have been studied in many aspects during the last 60 years. They have shown to be appropriate models for a wide variety of practical problems and, at the same time, they became numerically tractable even for very large scale instances. As standard formulation of a linear program—*acronym: LP*—we find problems like

$$\left.\begin{array}{rl} & \min c^{\mathrm{T}}x \\ \text{subject to} & Ax \propto b \\ & l \leq x \leq u, \end{array}\right\} \tag{1.1}$$

with the matrix $A \in \mathbb{R}^{m\times n}$, the objective's gradient $c \in \mathbb{R}^n$, the right–hand–side $b \in \mathbb{R}^m$, and the lower and upper bounds $l \in \mathbb{R}^n$ and $u \in \mathbb{R}^n$, respectively. If some x_i is unbounded below and/or above, this corresponds to $l_i = -\infty$ and/or $u_i = \infty$. A, b, c, l, u are assumed to be known fixed data in the above model. The relation '$\propto$' is to be replaced row-wise by one of the relations '$\leq$', '$=$', or '$\geq$'. Then the task is obviously to find the—or at least one—optimal feasible solution $x \in \mathbb{R}^n$. Alternatively, we often find also the LP-formulation

$$\left.\begin{array}{r} \min c^{\mathrm{T}}x \\ \text{subject to } Ax \propto b \\ x \geq 0, \end{array}\right\} \tag{1.2}$$

under the analogous assumptions as above. For these two LP types it holds obviously that, given a problem of one type, it may be reformulated into an equivalent problem of the other type. More precisely,

— given the LP in the formulation (1.2), by introducing the lower bounds $l = (0, \cdots, 0)^{\mathrm{T}}$ and the upper bounds $u = (\infty, \cdots, \infty)^{\mathrm{T}}$ (in computations rather markers $u = (M, \cdots, M)^{\mathrm{T}}$ with a sufficiently large number M, e.g. $M = 10^{20}$, just to indicate unboundedness), the problem is trivially of the type (1.1); and

— having the LP of type (1.1), introducing variables $x^+ \in \mathbb{R}^n_+$, $x^- \in \mathbb{R}^n_+$, inserting $x = x^+ - x^-$, $x^+ \geq 0, x^- \geq 0$, introducing the *slack variables* $y \in \mathbb{R}^n_+$ and $z \in \mathbb{R}^n_+$, and restating the conditions $l \leq x \leq u$ equivalently as

$$\begin{array}{rl} x^+ - x^- - y \quad\;\; &= l \\ x^+ - x^- \qquad\;\; + z &= u \\ y \qquad\quad &\geq 0 \\ z &\geq 0, \end{array}$$

the problem is transformed into the type (1.2).

In the same way it follows that every LP may be written as

$$\left.\begin{array}{r} \min c^{\mathrm{T}} x \\ \text{subject to } Ax = b \\ x \geq 0, \end{array}\right\} \tag{1.3}$$

i.e. as a special variant of (1.2).

Numerical methods known to be efficient in solving LP's belong essentially to one of the following classes:

— Pivoting methods, in particular the simplex and/or the dual simplex method;
— interior point methods for LP's with very sparse matrices;
— decomposition, dual decomposition and regularized decomposition approaches for LP's with special block structures of their coefficient matrices A.

In real life problems the fundamental assumption for linear programming, that the problem entries—except for the variables x—be known fixed data, does often happen not to hold. It either may be the case that (some of) the entries are constructed as statistical estimates from some observed real data, i.e. from some samples, or else that we know from the model design that they are random variables (like capacities, demands, productivities or prices). The standard approach to replace these random variables by their mean values—corresponding to the choice of statistical estimates mentioned before—and afterwards to solve the resulting LP may be justified only under special conditions; in general, it can easily be demonstrated to be dramatically wrong.

Assume, for instance, as a model for a diet problem the LP

$$\left.\begin{array}{r} \min c^{\mathrm{T}} x \\ \text{s. t. } Ax \geq b \\ Tx \geq h \\ x \geq 0, \end{array}\right\} \tag{1.4}$$

where x represents the quantities of various foodstuffs, and c is the corresponding price vector. The constraints reflect chemical or physiological requirements to be satisfied by the diet. Let us assume that the elements of A and b are fixed known data, i.e. deterministic, whereas at least some of the elements of T and/or h are random with a known joint probability distribution, which is not influenced by the choice of the decision x. Further, assume that the realizations of the random variables in T and h are not known before the decision on the diet x is taken, i.e. before the consumption of the diet. Then (1.4) is basically a *stochastic linear program—acronym: SLP*—for which it is not yet clear how a "solution" should be defined.

Replacing the random T and h by their expectations $\overline{T}$ and $\overline{h}$ and solving the resulting LP

$$\left.\begin{aligned} \min c^{\mathrm{T}}x \\ \text{s. t. } Ax \geq b \\ \overline{T}x \geq \overline{h} \\ x \geq 0, \end{aligned}\right\} \tag{1.5}$$

can result in a diet $\hat{x}$ violating the constraints in (1.4) very likely and hence with a probability much higher than feasible for the diet to serve successfully its medical purpose. Therefore, the medical experts would rather require a decision on the diet which satisfies all constraints jointly with a rather high probability, as 95% say, such that the problem to solve were

$$\left.\begin{aligned} &\min c^{\mathrm{T}}x \\ \text{s. t.} \quad & \qquad Ax \geq b \\ & P(Tx \geq h) \geq 0.95 \\ & \qquad\quad x \geq 0, \end{aligned}\right\} \tag{1.6}$$

a *single-stage stochastic linear program—acronym: SSLP*—with joint probabilistic constraints. Here we had at the starting point the LP (1.4) as model for our diet problem. However, the (practical) requirement to satisfy—besides the deterministic constraints $Ax \geq b$—also the reliability constraint $P(Tx \geq h) \geq 0.95$, yields with (1.6) a nonlinear program—*acronym: NLP*. This is due to the fact, that in general the probability function $G(x) := P(Tx \geq h)$ is clearly nonlinear.

As another example, let some production problem be formulated as

$$\left.\begin{aligned} \min c^{\mathrm{T}}x \\ \text{s. t. } Ax = b \\ Tx = h \\ x \geq 0, \end{aligned}\right\} \tag{1.7}$$

where T and h may contain random variables (productivities, demands, capacities, etc.) with a joint probability distribution (independent again of the choice of x), and the decision on x has to be taken before the realization of the random variables is known. Consequently, the decision x will satisfy the constraints $Ax = b,\ x \geq 0$; but after the observation of the random variables' realization it may turn out that $Tx \neq h$, i.e. that part of the target (like satisfying the demand for some of the products, capacity constraints, etc.) is not properly met. However, it may be necessary—by a legal commitment, the strong intention to maintain goodwill, or similar reasons—to compensate for the deficiency, i.e. for $h - Tx$, after its observation. One possibility to cope with this obligation may be the introduction of recourse by defining the constraints $Wy = h - Tx,\ y \geq 0$, for instance as model of an emergency production process or simply as the measurement of the absolute values of the deficiencies (represented by $W = (I, -I)$, with I the identity matrix). Let us assume W to be deterministic, and assume the recourse costs to be given as linear by $q^{\mathrm{T}}y$, say. Obviously we want to achieve this compensation with minimal costs. Hence we have the recourse problem

$$\left.\begin{array}{l} Q(x;T,h) := \min q^{\mathrm{T}}y \\ \text{s. t. } Wy = h - Tx \\ \qquad y \geq 0. \end{array}\right\} \tag{1.8}$$

For any x, feasible to the *first stage constraints* $Ax = b,\ x \geq 0$, the recourse function, i.e. the optimal value $Q(x;T,h)$ of the *second stage problem* (1.8), depends on T and h and is therefore a random variable. In many applications, e.g. in cases where the production plan x has to be implemented periodically (daily or weekly, for instance), it may be meaningful to choose x in such a way that the average overall costs, i.e. the sum of the first stage costs $c^{\mathrm{T}}x$ and the expected recourse costs $\mathbb{E}\, Q(x;T,h)$, are minimized. Hence we have the problem

$$\left.\begin{array}{l} \min c^{\mathrm{T}}x + \mathbb{E}\, Q(x;T,h) \\ \text{s. t. } Ax = b \\ \qquad x \geq 0, \end{array}\right\} \tag{1.9}$$

a *two-stage stochastic linear program—acronym: TSLP—with fixed recourse.*

Also in this case, although our starting point was the LP (1.7), the resulting problem (1.9) will be an NLP if the random variables in T and h have a continuous-type joint distribution (i.e. a distribution defined by a density function).

If, however, the random variables in T and h have a joint discrete distribution, defined by the realizations (T^j, h^j) with the probabilities $p_j,\ j = 1, \cdots, S$ (with $p_j > 0$ and $\sum_{j=1}^{S} p_j = 1$), problem (1.9) is easily seen to be equivalent to

$$\left.\begin{array}{rl} \min c^{\mathrm{T}}x + \sum_{j=1}^{S} p_j q^{\mathrm{T}} y^j & \\ \text{s. t. } \quad Ax \qquad\qquad & = b \\ T^j x + W y^j & = h^j,\ j = 1, \cdots, S \\ x \qquad\qquad & \geq 0 \\ y^j & \geq 0, \end{array}\right\} \tag{1.10}$$

such that under the discrete distribution assumption we get an LP again, with the special data structure indicated in Fig. 1.1.

In applications we observe an increasing need to deal with a generalization of the two-stage SLP with recourse (1.9) and (1.10), respectively. At this point we just give a short description as follows: In a first stage, a decision x_1 is chosen to be feasible with respect to some deterministic first stage constraints. Later on, after the realization of a random vector ξ_2, a deficiency in some second stage constraints has

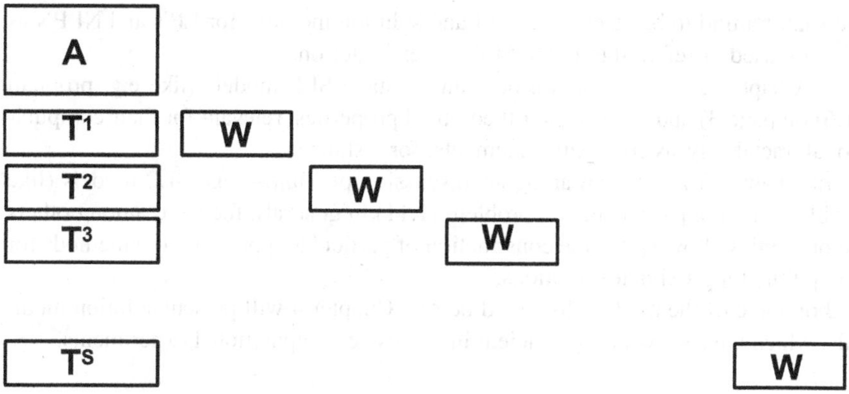

Fig. 1.1 Dual decomposition structure.

to be compensated for by an appropriate recourse decision $x_2(\xi_2)$. Then after the realization of a further random vector ξ_3, the former decisions x_1 and $x_2(\xi_2)$ may not be feasible with respect to some third stage constraints, and a further recourse decision $x_3(\xi_2,\xi_3)$ is needed, and so on, until a final stage T is reached. Again, we assume that, besides the first stage costs $c_1^{\mathrm{T}} x_1$, the recourse decisions $x_t(\zeta_t)$, $t \geq 2$, imply additional linear costs $c_t^{\mathrm{T}} x_t(\zeta_t)$, where $\zeta_t = (\xi_2,\cdots,\xi_t)$. Then the *multi-stage SLP—acronym: MSLP—with fixed recourse* is formulated as

$$\left.\begin{array}{l}
\min\left\{c_1^{\mathrm{T}} x_1 + \mathbb{E}_{\zeta_T}\left[\sum\limits_{t=2}^{T} c_t^{\mathrm{T}} x_t(\zeta_t)\right]\right\} \\
\text{subject to} \\
A_{11}x_1 \qquad\qquad\qquad\qquad\qquad\quad = b_1 \\
A_{t1}(\zeta_t)x_1 + \sum\limits_{\tau=2}^{t} A_{t\tau}(\zeta_t)x_\tau(\zeta_\tau) = b_t(\zeta_t), \text{ a.s.}, t = 2,\cdots,T \\
\qquad\qquad x_1 \geq 0,\ x_t(\zeta_t) \quad \geq 0, \qquad \text{a.s.}, t = 2,\cdots,T,
\end{array}\right\} \tag{1.11}$$

where, in general, we shall assume $A_{tt}(\zeta_t)$, $t \geq 2$, the matrices on the diagonal, to be deterministic, i.e. $A_{tt}(\zeta_t) \equiv A_{tt}$. It will turn out that, for general probability distributions, this problem—an NLP again—is much more difficult than the two-stage SLP (1.9), and methods to approximate a solution are just at their beginning phase, at best. However, under the assumption of discrete distributions of the random vectors ζ_t, problem (1.11) can also be reformulated into an equivalent LP, which in general is of (very) large scale, but again with a special data structure to be of use for solution procedures.

From this short sketch of the subject called SLP, which is by far not complete with respect to the various special problem formulations to be dealt with, we may already conclude that a basic toolkit of linear and nonlinear programming methods cannot be waived if we want to deal with the computational solution of SLP problems. To secure the availability of these resources, in the following sections of this chapter we shall remind to basic properties of and solution methods for LP's and NLP's as they are used or referred to in the SLP context, later on.

In Chapter 2, we present various Single–stage SLP models (like e.g. problem (1.6) on page 3) and discuss their theoretical properties, relevant for their computational tractability, as convexity statements, for instance.

In Chapter 3 follows an anlogous discussion of Multi–stage SLP models (like problem (1.9) in particular, and problem (1.11) in general), focussed among others on properties allowing for the construction of particular approximation methods for computing (approximate) solutions.

For some of the models discussed before, Chapter 4 will present solution methods, which have shown to be efficient in extensive computational experiments.

Exercises

1.1. Convert the following two LP's to the standard formulation used in (1.3):

(a)
$$\begin{aligned} \min\, & d^{\mathrm{T}} y \\ Ax - y &\geq b \\ x &\geq 0 \end{aligned}$$

(b)
$$\begin{aligned} \max\, & f^{\mathrm{T}} x - g^{\mathrm{T}} y \\ Ax + By &\leq d \\ Cx &= e \\ x &\leq 0 \end{aligned}$$

1.2. For the linear program—think of a production problem with the quantities x, y of two factors used to fabricate two products to meet their demand, aiming to minimize production costs $3x + 2y$—

$$\begin{aligned} \min\, & 3x + 2y \\ \text{s.t. } 2x + y &\geq 4 \\ x + y &\geq \xi \\ x, y &\geq 0, \end{aligned}$$

the demand ξ of the second product is only known to vary randomly in the interval $[2,4]$. Due to the lack of more precise information, ξ is assumed to be uniformly distributed, implying the density $\varphi(\xi) \equiv \frac{1}{2}$, $\xi \in [2,4]$, and the mean value $\bar{\xi} = \mathbb{E}[\xi] = 3$.

(a) Determine (e.g. graphically) the solution $(\hat{x}, \hat{y})$ of the above LP assuming that for the second product at least the mean demand $\bar{\xi}$ is to be met. Compute the expected supply shortage $\mathbb{E}[(\xi - \hat{x} - \hat{y})^+] = \frac{1}{2}\int_3^4 (\xi - \hat{x} - \hat{y})d\xi$.

(b) Consider instead this production problem under the condition that the random demand of the second product has to be met at least with a probability of $p = 95\%$. Determine the optimal factor combination and the minimal costs.

(c) Assume now, that the expected supply shortage of the second product is restricted to $0.05 \cdot \bar{\xi}$. Formulate the production model accordingly.

1.2 Linear Programming Prerequisites

In this section we briefly present the basic concepts in linear programming and, for various types of solution methods, the conceptual algorithms.

As mentioned on page 2 we may use the following standard formulation of an LP:

$$\left.\begin{array}{r} \min c^{\mathrm{T}}x \\ \text{s. t. } Ax = b \\ x \geq 0. \end{array}\right\} \tag{1.3}$$

With A being an $(m \times n)$-matrix, and b and c having corresponding dimensions, we know from linear algebra that the system of equations

$$Ax = b \ \text{ is solvable if and only if } \ \mathrm{rank}(A,b) = \mathrm{rank}(A).$$

Therefore, solvability of the system $Ax = b$ implies that

- either $\mathrm{rank}(A) = m$,
- or the system contains redundant equations which may be omitted, such that for the remaining system $\tilde{A}x = \tilde{b}$ we have the same set of solutions as for the original system, and that, for the $(m_1 \times n)$-matrix $\tilde{A}$, $m_1 < m$, the condition $\mathrm{rank}(\tilde{A}) = m_1$ holds.

Observing this well known fact, we henceforth assume without loss of generality, that $\mathrm{rank}(A) = m$ $(\leq n)$ for the $(m \times n)$-matrix A.

1.2.1 Algebraic concepts and properties

Solving the LP (1.3) obviously requires to find an extreme (minimal in our formulation) value of a linear function on a *feasible set* described as the intersection of a linear manifold, $\{x \mid Ax = b\}$, and finitely many halfspaces, $\{x \mid x_j \geq 0\}$, $j = 1, \cdots, n$, suggesting that this problem may be discussed in algebraic terms.

Definition 1.1. *Any feasible solution $\hat{x}$ of (1.3) is called a feasible basic solution if, for $I(\hat{x}) = \{i \mid \hat{x} > 0\}$, the set $\{A_i,\ i \in I(\hat{x})\}$ of columns in A is linearly independent.*

According to this definition, for any feasible basic solution $\hat{x}$ of (1.3) holds

$$\hat{x}_i > 0 \text{ for } i \in I(\hat{x}),\ \hat{x}_j = 0 \text{ for } j \notin I(\hat{x}), \ \text{ and } \sum_{i \in I(\hat{x})} A_i \hat{x}_i = b.$$

Furthermore, with $|I(\hat{x})|$ being the cardinality of this set (i.e. the number of its elements), if $|I(\hat{x})| < m$ such that the basic solution $\hat{x}$ contains less than m strictly positive components, then due to our rank assumption on A there is a superset $I_B(\hat{x})$

with $I_B(\hat{x}) \supset I(\hat{x})$ and $|I_B(\hat{x})| = m$ such that the column set $\{A_i,\ i \in I_B(\hat{x})\}$ is linearly independent or equivalently, that the $(m \times m)$-matrix $B = (A_i \mid i \in I_B(\hat{x}))$ is nonsingular. Introducing, with $I_B(\hat{x}) = \{i_1, \cdots, i_m\}$ and $I_N(\hat{x}) = \{1, \cdots, n\} \setminus I_B(\hat{x}) = \{j_1, \cdots, j_{n-m}\}$, the vectors $x^{\{B\}} \in \mathbb{R}^m$ —the *basic variables*— and $x^{\{N\}} \in \mathbb{R}^{n-m}$ —the *nonbasic variables*—according to

$$\begin{aligned} x_k^{\{B\}} &= x_{i_k},\ i_k \in I_B(\hat{x}) \text{ for } k = 1, \cdots, m; \\ x_l^{\{N\}} &= x_{j_l},\ j_l \in I_N(\hat{x}) \text{ for } l = 1, \cdots, n-m, \end{aligned}$$

then, with the $(m \times (n-m))$-matrix $N = (A_j \mid j \in I_N(\hat{x}))$ the system $Ax = b$ is, up to a possible rearrangement of columns and variables, equivalent to the system

$$Bx^{\{B\}} + Nx^{\{N\}} = b.$$

Therefore, up to the mentioned rearrangement of variables, the former feasible basic solution $\hat{x}$ corresponds to $(\hat{x}^{\{B\}} = B^{-1}b \geq 0,\ \hat{x}^{\{N\}} = 0)$, and the submatrix B of A is called a *feasible basis* . With the same rearrangement of the components of the vector c into the two vectors $c^{\{B\}}$ and $c^{\{N\}}$ we may rewrite problem (1.3) as

$$\begin{array}{lrcl} \min & c^{\{B\}^{\mathrm{T}}} x^{\{B\}} + c^{\{N\}^{\mathrm{T}}} x^{\{N\}} & & \\ \text{s. t.} & Bx^{\{B\}} + Nx^{\{N\}} & = & b \\ & x^{\{B\}} & \geq & 0 \\ & x^{\{N\}} & \geq & 0. \end{array}$$

Solving the system of equations for $x^{\{B\}}$ we get $x^{\{B\}} = B^{-1}b - B^{-1}Nx^{\{N\}}$ such that —with $\gamma_B := c^{\{B\}^{\mathrm{T}}} B^{-1}b$ the objective value of the feasible basic solution $(\hat{x}^{\{B\}} = B^{-1}b \geq 0,\ \hat{x}^{\{N\}} = 0)$—problem (1.3) is equivalent to

$$\left.\begin{array}{l} \min \gamma_B + \left(c^{\{N\}^{\mathrm{T}}} - c^{\{B\}^{\mathrm{T}}} B^{-1}N\right) x^{\{N\}} \\ \text{s. t. } x^{\{B\}} = B^{-1}b - B^{-1}Nx^{\{N\}} \geq 0 \\ \qquad\qquad\qquad\qquad\quad x^{\{N\}} \geq 0. \end{array}\right\} \tag{1.12}$$

For computational purposes (1.12) is usually represented by the *simplex tableau*

$$\begin{array}{|c|c|} \hline \zeta & d^{\mathrm{T}} \\ \hline \beta & D \\ \hline \end{array} = \begin{array}{|c|ccc|} \hline \zeta & \delta_1 & \cdots & \delta_{n-m} \\ \hline \beta_1 & \alpha_{11} & \cdots & \alpha_{1n-m} \\ \vdots & \vdots & & \vdots \\ \beta_m & \alpha_{m1} & \cdots & \alpha_{mn-m} \\ \hline \end{array} \tag{1.13}$$

such that the objective and the equality constraints of (1.12) are rewritten as

$$\left.\begin{array}{lcl} z & := & \zeta - d^{\mathrm{T}} x^{\{N\}} \\ x^{\{B\}} & = & \beta - D\, x^{\{N\}} \end{array}\right\} \tag{1.14}$$

with $\zeta = \gamma_B = c^{\{B\}^{\mathrm{T}}} B^{-1}b$, $\beta = (\beta_1, \cdots, \beta_m)^{\mathrm{T}} = B^{-1}b$, and furthermore

$$D = \begin{pmatrix} \alpha_{11} & \cdots & \alpha_{1n-m} \\ \vdots & & \vdots \\ a_{m1} & \cdots & a_{mn-m} \end{pmatrix} = B^{-1}N$$

and

$d^{\mathrm{T}} = (\delta_1, \cdots, \delta_{n-m}) = \left(c^{\{B\}^{\mathrm{T}}} B^{-1} N - c^{\{N\}^{\mathrm{T}}}\right) = \left(c^{\{B\}^{\mathrm{T}}} D - c^{\{N\}^{\mathrm{T}}}\right)$. Although not written down explicitly, we assume that also for the reformulation (1.13) and (1.14) the nonnegativity constraints $x^{\{B\}} \geq 0$, $x^{\{N\}} \geq 0$ have to hold.

To justify the simplex algorithm as a solution method for (1.3) the following statements are essential.

Proposition 1.1. *Provided that the LP (1.3) is feasible, i.e. that the feasible set* $\mathscr{B} := \{x \mid Ax = b,\ x \geq 0\} \neq \emptyset$, *there exists at least one feasible basic solution.*

Proposition 1.2. *If the LP (1.3) is solvable with the optimal value* $\hat{\gamma}$, *then there exists at least one feasible basis* $\hat{B}$, *yielding* $c^{\{\hat{B}\}^{\mathrm{T}}} \hat{B}^{-1} b = \hat{\gamma}$.

Definition 1.2. *Assume that* $\mathrm{rank}(A) = m$. *If for a feasible basis B and the corresponding feasible basic solution* $\hat{x}$ *with* $(\hat{x}^{\{B\}} = B^{-1}b, \hat{x}^{\{N\}} = 0)$ *it happens that* $|I(\hat{x})| < m$, *i.e. that less than m of the basic variables are strictly positive, then the basic solution* $\hat{x}$ *is called degenerate.*

Finally, if we have a feasible basis B such that $d^{\mathrm{T}} \leq 0$, than obviously this basis is optimal, i.e. $(\hat{x}^{\{B\}} = \beta,\ \hat{x}^{\{N\}} = 0)$ solves (1.3), since by (1.14) $z = \zeta - d^{\mathrm{T}} x^{\{N\}} \geq \zeta\ \ \forall x^{\{N\}} \geq 0$. On the other hand, assume that (1.3) is solvable, and that in addition all feasible basic solutions are nondegenerate. Then for an optimal feasible basis B, existing due to Prop. 1.2., $d^{\mathrm{T}} \leq 0$ has to hold due to the following argument:

If, for any feasible basis, $d_j > 0$ would hold for some $j \in \{1, \cdots, n-m\}$, due to $\beta > 0$ by the assumed nondegeneracy, we could choose $x^{\{N\}} = \tau e_j$ (e_j the j-th unit vector in $\mathbb{R}^{n-m}$) with some $\tau > 0$, such that according to (1.14) would follow

$$x^{\{B\}} = \beta - \tau D e_j = \beta - \tau D_j \geq 0 \text{ and } z = \zeta - \tau d^{\mathrm{T}} e_j = \zeta - d_j < \zeta.$$

Hence, the basis at hand would not be optimal.

Even without the nondegeneracy assumption the above optimality condition, also known as the *simplex criterion,* can be shown to hold true.

Proposition 1.3. *The LP (1.3) is solvable if and only if there exists an optimal feasible basis B such that the condition*

$$d^{\mathrm{T}} = \left(c^{\{B\}^{\mathrm{T}}} B^{-1} N - c^{\{N\}^{\mathrm{T}}}\right) = \left(c^{\{B\}^{\mathrm{T}}} D - c^{\{N\}^{\mathrm{T}}}\right) \leq 0 \qquad (1.15)$$

is satisfied.

The proof of the above statements may be found in the literature, among others in Dantzig [50], Maros [219], or Vanderbei [337].

1.2.2 Geometric interpretation

Besides the algebraic formulation of LP's, it is sometimes intuitively helpful to have in mind their geometric interpretation. To this end we need the concepts of a *convex polyhedron* and of a *convex polyhedral cone*.

Definition 1.3. *Given finitely many vectors $x^{(1)},\cdots,x^{(r)} \in \mathbb{R}^n$, then their convex hull*

$$\begin{aligned}\mathscr{P} &= \operatorname{conv}\{x^{(1)},\cdots,x^{(r)}\} \\ &:= \{x \mid x = \sum_{j=1}^{r} \lambda_j x^{(j)} \text{ with } \sum_{j=1}^{r} \lambda_j = 1,\ \lambda_j \geq 0\ \forall j\}\end{aligned}$$

is called a convex polyhedron, and their positive hull

$$\mathscr{C} = \operatorname{pos}\{x^{(1)},\cdots,x^{(r)}\} := \{y \mid y = \sum_{j=1}^{r} \mu_j x^{(j)} \text{ with } \mu_j \geq 0\ \forall j\}$$

is called a convex polyhedral cone.
Finally, $\mathscr{P}+\mathscr{C} = \{z \mid z = x+y:\ x \in \mathscr{P},\ y \in \mathscr{C}\}$ is called a convex polyhedral set.

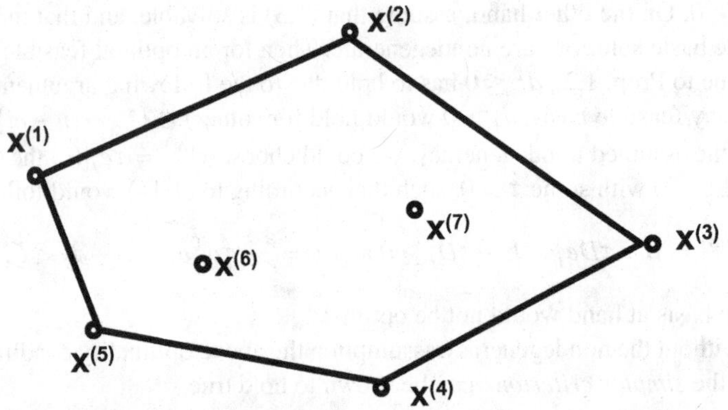

Fig. 1.2 Polyhedron $\hat{\mathscr{P}} = \operatorname{conv}\{x^{(1)},\cdots,x^{(7)}\}$.

To generate the polyhedron $\hat{\mathscr{P}}$ of Fig. 1.2, the elements $x^{(6)}$ and $x^{(7)}$ are obviously redundant, i.e. omitting these elements would result in the same polyhedron $\hat{\mathscr{P}}$, whereas no one of the elements $x^{(1)},\cdots,x^{(5)}$ can be deleted without changing the polyhedron essentially. The simple reason is that a polyhedron is uniquely determined by its vertices.

Definition 1.4. *Given a convex polyhedron $\mathscr{P}$, an element $y \in \mathscr{P}$ is a vertex if and only if there are no two further elements $v, w \in \mathscr{P}$ such that $v \neq y \neq w$ and $y = \lambda v + (1-\lambda)w$, $\lambda \in (0,1)$.*

Similarly, for a convex polyhedral cone not all of the generating elements mentioned in Def. 1.3. might be really needed to represent the cone. More precisely, whenever one of the generating elements equals a nonnegative linear combination of the other generating elements, it can be deleted without changing the cone.

With the LP (1.3) the set $\mathscr{C} = \{y \mid Ay = 0,\ y \geq 0\}$ can be associated.

Proposition 1.4. *The set $\mathscr{C} = \{y \mid Ay = 0,\ y \geq 0\}$ is a convex polyhedral cone,*

- *generated either trivially by $\{0\}$, if $\mathscr{C} = \{0\}$,*
- *or, if $\exists y \in \mathscr{C} : y \neq 0$, generated for instance by $\{y^{(1)}, \cdots, y^{(s)}\}$, the set of feasible basic solutions of the system*

$$\begin{aligned} Ay &= 0 \\ e^{\mathrm{T}}y &= 1 \quad \{\text{with } e^{\mathrm{T}} = (1,\cdots,1)\}, \\ y &\geq 0. \end{aligned}$$

With these concepts we may describe the feasible set

$$\mathscr{B} = \{x \mid Ax = b,\ x \geq 0\} \tag{1.16}$$

as follows:

Proposition 1.5. *For the feasible set $\mathscr{B} \neq \emptyset$ holds*

$$\mathscr{B} = \mathscr{P} + \mathscr{C} = \{z \mid z = x + y \text{ with } x \in \mathscr{P} \text{ and } y \in \mathscr{C}\},$$

where $\mathscr{C} = \{y \mid Ay = 0,\ y \geq 0\}$ and $\mathscr{P} = \operatorname{conv}\{x^{(1)}, \cdots, x^{(r)}\}$, with $\{x^{(1)}, \cdots, x^{(r)}\}$ being the set of feasible basic solutions of $\mathscr{B}$.

The set of feasible basic solutions of $\mathscr{B}$ can be shown to coincide with the set of vertices of $\mathscr{P}$ (and $\mathscr{B}$). The proofs of these statements may be found in the standard LP literature, or else in Kall–Wallace [172].

Definition 1.5. *For any nonempty set $\mathscr{M} \subset \mathbb{R}^n$ its polar cone is the set*

$$\mathscr{M}^P := \{z \in \mathbb{R}^n \mid z^{\mathrm{T}}x \leq 0\ \forall x \in \mathscr{M}\}.$$

An obvious consequence of this definition is

Proposition 1.6. *For any nonempty set $\mathscr{M} \subset \mathbb{R}^n$ its polar cone $\mathscr{M}^P \subset \mathbb{R}^n$ is a closed convex cone, i.e. $\mathscr{M}^P \neq \emptyset$ is a closed set such that for any two $z^{(i)} \in \mathscr{M}^P, i = 1,2$, holds $\lambda_1 z^{(1)} + \lambda_2 z^{(2)} \in \mathscr{M}^P\ \forall \lambda_i \geq 0$. In particular, for any convex polyhedral cone $\mathscr{C}$ its polar cone $\mathscr{C}^P$ is a convex polyhedral cone as well.*

Proof: Obviously, $0 \in \mathscr{M}^P$ and hence $\mathscr{M}^P \neq \emptyset$ is a convex cone. For $\{z^{(\nu)} \in \mathscr{M}^P, \nu \in \mathbb{N}\}$ converging to $\hat{z}$ we have for any arbitrary $\hat{x} \in \mathscr{M}$ that ${z^{(\nu)}}^{\mathrm{T}}\hat{x} \le 0 \;\forall \nu \in \mathbb{N}$ and hence $\hat{z}^{\mathrm{T}}\hat{x} = \lim_{\nu\to\infty} {z^{(\nu)}}^{\mathrm{T}}\hat{x} \le 0$, such that $\hat{z} \in \mathscr{M}^P$, i.e. $\mathscr{M}^P$ is closed.

If $\mathscr{C}$ is a convex polyhedral cone generated by $\{d^{(1)},\cdots,d^{(r)}\}$, with the matrix $D = (d^{(1)},\cdots,d^{(r)})$ the polar cone of $\mathscr{C}$ is given as $\mathscr{C}^P = \{z \mid D^{\mathrm{T}}z \le 0\}$ which, in analogy to Prop. 1.4., is a convex polyhedral cone. □

According to Proposition 1.5., using the set of feasible basic solutions $\{x^{(1)},\cdots,x^{(r)}\}$, i.e. the vertices of $\mathscr{P}$, and the generating set $\{y^{(1)},\cdots,y^{(s)}\}$ of $\mathscr{C}$ as described in Prop. 1.4., the LP (1.3) can now be rewritten as

$$\left.\begin{aligned} \min \sum_{i=1}^{r} \lambda_i c^{\mathrm{T}}x^{(i)} &+ \sum_{j=1}^{s} \mu_j c^{\mathrm{T}}y^{(j)} \\ \text{s. t.} \sum_{i=1}^{r} \lambda_i &= 1 \\ \lambda_i &\ge 0 \;\forall i \\ \mu_j &\ge 0 \;\forall j. \end{aligned}\right\} \tag{1.17}$$

This representation implies the following extension of Prop. 1.2..

Proposition 1.7. *Provided that $\mathscr{B} \neq \emptyset$, the LP (1.3) is solvable if and only if $c^{\mathrm{T}}y \ge 0 \;\forall y \in \mathscr{C}$, i.e. $-c \in \mathscr{C}^P$; in this case an optimal solution can be chosen as a vertex $x^{(i_0)}$ of $\mathscr{B}$ (a feasible basic solution of $\mathscr{B}$) such that $c^{\mathrm{T}}x^{(i_0)} = \min_{i\in\{1,\cdots,r\}} c^{\mathrm{T}}x^{(i)}$.*

Proof: The assumption, that $c^{\mathrm{T}}y \ge 0 \;\forall y \in \mathscr{C}$, is equivalent to the requirement that $c^{\mathrm{T}}y^{(j)} \ge 0,\; j = 1,\cdots,s$. If this condition is violated for at least one $y^{(j)}$ (e.g. for j_1), then according to (1.17) for $\mu_{j_1} \to \infty$ follows for the objective $z \to -\infty$, such that the LP is unsolvable.
If, on the other hand, the condition is satisfied, then—to solve (1.17)—we would choose $\mu_j = 0 \;\forall j$, which implies the assertion immediately. □

As a consequence we get

Proposition 1.8. *If $\mathscr{B} \neq \emptyset$, and if $c^{\mathrm{T}}x \ge \gamma \;\forall x \in \mathscr{B}$ for some $\gamma \in \mathbb{R}$, then the LP $\min\{c^{\mathrm{T}}x \mid x \in \mathscr{B}\}$ is solvable.*

Proof: For any fixed $\hat{x} \in \mathscr{B}$ and an arbitrary $y \in \mathscr{C}$ it holds true that $\hat{x} + \mu y \in \mathscr{B} \;\forall \mu > 0$, and by assumption we have $c^{\mathrm{T}}\hat{x} + \mu c^{\mathrm{T}}y \ge \gamma$, which implies that $c^{\mathrm{T}}y \ge 0$ is satisfied for each $y \in \mathscr{C}$; hence the assertion follows from Prop. 1.7.. □

1.2.3 Duality statements

To the *primal* LP in its standard formulation

$$\left.\begin{aligned} \min c^{\mathrm{T}}x \\ \text{s. t. } Ax = b \\ x \geq 0 \end{aligned}\right\} \tag{1.3}$$

another LP, called its *dual*, is assigned as

$$\left.\begin{aligned} \max b^{\mathrm{T}}u \\ \text{s. t. } A^{\mathrm{T}}u \leq c. \end{aligned}\right\} \tag{1.18}$$

The technical rules according to which the dual LP (1.18) is constructed from the primal LP (1.3) may roughly be stated as follows: To the equality constraints $Ax = b$ in (1.3) correspond the free variables $u \in \mathbb{R}^m$ in (1.18); to the nonnegative variables $x \in \mathbb{R}^n_+$ correspond the inequality contraints $A^{\mathrm{T}}u \leq c$ with the transpose of A as the matrix of coefficients; the right–hand–side b of the primal program yields the objective's gradient of the dual program, whereas the objective's gradient c of the primal LP turns into the right–hand–side of the dual LP; finally, to the minimization in (1.3) corresponds the maximization in (1.18).

Rewriting (1.18) into the standard form, we want to solve the problem

$$\begin{aligned} \gamma := \max\{b^{\mathrm{T}}u^+ - b^{\mathrm{T}}u^-\} &= -\min\{-b^{\mathrm{T}}u^+ + b^{\mathrm{T}}u^-\} \\ \text{s.t. } A^{\mathrm{T}}u^+ \; -A^{\mathrm{T}}u^- \; +v &= c \\ u^+, \quad u^-, \quad v &\geq 0. \end{aligned}$$

To this LP we assign analogously the dual LP

$$\begin{aligned} -\max c^{\mathrm{T}}z \\ \text{s. t. } \quad Az \leq -b \\ -Az \leq b \\ z \leq 0 \end{aligned}$$

which, using $x := -z$, yields

$$\begin{aligned} -\max -c^{\mathrm{T}}x = \min c^{\mathrm{T}}x \\ \text{s. t. } Ax = b \\ x \geq 0 \end{aligned}$$

coinciding with (1.3) again. Hence, the dual of the dual LP is the primal program again and we therefore can speak of a pair of dual LP's.

There are further relations between the primal and the dual LP which are less obvious. First, we have the *weak duality theorem.*

Proposition 1.9. *For any pair of feasible solutions $\tilde{x}$ and $\tilde{u}$ of (1.3) and (1.18), respectively, it holds that $b^{\mathrm{T}}\tilde{u} \leq c^{\mathrm{T}}\tilde{x}$.*

Proof: According to the assumed feasibilities $A\tilde{x} = b$, $\tilde{x} \geq 0$, and $A^{\mathrm{T}}\tilde{u} \leq c$ it follows that

$$b^{\mathrm{T}}\tilde{u} = (A\tilde{x})^{\mathrm{T}}\tilde{u} = \tilde{x}^{\mathrm{T}}(A^{\mathrm{T}}\tilde{u}) \leq \tilde{x}^{\mathrm{T}}c.$$

□

Moreover, there is the following relation between pairs of dual LP's.

Proposition 1.10. *If both of the dual LP's (1.3) and (1.18) are feasible, then both of them are solvable.*

Proof: Let $\hat{u}$ be feasible for (1.18). Then, by the weak duality theorem, $c^{\mathrm{T}}x \geq b^{\mathrm{T}}\hat{u}\ \forall x \in \mathscr{B}$. Hence Prop. 1.8. yields the solvability of (1.3). The solvability of (1.18) follows analogously. □

Finally, we have the *strong duality theorem.*

Proposition 1.11. *If the primal problem is solvable, then so is the dual problem, and the optimal values of the two problems coincide.*

Proof: According to Prop. 1.3. the LP (1.3) is solvable if and only if there exists an optimal feasible basis B such that the simplex criterion (1.15)

$$d^{\mathrm{T}} = \left(c^{\{B\}^{\mathrm{T}}} B^{-1} N - c^{\{N\}^{\mathrm{T}}}\right) = \left(c^{\{B\}^{\mathrm{T}}} D - c^{\{N\}^{\mathrm{T}}}\right) \leq 0$$

is satisfied. Since, up to a rearrangement of columns of A, we have that $(B,N) = A$, it follows that for $\hat{u} = B^{-1^{\mathrm{T}}} c^{\{B\}}$ it holds that

$$\begin{aligned} B^{\mathrm{T}}\hat{u} &= c^{\{B\}} \\ N^{\mathrm{T}}\hat{u} &\leq c^{\{N\}}. \end{aligned}$$

Hence, $\hat{u}$ is feasible for the dual program, and its (dual) objective value is

$$b^{\mathrm{T}}\hat{u} = b^{\mathrm{T}} B^{-1^{\mathrm{T}}} c^{\{B\}} = c^{\{B\}^{\mathrm{T}}} B^{-1} b,$$

thus coinciding with the primal optimal value. □

Proposition 1.12. *Both of the pair of dual LP's (1.3) and (1.18) are solvable if and only if there exist feasible solutions $x^\star$ and $u^\star$ such that the complementarity conditions*

$$x^{\star\mathrm{T}}(A^{\mathrm{T}}u^\star - c) = 0 \tag{1.19}$$

hold. Then, $x^\star$ and $u^\star$ are primal and dual optimal solutions, respectively.

Proof: If both of the LP's are solvable then there exist optimal feasible solutions $x^\star$ and $u^\star$ such that, by feasibility and strong duality,

$$0 = c^{\mathrm{T}}x^\star - b^{\mathrm{T}}u^\star = c^{\mathrm{T}}x^\star - x^{\star\mathrm{T}}A^{\mathrm{T}}u^\star = x^{\star\mathrm{T}}(c - A^{\mathrm{T}}u^\star).$$

On the other hand, from feasibility, complementarity and weak duality follows

$$0 \leq c^{\mathrm{T}}x^\star - b^{\mathrm{T}}u^\star = x^{\star\mathrm{T}}(c - A^{\mathrm{T}}u^\star) = 0$$

and hence the optimality of $x^\star$ and $u^\star$. □

The strong duality theorem implies a necessary and sufficient condition for the feasibility of a system of linear constraints, the *Farkas lemma*:

Proposition 1.13. *It holds*

$$\{x \mid Ax = b,\ x \geq 0\} \neq \emptyset \text{ if and only if } A^{\mathrm{T}}u \leq 0 \text{ implies } b^{\mathrm{T}}u \leq 0.$$

Proof: Assume that $A\hat{x} = b$ holds for some $\hat{x} \geq 0$. Then for any $\tilde{u}$ with $A^{\mathrm{T}}\tilde{u} \leq 0$ follows

$$b^{\mathrm{T}}\tilde{u} = (A\hat{x})^{\mathrm{T}}\tilde{u} = (\hat{x}^{\mathrm{T}}A^{\mathrm{T}})\tilde{u} = \hat{x}^{\mathrm{T}}(A^{\mathrm{T}}\tilde{u}) \leq 0.$$

On the other hand, assume that $A^{\mathrm{T}}u \leq 0$ always implies $b^{\mathrm{T}}u \leq 0$. For an arbitrary $\hat{u} \neq 0$ define $c := A^{\mathrm{T}}\hat{u}$. Then Prop. 1.7. implies that the LP $\max\{b^{\mathrm{T}}u \mid A^{\mathrm{T}}u \leq c\}$ is solvable. By the strong duality theorem, Prop. 1.11., its dual, $\min\{c^{\mathrm{T}}x \mid Ax = b,\ x \geq 0\}$, is then solvable as well, and hence feasible. □

Finally we mention, for later use, that the simplex criterion (1.15) is associated with a dual feasible basic solution.

Proposition 1.14. *Assume that from the LP (1.3) with some (not necessarily feasible) basis B the simplex tableau (1.13) has been derived, which satisfies the simplex criterion* $d^{\mathrm{T}} = \left(c^{\{B\}^{\mathrm{T}}}B^{-1}N - c^{\{N\}^{\mathrm{T}}}\right) \leq 0$. *Then with the primal basis B a dual feasible basis, i.e. a feasible basis of the dual program (1.18), is associated.*

Proof: Using the basis B, the matrix of the primal LP can be rewritten as $A = (B,N)$. Then the dual constraints read as

$$\left\{\begin{array}{l} B^{\mathrm{T}}u \leq c^{\{B\}} \\ N^{\mathrm{T}}u \leq c^{\{N\}} \end{array}\right\} \text{ or else } \left\{\begin{array}{llll} B^{\mathrm{T}}u & + I_m v & & = c^{\{B\}} \\ N^{\mathrm{T}}u & & + I_{n-m}w & = c^{\{N\}} \\ & v, & & w \geq 0 \end{array}\right\}$$

with unit matrices I_m and I_{n-m} of the indicated order. With $\hat{u} = {B^{\mathrm{T}}}^{-1}c^{\{B\}}$ it follows immediately, that, with $\hat{v} = 0$ and $\hat{w} = c^{\{N\}} - N^{\mathrm{T}}\hat{u} \geq 0$ due to the simplex criterion, $B^{\mathrm{T}}\hat{u} = c^{\{B\}}$ and $N^{\mathrm{T}}\hat{u} + I_{n-m}\hat{w} = c^{\{N\}}$. Hence

$$\begin{pmatrix} B^{\mathrm{T}} & 0 \\ N^{\mathrm{T}} & I_{n-m} \end{pmatrix}$$

is a dual feasible basis. □

Due to this relationship, any simplex tableau (1.13) for the primal LP (1.3), whether feasible or not, is called *dual feasible* if the simplex criterion (1.15) is satisfied.

1.2.4 The Simplex method

With this background we formulate the

Simplex Algorithm

S 1 Initialization
Find a feasible basis B for the LP (1.3).

S 2 Optimality Test
If $d^{\mathrm{T}} = \left(c^{\{B\}^{\mathrm{T}}} B^{-1}N - c^{\{N\}^{\mathrm{T}}}\right) \leq 0$, then stop ($B$ is optimal). Otherwise, continue.

S 3 Choice of Pivot Column
Choose $\rho \in \{1,\cdots,n-m\}$ such that $d_\rho > 0$, and let D_ρ be the corresponding column of D. If $D_\rho \leq 0$, then stop (LP (1.3) is unsolvable, since $x^{\{B\}} = B^{-1}b - D_\rho x_\rho^{\{N\}} \geq 0\ \forall x_\rho^{\{N\}} \geq 0$ and hence $\inf\{c^{\mathrm{T}}x \mid x \in \mathscr{B}\} = -\infty$); otherwise continue.

S 4 Choice of Pivot Row
The maximal increase $\tau \geq 0$ for $x_\rho^{\{N\}}$ such that $x^{\{B\}} = \beta - D_\rho \tau \geq 0$ remains satisfied implies choosing a row μ such that

$$\frac{\beta_\mu}{\alpha_{\mu\rho}} = \min\left\{ \frac{\beta_i}{\alpha_{i\rho}} \,\middle|\, i \in \{1,\cdots,m\};\ \alpha_{i\rho} > 0 \right\}.$$

S 5 Pivot Step
Exchange the roles of $x_\mu^{\{B\}}$ and $x_\rho^{\{N\}}$ such that $x_\mu^{\{B\}}$ becomes nonbasic and $x_\rho^{\{N\}}$ becomes basic, i.e. transform B and N into $\tilde{B}$ and $\tilde{N}$ according to

$$\begin{aligned} B = (A_{i_1},\cdots,A_{i_\mu},\cdots,A_{i_m}) &\longrightarrow \tilde{B} = (A_{i_1},\cdots,A_{j_\rho},\cdots,A_{i_m}) \\ N = (A_{j_1},\cdots,A_{j_\rho},\cdots,A_{j_{n-m}}) &\longrightarrow \tilde{N} = (A_{j_1},\cdots,A_{i_\mu},\cdots,A_{j_{n-m}}). \end{aligned}$$

With $B := \tilde{B}$ and $N := \tilde{N}$, and the implied adjustments of $x^{\{B\}}$, $x^{\{N\}}$, ζ, β, d and D, as well as of $I_B(x)$ and $I_N(x)$, return to step $S\,2$.

□

Remark 1.1. *In case that—in step S 3—$D_\rho \leq 0$, we may compute a generating element of the cone $\mathscr{C} = \{y \mid Ay = 0,\ y \geq 0\}$ from the present tableau as follows: Rearranging the components of y into $(y^{\{B\}}, y^{\{N\}})$, analogously to the corresponding rearrangement of the components of x, we get for $\hat{y}^{\{B\}} = -D_\rho$ and $\hat{y}^{\{N\}} = 1 \cdot e_\rho$ that $(\hat{y}^{\{B\}}, \hat{y}^{\{N\}}) \geq 0$ and $B\hat{y}^{\{B\}} + N\hat{y}^{\{N\}} = -BD_\rho + N_\rho = 0$ due to $D_\rho = B^{-1}N_\rho$. Hence, $\hat{y}$ is a (nontrivial) generating element of the cone $\mathscr{C} = \{y \mid Ay = 0,\ y \geq 0\}$ according to Prop. 1.4. (p. 11).*

□

Denoting one sequence of the steps $S\,2 - S\,5$ as a cycle, or else as a *pivot step*, we may easily prove

Proposition 1.15. *Provided that, after step S 1, we have a first feasible basis, and that all feasible basic solutions of LP (1.3) are nondegenerate, the simplex algorithm will terminate after finitely many cycles, either with an optimal feasible basis or with the information that* $\inf\{c^{\mathrm{T}}x \mid x \in \mathscr{B}\} = -\infty$.

Proof: As long as $D_\rho \not\leq 0$ in step $S\,3$, we shall get in step $S\,4$ that $\dfrac{\beta_\mu}{\alpha_{\mu\rho}} > 0$ due to the assumed nondegeneracy. Observing that

$$\frac{\beta_\mu}{a_{\mu\rho}} = \max\{\tau \mid \beta - \tau D_\rho \geq 0\} =: \hat{\tau}$$

we see that the pivot step in $S\,5$ yields

$$x^{\{B\}} = \beta - \hat{\tau} D_\rho,\; x_\rho^{\{N\}} = \hat{\tau}, \text{ and in particular } z = \zeta - \hat{\tau} d_\rho < \zeta, \tag{1.20}$$

according to the choice of ρ in step $S\,3$. Finally, since $D_\rho = B^{-1}N_\rho$ is equivalent with the solution of $BD_\rho = \sum_{i=1}^{m} B_i \alpha_{i\rho} = N_\rho$, where N_ρ depends nontrivially on the column B_μ (it holds $a_{\mu\rho} > 0$), it is well known from linear algebra that replacing column B_μ by the column N_ρ as in step $S\,5$, yields again a basis $\tilde{B}$ which is feasible due to the rule for selecting the pivot row in step $S\,4$. Hence, after one cycle we get another feasible basic solution with a strictly decreased objectiv value. Therefore, no feasible basis can show up more than once in this procedure, and the number of feasible bases of an LP is obviously finite. □

The nondegeneracy assumption is crucial for this proof. If there exist degenerate feasible basic solutions, it can happen in some finite sequence of cycles, that $\hat{\tau} = 0$ for each cycle, and hence the objective is not decreased (in contrast to (1.20)), and that at the end of this sequence we get the same basis with which the sequence began. Obviously, this may be repeated infinitely often, without any decrease of the objective and with nonoptimal bases. We then say that the procedure is cycling. However, even if degenerate feasible basic solutions exist, we can avoid cycling of the simplex algorithm by introducing additional rules in $S\,3$ and/or $S\,4$, the choice of the pivot column and/or the pivot row. Common approaches are lexicographic rules applied in every pivot step,

- either to the choice of variables entering the basis ($S\,3$) as well as of variables leaving the basis ($S\,4$), if they are not uniquely determined; the strategy to choose in either case the variable with the smallest index, is called *Bland's rule*,
- or else to the choice of the variables leaving the basis ($S\,4$) only, called the *lexicographic method*.

Proposition 1.16. *Provided that, after step S 1, we have a first feasible basis, and that either the lexicographic method or Bland's rule is used if the respective choice* *of variables in step S 3 and/or S 4 is not uniquely determined, the simplex algorithm will terminate after finitely many cycles, either with an optimal feasible basis or with the information that* $\inf\{c^{\mathrm{T}}x \mid x \in \mathscr{B}\} = -\infty$.

For the proof of Prop. 1.16. see e.g. Vanderbei [337], Theorem 3.2 and Theorem 3.3.

Obviously the pivot step *S 5* implies an update of the simplex tableau (1.13) which may be easily derived from the equivalent system (1.14) as follows: For simplicity rewrite the tableau (1.13) as

$$\begin{array}{|c|ccc|}\hline \zeta & \delta_1 & \cdots & \delta_{n-m} \\ \hline \beta_1 & \alpha_{11} & \cdots & \alpha_{1n-m} \\ \vdots & \vdots & & \vdots \\ \beta_m & \alpha_{m1} & \cdots & \alpha_{mn-m} \\ \hline \end{array} = \begin{array}{|c|ccc|}\hline \alpha_{00} & \alpha_{01} & \cdots & \alpha_{0n-m} \\ \hline \alpha_{10} & \alpha_{11} & \cdots & \alpha_{1n-m} \\ \vdots & \vdots & & \vdots \\ \alpha_{m0} & \alpha_{m1} & \cdots & \alpha_{mn-m} \\ \hline \end{array}$$

and hence the system (1.14), with $x_0^{\{B\}} := z$, as

$$\left.\begin{aligned} x_0^{\{B\}} &= \alpha_{00} - \sum_{j=1}^{n-m} \alpha_{0j} x_j^{\{N\}} \\ x_1^{\{B\}} &= \alpha_{10} - \sum_{j=1}^{n-m} \alpha_{1j} x_j^{\{N\}} \\ &\vdots \\ x_m^{\{B\}} &= \alpha_{m0} - \sum_{j=1}^{n-m} \alpha_{mj} x_j^{\{N\}} \end{aligned}\right\} \tag{1.14}$$

In *S 5* the μ-th equation ($\mu \geq 1$), $x_\mu^{\{B\}} = \alpha_{\mu 0} - \sum_{j=1}^{n-m} \alpha_{\mu j} x_j^{\{N\}}$, is solved for $x_\rho^{\{N\}}$ ($\rho \geq 1$)—requiring that $\alpha_{\mu\rho} \neq 0$ as given by *S 4*—and the resulting expression for $x_\rho^{\{N\}}$ is inserted into all other relations of (1.14). Under the assumption that (1.13) is a primal feasible tableau (i.e. $\alpha_{i0} \geq 0\ \forall i \geq 1$), μ and ρ are chosen in *S 3* and *S 4* in such a way, that $\alpha_{0\rho} > 0$ and that with the increase of $x_\rho^{\{N\}}$ to $\dfrac{\alpha_{\mu 0}}{\alpha_{\mu\rho}} \geq 0$ all basic variables stay nonnegative, and in particular $x_\mu^{\{B\}} \to 0$. The exchange of $x_\rho^{\{N\}}$ and $x_\mu^{\{B\}}$ yields a new tableau with the elements $\alpha_{ij}^\star$; $i = 0, \cdots, m$; $j = 0, \cdots, n-m$, to be computed as

$$\alpha_{ij}^{\star} = \begin{cases} \dfrac{1}{\alpha_{\mu\rho}} & i=\mu,\ j=\rho \\ \dfrac{\alpha_{\mu j}}{\alpha_{\mu\rho}} & i=\mu,\ j\neq\rho \\ -\dfrac{\alpha_{i\rho}}{\alpha_{\mu\rho}} & i\neq\mu,\ j=\rho \\ \alpha_{ij}-\dfrac{\alpha_{i\rho}\,\alpha_{\mu j}}{\alpha_{\mu\rho}} & i\neq\mu,\ j\neq\rho\,. \end{cases} \tag{1.21}$$

Instead of the primal pivot step, where with a primal feasible tableau we look for a pivot column ρ violating the simplex criterion and then for a pivot row μ such that the exchange of $x_{\rho}^{\{N\}}$ and $x_{\mu}^{\{B\}}$ yields again a primal feasible tableau, we also may consider the reverse situation: Given a dual feasible tableau, i.e. $\alpha_{0j} \le 0,\ j = 1,\cdots,n-m$, we may look for a pivot row μ violating primal feasibility, i.e. $\alpha_{\mu 0} < 0$, and then for a pivot column ρ such that after the exchange of $x_{\mu}^{\{B\}}$ and $x_{\rho}^{\{N\}}$ we get again a dual feasible tableau. Since the related transformation of the tableau is obviously given again by (1.21), according to these formulae it is obvious that now necessarily $\alpha_{\mu\rho} < 0$ has to hold to maintain $\alpha_{0\rho}^{\star} \le 0$, and that furthermore, to ensure also $\alpha_{0j}^{\star} \le 0$ for all other $j \ge 1$, the pivot column ρ has to be chosen such that

$$\frac{\alpha_{0\rho}}{\alpha_{\mu\rho}} = \min\left\{ \frac{\alpha_{0j}}{\alpha_{\mu j}} \,\middle|\, j \in \{1,\cdots,n-m\};\ \alpha_{\mu j} < 0 \right\}.$$

Transforming now the tableau according to (1.21) terminates a *dual pivot step*.

At this point we may present one method, out of several others described in the literature, to realize *S 1* of the simplex algorithm as follows:

a) Solve the system $Ax = b$ successively for m variables yielding, with some basis B, the tableau

α_{10}	α_{11}	$\cdots$	α_{1n-m}
$\vdots$	$\vdots$		$\vdots$
α_{m0}	α_{m1}	$\cdots$	α_{mn-m}

corresponding to $x^{\{B\}} = B^{-1}b - B^{-1}Nx^{\{N\}}$. If $B^{-1}b \ge 0$, this tableau is primal feasible, and we just have to fill in its first row, $c^{\{B\}^{\mathrm{T}}}B^{-1}b$ and $c^{\{B\}^{\mathrm{T}}}B^{-1}N - c^{\{N\}^{\mathrm{T}}}$. Otherwise:

b) Define the first row as $(0, -e^{\mathrm{T}})$ (with $e^{\mathrm{T}} = (1,\cdots,1)$) corresponding to the artificial objective $z = e^{\mathrm{T}}x^{\{N\}} =: h^{\mathrm{T}}x$, for which we now have a dual feasible tableau. As long as this tableau is primal infeasible, continue with dual pivot steps (if necessary with one of the additional lexicographic rules mentioned earlier).

c) When a primal feasible tableau—with the feasible basis $\hat{B}$, the corresponding nonbasic part $\hat{N}$ of A, and the artificial objective—has been found, then replace the first row of the tableau by $c^{\{\hat{B}\}^{\mathrm{T}}}\hat{B}^{-1}b$ and $c^{\{\hat{B}\}^{\mathrm{T}}}\hat{B}^{-1}\hat{N} - c^{\{\hat{N}\}^{\mathrm{T}}}$.

If $\mathscr{B} = \{x \mid Ax = b,\ x \geq 0\} \neq \emptyset$ then, due to Prop. 1.7., our artificial problem $\min\{h^{\mathrm{T}}x \mid x \in \mathscr{B}\}$ is solvable such that the above procedure will yield a first primal feasible tableau for our original problem $\min\{c^{\mathrm{T}}x \mid x \in \mathscr{B}\}$ after finitely many dual pivot steps.

1.2.5 The dual Simplex method

As mentioned in Prop. 1.14., a (primal) simplex tableau for the primal LP (1.3) being dual feasible is strongly related to a feasible basic solution of the dual LP (1.18). Hence, applying successively dual pivot steps to dual feasible tableaus of the primal LP (1.3) leads to

The Dual Simplex Algorithm

S 1 Initialization
Find a dual feasible primal tableau (with primal basis B) for the LP (1.3).

S 2 Feasibility Test
If $\beta = B^{-1}b \geq 0$, then stop (B is an optimal feasible basis). Otherwise, continue.

S 3 Choice of Pivot Row
Choose $\mu \in \{1, \cdots, m\}$ such that $\beta_\mu < 0$ and the corresponding μ-th row of D, i.e. $\alpha_{\mu\cdot} = (\alpha_{\mu 1}, \cdots, \alpha_{\mu\, n-m})$. If $\alpha_{\mu\cdot} \geq 0$, then stop (LP (1.3) is unsolvable, since $\mathscr{B} = \emptyset$). Otherwise, continue.

S 4 Choice of Pivot Column
The maximal increase $\tau \geq 0$ for $x_\mu^{\{B\}}$, such that $d^{\mathrm{T}} - \tau \cdot \alpha_{\mu\cdot} \leq 0$ remains satisfied, implies choosing a column ρ such that

$$\frac{d_\rho}{\alpha_{\mu\rho}} = \min\left\{ \frac{d_j}{\alpha_{\mu j}} \,\middle|\, j \in \{1, \cdots, n-n\};\ \alpha_{\mu j} < 0 \right\}.$$

S 5 Pivot Step
Exchange the roles of $x_\mu^{\{B\}}$ and $x_\rho^{\{N\}}$ such that $x_\mu^{\{B\}}$ becomes nonbasic and $x_\rho^{\{N\}}$ becomes basic, i.e. transform B and N into $\tilde{B}$ and $\tilde{N}$ according to

$$B = (A_{i_1}, \cdots, A_{i_\mu}, \cdots, A_{i_m}) \longrightarrow \tilde{B} = (A_{i_1}, \cdots, A_{j_\rho}, \cdots, A_{i_m})$$
$$N = (A_{j_1}, \cdots, A_{j_\rho}, \cdots, A_{j_{n-m}}) \longrightarrow \tilde{N} = (A_{j_1}, \cdots, A_{i_\mu}, \cdots, A_{j_{n-m}}).$$

With $B := \tilde{B}$ and $N := \tilde{N}$, and the implied adjustments of $x^{\{B\}}, x^{\{N\}}, \zeta, \beta$, d and D, as well as of $I_B(x)$ and $I_N(x)$, return to step *S 2*.

□

Proposition 1.17. *Given a first dual feasible tableau after step S 1, the dual simplex algorithm—if necessary with the dual version of one of the lexicographic rules mentioned above—yields after finitely many dual pivot steps either the optimal primal solution or else the information, that* $\mathcal{B} = \emptyset$.

Proof: By Prop. 1.14. any dual feasible primal tableau—with $A = (B,N)$—corresponds for the dual LP

$$\left\{\begin{array}{l}\min b^{\mathrm{T}}u\\ \text{subject to}\\ B^{\mathrm{T}}u \le c^{\{B\}}\\ N^{\mathrm{T}}u \le c^{\{N\}}\end{array}\right\} \text{ or else } \left\{\begin{array}{lll}\min b^{\mathrm{T}}u & & \\ \text{subject to} & & \\ B^{\mathrm{T}}u + I_m v & & = c^{\{B\}}\\ N^{\mathrm{T}}u & + I_{n-m}w & = c^{\{N\}}\\ \qquad v, & w & \ge 0\end{array}\right\}$$

to the dual feasible basis

$$\begin{pmatrix} B^{\mathrm{T}} & 0\\ N^{\mathrm{T}} & I_{n-m}\end{pmatrix} \text{ with the inverse } \begin{pmatrix} {B^{\mathrm{T}}}^{-1} & 0\\ -N^{\mathrm{T}}{B^{\mathrm{T}}}^{-1} & I_{n-m}\end{pmatrix}.$$

Hence we have for the dual constraints

$$\begin{pmatrix} u\\ w\end{pmatrix} = \begin{pmatrix} {B^{\mathrm{T}}}^{-1} & 0\\ -N^{\mathrm{T}}{B^{\mathrm{T}}}^{-1} & I_{n-m}\end{pmatrix}\begin{pmatrix} c^{\{B\}}\\ c^{\{N\}}\end{pmatrix} - \begin{pmatrix} {B^{\mathrm{T}}}^{-1} & 0\\ -N^{\mathrm{T}}{B^{\mathrm{T}}}^{-1} & I_{n-m}\end{pmatrix}\begin{pmatrix} v\\ 0\end{pmatrix}$$

or else, together with the objective $\eta = b^{\mathrm{T}}u$ and, as before, $B^{-1}b = \beta$ and $B^{-1}N = D$,

$$\begin{array}{rlcl}\eta &= \beta^{\mathrm{T}}c^{\{B\}} & - & \beta^{\mathrm{T}}v\\ u &= {B^{\mathrm{T}}}^{-1}c^{\{B\}} & - & {B^{\mathrm{T}}}^{-1}v\\ w &= -D^{\mathrm{T}}c^{\{B\}} + c^{\{N\}} & - & (-D^{\mathrm{T}})v \ge 0\\ v &\ge 0. & & \end{array}$$

From these formulae we see immediately that

- with dual feasibility after step *S 1*, i.e. $-D^{\mathrm{T}}c^{\{B\}} + c^{\{N\}} \ge 0$, from $\beta \ge 0$ follows dual optimality for $v = 0$ (*S 2*);
- with $\beta_\mu < 0$ and $D_\mu^{\mathrm{T}} \ge 0$, the dual nonbasic variable v_μ can grow arbitrarily and hence the objective $\eta \to \infty$ on the dual feasible set such that according to the weak duality theorem Prop. 1.9. there cannot exist any primal feasible solution (*S 3*);
- the requirement to maintain dual feasibility, i.e. $w \ge 0$ when increasing the nonbasic v_μ, results in the rule for choosing the pivot column (*S 4*).

Observing that now $\alpha_{\mu\rho} < 0$ implies again, as in the proof of Prop. 1.15., that the exchange of the nonbasic column N_ρ with the basic column B_μ yields a basis again, dual feasible by construction. □

Given the LP (1.3) with the $(m \times n)$-matrix A, the question may arise why we deal with the dual simplex method, carried out on the primal simplex tableau, instead of just applying the simplex method to the dual LP (1.18) and its associated tableau. One rather heuristic argument is the size of the tableaus which have to be updated in every pivot step: Whereas the primal tableau is of the order $m \times (n-m)$, we obviously have for the dual tableau the order $n \times m$, exceeding the former number of elements to be transformed by m^2, which can be large for realistic problems. In addition, professional implementations of the simplex method do not perform the pivot transformations on the respective simplex tableaus but essentially just on the corresponding basis inverses (in appropriate representations, e.g. in product form). But the basis inverses are of order $(m \times m)$ for the primal basis and of $(n \times n)$ for the dual, the latter being substantially greater if, as it is commonly the case in applications, $n \gg m$.

Although these considerations are not a strict argument for the advantage of the above dual simplex algorithm, they may serve as a heuristic explanation for the dual simplex method, as presented in this section, being often observed to be more efficient than the primal method. For more details on the implementation of the simplex method (and its different variants) we may refer for instance to Maros [219].

Exercises

1.3. How can you prove Prop. 1.1. (page 9)?

1.4. Find an argument showing that Prop. 1.2. (page 9) holds.

1.5. Consider the LP

$$(PP) \qquad \begin{array}{rrcrl} & \gamma := \min\{-10x_1 - 2x_2\} & & & \\ \text{s.t.} & -x_1 + 2x_2 & \geq & -4 & \\ & -3x_1 + 9x_2 & \geq & -12 & \\ & x_j & \geq & 0 & j = 1,2 \end{array}$$

as primal together with its dual program (DP). What can be said

(a) about the feasibility of this dual pair of LP's;
(b) about the solvability of (PP) and (DP)?

1.6. For any LP one of the following cases applies:

- FS: "feasible and finitely solvable",
- FU: "feasible and unbounded", and
- NF: "infeasible".

For any pair (PP) and (DP) of dual LP's mark with **YES** and/or **NO** the possible and impossible situations, respectively.

(DP) / (PP)	FS	FU	NF
FS			
FU			
NF			

Argue either with statements presented in Subsect. 1.2.3 or with examples to back your answers.

1.7. Use the simplex algorithm to compute a solution $(\hat{x}_1, \hat{x}_2)$ of the following LP. (Observe that in this case, with the slack variables as basic variables, a first feasible simplex tableau is immediately given!)

$$(PP) \qquad \begin{array}{rl} \gamma := & \max\{2x_1 + x_2\} \\ \text{s.t.} & x_1 + x_2 \le 2 \\ & 3x_1 + x_2 \le 3 \\ & x_2 \le 2 \\ & x_j \ge 0 \quad j = 1,2. \end{array}$$

With this solution $(\hat{x}_1, \hat{x}_2)$, use the complementarity conditions to determine a solution of (DP), the corresponding dual LP.

To verify your results, edit this model in SLP-IOR and run any LP solver (at your disposal with SLP-IOR).

1.8. Consider the LP

$$(PP) \qquad \begin{array}{rl} \gamma := & \min\{-x_1 + x_2\} \\ \text{s.t.} & x_1 + 2x_2 \ge 3 \\ & 2x_1 + x_2 \le 4 \\ & x_j \ge 0 \quad j = 1,2. \end{array}$$

(a) Augment the system of constraints by slack variables $y_i \ge 0$, $i = 1,2$. Choosing x_1 and y_1 as basic variables and solving the system for x_1, y_1 (as dependent variables) yields a dual feasible primal tableau.

(b) Beginning with this tableau, apply the dual simplex method to get a solution of (PP).

1.2.6 Dual decomposition method

As mentioned in the Introduction, in case of a discrete distribution we get for the two-stage SLP with recourse the LP (1.10) with the special data structure illustrated in Fig. 1.1 on page 5. This structure may be used according to an idea first presented in Benders [14], originally applied to mixed-integer NLP's. For simplicity we present the procedure for the special case of $S = 1$ realizations in (1.10), i.e. for

$$\left.\begin{array}{rl} & \min\{c^{\mathrm{T}}x + q^{\mathrm{T}}y\} \\ \text{s. t.} & Ax \qquad\quad = b \\ & Tx + Wy = h \\ & x \qquad\quad \geq 0 \\ & \qquad\quad y \geq 0. \end{array}\right\} \tag{1.22}$$

The extension of the method for $S > 1$ realizations is then immediate, although several variants and tricks can be involved.

We assume that the LP (1.22) is solvable and, in addition, that the first stage feasible set $\{x \mid Ax = b,\ x \geq 0\}$ is bounded. According to Prop. 1.7. the solvability of (1.22) implies that

$$\{(x,y) \mid Ax = b,\ Tx + Wy = h,\ x \geq 0, y \geq 0\} \neq \emptyset$$

and

$$c^{\mathrm{T}}\xi + q^{\mathrm{T}}\eta \geq 0\ \forall (\xi,\eta) \in \{(\xi,\eta) \mid A\xi = 0,\ T\xi + W\eta = 0,\ \xi \geq 0,\ \eta \geq 0\},$$

and therefore in particular, for $\xi = 0$,

$$q^{\mathrm{T}}\eta \geq 0\ \forall \eta \in \{\eta \mid W\eta = 0,\ \eta \geq 0\},$$

such that the second stage optimum, also called the *recourse function*,

$$f(x) := \min\{q^{\mathrm{T}}y \mid Wy = h - Tx,\ y \geq 0\}$$

is finite if the recourse constraints are feasible. Otherwise, we define the recourse function as $f(x) = \infty$ if $\{y \mid Wy = h - Tx,\ y \geq 0\} = \emptyset$. Then we have

Proposition 1.18. *The recourse function $f(x)$, defined on the bounded set $\{x \mid Ax = b,\ Tx + Wy = h,\ x \geq 0, y \geq 0\} \neq \emptyset$, is piecewise linear, convex, and bounded below.*

Proof: By our assumptions, with $\mathscr{B}_1 = \{x \mid Ax = b,\ x \geq 0\}$ it follows that $\mathscr{B} := \mathscr{B}_1 \cap \{x \mid \exists y \geq 0 : Wy = h - Tx\} \neq \emptyset$ is bounded. Since $\{x \mid \exists y \geq 0 : Wy = h - Tx\} \neq \emptyset$ is the projection of the convex polyhedral set $\{(x,y) \mid Tx + Wy = h,\ y \geq 0\}$ in (x,y)-space into x-space, it is convex polyhedral. Hence, $\mathscr{B}$ as the intersection of a convex polyhedron with a convex polyhedral set is a convex polyhedron, and it holds for $x \in \mathscr{B}$

$$f(x) = q^{\{B_W\}^{\mathrm{T}}} B_W^{-1}(h - Tx) \text{ if } B_W^{-1}(h - Tx) \geq 0,$$

where B_W (out of W) is an appropriate optimal basis, chosen from the finitely many feasible bases of W. Hence, $f(x)$ is piecewise linear and bounded below in $x \in \mathscr{B}$. Finally, with $x^1 \in \mathscr{B}$ and $x^2 \in \mathscr{B}$ such that $f(x^i)$, $i = 1,2$, is finite, and with corresponding recourse solutions y^1, y^2 satisfying

$$f(x^i) = q^{\mathrm{T}} y^i,\ i = 1,2, \ \text{ and } \ Wy^i = h - Tx^i,\ y^i \geq 0,\ i = 1,2,$$

for arbitrary $\lambda \in (0,1)$ and $\tilde{x} = \lambda x^1 + (1-\lambda)x^2$ it follows that

$$\lambda y^1 + (1-\lambda)y^2 \in \{y \mid Wy = h - T\tilde{x},\ y \geq 0\}$$

and hence that

$$\begin{aligned} f(\tilde{x}) &= \min\{q^{\mathrm{T}} y \mid Wy = h - T\tilde{x},\ y \geq 0\} \\ &\leq q^{\mathrm{T}}(\lambda y^1 + (1-\lambda)y^2) = \lambda f(x^1) + (1-\lambda) f(x^2), \end{aligned}$$

demonstrating the convexity of $f(x)$ on its effective domain $\operatorname{dom} f = \mathscr{B}$. □

Obviously, with the recourse function $f(x)$, the LP (1.22) can be rewritten equivalently as the NLP

$$\begin{aligned} &\min\{c^{\mathrm{T}} x + f(x)\} \\ \text{s. t. } & Ax = b \\ & x \geq 0, \end{aligned}$$

restricting x implicitly to the effective domain of f, or else as

$$\left.\begin{aligned} &\min\{c^{\mathrm{T}} x + \theta\} \\ \text{s. t.} \quad & Ax = b \\ & \theta - f(x) \geq 0 \\ & x \geq 0. \end{aligned}\right\} \tag{1.23}$$

However, this may not yet help a lot since, in general, we do not know the (convex polyhedral) recourse function $f(x)$ explicitly. To say it in other terms: $f(x)$ being bounded below, piecewise linear and convex on $\mathscr{B}$ implies the existence of finitely many linear functions $\varphi_\nu(x)$, $\nu = 1, \cdots, L$, such that, on $\operatorname{dom} f = \mathscr{B}$, it holds that $f(x) = \max_{\nu \in \{1,\cdots,L\}} \varphi_\nu(x)$. Hence, to reduce the feasible set of (1.23) to the effective domain $\mathscr{B}$ of f, it may be necessary to add some further linear constraints $\psi_1(x), \cdots, \psi_K(x)$ (observe that the polyhedron $\mathscr{B}$ is defined by finitely many linear constraints) to achieve feasibility of the recourse problem, such that instead of (1.23) we get the equivalent LP

$$\left.\begin{aligned} &\min\{c^{\mathrm{T}} x + \theta\} \\ \text{s. t.} \quad & Ax = b \\ & \theta - \varphi_\nu(x) \geq 0 \quad \nu = 1, \cdots, L \\ & \psi_\mu(x) \geq 0 \quad \mu = 1, \cdots, K \\ & x \geq 0. \end{aligned}\right\} \tag{1.24}$$

Also in this case, we do not know in advance the linear constraints needed for the complete coincidence of this problem with the original LP (1.22). Therefore, the idea of the following procedure—also called Benders decomposition— is to generate successively those additional constraints needed to approximate (and finally to hit) the solution of the original LP (1.22).

The Dual Decomposition Algorithm

S 1 Initialization

Find a lower bound θ_0 for

$$\min\{q^{\mathrm{T}}y \mid Ax = b,\ Tx + Wy = h,\ x \geq 0, y \geq 0\}$$

and solve the LP

$$\min\{c^{\mathrm{T}}x + \theta \mid Ax = b,\ x \geq 0,\ \theta \geq \theta_0\}$$

yielding the solution $(\hat{x}, \hat{\theta})$. Define

$$\mathscr{B}_0 = \{(x, \theta) \mid Ax = b,\ x \geq 0,\ \theta \in \mathbb{R}\} \text{ and}$$
$$\mathscr{B}_1 = \{\mathbb{R}^n \times \{\theta\} \mid \theta \geq \theta_0\}.$$

S 2 Evaluate the recourse function

To get $f(\hat{x}) = \min\{q^{\mathrm{T}}y \mid Wy = h - T\hat{x},\ y \geq 0\}$, solve the dual LP

$$f(\hat{x}) = \max\{(h - T\hat{x})^{\mathrm{T}}u \mid W^{\mathrm{T}}u \leq q\}.$$

If $f(\hat{x}) = +\infty$, then go to step *S 3*, else to *S 4*.

S 3 The feasibility cut

$\hat{x}$ is infeasible for (1.22). In this case by Prop. 1.7. (p. 12) there exists an unbounded growth direction $\tilde{u}$ (to be revealed in step *S 3* of the simplex algorithm as one of the finitely many generating elements of the cone $\{u \mid W^{\mathrm{T}}u \leq 0\}$; see Remark 1.1. on page 16) such that $W^{\mathrm{T}}\tilde{u} \leq 0$ and $(h - T\hat{x})^{\mathrm{T}}\tilde{u} > 0$, whereas for any feasible x of (1.22) there exists some $y \geq 0$ such that $Wy = h - Tx$. Multiplying this equation by $\tilde{u}$ yields the inequality

$$\tilde{u}^{\mathrm{T}}(h - Tx) = \tilde{u}^{\mathrm{T}}Wy \leq 0,$$

which has to hold for any feasible x but is violated by $\hat{x}$. Therefore we redefine $\mathscr{B}_1 := \mathscr{B}_1 \cap \{(x, \theta) \mid \tilde{u}^{\mathrm{T}}(h - Tx) \leq 0\}$ such that the infeasible $\hat{x}$ is cut off, and go on to step *S 5*.

S 4 The optimality cut

Since $f(\hat{x})$ is finite, by Prop. 1.3. there exists for the recourse problem a dual optimal feasible basic solution $\hat{u}$, determined in step *S 2* above, such that

$$f(\hat{x}) = (h - T\hat{x})^{\mathrm{T}}\hat{u},$$

whereas for any arbitrary x we have

$$\begin{aligned} f(x) &= \sup\{(h - Tx)^{\mathrm{T}}u \mid W^{\mathrm{T}}u \leq q\} \\ &\geq (h - Tx)^{\mathrm{T}}\hat{u}. \end{aligned}$$

Therefore, the inequality $\theta - f(x) \geq 0$ in (1.23) implies the linear constraint

$$\theta \geq \hat{u}^{\mathrm{T}}(h - Tx).$$

If this constraint is satisfied for $(\hat{x}, \hat{\theta})$, i.e. if $f(\hat{x}) \leq \hat{\theta}$, stop the procedure, since $x^\star := \hat{x}$ is an optimal first stage solution; otherwise redefine the set of constraints as $\mathscr{B}_1 := \mathscr{B}_1 \cap \{(x, \theta) \mid \theta \geq \hat{u}^{\mathrm{T}}(h - Tx)\}$, thus cutting off the nonoptimal $(\hat{x}, \hat{\theta})$, and go on to step *S 5*.

S 5 Solve the updated LP, called the *master program*,

$$\min\{c^{\mathrm{T}}x + \theta \mid (x, \theta) \in \mathscr{B}_0 \cap \mathscr{B}_1\}$$

yielding the optimal solution $(\tilde{x}, \tilde{\theta})$.
With $(\hat{x}, \hat{\theta}) := (\tilde{x}, \tilde{\theta})$ return to step *S 2*.

Proposition 1.19. *Given the above assumptions, the dual decomposition algorithm yields an optimal first stage solution $x^\star$ of (1.22) after finitely many cycles.*

Proof: According to Prop. 1.18. the lower bound θ_0 of *S 1* exists (for instance, by weak duality for any $(w, u) \in \{(w, u) \mid A^{\mathrm{T}}w + T^{\mathrm{T}}u \leq 0,\ W^{\mathrm{T}}u \leq q\}$, the value $b^{\mathrm{T}}w + h^{\mathrm{T}}u$ could be chosen as θ_0).

Due to the solvability of (1.22) the dual constraints $W^{\mathrm{T}}u \leq q$ are feasible and independent of x. Hence the dual representation of $f(\hat{x})$ in *S 2* is always feasible implying that $f(\hat{x})$ is either finite or equal to $+\infty$, the latter indicating primal infeasibility.

If $f(\hat{x}) = +\infty$, such that $\hat{x}$ is infeasible for (1.22), due to Prop. 1.7. there is a $\tilde{u}:\ W^{\mathrm{T}}\tilde{u} \leq 0$ and $(h - T\hat{x})^{\mathrm{T}}\tilde{u} > 0$. We may assume that $\tilde{u}$ is one of finitely many generating elements of the cone $\{u \mid W^{\mathrm{T}}u \leq 0\}$, as we get it in step *S 3* of the simplex algorithm (see Remark 1.1. on page 16). Since the cone $\{u \mid W^{\mathrm{T}}u \leq 0\}$ is finitely generated, we shall add at most finitely many constraints of the type $\tilde{u}^{\mathrm{T}}(h - Tx) \leq 0$ before we have finite recourse in all further cycles.

If $f(\hat{x}) = (h - T\hat{x})^{\mathrm{T}}\hat{u}$ is finite, we assume $\hat{u}$ to be an optimal dual feasible basic solution (as delivered by the simplex algorithm). Since there are only finitely many dual feasible basic solutions and hence finitely many constraints of the type $\theta \geq \hat{u}^{\mathrm{T}}(h - Tx)$ to be added at most, after finitely many cycles, with the solution of the updated LP in *S 5*, we must get in the subsequent step *S 4* that $\hat{\theta} \geq \hat{u}^{\mathrm{T}}(h - T\hat{x}) = f(\hat{x})$. Due to the facts that

a) the feasible set of (1.23) is contained in the feasible set $\mathscr{B}_0 \cap \mathscr{B}_1$ of the last master program in the previous step *S 5*, solved by $(\hat{x}, \hat{\theta})$, and that
b) this solution $(\hat{x}, \hat{\theta})$ is obviously feasible for (1.23),

it follows for any solution $(x^\star, \theta^\star)$ of (1.23) that

$$\begin{aligned} c^{\mathrm{T}}x^\star + \theta^\star &= c^{\mathrm{T}}x^\star + f(x^\star) \\ &\geq c^{\mathrm{T}}\hat{x} + \hat{\theta} && \text{due to a)} \\ &\geq c^{\mathrm{T}}x^\star + \theta^\star && \text{due to b).} \end{aligned}$$

Hence, $\hat{x}$ is a first stage solution of (1.22). □

Observe that whenever we have that $\hat{x} \in \operatorname{dom} f$ with the stopping criterion not satisfied, we have to add in $S4$ a linear constraint of the type $\theta \geq \phi(x) := \hat{\gamma} + \hat{g}^{\mathrm{T}}x$, where $\hat{\gamma} = \hat{u}^{\mathrm{T}}h$ and $\hat{g} = -T^{\mathrm{T}}\hat{u} \in \partial f(\hat{x})$, the subdifferential of f in $\hat{x}$ (see Def. 1.10. on page 54 below). Hence $\phi(x)$ is a linear lower bound of f in $x \in \operatorname{dom} f$ such that $\phi(\hat{x}) = f(\hat{x})$. This is illustrated in Fig. 1.3.

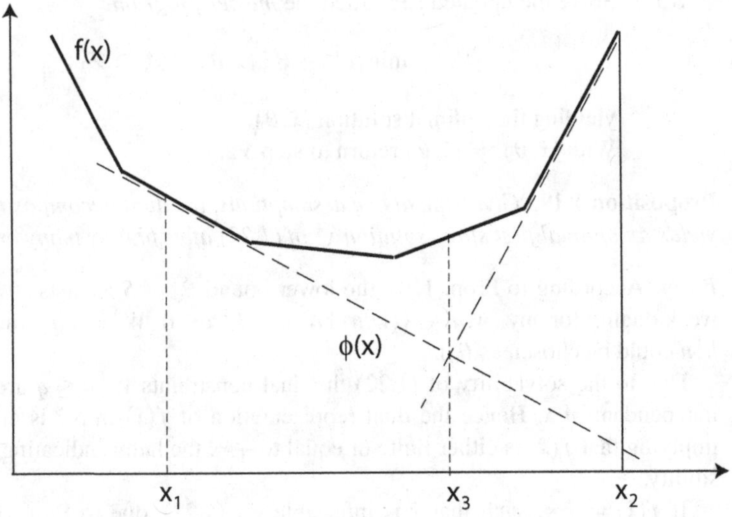

Fig. 1.3 Dual decomposition: Optimality cuts.

Let us consider now, instead of (1.22), the two-stage SLP (1.10) with $S > 1$ realizations, given as

$$\left.\begin{array}{rl} & \min c^{\mathrm{T}}x + \sum_{j=1}^{S} p_j q^{\mathrm{T}} y^j \\ \text{s. t.} & Ax \qquad\quad = b \\ & T^j x + W y^j = h^j,\ j = 1, \cdots, S \\ & x \qquad\quad \geq 0 \\ & \qquad\quad y^j \geq 0,\ j = 1, \cdots, S. \end{array}\right\}$$

This is equivalent to the NLP

$$\left.\begin{array}{rl} & \min\{c^{\mathrm{T}}x + \sum_{j=1}^{S} p_j \theta_j\} \\ \text{s. t.} & Ax = b \\ & \theta_j - f_j(x) \geq 0,\ j = 1, \cdots, S \\ & x \geq 0 \end{array}\right\} \qquad (1.25)$$

with the recourse functions

$$f_j(x) = \min\{q^{\mathrm{T}}y^j \mid Wy^j = h^j - T^j x,\ y^j \geq 0\},\ j = 1,\cdots,S.$$

Then we can modify the above dual decomposition algorithm as follows:

Dual Decomposition – Multicut Version

S 1 Initialization

Find, for $j = 1,\cdots,S$, lower bounds $\tilde{\theta}_j$ for

$$\min\{q^{\mathrm{T}}y^j \mid Ax = b,\ T^j x + Wy^j = h^j,\ x \geq 0, y^j \geq 0\}$$

and, with $p = (p_1,\cdots,p_S)^{\mathrm{T}}$, $\theta = (\theta_1,\cdots,\theta_S)^{\mathrm{T}}$ and $\tilde{\theta} = (\tilde{\theta}_1,\cdots,\tilde{\theta}_S)^{\mathrm{T}}$ solve the LP

$$\min\{c^{\mathrm{T}}x + p^{\mathrm{T}}\theta \mid Ax = b,\ x \geq 0, \theta \geq \tilde{\theta}\},$$

yielding the solution $(\hat{x},\hat{\theta})$. Define

$$\mathscr{B}_0 = \{(x,\theta) \mid Ax = b,\ x \geq 0,\ \theta \in \mathbb{R}^S\} \text{ and}$$
$$\mathscr{B}_1 = \{\mathbb{R}^n \times \{\theta\} \mid \theta \geq \tilde{\theta}\}.$$

S 2 Evaluate the recourse functions

To get $f_j(\hat{x}) = \min\{q^{\mathrm{T}}y^j \mid Wy^j = h^j - T^j\hat{x},\ y^j \geq 0\}$, solve the dual LP's

$$f_j(\hat{x}) = \max\{(h^j - T^j\hat{x})^{\mathrm{T}}u^j \mid W^{\mathrm{T}}u^j \leq q\},\ j = 1,\cdots,S.$$

If $J := \{j \mid f_j(\hat{x}) = +\infty\} \neq \emptyset$, then go to step *S 3*, else to *S 4*.

S 3 Feasibility cuts

We have $f_j(\hat{x}) = +\infty$ for $j \in J \neq \emptyset$ implying that $\hat{x}$ is infeasible for (1.25). In this case by Prop. 1.7. there exist unbounded growth directions $\tilde{u}^j,\ j \in J$ (to be revealed in step *S 3* of the simplex algorithm; see Remark 1.1. on page 16) such that $\forall j \in J$ holds $W^{\mathrm{T}}\tilde{u}^j \leq 0$ and $(h^j - T^j\hat{x})^{\mathrm{T}}\tilde{u}^j > 0$, whereas for any feasible x of (1.25) there exist some $y^j \geq 0$ such that $Wy^j = h^j - T^j x$. Multiplying these equations by $\tilde{u}^j$ yields the inequalities

$$\tilde{u}^{j\mathrm{T}}(h^j - T^j x) = \tilde{u}^{j\mathrm{T}}Wy^j \leq 0,$$

which have to hold for any feasible x but are violated by $\hat{x}$ for $j \in J$. Therefore we redefine $\mathscr{B}_1 := \mathscr{B}_1 \cap \{(x,\theta) \mid \tilde{u}^{j\mathrm{T}}(h^j - T^j x) \leq 0,\ j \in J\}$ such that the infeasible $\hat{x}$ is cut off, and go on to step *S 5*.

S 4 Optimality cuts

Since $f_j(\hat{x})$ is finite for all $j = 1,\cdots,S$, by Prop. 1.3. there exist for the recourse problems dual optimal feasible basic solutions $\hat{u}^j$, determined in step *S 2* above, such that

$$f_j(\hat{x}) = (h^j - T^j\hat{x})^{\mathrm{T}}\hat{u}^j,$$

whereas for any arbitrary x we have

$$\begin{aligned} f_j(x) &= \sup\{(h^j - T^j x)^{\mathrm{T}} u^j \mid W^{\mathrm{T}} u^j \leq q\} \\ &\geq (h^j - T^j x)^{\mathrm{T}} \hat{u}^j. \end{aligned}$$

Therefore, the inequalities $\theta_j - f_j(x) \geq 0$ in (1.25) imply the linear constraints

$$\theta_j \geq \hat{u}^{j\mathrm{T}}(h^j - T^j x).$$

If these constraints are satisfied for $(\hat{x}, \hat{\theta})$, i.e. if $f_j(\hat{x}) \leq \hat{\theta}_j\ \forall j$, stop the procedure, since $x^\star := \hat{x}$ is an optimal first stage solution; otherwise, if $f_j(\hat{x}) > \hat{\theta}_j$ for $j \in J \neq \emptyset$, redefine the set of constraints as $\mathscr{B}_1 := \mathscr{B}_1 \cap \{(x, \theta) \mid \theta_j \geq \hat{u}^{j\mathrm{T}}(h^j - T^j x) \text{ for } j \in J\}$, thus cutting off the nonoptimal $(\hat{x}, \hat{\theta})$, and go on to step *S 5*.

S 5 Solve the updated master program

$$\min\{c^{\mathrm{T}} x + \theta \mid (x, \theta) \in \mathscr{B}_0 \cap \mathscr{B}_1\}$$

yielding the optimal solution $(\tilde{x}, \tilde{\theta})$.
With $(\hat{x}, \hat{\theta}) := (\tilde{x}, \tilde{\theta})$ return to step *S 2*.

This *multicut* version of the dual decomposition method for solving the two-stage SLP (1.10) or its equivalent NLP (1.25) is due to Birge and Louveaux (see Birge–Louveaux [25]). Similarly to Prop. 1.19., the multicut method can also be shown to yield an optimal first stage solution after finitely many cycles.

Instead of introducing S variables θ_j as in the multicut version, we may also get along with just one additional variable θ: Instead of (1.25) we deal, again equivalently to the SLP (1.10), with the NLP

$$\left.\begin{array}{rr} & \min\{c^{\mathrm{T}} x + \theta\} \\ \text{s. t.} & Ax = b \\ & \theta - \sum_{j=1}^{S} p_j f_j(x) \geq 0, \\ & x \geq 0. \end{array}\right\} \tag{1.26}$$

In step *S 3* we add feasibility cuts to $\mathscr{B}_1$ as long as we find $f_j(\hat{x}) = +\infty$ for at least one j. In step *S 4*, where all recourse function values are finite with $f_j(\hat{x}) = (h^j - T^j \hat{x})^{\mathrm{T}} \hat{u}^j$, we

- either add the optimality cut $\theta \geq \sum_{j=1}^{S} p_j \hat{u}^{j\mathrm{T}}(h^j - T^j x)$ to $\mathscr{B}_1$ if $\hat{\theta} < \sum_{j=1}^{S} p_j \hat{u}^{j\mathrm{T}}(h^j - T^j \hat{x})$, and then go on to the master program in step *S 5*;

– or else, if $\hat{\theta} \geq \sum_{j=1}^{S} p_j \hat{u}^{j\mathrm{T}}(h^j - T^j \hat{x})$, we stop with $\hat{x}$ as an optimal first stage solution.

This *L–shaped method* was introduced by Van Slyke–Wets [336]. Both variants of Benders' decomposition are described in detail in Birge–Louveaux [26].

1.2.7 Nested decomposition

This section is devoted to an extension of the dual decomposition method to multi–stage SLP problems (MSLP's). In (1.11) on page 5, we have introduced a general MSLP with fixed recourse. Now we will allow randomness of the recourse matrices and objective coefficients, too. Due to Remark 3.10., pages 265–266, we may restrict MSLP's to the widely used staircase formulation:

$$\left.\begin{array}{lll}
\min \left\{ c_1^{\mathrm{T}} x_1 + \mathbb{E}_{\zeta_T} \left[\sum_{t=2}^{T} c_t^{\mathrm{T}}(\zeta_t) x_t(\zeta_t) \right] \right\} & & \\
W_1 x_1 & & = b_1 \\
T_2(\zeta_2) x_1 & + W_2(\zeta_2) x_2(\zeta_2) & = b_2(\zeta_2), \text{ a.s.,} \\
T_3(\zeta_3) x_2(\zeta_2) & + W_3(\zeta_3) x_3(\zeta_3) & = b_3(\zeta_3), \text{ a.s.,} \\
\vdots & & \\
T_T(\zeta_T) x_{T-1}(\zeta_{T-1}) & + W_T(\zeta_T) x_T(\zeta_T) & = b_T(\zeta_T), \text{ a.s.,} \\
 & x_1,\ x_t(\zeta_t) & \geq 0, \text{ a.s. } \forall t,
\end{array}\right\} \quad (1.27)$$

where $\xi_2, \cdots, \xi_T$ and therefore also $\zeta_t = (\xi_2, \cdots, \xi_t)$, $t = 2, \cdots, T$, are random vectors with given distributions. Furthermore, since in stage t with $2 \leq t \leq T$ the constraint

$$T_t(\zeta_t) x_{t-1}(\zeta_{t-1}) + W_t(\zeta_t) x_t(\zeta_t) = b_t(\zeta_t)$$

has to hold a.s., it should be obvious that for almost every realization $\widehat{\zeta}_t = (\widehat{\zeta}_{t-1}, \widehat{\xi}_t)$, with $x_{t-1}(\cdot)$ being the decision $x_{t-1}(\widehat{\zeta}_{t-1})$ taken for the corresponding sub–path of $\widehat{\zeta}_t$, the decision $x_t(\widehat{\zeta}_t)$ in stage t has to satisfy the constraints $W_t(\widehat{\zeta}_t) x_t(\widehat{\zeta}_t) = b_t(\widehat{\zeta}_t) - T_t(\widehat{\zeta}_t) x_{t-1}(\widehat{\zeta}_{t-1})$, $x_t(\widehat{\zeta}_t) \geq 0$.

If in particular the random vector $\xi := (\xi_2, \cdots, \xi_T)$ (and hence all the vectors ξ_t and ζ_t) has a finite discrete distribution, defined by realizations and corresponding probabilities as $\{\widehat{\xi}^s,\ \mathbb{P}_\xi(\xi = \widehat{\xi}^s) = q_s;\ s \in \mathscr{S} := \{1, \cdots, S\}\}$, we can represent the process $\{\zeta_t;\ t = 2, \cdots, T\}$ on a *scenario tree* as follows:
Node $n = 1$ in stage 1 corresponds to the assumed deterministic state at the beginning of the process;
in stage 2 we have the nodes $n = 2, \cdots, K_2$, each one corresponding to one of the different sub–paths $\widehat{\zeta}_2^{\rho(n)}$ contained in the scenarios $\widehat{\xi}^1, \cdots, \widehat{\xi}^S$, endowed with the

probability $p_n = \sum_{s \in \mathscr{S}} \left\{ q_s \mid \widehat{\zeta}_2^s = \widehat{\zeta}_2^{\rho(n)} \right\}$; in stage 3 there are then the nodes $n = K_2 + 1, \cdots, K_3$ corresponding to one of the different sub–paths $\widehat{\zeta}_3^{\rho(n)}$ contained in $\{\widehat{\xi}^s;\ s \in \mathscr{S}\}$, with the probabilities $p_n = \sum_{s \in \mathscr{S}} \left\{ q_s \mid \widehat{\zeta}_3^s = \widehat{\zeta}_3^{\rho(n)} \right\}$; and so on. As an example of a scenario tree see the four-stage case in Fig. 1.4 with 10 scenarios.

Scenarios are the different realizations of ζ_T, they correspond to the root–to–leaf paths in the tree. The superscript $\rho(n)$ denotes the first scenario which passes through node n, in a fixed ordering of scenarios. In Fig. 1.4 we have, e.g., $\rho(2) = 1$ and $\rho(8) = 8$. For further details on this notation see the section about notations.

By construction, any node n in some stage $t_n \geq 2$ has exactly one predecessor (node) h_n in stage $t_n - 1$, whereas each node n in stage $t_n < T$ has a nonempty finite set $\mathscr{C}(n)$ of successors (nodes in stage $t_n + 1$), also called the children of n. For any node n in stage $t_n \geq 2$ (i.e. $K_{t_n - 1} < n \leq K_{t_n}$) we shall use the short-hand T_n, W_n, x_n, b_n, c_n instead of $T_{t_n}(\widehat{\zeta}_{t_n}^{\rho(n)})$, $W_{t_n}(\widehat{\zeta}_{t_n}^{\rho(n)})$, $x_{t_n}(\widehat{\zeta}_{t_n}^{\rho(n)})$, $b_{t_n}(\widehat{\zeta}_{t_n}^{\rho(n)})$, and $c_{t_n}(\widehat{\zeta}_{t_n}^{\rho(n)})$, respectively.

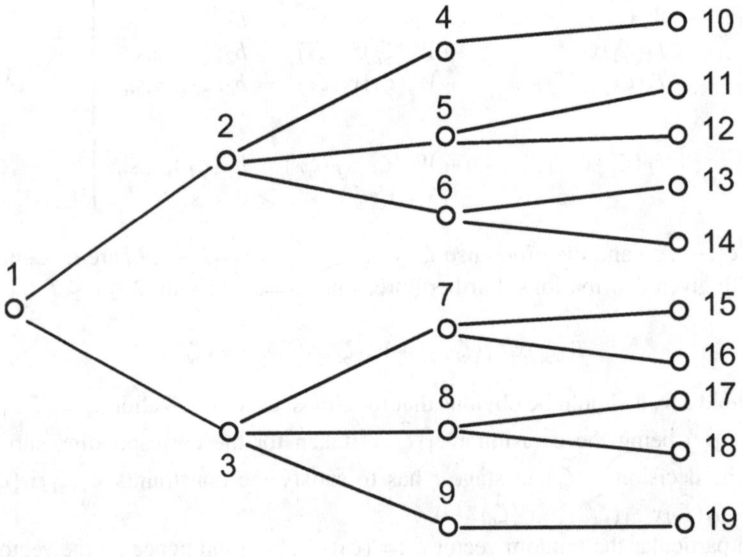

Fig. 1.4 Four-stage scenario tree.

For later use we introduce some further notations. $\mathscr{N}$ denotes the set of nodes of the scenario tree and $\mathscr{N}_t$ stands for the set of nodes in stage t, i.e., $\mathscr{N}_1 = \{1\}$ and $\mathscr{N}_t = \{K_{t-1} + 1, \ldots, K_t\}$ for $t > 1$. The set of nodes of the subtree rooted at $n \in \mathscr{N}$ will be denoted by $\mathscr{G}(n)$. In our example in Fig. 1.4 we have, e.g., $\mathscr{N}_3 = \{4, \ldots, 9\}$ and $\mathscr{G}(2) = \{2; 4, \ldots, 6; 10, \ldots, 14\}$.

Now we can rewrite problem (1.27) as the following optimization problem on the corresponding scenario tree:

$$\left.\begin{array}{l}\min\left\{c_1^{\mathrm{T}}x_1+\sum\limits_{n=2}^{K_2}p_nc_n^{\mathrm{T}}x_n+\sum\limits_{n=K_2+1}^{K_3}p_nc_n^{\mathrm{T}}x_n+\cdots\right.\\ \qquad\qquad\left.\cdots+\sum\limits_{n=K_{T-1}+1}^{K_T}p_nc_n^{\mathrm{T}}x_n\right\}\\[2ex] W_1x_1 \qquad\qquad\qquad\qquad = b_1\\ T_nx_1+W_nx_n \qquad\qquad = b_n,\ n=2,\cdots,K_2\\ \qquad T_nx_{h_n}+W_nx_n \qquad = b_n,\ n=K_2+1,\cdots,K_3\\ \qquad\qquad\ddots\\ \qquad\qquad T_nx_{h_n}+W_nx_n = b_n,\ n=K_{T-1}+1,\cdots,K_T\\ \qquad\qquad\qquad x_n\geq 0,\ n=1,\cdots,K_T\,.\end{array}\right\} \tag{1.28}$$

The above problem can compactly be written as follows:

$$\left.\begin{array}{rl}F_1=\min\ c_1^{\mathrm{T}}x_1+&\sum\limits_{\nu\in\mathcal{N}\setminus\{1\}}p_\nu c_\nu^{\mathrm{T}}x_\nu\\ \text{s.t.}\qquad W_1x_1&=b_1\\ T_\nu x_{h_\nu}+W_\nu x_\nu&=b_\nu,\ \nu\in\mathcal{N}\setminus\{1\}\\ x_\nu&\geq 0,\ \nu\in\mathcal{N}.\end{array}\right\} \tag{1.29}$$

For the ease of presentation we make the following assumption:

Assumption 1.1. *There exist individual upper bounds $x_n\leq U_n$, $U_n\geq 0$, $\forall n\in\mathcal{N}$.*

Note that U_n does also depend on $n\in\mathcal{N}$, i.e., multistage models involving individual stochastic upper bounds are incorporated in the above formulation.

Assumption 1.1., implying that (1.29) is either infeasible or otherwise has an optimal solution, may be included into (1.29) by adding the constraints $y_\nu\geq 0$, $I_\nu x_\nu+y_\nu=U_\nu$, $\forall\nu\in\mathcal{N}$ (I_ν identity matrices of the order $\operatorname{ord}I_\nu=\dim x_\nu$). Aggregating the sets $\{x_\nu,\ \nu\in\mathcal{N}\}$ and $\{y_\nu,\ \nu\in\mathcal{N}\}$ into one vector x and y each, and consequently condensing all the equality constraints into the two blocks $Ax=b$ and $Ix+y=U$, a further straightforward implication of assumption 1.1. is that (1.29) is dual feasible: considering the resulting (condensed) primal–dual pair of LP problems

$$\left\{\begin{array}{lll}\min\ c^{\mathrm{T}}x&&\\ \text{s.t.}\ \ Ax&&=b\\ \quad\ \ Ix+y&&=U\\ \quad\ \ x&&\geq 0\\ &&y\geq 0,\end{array}\right\},\quad\left\{\begin{array}{lll}\max\ b^{\mathrm{T}}v+U^{\mathrm{T}}w&&\\ \text{s.t.}\ \ A^{\mathrm{T}}v&+w&\leq c\\ &w&\leq 0,\end{array}\right\} \tag{1.30}$$

it is obvious that, to an arbitrarily chosen v, there exists w such that (v,w) is a feasible solution of the dual.

Problem (1.28) corresponds to the following sequence of programs: For node $n=1$

$$\left.\begin{aligned} F_1 = \min c_1^{\mathrm{T}} x_1 + \sum_{n=2}^{K_2} p_n F_n(x_1) \\ W_1 x_1 = b_1 \\ x_1 \geq 0; \end{aligned}\right\} \tag{1.31}$$

then for each node in stage 2, i.e., for $n \in \mathscr{N}_2 = \{2,\cdots,K_2\}$

$$\left.\begin{aligned} F_n(x_1) = \min c_n^{\mathrm{T}} x_n + \sum_{m\in\mathscr{C}(n)} \frac{p_m}{p_n} F_m(x_n) \\ W_n x_n = b_n - T_n x_1 \\ x_n \geq 0; \end{aligned}\right\} \tag{1.32}$$

and in general for any node $n \in \mathscr{N}_t = \{K_{t-1}+1,\ldots,K_t\}$ in stage $t_n \in \{3,\cdots,T-1\}$

$$\left.\begin{aligned} F_n(x_{h_n}) = \min c_n^{\mathrm{T}} x_n + \sum_{m\in\mathscr{C}(n)} \frac{p_m}{p_n} F_m(x_n) \\ W_n x_n = b_n - T_n x_{h_n} \\ x_n \geq 0. \end{aligned}\right\} \tag{1.33}$$

Finally, for nodes n in stage $t_n = T$, i.e., $n \in \mathscr{N}_T = \{K_{T-1}+1,\cdots,K_T\}$, we get

$$\left.\begin{aligned} F_n(x_{h_n}) = \min c_n^{\mathrm{T}} x_n \\ W_n x_n = b_n - T_n x_{h_n} \\ x_n \geq 0. \end{aligned}\right\} \tag{1.34}$$

For n with $t_n = T$ it is obvious from (1.34) that $F_n(x_{h_n})$ is piecewise linear and convex in x_{h_n} for all $n \in \{K_{T-1}+1,\cdots,K_T\}$ (see Prop. 1.18., p. 24). Then, going backwards through stages $T-1, T-2,\cdots,2$, it follows immediately from (1.33), that the additive terms $\sum_{m\in\mathscr{C}(n)} \frac{p_m}{p_n} F_m(x_n)$ are piecewise linear and convex in x_n implying that also the functions $F_n(x_{h_n})$ are piecewise linear and convex, such that by (1.32) also the additive term $\sum_{n=2}^{K_2} p_n F_n(x_1)$ in (1.31) is piecewise linear and convex in x_1.

Note that for $t_n > 1$ the NLP (1.33) can also be written in the following equivalent LP form:

$$\left.\begin{array}{rrl} F_n(x_{h_n}) = \min\ c_n^{\mathrm{T}} x_n + & \displaystyle\sum_{\nu \in \mathscr{G}(n)\setminus\{n\}} \frac{p_\nu}{p_n} c_\nu^{\mathrm{T}} x_\nu & \\ \text{s.t.} & W_n x_n & = b_n - T_n x_{h_n} \\ & T_\nu x_{h_\nu} + W_\nu x_\nu & = b_\nu,\ \nu \in \mathscr{G}(n)\setminus\{n\} \\ & x_\nu & \geq 0,\ \nu \in \mathscr{G}(n), \end{array}\right\} \tag{1.35}$$

with the parameter vector x_{h_n} and optimal–value function F_n. As usual, $F_n(x_{h_n}) := +\infty$ is taken in the infeasible case. The LP problem (1.35) will be called a *descendant recourse problem* and will be denoted by `LPDesc` (n, x_{h_n}). In this context, (1.33) is the recursive NLP formulation of `LPDesc` (n, x_{h_n}). For the sake of uniform notation we introduce a virtual node, indexed by 0, as the ancestor of the root node. Since the virtual node merely serves for simplifying notation, it is not added to $\mathscr{N}$. We define a matrix T_1 as an $(m_1 \times 1)$ zero matrix (a column vector), where m_1 is the number of rows of W_1. Interpreting x_{h_1} as an arbitrary number, the original multistage problem (1.29) is included into this notation, resulting in a constant optimal–value function $F_1(x_{h_1}) \equiv F_1$.

Assumption 1.1. implies that `LPDesc` (n, x_{h_n}) is either infeasible, or otherwise it has an optimal solution, furthermore, it is dual feasible for all $n \in \mathscr{N}$ and all x_{h_n}.

For a fixed x_{h_n} and $t_n < T$, (1.35) is the LP–equivalent of a $(T - t_n + 1)$–stage recourse problem, corresponding to the following scenario tree: take the subtree of the original scenario tree, which is rooted at $n \in \mathscr{N}$, and divide by p_n all probabilities associated with the nodes of the subtree. In particular, for $t_n = T - 1$ the LP problems `LPDesc` (n, x_{h_n}) are LP equivalents of two–stage recourse problems. For $n \in \mathscr{N}_T$ n is a leaf of the scenario tree and the LP `LPDesc` (n, x_{h_n}) is an ordinary one–stage LP problem. Nevertheless, for the sake of simplicity, we call also these LP's descendant recourse problems.

Above we have derived the piecewise linearity of $F_n(x_{h_n})$ using backward induction. An alternative way of showing this consists of considering the LP (1.35) for which Prop. 1.18. (p. 24) directly applies.

Consider problem (1.33) for some n with $t_n < T$. In analogy to (1.23) on page 25, we introduce an upper bound θ_n to replace the additive term $\displaystyle\sum_{m \in \mathscr{C}(n)} \frac{p_m}{p_n} F_m(x_n)$ in the objective function. Due to the piecewise linearity of the latter term, the upper bound θ_n has to satisfy some additional linear constraints

$$d_{nk}^{\mathrm{T}} x_n + \theta_n \geq \delta_{nk}, k = 1, \cdots, S_n .$$

In addition, some further linear constraints

$$a_{nj}^{\mathrm{T}} x_n \geq \alpha_{nj},\ j = 1, \cdots, R_n ,$$

may be necessary to ensure the feasibility (i.e., the finiteness of F_m) of the LP's for the nodes $m \in \mathscr{C}(n)$, such that (1.33) is now replaced by

$$
\left.\begin{array}{llll}
F_n(x_{h_n}) = \min c_n^{\mathrm{T}} x_n + \theta_n & & & \\
\quad W_n x_n & = b_n - T_n x_{h_n} & & \\
\quad a_{nj}^{\mathrm{T}} x_n & \geq \alpha_{nj}, & j = 1, \cdots, R_n & \\
\quad d_{nk}^{\mathrm{T}} x_n + \theta_n & \geq \delta_{nk}, & k = 1, \cdots, S_n & \\
\quad x_n & \geq 0. & &
\end{array}\right\} \quad (1.36)
$$

As discussed in connection with (1.24) on page 25, the main idea in dual decomposition is to solve a sequence of successively built *relaxed master problems*

$$
\left.\begin{array}{llll}
\tilde{F}_n(x_{h_n}) = \min c_n^{\mathrm{T}} x_n + \theta_n & & & \\
\quad W_n x_n & = b_n - T_n x_{h_n} & & \\
\quad a_{nj}^{\mathrm{T}} x_n & \geq \alpha_{nj}, & j = 1, \cdots, r_n & \\
\quad d_{nk}^{\mathrm{T}} x_n + \theta_n & \geq \delta_{nk}, & k = 1, \cdots, s_n & \\
\quad x_n & \geq 0, & &
\end{array}\right\} \quad (1.37)
$$

with parameter vector x_{h_n} and optimal–value function $\tilde{F}_n$. Similarly as for descendant recourse problems, for the root node holds $\tilde{F}_1(x_{h_1}) \equiv \tilde{F}_1$, with a constant value $\tilde{F}_1$.

The LP (1.37) will be denoted by `LPMast` (n, x_{h_n}) and will be called a *relaxed master problem*. Constraints in the second and third group of constraints will be called *feasibility cuts* and *optimality cuts*, respectively. In the nested decomposition (ND) algorithm these cuts will be added in a one–by–one manner to `LPMast` (n, x_{h_n}), as it will be discussed later. $r_n = 0$ or $s_n = 0$ means that the corresponding group of constraints is missing. Furthermore, if $s_n = 0$, then we assume that the variable θ_n is fixed by an additional constraint $\theta_n = 0$. Finally, we will use the above notation also for the leaves ($t_n = T$), by keeping $r_n = 0$ and $s_n = 0$ throughout.

Due to Assumption 1.1., `LPMast` (n, x_{h_n}) is either infeasible, or otherwise it has an optimal solution. It is also dual feasible, $\forall n \in \mathcal{N}$, $\forall x_{h_n}$.

Assume that $s_n > 0$ holds. In this case the third group of inequalities in the relaxed master problem (1.37) can equivalently be written as the following single inequality constraint:

$$
\theta_n \geq \max_{1 \leq k \leq s_n} (\delta_{nk} - d_{nk}^{\mathrm{T}} x_n).
$$

This can be put into the objective function thus leading to an equivalent formulation of the relaxed master problem

$$
\left.\begin{array}{lll}
\tilde{F}_n(x_{h_n}) = \min \left[c_n^{\mathrm{T}} x_n + \max\limits_{1 \leq k \leq s_n} (\delta_{nk} - d_{nk}^{\mathrm{T}} x_n) \right] & & \\
\quad W_n x_n = b_n - T_n x_{h_n} & & \\
\quad a_{nj}^{\mathrm{T}} x_n \geq \alpha_{nj}, & j = 1, \cdots, r_n & \\
\quad x_n \geq 0, & &
\end{array}\right\} \quad (1.38)
$$

as an NLP problem with a piecewise linear convex objective function. Note that for descendant recourse problems we had both an LP formulation (1.35) and an NLP formulation (1.33). These have their counterparts (1.37) and (1.38) concerning relaxed master problems.

In the multistage case with $T > 2$, for $1 \leq t_n < T-1$ two new features appear in comparison with the two–stage case, which have to be dealt with. On the one hand, both the descendant recourse problem (1.33) and the relaxed master problem (1.37) depend on a parameter vector x_{h_n}. On the other hand, the terms F_m in the objective function in (1.33) are defined by multistage (at least two–stage) problems. We have to explain how in this situation valid cuts can be constructed.

Let us consider a node $n \in \mathcal{N}_t$, $t < T$, and its child–nodes $m \in \mathscr{C}(n)$. We assume that the current relaxed master `LPMast` (n, x_{h_n}) has a solution $(\hat{x}_n, \hat{\theta}_n)$. The problems `LPMast` $(m, \hat{x}_n)$, assigned to the child–nodes $m \in \mathscr{C}(n)$, are either infeasible or have an optimal solution.

Feasibility cuts

If `LPMast` $(m, \hat{x}_n)$ is infeasible for an $m \in \mathscr{C}(n)$ then a feasibility cut will be added to `LPMast` (n, x_{h_n}). The infeasibility of (1.37) implies the objective of the corresponding dual

$$\left.\begin{array}{lrl} \max & (b_m - T_m\hat{x}_n)^{\mathrm{T}} u_m + \sum_{j=1}^{r_m} \alpha_{mj} v_{mj} + \sum_{k=1}^{s_m} \delta_{mk} w_{mk} & \\ \text{s.t.} & W_m^{\mathrm{T}} u_m + \sum_{j=1}^{r_m} a_{mj} v_{mj} + \sum_{k=1}^{s_m} d_{mk} w_{mk} & \leq c_n \\ & \sum_{k=1}^{s_m} w_{mk} & = 1 \\ & v_{mj} & \geq 0,\ \forall j \\ & w_{mk} & \geq 0,\ \forall k \end{array}\right\} \tag{1.39}$$

to be unbounded from above (note that (1.39) is feasible, due to Assumption 1.1.). Hence there exists a cone–generating vector $(\tilde{u}_m, \tilde{v}_m)$ with $\tilde{v}_m \geq 0$ satisfying

$$\begin{array}{rl} W_m^{\mathrm{T}} \tilde{u}_m + \sum_{j=1}^{r_m} a_{mj} \tilde{v}_{mj} \leq 0 & \text{and} \\ (b_m - T_m\hat{x}_n)^{\mathrm{T}} \tilde{u}_m + \sum_{j=1}^{r_m} \alpha_{mj} \tilde{v}_{mj} > 0. & \end{array} \tag{1.40}$$

In (1.40) $\tilde{w}_{mk}$ is missing since any cone–generating $(\tilde{u}_m, \tilde{v}_m, \tilde{w}_m)$ of (1.39) has to satisfy $\sum_{k=1}^{s_m} \tilde{w}_{mk} = 0$, $\tilde{w}_{mk} \geq 0 \forall k$, such that $\tilde{w}_{mk} = 0 \forall k$.

Thus, analogously to Benders' algorithm, for cutting off $\hat{x}_n$ which has led to the infeasible relaxed master `LPMast` $(m,\hat{x}_n)$, the following feasibility cut is added to `LPMast` (n,x_{h_n}):

$$(b_m - T_m x_n)^{\mathrm{T}} \tilde{u}_m + \sum_{j=1}^{r_m} \alpha_{mj} \tilde{v}_{mj} \leq 0, \tag{1.41}$$

or equivalently

$$a_m^{\mathrm{T}} x_n \geq \alpha_m \tag{1.42}$$

where $a_m := T_n^{\mathrm{T}} \tilde{u}_m$ and $\alpha_n := b_m^{\mathrm{T}} \tilde{u}_m + \sum_{j=1}^{r_m} \alpha_{mj} \tilde{v}_{mj}$ hold.

Notice that $\hat{x}_n$ is infeasible also for the descendant linear programming problem `LPDesc` (n,x_{h_n}) (see 1.35) in the following sense: it can not be extended to a feasible solution of this problem. It makes sense therefore to cut off this point.

Definition 1.6. *A feasibility cut* $a_{nj}^{\mathrm{T}} x_n \geq \alpha_{nj}$ *in* `LPMast` (n,x_{h_n}) *will be called valid, if for any feasible solution* $(\bar{x}_\nu,\ \nu \in \mathscr{G}(n))$ *of the descendant recourse problem* `LPDesc` (n,x_{h_n})*, the inequality* $a_{nj}^{\mathrm{T}} \bar{x}_n \geq \alpha_{nj}$ *holds.*

Validity of a feasibility cut means for any feasible solution of `LPDesc` (n,x_{h_n}), that the piece of this solution which corresponds to node n will not be cut off. In context of the NLP formulation (1.33), the jth feasibility cut in `LPMast` (n,x_{h_n}) is valid, if and only if for any feasible solution $\bar{x}_n$ of (1.33) for which $F_m(\bar{x}_n) < +\infty$ holds for all $m \in \mathscr{C}(n)$, the inequality $a_{nj}^{\mathrm{T}} \bar{x}_n \geq \alpha_{nj}$ holds.

Proposition 1.20. *The following assertions hold:*

(i) *Let* $n \in \mathscr{N}$ *be an arbitrary node. If in* `LPMast` (n,x_{h_n}) *either* $r_n = 0$ *holds or otherwise all feasibility cuts are valid then for any feasible solution* $(\bar{x}_\nu,\ \nu \in \mathscr{G}(n))$ *of* `LPDesc` (n,x_{h_n})*,* $\bar{x}_n$ *is a feasible solution of* `LPMast` (n,x_{h_n}).

(ii) *Let* $n \in \mathscr{N}$ *and* $m \in \mathscr{C}(n)$ *be the nodes which have been considered for generating the feasibility cut. Provided that in* `LPMast` $(m,\hat{x}_n)$ *either* $r_m = 0$ *holds or otherwise all feasibility cuts are valid, the new cut is a valid feasibility cut in* `LPMast` (n,x_{h_n}).

Proof:

(i): If $r_n = 0$ then *(i)* is obviously true. Otherwise the assertion is an immediate consequence of the definition of validity.

(ii): To see this, assume that $(\bar{x}_\nu,\ \nu \in \mathscr{G}(n))$ is a feasible solution of the descendant recourse problem `LPDesc` (n,x_{h_n}). In particular for node $m \in \mathscr{C}(n)$, from which the cut has been generated, we have:

$$\begin{aligned} W_m \bar{x}_m &= b_m - T_m \bar{x}_n \\ a_{mj}^{\mathrm{T}} \bar{x}_m &\geq \alpha_{mj},\ j = 1,\ldots,r_m \end{aligned}$$

where the second inequality holds for the following reason: $(\bar{x}_\mu, \mu \in \mathscr{G}(m))$ is obviously a feasible solution of `LPDesc` $(m,\bar{x}_n)$ and then the inequality follows from the assumption concerning feasibility cuts in `LPMast` $(m,\hat{x}_n)$. Multiplying by $\tilde{u}_m$, $\tilde{v}_m$, and summing up, we get

$$(b_m - T_m\bar{x}_n)^{\mathrm{T}}\tilde{u}_m + \sum_{j=1}^{r_m} \alpha_{mj}\tilde{v}_{mj} \le (W_m^{\mathrm{T}}\tilde{u}_m + \sum_{j=1}^{r_m} a_{mj}\tilde{v}_{mj})\bar{x}_m \le 0$$

where the last inequality follows from (1.40) and from the nonnegativity of $\bar{x}_m$. This shows (see (1.41)) that for $\bar{x}_n$ the newly added inequality holds.

□

Optimality cuts

If `LPMast` $(m,\hat{x}_n)$ has a solution for all $m \in \mathscr{C}(n)$ then we consider appending an optimality cut to `LPMast` (n,x_{h_n}). Let $(\hat{x}_m, \hat{\theta}_m)$ be an optimal solution of `LPMast` $(m,\hat{x}_n)$ and $(\hat{u}_m, \hat{v}_m, \hat{w}_m)$ be an optimal solution of its dual (1.39), then we have

$$\begin{aligned} \tilde{F}_m(\hat{x}_n) &= c_m^{\mathrm{T}}\hat{x}_m + \hat{\theta}_m \\ &= (b_m - T_m\hat{x}_n)^{\mathrm{T}}\hat{u}_m + \sum_{j=1}^{r_m} \alpha_{mj}\hat{v}_{mj} + \sum_{k=1}^{s_m} \delta_{mk}\hat{w}_{mk}, \end{aligned} \tag{1.43}$$

for all $m \in \mathscr{C}(n)$. The key observation concerning optimality cuts is the following: The feasible domain of the dual problem (1.39) does not depend on $\hat{x}_n$. Consequently, due to weak duality in LP, we have that

$$\tilde{F}_m(x_n) \ge (b_m - T_m x_n)^{\mathrm{T}}\hat{u}_m + \sum_{j=1}^{r_m} \alpha_{mj}\hat{v}_{mj} + \sum_{k=1}^{s_m} \delta_{mk}\hat{w}_{mk} \tag{1.44}$$

holds for any x_n. Therefore we consider adding the following optimality cut to `LPMast` (n,x_{h_n}):

$$\theta_n \ge \sum_{m \in \mathscr{C}(n)} \frac{p_m}{p_n} \left[(b_m - T_m x_n)^{\mathrm{T}}\hat{u}_m + \sum_{j=1}^{r_m} \alpha_{mj}\hat{v}_{mj} + \sum_{k=1}^{s_m} \delta_{mk}\hat{w}_{mk} \right]. \tag{1.45}$$

If the above inequality holds for $(\hat{x}_n, \hat{\theta}_n)$, which is the current solution of `LPMast` (n,x_{h_n}), then the new constraint would be redundant, otherwise the optimality cut will be added to `LPMast` (n,x_{h_n}).

The optimality cut can equivalently be written as

$$d_{nk}^{\mathrm{T}} x_n + \theta_n \ge \delta_{nk},\ k = s_n + 1 \tag{1.46}$$

with

$$d_{nk} := \sum_{m \in \mathscr{C}(n)} \frac{p_m}{p_n} T_m^{\mathrm{T}} \hat{u}_m \quad \text{and}$$

$$\delta_{nk} := \sum_{m \in \mathscr{C}(n)} \frac{p_m}{p_n} \left[b_m^{\mathrm{T}} \hat{u}_m + \sum_{j=1}^{r_m} \alpha_{mj} \hat{v}_{mj} + \sum_{k=1}^{s_m} \delta_{mk} \hat{w}_{mk} \right].$$

With the notation just introduced, (1.44) implies that

$$\sum_{m \in \mathscr{C}(n)} \frac{p_m}{p_n} \tilde{F}_m(x_n) \geq \delta_{nk} - d_{nk}^{\mathrm{T}} x_n, \quad \forall x_n, \tag{1.47}$$

holds for $k = s_n + 1$ and for arbitrary x_n. In deriving this inequality we have also used the fact, that for the scenario tree $\sum_{m \in \mathscr{C}(n)} p_m = p_n$ holds.

The optimality cut clearly cuts off the current solution $(\hat{x}_n, \hat{\theta}_n)$.

Definition 1.7. *An optimality cut $d_{nk}^{\mathrm{T}} x_n + \theta_n \geq \delta_{nk}$ in* `LPMast` (n, x_{h_n}) *is called valid, if the inequality $\delta_{nk} - d_{nk}^{\mathrm{T}} x_n \leq \sum_{m \in \mathscr{C}(n)} \frac{p_m}{p_n} F_m(x_n)$ holds for arbitrary x_n.*

Comparing the NLP formulations (1.38) and (1.33) of `LPMast` (n, x_{h_n}) and `LPDesc` (n, x_{h_n}), respectively, we observe the reason for this requirement: We wish to achieve that the objective function of the relaxed master problem yields a lower bound to the objective function of the descendant recourse problem.

Proposition 1.21. *The following assertions hold:*

(i) *Let $n \in \mathscr{N}$ be an arbitrary node and assume that all feasibility cuts are valid in* `LPMast` (n, x_{h_n}). *If either $n \in \mathscr{N}_T$, or in* `LPMast` (n, x_{h_n}) $s_n > 0$ *holds and all optimality cuts are valid then $\tilde{F}_n(x_{h_n}) \leq F_n(x_{h_n})$ holds for any x_{h_n}.*

(ii) *Let $n \in \mathscr{N}$ be the node considered in the discussion on optimality cuts. If either $n \in \mathscr{N}_{T-1}$, or for all $m \in \mathscr{C}(n)$ $s_m > 0$ holds and all feasibility– and optimality cuts are valid in* `LPMast` $(m, \hat{x}_n)$, *then the new cut is a valid optimality cut in* `LPMast` (n, x_{h_n}).

Proof:

(i): In the case $n \in \mathscr{N}_T$ the problems `LPMast` (n, x_{h_n}) and `LPDesc` (n, x_{h_n}) are identical and therefore we have $\tilde{F}_n(x_{h_n}) = F_n(x_{h_n})$ for all x_{h_n}. Assume $n \in \mathscr{N}_t$ with $t \leq T-1$. Our assumption implies the inequality

$$c_n^{\mathrm{T}} x_n + \max_{1 \leq k \leq s_n} (\delta_{nk} - d_{nk}^{\mathrm{T}} x_n) \leq c_n^{\mathrm{T}} x_n + \sum_{m \in \mathscr{C}(n)} \frac{p_m}{p_n} F_m(x_n) \tag{1.48}$$

for arbitrary x_n. We consider the NLP formulations (1.33) and (1.38) of `LPDesc` (n, x_{h_n}) and `LPMast` (n, x_{h_n}), respectively. If (1.38) is infeasible ($\tilde{F}_n(x_{h_n}) = +\infty$) then due to Proposition 1.20. *(i)*, `LPDesc` (n, x_{h_n}) is also infeasible and consequently $F_n(x_{h_n}) = +\infty$ holds. Thus we have $\tilde{F}_n(x_{h_n}) =$

$F_n(x_{h_n}) = +\infty$. Assume that (1.38) is feasible. For any feasible solution in (1.33), which is infeasible in (1.38), at least one feasibility–cut constraint in the latter is violated. The validity of this feasibility cut implies that the right–hand–side in (1.48) is $+\infty$. Thus taking minima on both sides of (1.48) over the feasible domain of (1.38) results in our inequality.

(ii): By part *(i)* of this proposition, our assumption implies that $\tilde{F}_m(x_n) \le F_m(x_n)$ holds for all $m \in \mathscr{C}(n)$ and arbitrary x_n. Utilizing (1.47) we get the inequality

$$\delta_{nk} - d_{nk}^{\mathrm{T}} x_n \le \sum_{m \in \mathscr{C}(n)} \frac{p_m}{p_n} \tilde{F}_m(x_n) \le \sum_{m \in \mathscr{C}(n)} \frac{p_m}{p_n} F_m(x_n)$$

which proves *(ii)*.

□

Now we are prepared to describe the nested decomposition (ND) algorithm. This consists of carrying out the following three basic operations in an iterative fashion.

Starting with stage t_0, a *forward pass* consists of an attempt of solving all relaxed master problems in stages $t \ge t_0$, in a stage–by–stage manner. The solutions obtained in stage t are used to set up the relaxed master problems in stage $t+1$. A forward pass terminates either by encountering a node n such that `LPMast` $(n,\hat{x}_{h_n})$ is infeasible, or by obtaining a solution $\hat{x}_n$ for all nodes n with $t_n \ge t_0$. The solutions obtained this way are consistent in the following sense: for any node n with $t_n > t_0$, before setting up and solving `LPMast` $(n,\hat{x}_{h_n})$ the relaxed master problem associated with the ancestor node has been already solved and the solution of the ancestor problem is used to set up `LPMast` $(n,\hat{x}_{h_n})$. In particular, this implies that for any node n with $t_n \ge t_0$, $(\hat{x}_\nu, \nu \in \mathscr{G}(n))$ is a feasible solution of the descendant recourse problem `LPDesc` $(n,\hat{x}_{h_n})$.

Backtracking starts with a node n, for which `LPMast` $(n,\hat{x}_{h_n})$ is infeasible. The following steps are carried out along the unique path from n to the root. First a feasibility cut is added to the ancestor's relaxed master problem. The relaxed master of the ancestor is solved next. If this turns out to be infeasible then the procedure is repeated with the ancestor node being the current node. Backtracking terminates either by finding a node along the path with a feasible relaxed master problem, or by reaching the root node with an infeasible associated relaxed master problem. In the latter case the multistage problem is infeasible, the overall procedure terminates.

A *backward pass* presupposes that `LPMast` $(n,\hat{x}_{h_n})$ is feasible with an optimal solution $\hat{x}_n$ for all $n \in \mathscr{N}$. Starting with $t = T-1$, an attempt is made to add optimality cuts to all relaxed master problems in stage t. Relaxed master problems with added optimality cuts are solved. Afterwards this is repeated with stage $t = T-2$, and so on, in a backward stage–by–stage manner. Since adding an optimality cut does not render a feasible relaxed master problem infeasible, the backward pass terminates by reaching the root node. If during a whole backward pass no optimality cuts have been added then the current solution is optimal and the overall procedure terminates.

Note that if for any node n with $t_n < T$ the solution $\hat{x}_n$ changes then the current solutions (if any) associated with the nodes in $\mathcal{G}(n) \setminus \{n\}$ become invalid in the overall procedure, in general. The reason is that changing $\hat{x}_n$ implies changing the parameter in `LPMast` $(m, \hat{x}_n)$ for all $m \in \mathcal{C}(n)$ which may result in changing the solution $\hat{x}_m$. This in turn implies changes in the parametrization of the relaxed master problems associated with the child-nodes of $m \in \mathcal{C}(n)$, and so on.

Next we formulate the nested decomposition (ND) algorithm.

Nested Decomposition Algorithm

S 0 Initialization
Let $r_n = 0,\ s_n = 0,\ \gamma_n =$ False
and add the constraint $\theta_n = 0$ to `LPMast` (n, x_{h_n}), $\forall n \in \mathcal{N}$.
Set $t := 1$ and for formal reasons set $\hat{x}_{h_1} = 0$.
The Boolean variable γ_n will be used for the following purpose: $\gamma_n = True$ indicates that the current relaxed master `LPMast` $(n, \hat{x}_{h_n})$ has a solution and it is legitimate to use the current solution $(\hat{x}_n, \hat{\theta}_n)$ when node n is encountered during the subsequent iterations. $\gamma_n = False$ indicates that `LPMast` $(n, \hat{x}_{h_n})$ is to be solved whenever node n is encountered. (Observe that for $n \in \mathcal{N}_T$ we'll have $r_n = s_n = 0$ as well as $\theta_n \equiv 0$ throughout the procedure.)

S 1 Select Direction
If $t < T$ then go to *S 2* (forward pass), otherwise go to *S 3* (backward pass).

S 2 Forward Pass
For $n \in \mathcal{N}_t$ for which $\gamma_n =$ False in turn do:

- Solve `LPMast` $(n, \hat{x}_{h_n})$. If infeasible then store $(\tilde{u}_n, \tilde{v}_n)$ which fulfills (1.40) and continue with *S 4* (backtracking). Otherwise continue this loop with the next step.
- Store the solution $(\hat{x}_n, \hat{\theta}_n)$; if $t = T$ then store also the dual solution $(\hat{u}_n, \hat{v}_n, \hat{w}_n)$;
- set $\gamma_n =$ True and $\gamma_\nu :=$ False for all $\nu \in \mathcal{G}(n) \setminus \{n\}$;
- take the next node in $\mathcal{N}_t$.

If this loop goes through without jumping to *S 4* then proceed as follows: if $t = T$ then go to *S 1*, otherwise set $t := t + 1$ and repeat *S 2*.

S 3 Backward pass
Set $\gamma :=$ True. This Boolean variable is only used in the present backward pass. $\gamma = True$ indicates that no optimality cuts have been added so far.
For $n \in \mathcal{N}_{t-1}$ in turn do:

- Check whether (1.45) holds for the current solution $(\hat{x}_n, \hat{\theta}_n)$;
- if yes, then take the next node in $\mathcal{N}_{t-1}$, otherwise
- add an optimality cut:
  - Set $\gamma :=$ False;
  - if $s_n = 0$ then drop the constraint $\theta_n = 0$;

- add the optimality cut (1.46) to `LPMast` $(n, \hat{x}_{h_n})$ with $k := s_n + 1$; set $s_n := s_n + 1$;
- solve `LPMast` $(n, \hat{x}_{h_n})$ and temporarily store the dual solution $(\hat{u}_n, \hat{v}_n, \hat{w}_n)$.

Note that this loop always goes through: adding an optimality cut does not render a previously feasible relaxed master problem infeasible. After this loop has gone through check for optimality: If $t = 1$ and $\gamma =$ True then no optimality cut has been added through a whole backward cycle. In this case the current solution is optimal, **Stop**. Otherwise if $t > 1$ then set $t := t - 1$ and repeat *S 3*, else return to *S 1*.

S 4 Backtracking

- If $n = 1$ then the multistage problem is infeasible, **Stop**. Otherwise
- make the predecessor of n the current node, i.e., set $m := n$ and subsequently $n := h_m$.
- Add a feasibility cut to `LPMast` $(n, \hat{x}_{h_n})$ according to (1.42);
- set $\gamma_v :=$ False for all $v \in \mathscr{G}(n)$;
- solve `LPMast` $(n, \hat{x}_{h_n})$. If infeasible then compute $(\tilde{u}_n, \tilde{v}_n)$ which fulfills (1.40) and repeat *S 4*. Otherwise set $\gamma_n :=$ True, store a solution $(\hat{x}_n, \hat{\theta}_n)$ and return to *S 1*.

Proposition 1.22. *The following assertions hold:*

(i) *The feasibility cuts generated by the algorithm are valid.*
(ii) *The optimality cuts generated by the algorithm are also valid. Furthermore, $\tilde{F}_n(x_{h_n}) \le F_n(x_{h_n})$ holds for all $n \in \mathscr{N}$ and all x_{h_n}.*
(iii) *The algorithm terminates in a finite number of iterations.*
(iv) *If the algorithm terminates in S 4 then the multistage problem is infeasible; if termination occurs in S 3 then the current solution $(\hat{x}_n,\ n \in \mathscr{N})$ is optimal.*

Proof:

(i) Feasibility cuts are generated along backward chains in *S 4*. If $r_n = 0$ holds for `LPMast` $(n, \hat{x}_{h_n})$, belonging to the starting node n of the chain (the node in the highest stage), then Proposition 1.20. implies that all feasibility cuts added along the chain are valid. This is the case in the initial phase of the method. If later on the starting node already has feasibility cuts, they are valid, therefore again Proposition 1.20. applies thus ensuring the validity of the newly generated cuts.

(ii) The validity of the optimality cuts follows immediately from Proposition 1.21.. For the inequality we observe that $\tilde{F}(x_{h_n}) = F(x_{h_n})$ holds for the leaves $n \in \mathscr{N}_T$, therefore our inequality follows from Proposition 1.21. by backward induction.

(iii) Due to the construction of the algorithm, none of the cone–generating elements and dual feasible basic solutions of `LPMast` $(m, \hat{x}_n)$ $(m \in \mathscr{C}(n))$ is

used repeatedly for adding cuts to `LPMast` (n, x_{h_n}). Consequently, for finite termination it is sufficient to show that for any node $n \in \mathcal{N}$ there exist finitely many different cone–generating elements and dual feasible basic solutions of relaxed master problems associated with the child–nodes. This is a consequence of the fact that the dual feasible region of `LPMast` $(m, \hat{x}_n)$ does not depend on $\hat{x}_n$ (see also the discussion on page 39).
For nodes n with $t_n = T$ (leaves), both the set of cone–generating elements and the set of feasible basic dual solutions are obviously finite. Let us consider a node n with $t_n = T - 1$. Both types of cuts for this node are generated either on the basis of cone–generating elements or on the basis of dual basic feasible solutions of `LPMast` $(m, \hat{x}_n)$ with $m \in \mathscr{C}(n)$. Consequently, the number of different feasibility– and optimality cuts in `LPMast` $(n, \hat{x}_{h_n})$ is finite and the set of possible cuts is independent on the specific value of x_{h_n}. This implies that for `LPMast` (n, x_{h_n}) the number of different dual feasible sets is also finite. Consequently, for each node n with $t_n = T - 1$, the number of cone–generating elements and dual basic feasible solutions is finite. These are used for generating cuts for nodes n with $t_n = T - 2$. Using backward induction according to stages, it follows that, for any node $n \in \mathcal{N}$, there are finitely many different feasibility– and optimality cuts. This proves *(iii)*.

(iv) If the algorithm terminates in *S* 4 then `LPMast` $(1, \hat{x}_{h_1})$ is infeasible. Then, due to assertion (i), `LPDesc` $(1, \hat{x}_{h_1})$ is also infeasible. The latter being the original multistage problem this proves the first statement.
For any node $n \in \mathcal{N}$, by successively applying (1.45) and (1.43) we get

$$
\begin{aligned}
\tilde{F}_n(x_{h_n}) &= c_n^{\mathrm{T}} \hat{x}_n + \hat{\theta}_n \\
&\geq c_n^{\mathrm{T}} \hat{x}_n + \sum_{m \in \mathscr{C}(n)} \frac{p_m}{p_n} (c_m^{\mathrm{T}} \hat{x}_m + \hat{\theta}_m) \\
&\geq c_n^{\mathrm{T}} \hat{x}_n + \sum_{m \in \mathscr{C}(n)} \frac{p_m}{p_n} \left(c_m^{\mathrm{T}} \hat{x}_m + \sum_{\mu \in \mathscr{C}(m)} \frac{p_\mu}{p_m} (c_\mu^{\mathrm{T}} \hat{x}_\mu + \hat{\theta}_\mu) \right) \\
&= c_n^{\mathrm{T}} \hat{x}_n + \sum_{m \in \mathscr{C}(n)} \frac{p_m}{p_n} c_m^{\mathrm{T}} \hat{x}_m + \sum_{m \in \mathscr{C}(n)} \sum_{\mu \in \mathscr{C}(m)} \frac{p_\mu}{p_n} (c_\mu^{\mathrm{T}} \hat{x}_\mu + \hat{\theta}_\mu) \\
&\;\;\vdots \\
&\geq c_n^{\mathrm{T}} \hat{x}_n + \sum_{\nu \in \mathscr{G}(n)} \frac{p_\nu}{p_n} c_\nu^{\mathrm{T}} \hat{x}_\nu \\
&\geq F_n(x_{h_n}),
\end{aligned}
\tag{1.49}
$$

where the last inequality follows from the fact, that $((\hat{x}_\nu, \hat{\theta}_\nu)), \nu \in \mathscr{G}(n))$ is a feasible solution of `LPDesc` $(n, \hat{x}_{h_n})$. The full proof follows by an obvious induction. Applying this for $n = 1$, together with assertion *(ii)*, the result follows. The above proof also shows that at optimality (1.45) is fulfilled as an equality throughout. □

Regarding (1.36) and (1.37), we took the liberty of using in both problems the same notation for the cuts. For $T > 2$, however, the nested decomposition method generates optimality cuts for `LPMast` (n, x_{h_n}) which are not necessarily among the optimality cuts of `LPDesc` (n, x_{h_n}), not even at points of differentiability of the objective function in (1.33).

For the dual decomposition method, master problems can be considered as relaxations of the full representation (1.24) and the algorithm can be interpreted as building the set of additional constraints in a step–by–step fashion (see page 25). As indicated above, this interpretation is no more valid in the multistage case. The reason is that, for $T > 2$ and $n \in \mathcal{N}_t$ with $1 \leq t \leq T-2$, optimality cuts are based on relaxed master problems which are themselves in the process of being built up. Therefore, optimality cuts do not provide necessarily supporting hyperplanes to the true optimal–value function. An example for this behavior can be found in Birge–Louveaux [26], Section 7.1. For indicating this distinctive feature, we used the term "*relaxed* master problem" whereas in Section 1.2.6 on dual decomposition the term "master problem" has been employed.

As in the dual decomposition method, after a backward pass the current value of $\tilde{F}_1$ clearly provides a lower bound on the optimal objective value of the multistage problem. After a complete forward pass, i.e. if during a forward pass all relaxed master problems turn out to be feasible, the current solution $(\hat{x}_n, n \in \mathcal{N})$ is a feasible solution of the multistage problem (1.28). Thus, computing the corresponding objective value results in an upper bound on the optimal objective value of the multistage problem.

Finally let us remark that, based on Propositions 1.20. and 1.21., several different variants of ND can be built, which differ on the *sequencing protocol*, the latter meaning the sequence in which nodes are processed (relaxed master problems are considered) in the algorithm. The variant which has been discussed in this section implements the *FFFB* (fast–forward–fast–backward) protocol, which has been found in empirical studies by Gassmann [112] to be the best variant.

Nested decomposition for deterministic LP's with a staircase structure has been studied by Abrahamson [2], Dantzig [51], and Wittrock [345], [346]. The generalization of the dual decomposition to a nested decomposition scheme for multistage problems is due to Birge [23]. The method is also called nested *L*–shaped method, see Birge–Louveaux [26].

Finally let us mention that multi–cut versions of the ND method can also be built analogously as for two–stage problems, see Section 1.2.6.

1.2.8 Regularized decomposition

To reduce the notation, we may write the k-th master problem for the multicut method as

$$\min\left\{ c^{\mathrm{T}}x+\sum_{j=1}^{S} p_j\theta_j \,\middle|\, (x,\theta_1,\cdots,\theta_S)\in\mathcal{D}_k \right\}, \tag{1.50}$$

where $\mathcal{D}_k$ is the feasible set associated with the set $\mathcal{G}_k$ of constraints required in this master program. Hence, instead of minimizing

$$\Phi(x)=c^{\mathrm{T}}x+\sum_{j=1}^{S} p_j f_j(x)$$

we minimize, with respect to x,

$$\hat{\Phi}_k(x)=c^{\mathrm{T}}x+\min_{\theta}\left\{ \sum_{j=1}^{S} p_j\theta_j \,\middle|\, (x,\theta_1,\cdots,\theta_S)\in\mathcal{D}_k \right\},$$

a piecewise linear function supporting from below the piecewise linear objective function Φ of our original NLP (1.25). In particular, in the early cycles of the algorithm, this support function $\hat{\Phi}_k$ is likely not to represent very well the true function Φ in some neighborhood of the last iterate $\hat{x}^{(k)}$. This may imply, that even for an $\hat{x}^{(k)}$ close to the overall optimum of (1.25) we get from solving (1.50) an $\hat{x}^{(k+1)}$ far away from the optimal point. Hence, it is no surprise that, in real size problems, we often observe an "erratic jumping around" of the subsequent iterates $x^{(\ell)}$ without a substantial progress in the objective, even when starting from an overall feasible iterate $x^{(k)}$ close to the solution of the original NLP (1.25). This undesirable behaviour may be improved substantially by regularizing the master program with an additive quadratic term which shall avoid too big steps away from an overall feasible approximate solution $z^{(k)}$ within one iteration. Hence, with some control parameter $\rho>0$ and denoting the Euclidean norm as $\|\cdot\|$, we deal with master programs of the form

$$\min\left\{ \frac{1}{2\rho}\|x-z^{(k)}\|^2+c^{\mathrm{T}}x+\sum_{j=1}^{S} p_j\theta_j \,\middle|\, (x,\theta_1,\cdots,\theta_S)\in\mathcal{D}_k \right\} \tag{1.51}$$

to find a next trial point $x^{(k)}$, for which we have to decide by criteria to be mentioned in the presentation of the algorithm, whether it is accepted as the next approximate or whether we continue with the current approximate, $z^{(k)}$.

We restrict ourselves to just giving a sketch of the modified algorithm. For simplicity, degeneracy in the constraints $\mathcal{G}_k$ of (1.50) is excluded by assumption, such that every vertex of the feasible set $\mathcal{D}_k\subset\mathbb{R}^{n+S}$ is determined by exactly $n+S$ active constraints (including the first stage equations $Ax = b$ and active nonnegativity conditions, i.e. $x_i = 0$ in case). Now we can present a sketch of the

Regularized Decomposition Algorithm QDECOM

S1 Determine a first approximate $z^{(1)}$, overall feasible for (1.25); let $k := 1$, and define $\mathscr{D}_1$ as the feasible set determined by the constraint set

$$\mathscr{G}_1 := \{Ax = b\} \cup \{\text{all optimality cuts at } z^{(1)}\}.$$

S2 Solve (1.51) for $x^{(k)}$ as first stage trial point and $\theta^{(k)} = (\theta_1^{(k)}, \cdots, \theta_S^{(k)})^{\mathrm{T}}$ as recourse approximates.
If $\Phi(z^{(k)}) = \hat{\Phi}(x^{(k)})$ $(= c^{\mathrm{T}}x^{(k)} + \sum_{j=1}^{S} p_j \theta_j^{(k)})$, then stop; $z^{(k)}$ is an optimal first stage solution for (1.25). Otherwise continue.

S3 Delete from the constraint set $\mathscr{G}_k$ of (1.51) constraints being inactive at $(x^{(k)}, \theta^{(k)})$, such that no more than $n+S$ constraints are left.

S4 If $x^{(k)}$ satisfies all first stage constraints (i.e. in particular $x^{(k)} \geq 0$), then go to step *S5*; otherwise add to $\mathscr{G}_k$ no more than S violated first stage constraints (nonnegativity conditions $x_i \geq 0$), yielding $\mathscr{G}_{k+1}$; let $z^{(k+1)} := z^{(k)}$, $k := k+1$, and go to step *S2*.

S5 Determine $f_j(x^{(k)})$, $j = 1, \cdots, S$.
If $f_j(x^{(k)}) = +\infty$ then add a feasibility cut to $\mathscr{G}_k$,
else if $f_j(x^{(k)}) > \theta_j^{(k)}$ then add an optimality cut to $\mathscr{G}_k$.

S6 If $f_j(x^{(k)}) = +\infty$ for at least one j then let $z^{(k+1)} := z^{(k)}$ and go to step *S8*; otherwise go to step *S7*.

S7 If $\Phi(x^{(k)}) = \hat{\Phi}(x^{(k)})$,
or else if $\Phi(x^{(k)}) \leq \mu\Phi(z^{(k)}) + (1-\mu)\hat{\Phi}(x^{(k)})$ for some parameter $\mu \in (0,1)$ and exactly $n+S$ constraints were active at $(x^{(k)}, \theta^{(k)})$,
then let $z^{(k+1)} := x^{(k)}$;
otherwise, let $z^{(k+1)} := z^{(k)}$.

S8 Let $\mathscr{G}_{k+1}$ be the constraint set resulting from $\mathscr{G}_k$ after deleting and adding constraints due to steps *S3* and *S5*, respectively. With $\mathscr{D}_{k+1}$ the corresponding feasible set and $k := k+1$ return to step *S2*.

The parameters $\rho > 0$ and $\mu \in (0,1)$ can be chosen adaptively between fixed bounds in order to improve the progress of the algorithm.

As we see immediately, during this algorithm all approximates $z^{(k)}$ are overall feasible since the change $z^{(k+1)} := x^{(k)}$ only takes place in step *S7*,

— either if $\Phi(x^{(k)}) = \hat{\Phi}(x^{(k)})$, which means that the piecewise linear support $\hat{\Phi}$ of Φ coincides with Φ in $x^{(k)}$, as well as obviously in $z^{(k)}$, such that, since $(x^{(k)}, \theta^{(k)})$ minimizes (1.51), we have the inequality $\hat{\Phi}(x^{(k)}) \leq \hat{\Phi}(z^{(k)})$ implying $\Phi(x^{(k)}) < +\infty$ and hence the overall feasibility of $x^{(k)}$, and continuing with the unchanged approximate $z^{(k)}$ would block the procedure;

— or if $(x^{(k)}, \theta^{(k)})$ is a vertex of $\mathscr{D}_k$ (corresponding to $\hat{\Phi}$ having a kink in $x^{(k)}$) and the decrease of Φ from $z^{(k)}$ to $x^{(k)}$

$$\begin{aligned}\Phi(x^{(k)}) - \Phi(z^{(k)}) &\le (1-\mu)(\hat{\Phi}(x^{(k)}) - \Phi(z^{(k)})) \\ &= (1-\mu)(\hat{\Phi}(x^{(k)}) - \hat{\Phi}(z^{(k)})) < 0\end{aligned}$$

is substantial with respect to the corresponding decrease of $\hat{\Phi}$ and implies, due to $\Phi(x^{(k)}) - \Phi(z^{(k)}) < 0$ and therefore $\Phi(x^{(k)}) < +\infty$, again the overall feasibility of $x^{(k)}$. As an example see Fig. 1.3, with the correspondences $\Phi \widehat{=} f$ and $\hat{\Phi} \widehat{=} \phi$. Here, starting from $z^{(1)} = x^{(1)}$ with the related optimality cut, we find $x^{(2)}$ according to the feasibility cut being active there. Then we add a new optimality cut in $x^{(2)}$ due to step $S5$, but keep $z^{(2)} := z^{(1)}$ since $\Phi(x^{(2)}) > \Phi(z^{(1)})$. Hence we get next the trial point $x^{(3)}$ which—depending on the choice of μ—could be a candidate for the next approximate $z^{(3)}$.

The algorithm QDECOM was proposed by Ruszczyński [293], where the details including the proof of its finiteness can be found. The same author also provided an implementation of QDECOM which for a very large variety of test problems has shown to be highly reliable as well as efficient.

Remark 1.2. *It should be pointed out that in the above examples of decomposition algorithms just prototype variants for the generation of cuts were presented. For particular applications it might however be advantageous to take into consideration special data structures of the corresponding models when designing the cuts.* □

1.2.9 Interior Point Methods

For the primal LP (1.3) and its dual (1.18), introducing for the latter one the slack variables $s_i \ge 0$, $i = 1, \cdots, n$, we know from Prop. 1.12. that for a primal-dual pair of solutions the following system has to be satisfied:

$$\left.\begin{array}{rl} Ax & = b \\ A^{\mathrm{T}}u + s & = c \\ x & \ge 0 \\ s & \ge 0 \\ x^{\mathrm{T}}s & = 0. \end{array}\right\} \qquad (1.52)$$

Defining the diagonal matrices $X := \operatorname{diag}(x_i)$ and $S := \operatorname{diag}(s_i)$, the above system requires to find a solution (with $e = (1, \cdots, 1)^{\mathrm{T}}$) of

$$F(x,u,s) := \begin{pmatrix} A^{\mathrm{T}}u + Is - c \\ Ax - b \\ XSe \end{pmatrix} = 0 \tag{1.53}$$

such that

$$x \geq 0,\ s \geq 0. \tag{1.54}$$

For the Jacobian of F we have

$$\mathscr{J}(x,u,s) = \begin{pmatrix} 0 & A^{\mathrm{T}} & I \\ A & 0 & 0 \\ S & 0 & X \end{pmatrix}$$

which, due to our general assumption that $\operatorname{rank}(A) = m$ (see page 7), is nonsingular as long as $x_i > 0,\ s_i > 0,\ i = 1,\cdots,n$. Hence, having at hand a primal-dual feasible pair $(\hat{x},\hat{u},\hat{s})$ satisfying the condition

$$Ax = b,\ A^{\mathrm{T}}u + s = c,\ x > 0,\ s > 0, \tag{1.55}$$

called *strict feasibility* or else *interior–point* condition, we may uniquely determine the search direction of the Newton method for the solution of the system (1.53) with the conditions (1.54) by solving the linear equations

$$F(\hat{x},\hat{u},\hat{s}) + \mathscr{J}(\hat{x},\hat{u},\hat{s}) \begin{pmatrix} \Delta x \\ \Delta u \\ \Delta s \end{pmatrix} = \begin{pmatrix} 0 \\ 0 \\ 0 \end{pmatrix} \tag{1.56}$$

or equivalently

$$\begin{pmatrix} 0 & A^{\mathrm{T}} & I \\ A & 0 & 0 \\ \hat{S} & 0 & \hat{X} \end{pmatrix} \begin{pmatrix} \Delta x \\ \Delta u \\ \Delta s \end{pmatrix} = \begin{pmatrix} 0 \\ 0 \\ -\hat{X}\hat{S}e \end{pmatrix}. \tag{1.57}$$

Proposition 1.23. *Given the strict feasibility condition (1.55), for any $w \in \mathbb{R}^n$: $w_i > 0\ \forall i$, there are uniquely determined x,u,s satisfying*

$$Ax = b,\ x \geq 0,\ A^{\mathrm{T}}u + s = c,\ s \geq 0, \ \textit{and}\ \ x_i s_i = w_i,\ i = 1,\cdots,n.$$

A proof of this statement may be found in S.J. Wright [348], for instance. Due to this statement the concept of the *central path*, playing an important role in the field of interior point methods, can be introduced.

Definition 1.8. *For $\mu > 0$, the primal-dual central path is defined as*

$$\mathscr{C} := \left\{ (x^{\{\mu\}},u^{\{\mu\}},s^{\{\mu\}}) \,\middle|\, F(x^{\{\mu\}},u^{\{\mu\}},s^{\{\mu\}}) = \begin{pmatrix} 0 \\ 0 \\ \mu e \end{pmatrix},\ (x^{\{\mu\}},s^{\{\mu\}}) > 0 \right\}.$$

This definition suggests to drive $\mu \to 0$, due to the conjecture that the limit $(x^\star,u^\star,s^\star) = \lim_{\mu \to 0}(x^{\{\mu\}},u^{\{\mu\}},s^{\{\mu\}})$ (if it exists) yields a primal-dual pair of solutions according to (1.52). Now, starting again with a strictly feasible primal-dual pair $(\hat{x}, \hat{u}, \hat{s})$, we could, instead of (1.56), design a Newton search direction in order to drive the system towards the central path for $\hat{\mu} = \frac{\hat{x}^{\mathrm{T}}\hat{s}}{n}$, such that we had to deal with the system

$$F(\hat{x}, \hat{u}, \hat{s}) + \mathscr{J}(\hat{x}, \hat{u}, \hat{s}) \begin{pmatrix} \Delta x \\ \Delta u \\ \Delta s \end{pmatrix} = \begin{pmatrix} 0 \\ 0 \\ \hat{\mu} e \end{pmatrix}. \tag{1.58}$$

Finally, the two approaches (1.56) and (1.58) may be mixed by choosing for the latter one $\sigma\hat{\mu}$ instead of $\hat{\mu}$ with some $\sigma \in [0,1]$, where $\sigma = 0$ corresponds to (1.56), whereas $\sigma = 1$ reflects fully the goal to move towards the central path. Hence the Newton system becomes

$$F(\hat{x}, \hat{u}, \hat{s}) + \mathscr{J}(\hat{x}, \hat{u}, \hat{s}) \begin{pmatrix} \Delta x \\ \Delta u \\ \Delta s \end{pmatrix} = \begin{pmatrix} 0 \\ 0 \\ \sigma\hat{\mu} e \end{pmatrix}, \tag{1.59}$$

and for the corresponding search direction we have to solve the linear equations

$$\begin{pmatrix} 0 & A^{\mathrm{T}} & I \\ A & 0 & 0 \\ \hat{S} & 0 & \hat{X} \end{pmatrix} \begin{pmatrix} \Delta x \\ \Delta u \\ \Delta s \end{pmatrix} = \begin{pmatrix} 0 \\ 0 \\ -\hat{X}\hat{S}e + \sigma\hat{\mu} e \end{pmatrix}. \tag{1.60}$$

Thus we have the following conceptual

Primal-Dual (Interior Point) Algorithm

S 1 Find (x^0, u^0, s^0) satisfying the interior–point condition (1.55) and let $k := 0$.

S 2 For some $\sigma_k \in [0,1]$ and $\mu_k = \frac{{x^k}^{\mathrm{T}} s^k}{n}$ solve

$$\begin{pmatrix} 0 & A^{\mathrm{T}} & I \\ A & 0 & 0 \\ S^k & 0 & X^k \end{pmatrix} \begin{pmatrix} \Delta x^k \\ \Delta u^k \\ \Delta s^k \end{pmatrix} = \begin{pmatrix} 0 \\ 0 \\ -X^k S^k e + \sigma_k \mu_k e \end{pmatrix}.$$

S 3 Let

$$(x^{k+1}, u^{k+1}, s^{k+1}) := (x^k, u^k, s^k) + \alpha_k (\Delta x^k, \Delta u^k, \Delta s^k),$$

where α_k is chosen such that $(x^{k+1}, s^{k+1}) > 0$. If $X^{k+1} S^{k+1} e < \varepsilon e$ for some small tolerance ε, stop; else return to $S\,2$ with $k := k+1$.

In practice, the requirement of a strictly feasible (x^0, u^0, s^0) as a first iterate in the above algorithm may involve severe difficulties. Instead, it is possible—and also much easier—to start with an infeasible first iterate, more precisely with some $(\hat{x}, \hat{u}, \hat{s})$ such that $(\hat{x}, \hat{s}) > 0$ is satisfied, but the equality constraints are violated, i.e. $\hat{w}_p := A\hat{x} - b \neq 0$ and/or $\hat{w}_d := A^{\mathrm{T}}\hat{u} + I\hat{s} - c \neq 0$. Instead of the system (1.60) for the search direction we then have to begin the above algorithm in step $S\,2$ with the system

$$\begin{pmatrix} 0 & A^{\mathrm{T}} & I \\ A & 0 & 0 \\ \hat{S} & 0 & \hat{X} \end{pmatrix} \begin{pmatrix} \Delta x \\ \Delta u \\ \Delta s \end{pmatrix} = \begin{pmatrix} -\hat{w}_d \\ -\hat{w}_p \\ -\hat{X}\hat{S}e + \sigma\hat{\mu}e \end{pmatrix}. \tag{1.61}$$

As soon as the first iterate becomes strictly feasible (equivalently, as soon as we can choose $\alpha_k = 1$ in step $S\,3$), the subsequent iterates remain strictly feasible, such that (1.61) coincides with the original search direction (1.60) again. This modification of the above conceptual algorithm is referred to as *infeasible interior point method*.

The linear system (1.61) (and (1.60) as well), due to the special structure of its coefficient matrix, may be reformulated to more compact systems with symmetric nonsingular coefficient matrices. First we eliminate Δs using the last block of equations of (1.61),

$$\hat{S}\Delta x + \hat{X}\Delta s = -\hat{X}\hat{S}e + \sigma\hat{\mu}e,$$

yielding

$$\Delta s = -\hat{S}e + \sigma\hat{\mu}\hat{X}^{-1}e - \hat{X}^{-1}\hat{S}\Delta x, \tag{1.62}$$

such that for the two other blocks of equations of (1.61) we have

$$\begin{aligned} A^{\mathrm{T}}\Delta u + \Delta s &= -\hat{w}_d \\ A\Delta x \qquad &= -\hat{w}_p \end{aligned}$$

and hence due to (1.62)

$$\begin{pmatrix} 0 & A \\ A^{\mathrm{T}} & -\hat{X}^{-1}\hat{S} \end{pmatrix} \begin{pmatrix} \Delta u \\ \Delta x \end{pmatrix} = \begin{pmatrix} -\hat{w}_p \\ -\hat{w}_d + \hat{S}e - \sigma\hat{\mu}\hat{X}^{-1}e \end{pmatrix}. \tag{1.63}$$

Hence, to determine the search direction with this so called *augmented system*, we first solve (1.63) for Δu and Δx, and then insert Δx in (1.62) to get Δs. With the notation $S^{\frac{1}{2}} := \operatorname{diag}(\sqrt{s_i})$, $X^{\frac{1}{2}} := \operatorname{diag}(\sqrt{x_i})$, the system (1.63) contains, with $D := \hat{S}^{-\frac{1}{2}}\hat{X}^{\frac{1}{2}}$, the nonsingular diagonal matrix $-D^{-2}$ such that we can eliminate Δx from

$$A^{\mathrm{T}}\Delta u - D^{-2}\Delta x = -\hat{w}_d + \hat{S}e - \sigma\hat{\mu}\hat{X}^{-1}e$$

to get

$$\Delta x = D^2(A^{\mathrm{T}}\Delta u + \hat{w}_d - \hat{S}e + \sigma\hat{\mu}\hat{X}^{-1}e) \tag{1.64}$$

such that the first block of (1.63)) yields

$$A\Delta x = AD^2(A^{\mathrm{T}}\Delta u + \hat{w}_d - \hat{S}e + \sigma\hat{\mu}\hat{X}^{-1}e) = -\hat{w}_p,$$

leading, together with (1.64) and (1.62), to the *normal equations* system

$$\left.\begin{aligned} AD^2A^{\mathrm{T}}\Delta u &= -\hat{w}_p + A(-\hat{S}^{-1}\hat{X}\hat{w}_d + \hat{X}e - \sigma\hat{\mu}\hat{S}^{-1}e) \\ \Delta s &= -A^{\mathrm{T}}\Delta u - \hat{w}_d \\ \Delta x &= -\hat{S}^{-1}\hat{X}\Delta s - \hat{X}e + \sigma\hat{\mu}\hat{S}^{-1}e\,. \end{aligned}\right\} \quad (1.65)$$

The starting point for the field of interior point methods can be seen in the paper published by Dikin [70] in 1967 (see also the much later joint publication of this author and Roos [71] in 1997), although the very wide activity in this area only boiled up when the paper of Karmarkar [177] had appeared in 1984—accompanied by a rather unusual amount of public relation!

In particular among primal-dual interior point methods many variants were designed, depending on the adaptive choices of the parameter σ and of the steplengths α_k, and on modifications of the right–hand–sides of (1.61) (or the augmented or normal equations system derived thereoff), among others. As an alternative to the above attempt of driving the system towards the central path, we just mention a method aiming to approach so-called *analytic centers*, a concept originally introduced by Sonnevend [312]. For more details on the variety of interior point algorithms we refer to books especially devoted to this subject, for instance the monographs of D. den Hertog [134], Roos–Terlaky–Vial [290], Wright [348], and Ye [351], just to mention a few.

In order to get an efficient method in the frame of interior point algorihms, it is important to determine efficiently the search directions, to be evaluated in every iteration step. For this purpose it is certainly advantageous to have the reformulation (1.65), which amounts essentially to solve a system of linear equations

$$Mv = r,$$

with $M = AD^2A^{\mathrm{T}}$ being a symmetric positive definite matrix. Therefore, M allows for a Cholesky factorization $M = L \cdot L^{\mathrm{T}}$ with L being a nonsingular lower triangular matrix, such that the above linear equations can easily be dealt with by solving consecutively the two systems

$$Ly = r \text{ and then } L^{\mathrm{T}}v = y.$$

In general, interior point methods are said to be efficient for large scale LP's, in particular for those with (very) sparse coefficient matrices. However, this statement requires that with M being sparse also L will be sparse such that solving the two last systems involving L and L^{T} becomes very cheap. Unfortunately, this consequence does not always hold. In particular, if M is overall sparse, but nevertheless contains some dense columns, then very likely an undesired fill in of nonzeros into L may happen. Hence, several heuristics have been designed to deal with the submatrices with dense columns separately, in order to maintain efficiency first for the sparse part and finally also for the rest of the system. The success of these attempts seems to depend substantially on the data structure of the LP's considered. For instance, for two-stage SLP's with discrete distributions (and S large) we have—according to Fig. 1.1 on page 5 in the introduction—to expect dense columns in the leading band matrix containing the submatrices $T^1, \cdots, T^S$. Based on many of our computational experiments we have to say that various interior point solvers, including those general purpose variants implemented in several commercial LP software packages, either fail with this problem class or else are clearly ruled out by some efficient implementations based on the simplex method, on Benders' decomposition as the L-shaped method, or on regularized decomposition as the algorithm QDECOM presented in Section 1.2.8. On the other hand, there are interior point implementations designed especially with attention to the data structure of two-stage SLP's and behaving in many cases better than the simplicial or decomposition type methods tested. To mention just one of these, BPMPD implemented by Mészáros [233] behaves impressingly well. Not to be misunderstood: This does not mean that this solver is always the most efficient. It appears to be true with this class of problems that there are implemented solvers of various types, designed regarding our data structure, each of which may outperform the others on various subsets of problem instances.

Exercises

1.9. The dual decomposition algorithm (page 26) has been discussed under the

Assumption: The LP (1.22) is solvable implying that for the recourse problem $\min\{q^{\mathrm{T}}y \mid Wy = \zeta, y \geq 0\}$ follows its solvability, if it is feasible, and therefore that $\{u \mid W^{\mathrm{T}}u \leq q\} \neq \emptyset$; furthermore, the first stage feasible set $\{x \mid Ax = b, x \geq 0\}$ is assumed to be bounded.

(a) Show that the sequence of optimal values $\{c^{\mathrm{T}}\hat{x}^{\nu} + \hat{\theta}^{\nu}\}$ generated by the successive master programs in step $S\,5$ is monotonically increasing.

(b) Under the additional assumption that $\{y \mid Wy = \zeta, y \geq 0\} \neq \emptyset$ for any arbitrary ζ, in step $S\,2$ the case $f(\hat{x}) = \infty$ cannot happen, i.e. the feasibility cuts in step $S\,3$ are never used. Why?

1.10. Given the LP

$$(P) \qquad \left\{\begin{array}{ll} \min & -x_1 - x_2 \\ & x_1 + x_2 + x_3 = 1 \\ & x_1, x_2, x_3 \geq 0. \end{array}\right.$$

(a) Do (P) together with its dual program (D) satisfy the interior-point condition (1.55)?

(b) Determine the central path of the dual pair (P) and (D), given for $\lambda > 0$ due to Def. 1.8. as the set

$$\mathscr{C} = \left\{ (x(\lambda), u(\lambda), s(\lambda)) \,\middle|\, \begin{array}{c} Ax(\lambda) = b, A^{\mathrm{T}}u(\lambda) + s(\lambda) = c \\ (x(\lambda), s(\lambda)) > 0,\; x_j(\lambda) \cdot s_j(\lambda) = \lambda \;\forall j \end{array} \right\}.$$

1.11. Given the interior-point condition (1.55), according to Prop. 1.23. the system

$$\begin{pmatrix} A^{\mathrm{T}}u + Is - c \\ Ax - b \\ XSe \end{pmatrix} = \begin{pmatrix} 0 \\ 0 \\ \mu I \end{pmatrix}$$

has the central path $\{(x^{\mathrm{T}}(\mu), u^{\mathrm{T}}(\mu), s^{\mathrm{T}}(\mu))^{\mathrm{T}} \mid \mu > 0\}$ as unique solution.
The mapping $(x,u,s) : (0,\infty) \longrightarrow \mathbb{R}^n \times \mathbb{R}^m \times \mathbb{R}^n$ is continuously differentiable. Why?

1.12. For the symmetric matrix

$$D = \begin{pmatrix} 1\,2\,1 \\ 2\,8\,2 \\ 1\,2\,2 \end{pmatrix}$$

compute the Cholesky factorization $D = L \cdot L^{\mathrm{T}}$.
Then, with $d = (7,\, 18,\, 10)^{\mathrm{T}}$, solve the linear system $Dx = d$ by solving successively $Ly = d$ and $L^{\mathrm{T}}x = y$.

1.3 Nonlinear Programming Prerequisites

Considering for instance the chance constrained problem (1.6) on page 3 (under some additional assumptions), or else the regularized master program (1.51) on page 46, we shall encounter NLP's of the general form

$$\left.\begin{array}{r} \min f(x) \\ \text{s. t. } g_i(x) \le 0, \;\; i = 1, \cdots, m, \end{array}\right\} \tag{1.66}$$

where we henceforth assume the functions $f : \mathbb{R}^n \to \mathbb{R}$ and $g_i : \mathbb{R}^n \to \mathbb{R}$ to be convex.

Definition 1.9. *A set $\mathscr{C} \subseteq \mathbb{R}^n$ is convex if for arbitrary $x, y \in \mathscr{C}$ and for any $\lambda \in [0,1]$ holds $\lambda x + (1-\lambda)y \in \mathscr{C}$. Then a function $\varphi : \mathscr{C} \to \mathbb{R}$ is convex if*

$$\varphi(\lambda x + (1-\lambda)y) \le \lambda\varphi(x) + (1-\lambda)\varphi(y) \;\; \forall x, y \in \mathscr{C},\; \forall \lambda \in [0,1]\,.$$

This definition implies further properties. First,

Proposition 1.24. *If $\varphi : \mathbb{R}^n \longrightarrow \mathbb{R}$ is convex, then φ is continuous.*

Furthermore, a convex function need not be differentiable, but it is—under mild assumptions—*subdifferentiable* everywhere.

Definition 1.10. *A vector $g \in \mathbb{R}^n$ is a subgradient of a convex function φ at $x \in \mathbb{R}^n$, if*

$$g^{\mathrm{T}}(z - x) \le \varphi(z) - \varphi(x) \;\; \forall z \in \mathbb{R}^n\,.$$

The set of all subgradients of φ at x is the subdifferential of φ at x, denoted by $\partial\varphi(x)$.
If $\partial\varphi(x) \neq \emptyset$, then φ is called subdifferentiable at x.

A typical result for convex functions is referred to as

Proposition 1.25. *Given a convex function $\varphi : \mathbb{R}^n \longrightarrow \mathbb{R}$, then for any $x \in \mathbb{R}^n$ the set $\partial\varphi(x)$ is nonempty, convex, closed, and bounded.*
In addition, φ is differentiable in x with the gradient $\hat{g} = \nabla\varphi(x)$ if and only if $\partial\varphi(x) = \{\hat{g}\}$, i.e. $\partial\varphi(x)$ is a singleton.
Finally, given a convex function $\psi : \mathbb{R}^m \longrightarrow \mathbb{R}$ and a linear affine mapping $y : \mathbb{R}^n \longrightarrow \mathbb{R}^m$ defined by $y(x) := d + Dx$ with some $d \in \mathbb{R}^m$ and $D \in \mathbb{R}^{m\times n}$, then $f : \mathbb{R}^n \longrightarrow \mathbb{R}$ composed as $f(x) := \psi(y(x))$ is convex, and for its subdifferential holds the chain rule $\partial f(x) = D^{\mathrm{T}}\partial\psi(y(x))$, or equivalently

$$h \in \partial f(x) \Longleftrightarrow \exists g \in \partial\psi(y(x)) : h = D^{\mathrm{T}} g.$$

For more detailed statements on subdifferentiability of convex functions we refer to Rockafellar [281].

Continuing the discussion of problem (1.66), due to the convexity assumption on g_i we have that the feasible set of (1.66)

$$\mathscr{B} := \{x \mid g_i(x) \leq 0,\ i = 1, \cdots, m\}$$

is convex, and that any local minimum at $\hat{x} \in \mathscr{B}$ is at the same time a global minimum, i.e. $f(\hat{x}) = \min_{x\in\mathscr{B}} f(x)$.

Henceforth, in addition to convexity, we assume the functions describing problem (1.66), $f : \mathbb{R}^n \to \mathbb{R}$ and $g_i : \mathbb{R}^n \to \mathbb{R}$, to be continuously differentiable.

The fact of some continuously differentiable function $\varphi : \mathbb{R}^n \longrightarrow \mathbb{R}$ to be convex obviously implies the subgradient inequality of Def. 1.10. at any $x \in \mathbb{R}^n$ to hold with $g = \nabla\varphi(x)$; but now also the reverse conclusion is valid.

Proposition 1.26. *φ is convex if and only if*

$$(y-x)^{\mathrm{T}}\nabla\varphi(x) \leq \varphi(y) - \varphi(x) \quad \forall x, y \in \mathbb{R}^n . \tag{1.67}$$

Proof: Assume that (1.67) holds true. Then for arbitrary $y, z \in \mathbb{R}^n$, $\lambda \in (0,1)$, and $x = \lambda y + (1-\lambda)z$ follows

$$\begin{aligned}(y-x)^{\mathrm{T}}\nabla\varphi(x) &\leq \varphi(y) - \varphi(x)\\ (z-x)^{\mathrm{T}}\nabla\varphi(x) &\leq \varphi(z) - \varphi(x),\end{aligned}$$

implying

$$\underbrace{(\lambda y + (1-\lambda)z - x)}_{=0}{}^{\mathrm{T}}\nabla\varphi(x) \leq \lambda\varphi(y) + (1-\lambda)\varphi(z) - \varphi(x),$$

i.e. the convexity of φ.

Assume now φ to be convex. Then, for any $x,y \in \mathbb{R}^n$ and $\lambda \in (0,1)$, together with the mean value theorem we get, with $\theta_\lambda \in (0,1)$,

$$\begin{aligned}\varphi(y)-\varphi(x) &\geq \frac{1}{1-\lambda}\{\varphi(x+(1-\lambda)(y-x))-\varphi(x)\}\\ &= \frac{1}{1-\lambda}\{(1-\lambda)(y-x)^{\mathrm{T}}\nabla\varphi(x+\theta_\lambda(1-\lambda)(y-x))\}\\ &= (y-x)^{\mathrm{T}}\nabla\varphi(x+\theta_\lambda(1-\lambda)(y-x))\end{aligned}$$

yielding (1.67) for $\lambda \to 1$. □

To get optimality conditions for the NLP (1.66) assume that we have an optimal solution $\hat{x} \in \mathcal{B}$. Let $I(\hat{x}) := \{i \mid g_i(\hat{x}) = 0\}$. For $i \notin I(\hat{x})$, i.e. for $g_i(\hat{x}) < 0$, it follows that $g_i(\hat{x}+\alpha z) \leq 0$ for any $z \in \mathbb{R}^n$ if we choose $\alpha > 0$ small enough. On the other hand, for $\tilde{z} \in \mathbb{R}^n$ with $\tilde{z}^{\mathrm{T}}\nabla g_i(\hat{x}) < 0\ \forall i \in I(\hat{x})$, there exists an $\overline{\alpha} > 0$ such that for $i = 1,\cdots,m$ holds $g_i(\hat{x}+\alpha\tilde{z}) \leq 0\ \forall\alpha \in (0,\overline{\alpha})$ and hence $\hat{x}+\alpha\tilde{z} \in \mathcal{B}\ \forall\alpha \in (0,\overline{\alpha})$. For $\hat{x}$ to be a minimal point of f in $\mathcal{B}$ it follows that $f(\hat{x}+\alpha\tilde{z})-f(\hat{x}) \geq 0\ \forall\alpha \in (0,\overline{\alpha})$. For $\hat{x}$ to be a solution of (1.66) we have therefore the (necessary) condition

$\underline{\mathcal{RC}^\star}$ $\quad z^{\mathrm{T}}\nabla g_i(\hat{x}) < 0,\ i \in I(\hat{x})$ implies that $z^{\mathrm{T}}\nabla f(\hat{x}) \geq 0$.

Hence we know the requirements for all $z \in \mathbb{R}^n$ satisfying $z^{\mathrm{T}}\nabla g_i(\hat{x}) < 0\ \forall i \in I(\hat{x})$, but for $z \neq 0$ such that $z^{\mathrm{T}}\nabla g_i(\hat{x}) = 0$ for at least one $i \in I(\hat{x})$, it is not clear what we should expect. For technical reasons which will be apparent below, we strengthen the above condition $\underline{\mathcal{RC}^\star}$ slightly and state the somewhat voluntary modification as *regularity condition*

$\underline{\mathcal{RC}\ 0}$ For any optimal $\hat{x}$ of (1.66) holds that

$$z^{\mathrm{T}}\nabla g_i(\hat{x}) \leq 0\ \forall i \in I(\hat{x}) \text{ implies } z^{\mathrm{T}}\nabla f(\hat{x}) \geq 0.$$

Remark 1.3. *Observe that for linear constraints the regularity condition is always satisfied: Having*

$$g_i(\hat{x}) = b_i - a^{(i)^{\mathrm{T}}}\hat{x} = 0\ \forall i \in I(\hat{x})$$

implies that for any z such that $z^{\mathrm{T}}\nabla g_i(\hat{x}) = -a^{(i)^{\mathrm{T}}}z \leq 0$ it follows that

$$g_i(\hat{x}+\alpha z) = b_i - a^{(i)^{\mathrm{T}}}(\hat{x}+\alpha z) = \underbrace{b_i - a^{(i)^{\mathrm{T}}}\hat{x}}_{=0} - \alpha a^{(i)^{\mathrm{T}}}z \leq 0\ \forall\alpha > 0.$$

Hence, there is an $\overline{\alpha} > 0$ such that $\hat{x}+\alpha z \in \mathcal{B}\ \forall\alpha \in (0,\overline{\alpha})$, and due to the optimality of $\hat{x}$ follows $f(\hat{x}+\alpha z)-f(\hat{x}) \geq 0$; in view of the mean value theorem the last inequality implies for $\alpha \downarrow 0$, that $z^{\mathrm{T}}\nabla f(\hat{x}) \geq 0$, i.e. $\underline{\mathcal{RC}\ 0}$ is satisfied.

In the nonlinear case it may happen that the above regularity condition does not hold. Take for example the elementary problem

$$\min\{x \in \mathbb{R}^1 \mid x^2 \leq 0\}.$$

Since there is only one feasible solution, $\hat{x} = 0$, this is also the optimal solution of the problem. Here $z^{\mathrm{T}}\nabla g_i(\hat{x}) \leq 0$ means that $2 \cdot z \cdot \hat{x} \leq 0$, which is true for all $z \in \mathbb{R}^1$ since $\hat{x} = 0$, but $z^{\mathrm{T}}\nabla f(\hat{x}) = z \cdot 1 < 0\ \forall z < 0$, such that $\underline{\mathscr{RC}\ 0}$ is violated. □

To check the condition $\underline{\mathscr{RC}\ 0}$ seems to be almost impossible, in general, since it would require to know an optimal solution $\hat{x} \in \mathscr{B}$ in advance, which usually is not the case. However, there are various other regularity conditions which are easier to check and which imply the validity of $\underline{\mathscr{RC}\ 0}$. For convex problems (1.66) a very popular assumption is the *Slater condition*:

$\underline{\mathscr{RC}\ 1}$ For (1.66) there exists a feasible point $\tilde{x}$ such that $g_i(\tilde{x}) < 0\ \forall i$.

Similarly to Remark 1.3. the Slater condition needs to be required for nonlinear constraints only, whereas for linear constraints it may be abandonned. Without proof we mention that, for convex problems, $\underline{\mathscr{RC}\ 1}$ implies $\underline{\mathscr{RC}\ 0}$.

1.3.1 Optimality Conditions

We just have seen a particular condition, $\underline{\mathscr{RC}\ 0}$, which obviously is sufficient for the optimality of $\hat{x}$ in (1.66): For any direction z leading from $\hat{x}$ into $\mathscr{B}$, i.e. for $z \in \mathbb{R}^n$ such that $\hat{x} + \alpha z \in \mathscr{B}$ for sufficiently small $\alpha > 0$, and therefore in particular for which $g_i(\hat{x} + \alpha z) \leq 0 = g_i(\hat{x})\ \forall i \in I(\hat{x})$, it follows that $z^{\mathrm{T}}\nabla g_i(\hat{x}) \leq 0\ \forall i \in I(\hat{x})$, which by $\underline{\mathscr{RC}\ 0}$ implies $z^{\mathrm{T}}\nabla f(\hat{x}) \geq 0$. Hence, from Proposition 1.26. we get $f(\hat{x} + \alpha z) - f(\hat{x}) \geq \alpha z^{\mathrm{T}}\nabla f(\hat{x}) \geq 0$ for $\alpha > 0$ and therefore the optimality of $\hat{x}$ for (1.66). However as discussed above, $\underline{\mathscr{RC}\ 0}$ is anything but operational for finding optimal solutions. Nevertheless, it is useful for deriving more tractable optimality conditions, called the *Karush-Kuhn-Tucker conditions* (KKT):

Proposition 1.27. *Assume that for the convex program (1.66) the Slater condition $\underline{\mathscr{RC}\ 1}$ is satisfied. Then an $\hat{x} \in \mathscr{B}$ solves (1.66) if and only if there exists an $\hat{u} \in \mathbb{R}^m$ such that the following conditions hold:*

$$\left.\begin{array}{ll} i) & \nabla f(\hat{x}) + \sum_{i=1}^{m} \hat{u}_i \nabla g_i(\hat{x}) = 0 \\ ii) & g_i(\hat{x}) \leq 0\ \forall i \\ iii) & \hat{u}_i \cdot g_i(\hat{x}) = 0\ \forall i \\ iv) & \hat{u} \geq 0. \end{array}\right\} \tag{1.68}$$

Proof: To show that (1.68) is sufficient for $\hat{x}$ to be a solution of (1.66), we observe first that by (1.68) *ii*) the point $\hat{x}$ is feasible. Further, for all $i \in I(\hat{x})$ we have $g_i(\hat{x}) = 0$ and hence for arbitrary $y \in \mathscr{B}$ due to Proposition 1.26.

$$0 \geq g_i(y) - g_i(\hat{x}) \geq (y - \hat{x})^{\mathrm{T}}\nabla g_i(\hat{x})\ \ \forall i \in I(\hat{x}).$$

Using again Proposition 1.26. as well as (1.68) *iv*) and the *complementarity conditions iii*), it follows from condition *i*) in (1.68) that

$$f(y)-f(\hat{x}) \geq (y-\hat{x})^{\mathrm{T}}\nabla f(\hat{x}) = -\sum_{i\in I(\hat{x})} \hat{u}_i (y-\hat{x})^{\mathrm{T}}\nabla g_i(\hat{x}) \geq 0 \;\forall y \in \mathscr{B},$$

such that

$$f(\hat{x}) \leq f(y) \;\forall y \in \mathscr{B}.$$

To show the necessity of KKT assume that $f(\hat{x}) = \min_{x\in\mathscr{B}} f(x)$. Since with the assumed Slater condition $\underline{\mathscr{RC}\ 1}$ the regularity condition $\underline{\mathscr{RC}\ 0}$ holds as well at $\hat{x}$, we know that with the active set $I(\hat{x}) = \{i \mid g_i(\hat{x}) = 0\}$

$$z^{\mathrm{T}}\nabla g_i(\hat{x}) \leq 0 \;\forall i \in I(\hat{x}) \text{ implies } z^{\mathrm{T}}\nabla f(\hat{x}) \geq 0.$$

Then from the Farkas lemma (Proposition 1.13., page 15) it follows that

$$\{\hat{u}_i,\ i \in I(\hat{x}) \mid \sum_{i\in I(\hat{x})} \hat{u}_i \nabla g_i(\hat{x}) = -\nabla f(\hat{x}),\ \hat{u}_i \geq 0 \;\forall i \in I(\hat{x})\} \neq \emptyset,$$

such that with $\hat{u}_i = 0 \;\forall i \notin I(\hat{x})$ the conditions (1.68) *i*)–*iv*) are satisfied. □

Since there are, in addition to (1.66), various other NLP formulations, the KKT conditions have to be adapted correspondingly. If we have for instance the NLP

$$\min\{f(x) \mid g(x) \leq 0,\ x \geq 0\} \tag{1.69}$$

with the vector valued function $g(x) = (g_1(x), \cdots, g_m(x))^{\mathrm{T}}$ and all g_i and f being continuously differentiable and convex as before, we get immediately the KKT conditions

$$\left.\begin{array}{lr} i) & \nabla f(\hat{x}) + \sum_{i=1}^{m} \hat{u}_i \nabla g_i(\hat{x}) \geq 0 \\ ii) & \hat{x}^{\mathrm{T}}(\nabla f(\hat{x}) + \sum_{i=1}^{m} \hat{u}_i \nabla g_i(\hat{x})) = 0 \\ iii) & g(\hat{x}) \leq 0 \\ iv) & \hat{u}^{\mathrm{T}}\, g(\hat{x}) = 0 \\ v) & \hat{x} \geq 0 \\ vi) & \hat{u} \geq 0. \end{array}\right\} \tag{1.70}$$

To see this, deal with the additional constraints $-x \leq 0$ just as with $g(x) \leq 0$. Introducing additional multipliers $w \in \mathbb{R}^n_+$ (for $-x \leq 0$) and afterwards eliminating them again leads to the inequalities *i*) and the additional complementarity conditions *ii*).

Coming back to the original NLP (1.66), the corresponding KKT conditions (1.68) have an interpretation which may be of interest also with respect to solution methods. Defining the *Lagrange function* to (1.66) as

$$L(x,u) := f(x) + \sum_{i=1}^{m} u_i g_i(x),$$

it is obvious that for any fixed $\tilde{u} \geq 0$ the function $L(\cdot, \tilde{u})$ is convex in x. Considering (1.68) i) we have, with ∇_x being the gradient with respect to x,

$$L(x,\hat{u}) - L(\hat{x},\hat{u}) \geq (x-\hat{x})^{\mathrm{T}} \nabla_x L(\hat{x},\hat{u}) = 0 \; \forall x \in \mathbb{R}^n$$

such that $L(\hat{x},\hat{u}) = \min_{x \in \mathbb{R}^n} L(x,\hat{u})$. On the other hand, for any fixed $\tilde{x}$ the function $L(\tilde{x},\cdot)$ is linear affine and hence concave in u, resulting in the inverse inequality to (1.67) and implying with $\hat{u}^{\mathrm{T}} g(\hat{x}) = 0$ due to (1.68) ii) and iii) that

$$\begin{aligned} L(\hat{x},u) - L(\hat{x},\hat{u}) &\leq (u-\hat{u})^{\mathrm{T}} \nabla_u L(\hat{x},\hat{u}) \\ &= (u-\hat{u})^{\mathrm{T}} g(\hat{x}) = u^{\mathrm{T}} g(\hat{x}) \leq 0 \;\; \forall u \geq 0 \end{aligned}$$

such that $L(\hat{x},\hat{u}) = \max_{u \geq 0} L(\hat{x},u)$.

Hence, the KKT point $(\hat{x}, \hat{u} \geq 0)$ of Proposition 1.27. is a *saddle point* of the Lagrange function:

$$L(\hat{x},u) \leq L(\hat{x},\hat{u}) \leq L(x,\hat{u}) \;\; \forall u \geq 0, \; \forall x \in \mathbb{R}^n. \tag{1.71}$$

From (1.71) we get the following *saddle point theorem* which may also be interpreted as a strong duality theorem for nonlinear programming.

Proposition 1.28. (Saddle point theorem) *If for the Lagrange function to (1.66) there exists a saddle point* $(\hat{x}, \hat{u} \geq 0)$, *then*

$$\max_{u \geq 0} \inf_{x \in \mathbb{R}^n} L(x,u) = \min_{x \in \mathbb{R}^n} \sup_{u \geq 0} L(x,u), \tag{1.72}$$

and $(\hat{x}, \hat{u})$ *solves each of these two problems.*

For a proof of this statement see e.g. Luenberger [209].

By definition of $L(\cdot,\cdot)$ follows $\min_x \sup_{u \geq 0} L(x,u) = \min\{f(x) \mid g(x) \leq 0\}$. Therefore, $\min_x \sup_{u \geq 0} L(x,u)$ is considered as the primal problem, whereas, in contrast, $\max_{u \geq 0} \inf_x L(x,u)$ is its dual.

1.3.2 Solution methods

Several types of solution methods have been proposed for the numerical approximation of solutions for nonlinear programs (1.66). Many of these approaches may be found in the books of Bazaraa–Shetty [11], Bertsekas [17, 18], Geiger–

Kanzow [123], and McCormick [232], just to mention a few. Most of the methods dealt with in this literature belong to one of the following categories:

- cutting plane methods
  (e.g. Elzinga–Moore [86], Kelley [180], Kleibohm [187], Veinott [338])
- feasible direction methods (e.g. Topkis–Veinott [329], Zoutendijk [355])
- penalty methods (e.g. Fiacco–McCormick [93])
- Lagrangian methods (e.g. Bertsekas [17]).

In stochastic programming variants of cutting plane methods have first and foremost been used so far. The reason for this fact seems to be, that in all other classes of NLP solution methods, within any iteration there are iterative subcycles requiring the repeated evaluation of gradients of some modified objective functions containing integral functions (expected value functions or probability functions) which is expensive to perform, if not impossible at all.

Therefore, we restrict ourselves to a few prototypes of methods based on cutting planes (separating hyperplanes). A hyperplane $H = \{v \mid a^{\mathrm{T}}v = \alpha\}$ with $a \neq 0$ is *separating* two nonempty closed convex sets $C \subset \mathbb{R}^{\ell}$ and $D \subset \mathbb{R}^{\ell}$, if $\sup_C a^{\mathrm{T}}v \leq \alpha \leq \inf_D a^{\mathrm{T}}v$; then C and D have at most boundary points in common. The hyperplane H is said to separate C and D *strongly* if $\sup_C a^{\mathrm{T}}v < \inf_D a^{\mathrm{T}}v$, such that $\alpha > \sup_C a^{\mathrm{T}}v$ or else $\alpha < \inf_D a^{\mathrm{T}}v$, and hence $C \cap D = \emptyset$, hold true. For a singleton $D = \{\hat{u}\} \in \operatorname{bd} C$, the Hyperplane H separating C and D is called a *supporting hyperplane* of C at $\hat{u}$.

First, let us consider the following special variant of the NLP (1.66):

$$\begin{array}{c}\min\{c^{\mathrm{T}}x + \Phi(x)\} \\ \text{s.t. } x \in X\,,\end{array}$$

where the convex polyhedral set $X := \{x \mid Ax = b\,,\ x \geq 0\} \neq \emptyset$ is assumed to be compact and $\Phi : \mathbb{R}^n \longrightarrow \mathbb{R}$ is convex and hence continuous. Obviously, this problem can equivalently be rewritten as

$$\left.\begin{array}{rl}\min\{c^{\mathrm{T}}x + \Theta\} & \\ \text{s.t. } x & \in X \\ \Phi(x) - \Theta & \leq 0.\end{array}\right\} \tag{1.73}$$

The last constraint coincides with $F(x,\Theta) := \Phi(x) - \Theta \leq 0$, where F is convex and continuous in (x,Θ) as well. With Φ continuous and $X \neq \emptyset$ compact, the extrema $\hat{\Phi} = \min_{x\in X}\Phi(x)$ and $\tilde{\Phi} = \max_{x\in X}\Phi(x)$ obviously exist; hence for (x,Θ) feasible in (1.73) holds $\Theta \geq \hat{\Phi}$, and bounding Θ above by $\Theta \leq \tilde{\Phi} + \gamma,\ \gamma > 0$, does not affect the solution set of (1.73). Then, with some $L \leq \hat{\Phi}$ and $M \geq \tilde{\Phi} + \gamma$, the set $Z := \{z = (x,\Theta) \mid x \in X\,,\ L \leq \Theta \leq M\} \neq \emptyset$ is compact in $\mathbb{R}^{n+1}$, and (1.73) has a solution $\hat{z} = (\hat{x},\hat{\Theta})$, solving with $d^{\mathrm{T}} = (c^{\mathrm{T}},1)$ and $\hat{\zeta} = d^{\mathrm{T}}\hat{z}$ the equivalent problem

$$\min\{d^{\mathrm{T}}z \mid z \in Z,\ F(z) \leq 0\}. \tag{1.74}$$

Under the assumptions made for this problem, i.e. $Z \in \mathbb{R}^{n+1} \neq \emptyset$ convex polyhedral compact and $F : \mathbb{R}^{n+1} \longrightarrow \mathbb{R}$ convex, the following method was proposed by Kelley [180] and independently by Cheney–Goldstein [42].

Cutting Planes: Outer Linearization (Kelley)

S 1 With $\mathscr{P}_0 := Z$ let $\hat{z}_0 \in \mathscr{P}_0$ solve the problem $\hat{\zeta}_0 := \min_{z\in\mathscr{P}_0} d^{\mathrm{T}}z$ and set $k := 0$.

S 2 If $\hat{F}_k := F(\hat{z}_k) \leq 0$, stop; $\hat{z}_k$ solves (1.74).

S 3 Else, if $\hat{F}_k = F(\hat{z}_k) > 0$, choose a subgradient $g_k \in \partial F(\hat{z}_k)$ to define $h_k(z) := \hat{F}_k + g_k^{\mathrm{T}}(z - \hat{z}_k)$ $(\leq F(z)\ \forall z$ due to Def. 1.10.).
With $\mathscr{P}_{k+1} := \mathscr{P}_k \cap \{z \mid h_k(z) \leq 0\}$ determine a solution $\hat{z}_{k+1}$ of $\hat{\zeta}_{k+1} := \min_{z\in\mathscr{P}_{k+1}} d^{\mathrm{T}}z$, let $k := k+1$, and return to step *S 2*.

□

Proposition 1.29. *If the above procedure is not finite, then any accumulation point $\tilde{z}$ of the sequence $\{\hat{z}_k\}$ solves problem (1.74), yielding its optimal value as $\hat{\zeta} = d^{\mathrm{T}}\tilde{z}$, and for the sequence $\{\hat{\zeta}_k\}$ with $\hat{\zeta}_{k+1} \geq \hat{\zeta}_k\ \forall k$ follows that* $\lim_{k\to\infty}\hat{\zeta}_k = \hat{\zeta}$.

In *S 3*, by $h_k(z) \equiv 0$ the hyperplane $H_k := \{z \mid g_k^{\mathrm{T}}z = \delta_k\}$ with $\delta_k = g_k^{\mathrm{T}}\hat{z}_k - \hat{F}_k$ is defined, and $\forall z : F(z) \leq 0$ holds $g_k^{\mathrm{T}}z \leq \delta_k < g_k^{\mathrm{T}}\hat{z}_k$. Hence, H_k separates $C := \{z \mid F(z) \leq 0\}$ and $D := \{\hat{z}_k\}$ strongly, and the new constraint $g_k^{\mathrm{T}}z \leq \delta_k$ is cutting off $\hat{z}_k$ from the feasibility sets of the following steps. This suggests the term *cutting plane method* (see Fig. 1.5).

The proof presented here follows the line of reasoning in Blum–Oettli [30].

Proof: Obviously holds $\mathscr{P}_k \supset \mathscr{P}_{k+1} \supset \cdots \supset \{z \mid z \in Z,\ F(z) \leq 0\}$ for $k = 0, 1, 2, \cdots$, which implies that $\hat{\zeta}_k = d^{\mathrm{T}}\hat{z}_k \leq \hat{\zeta}_{k+1} = d^{\mathrm{T}}\hat{z}_{k+1} \leq \cdots \leq \hat{\zeta}$. Hence there exists $\tilde{\zeta} := \lim_{k\to\infty}\hat{\zeta}_k$, and $\tilde{\zeta} \leq \hat{\zeta}$. Furthermore, $\mathscr{P}_k \subset Z\ \forall k$ and hence $\hat{z}_k \in \mathscr{P}_k \subset Z$. Due to the compactness of Z follows that the sequence $\{\hat{z}_k\}$ has at least one accumulation point $\tilde{z}$, and some subsequence $\{\hat{z}_\kappa\} \subset \{\hat{z}_k\}$ converges to $\tilde{z} \in Z$. This implies $\hat{\zeta}_\kappa = d^{\mathrm{T}}\hat{z}_\kappa \to d^{\mathrm{T}}\tilde{z} = \tilde{\zeta} = \lim_{k\to\infty}\hat{\zeta}_k$.

For the subsequence $\{\hat{z}_\kappa\}$ holds $\hat{F}_\nu + g_\nu^{\mathrm{T}}(\hat{z}_\kappa - \hat{z}_\nu) \leq 0\ \forall \kappa > \nu$ and therefore $\hat{F}_\nu + g_\nu^{\mathrm{T}}(\tilde{z} - \hat{z}_\nu) \leq 0\ \forall \nu$. The subdifferentials $\partial F(z), z \in Z$, are uniformly bounded (see Rockafellar [281], Th. 24.7); hence there exists $\Gamma \in \mathbb{R}$ such that due to Schwarz' inequality $\hat{F}_\kappa \leq |\tilde{z} - \hat{z}_\kappa| \cdot \Gamma\ \forall \kappa$. From $\hat{F}_\kappa = F(\hat{z}_\kappa) \to F(\tilde{z})$ and $\hat{z}_\kappa \to \tilde{z}$ follows $F(\tilde{z}) \leq 0$ such that $\tilde{z}$ is feasible in (1.74) and $\tilde{\zeta} \leq \hat{\zeta}$; however, by the definition of $\hat{\zeta}$ follows $\hat{\zeta} \leq \tilde{\zeta}$. □

Let us now consider the original NLP (1.66)

$$\min_{x\in\mathbb{R}^n} \{f(x) \mid g_i(x) \leq 0,\ i = 1, \cdots, m\}$$

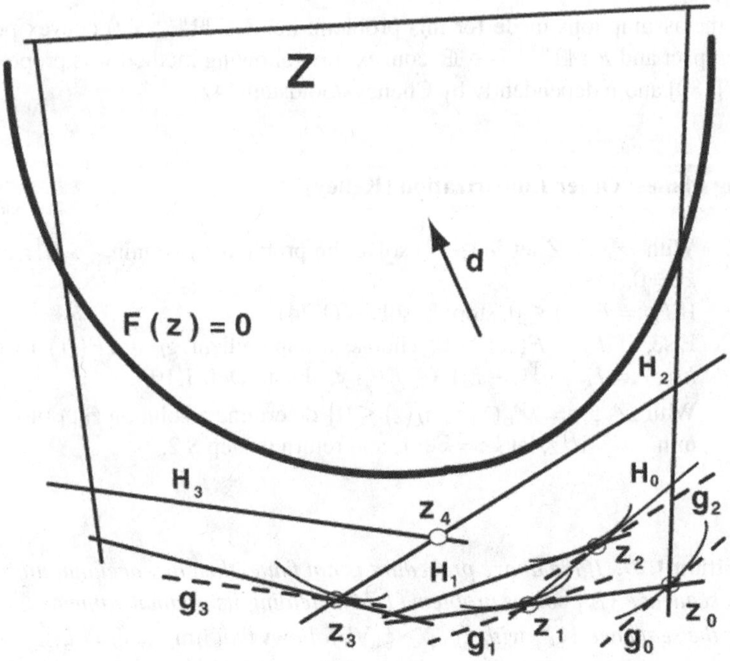

Fig. 1.5 Kelley's outer linearization with cutting planes $H_k := \{z \mid g_k^{\mathrm{T}} z = \delta_k\}$.

where in addition to the assumptions on pages 54/55 we require that

$$\mathscr{B} = \{x \mid g_i(x) \leq 0,\ i = 1, \cdots, m\}$$

be bounded and that $\exists \tilde{x} \in \operatorname{int} \mathscr{B}$, the latter condition being satisfied if there exists a Slater point $\tilde{x}$, for instance (see $\mathscr{RC}$ 1 on page 57). Then, problem (1.66) is equivalent to

$$\left.\begin{array}{ll} \min \theta & \\ \text{s.t. } g_i(x) & \leq 0,\ i = 1, \cdots, m, \\ \quad f(x) - \theta \leq 0. & \end{array}\right\} \tag{1.75}$$

Obviously the additional condition $\theta \leq f(\tilde{x}) + \gamma$ with some constant $\gamma > 0$ does not change the solution set. Hence, instead of this problem we may consider the minimization of the linear function $\varphi(x, \theta) \equiv \theta$ on the bounded convex set $\overline{\mathscr{B}} := \{(x, \theta) \mid x \in \mathscr{B},\ f(x) \leq \theta \leq f(\tilde{x}) + \gamma\}$, for which obviously a point $(\hat{x}, \hat{\theta}) \in \operatorname{int} \overline{\mathscr{B}}$ exists as well. Therefore, we may confine our considerations on NLP's of the type

$$\min\{c^{\mathrm{T}} x \mid x \in \mathscr{B}\} \tag{1.76}$$

with a bounded convex set $\mathscr{B}$ containing an interior point $\tilde{x}$. In this situation there exists a convex polyhedron $\mathscr{P}$ such that $\mathscr{P} \supset \mathscr{B}$. In other words, due to Section

1.2.2 there are linear constraints defining the feasible set $\mathscr{P}$, and it holds

$$\min_{x\in\mathscr{P}} c^{\mathrm{T}}x \le \min_{x\in\mathscr{B}} c^{\mathrm{T}}x.$$

Now we may describe a first method using supporting hyperplanes as proposed originally by Veinott [338] and discussed later by Kleibohm [187]:

Cutting Planes: Outer Linearization; Fixed Slater Point (Veinott)

S 1 Find a $\tilde{x} \in \mathrm{int}\mathscr{B}$ and a convex polyhedron $\mathscr{P}_0 \supset \mathscr{B}$; let $k := 0$.

S 2 Solve the LP $\min\{c^{\mathrm{T}}x \mid x \in \mathscr{P}_k\}$, yielding the solution $\hat{x}^{(k)}$.
If $\hat{x}^{(k)} \in \mathscr{B}$, stop; $\hat{x}^{(k)}$ solves (1.76).
Else, determine $z^{(k)} \in [\hat{x}^{(k)}, \tilde{x}] \cap \mathrm{bd}\mathscr{B}$ (with $[\hat{x}^{(k)}, \tilde{x}]$ the straight line between $\hat{x}^{(k)}$ and $\tilde{x}$, and $\mathrm{bd}\mathscr{B}$ the boundary of $\mathscr{B}$).

S 3 Determine a supporting hyperplane H_k of $\mathscr{B}$ in $z^{(k)}$, i.e. find $a^{(k)} \in \mathbb{R}^n$ and $\alpha_k = a^{(k)^{\mathrm{T}}} z^{(k)}$ such that

$$H_k := \{x \mid a^{(k)^{\mathrm{T}}} x = \alpha_k\} \quad \text{and} \quad a^{(k)^{\mathrm{T}}} \hat{x}^{(k)} > \alpha_k \ge a^{(k)^{\mathrm{T}}} x \ \forall x \in \mathscr{B}.$$

Define $\mathscr{P}_{k+1} := \mathscr{P}_k \cap \{x \mid a^{(k)^{\mathrm{T}}} x \le \alpha_k\}$, let $k := k+1$, and return to step *S 2*.

□

As for Kelley's algorithm, we may not expect the iterates $z^{(k)} \in \mathscr{B}$ or $\hat{x}^{(k)} \notin \mathscr{B}$ to converge, in general. However the following statement is easy to prove.

Proposition 1.30. *Under the above assumptions, the accumulation points of $\{\hat{x}^{(k)}\}$ as well as of $\{z^{(k)}\}$ solve (1.76). Furthermore, the objective values $\{c^{\mathrm{T}}\hat{x}^{(k)}\}$ and $\{c^{\mathrm{T}}z^{(k)}\}$ converge to $\min\{c^{\mathrm{T}}x \mid x \in \mathscr{B}\}$. Finally, in every iteration we have an error estimate with respect to the true optimal value δ of (1.76) as*

$$\Delta_k = \min_{l=1,\cdots,k} c^{\mathrm{T}}z^{(l)} - c^{\mathrm{T}}\hat{x}^{(k)}.$$

For the proof we refer to the NLP literature mentioned on page 59.

Remark 1.4. *Observe that due to $\mathscr{P}_{k+1} \subset \mathscr{P}_k \ \forall k$ it follows $c^{\mathrm{T}}\hat{x}^{(k+1)} \ge c^{\mathrm{T}}\hat{x}^{(k)}$ whereas the sequence $\{c^{\mathrm{T}}z^{(k)}\}$ need not be monotone. However, since $z^{(k)} \in \mathscr{B} \ \forall k$, we have $c^{\mathrm{T}}z^{(k)} \ge \delta \ \forall k$, whereas $c^{\mathrm{T}}\hat{x}^{(k)} \le \delta$ as long as $\hat{x}^{(k)} \notin \mathscr{B}$. Obviously, the above error estimate yields an additional stopping criterion in step* S 2 *according to $\Delta_k < \varepsilon$, with a predetermined tolerance $\varepsilon > 0$.*

As to the supporting hyperplane H_k: For the feasible set

$$\mathscr{B} = \{x \mid g_i(x) \le 0,\ i = 1, \cdots, m\} = \{x \mid G(x) \le 0\}$$

with $G(x) := \max_{1 \le i \le m} g_i(x)$ *we determine in* S 2 *the (unique) boundary point* $z^{(k)} \in [\hat{x}^{(k)}, \tilde{x}] \cap \{x \mid G(x) = 0\}$, *and afterwards we define the hyperplane* $H_k := \{x \mid {a^{(k)}}^{\mathrm{T}} x = \alpha_k\}$ *with* $a^{(k)} \in \partial G(z^{(k)})$, *which may be chosen e.g. as* $a^{(k)} = \nabla g_j(z^{(k)})$ *for any* $j:\ g_j(z^{(k)}) = G(z^{(k)})$, *and then let* $\alpha_k := {a^{(k)}}^{\mathrm{T}} z^{(k)}$. *Due to (1.67), page 55, it follows* ${a^{(k)}}^{\mathrm{T}} x \le \alpha_k\ \forall x \in \mathscr{B}$, *whereas* ${a^{(k)}}^{\mathrm{T}} \hat{x}^{(k)} > \alpha_k$. *Hence,with the inequality* ${a^{(k)}}^{\mathrm{T}} x \le \alpha_k$ *added in step* S 3 *all feasible points of* $\mathscr{B}$ *are maintained, and the outer approximate* $\hat{x}^{(k)}$ *is cut off (see Fig. 1.6).* □

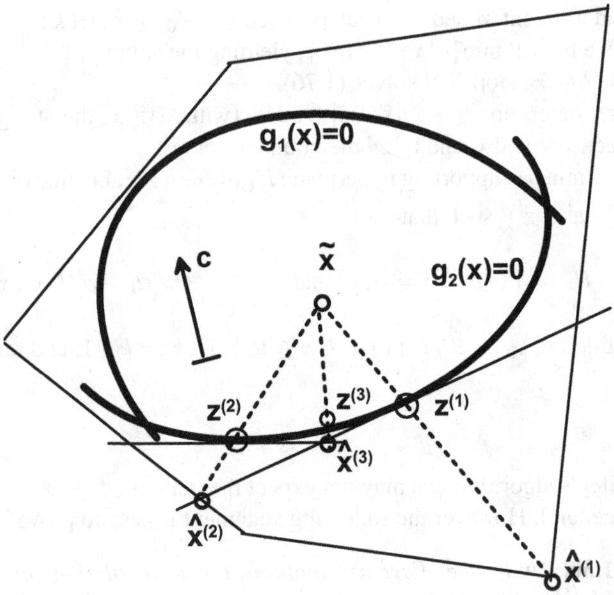

Fig. 1.6 Three cycles of Veinott's cutting plane method.

As mentioned already, in stochastic programming several cutting plane methods are used, as e.g. the above algorithms of Kelley (explicitely) or of Veinott (implicitly) to solve recourse problems, or else appropriate types for solving (explicitely) problems with probabilistic constraints as (1.6), for instance. In the latter case, we usually have special NLP's as

$$\left.\begin{array}{ll} \min c^{\mathrm{T}} x & \\ \text{s.t. } {a^{(i)}}^{\mathrm{T}} x \ge b_i, \ i = 1, \cdots, m, & \\ \quad F(x) \ge \alpha, & \end{array}\right\} \tag{1.77}$$

where $F(x) = P(\omega \mid Tx \ge h(\omega))$ with a given probability distribution P.

We shall briefly describe two further cutting plane approaches specialized to the problem type (1.77) under the following assumptions:

- F is a concave continuously differentiable function;

- $\mathscr{B}_{lin} := \{x \mid {a^{(i)}}^{\mathrm{T}} x \geq b_i,\ i = 1,\cdots,m\}$ is bounded, and hence this also holds for $\mathscr{B} = \mathscr{B}_{lin} \cap \{x \mid F(x) \geq \alpha\}$;
- $\exists x^S \in \mathscr{B}_{lin}$ being a Slater point for the nonlinear constraint, i.e. satisfying $F(x^S) > \alpha$.

The following method was originally proposed by Zoutendijk [355] and later on specialized for chance constrained programs in Szántai [320]. Obviously it is closely related to the above Veinott approach.

Cutting Planes: Outer Linearization; Moving Slater Points (Zoutendijk)

S 1 Let $y^{(1)} := x^S$, $\mathscr{B}_1 := \mathscr{B}_{lin}$, and $k := 1$.
S 2 Solve the LP $\min\{c^{\mathrm{T}} x \mid x \in \mathscr{B}_k\}$ yielding a solution $x^{(k)}$.
S 3 If $F(x^{(k)}) > \alpha - \varepsilon$ (for some predefined tolerance $\varepsilon > 0$), then stop; else add a feasibility cut according to the next step.
S 4 Determine $z^{(k)} \in [y^{(k)}, x^{(k)}] \cap \{x \mid F(x) = \alpha\}$;
$\mathscr{B}_{k+1} := \mathscr{B}_k \cap \{x \mid \nabla F(z^{(k)})^{\mathrm{T}} (x - z^{(k)}) \geq 0\}$;
$y^{(k+1)} := y^{(k)} + \dfrac{1}{k+1}(z^{(k)} - y^{(k)})$; $k := k+1$; return to step *S 2*.

□

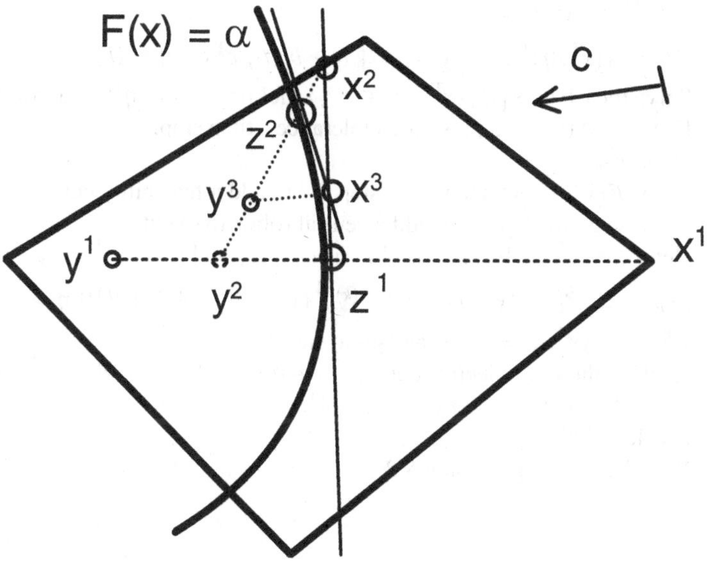

Fig. 1.7 Outer linearization with moving Slater points.

Under the above assumptions on problem (1.77) the same statements concerning convergence and error estimates as in Prop. 1.30. hold true for this algorithm.

Remark 1.5. *Whereas in the previous method the interior point $\tilde{x}$ was kept fixed throughout the procedure, in this variant the interior point of the set $\{x \mid F(x) \geq \alpha\}$ (originally $y^{(1)} = x^S$) is changed in each cycle as shown in Fig. 1.7. Since for any convex set $\mathscr{D}$ with some $y \in \operatorname{int}\mathscr{D}$ and any $z \in \operatorname{bd}\mathscr{D}$ it follows that $\lambda z + (1-\lambda)y \in \operatorname{int}\mathscr{D}\ \forall \lambda \in (0,1)$, we conclude that in step S 4 with $y^{(k)}$ interior to $\{x \mid F(x) \geq \alpha\}$ and $z^{(k)}$ on its boundary, we get $y^{(k+1)} \in \{x \mid F(x) > \alpha\}$ and hence again an interior point. However, these changes of the interior (Slater) points may improve the convergence rate of the algorithm.*

□

Again for problems of the type (1.77) with the above assumptions modified as

- $\exists x^S \in \operatorname{int}\mathscr{B}_{lin}$ being a Slater point for the nonlinear constraint, i.e. satisfying $F(x^S) > \alpha$, let U be such that $c^{\mathrm{T}}x \leq U\ \forall x \in \mathscr{B}$, and assume (normalize) c to satisfy $\|c\| = 1$,

we present the following method adapted by Mayer [230] from the central cutting plane method introduced by Elzinga–Moore [86] for general convex nonlinear programs. Similar methods have been investigated by Bulatov [38] as well as Zoutendijk [356] and Zukhovitskii–Primak [358].

A Central Cutting Plane Method (Elzinga–Moore)

S 1 Let $y^{(1)} := x^S$, $k := 1$, and
$$\mathscr{P}_1 := \{(x^{\mathrm{T}},\eta)^{\mathrm{T}} \mid {a^{(i)}}^{\mathrm{T}}x - \|a^{(i)}\|\,\eta \geq b_i\ \forall i,\ c^{\mathrm{T}}x + \eta \leq U\}.$$

S 2 Solve the LP $\max\{\eta \mid (x^{\mathrm{T}},\eta)^{\mathrm{T}} \in \mathscr{P}_k\}$ yielding $({x^{(k)}}^{\mathrm{T}},\eta^{(k)})^{\mathrm{T}}$ as a solution.

S 3 If $\eta^{(k)} < \varepsilon$ ($\varepsilon > 0$ a prescribed tolerance), then stop; otherwise
- if $F(x^{(k)}) < \alpha$, then go to step *S 4* to add a feasibility cut;
- else go to step *S 5* to add a central (objective) cut.

S 4 Determine $z^{(k)} \in [y^{(k)},x^{(k)}] \cap \{x \mid F(x) = \alpha\}$ and let
$$\mathscr{P}_{k+1} := \mathscr{P}_k \cap \{(x^{\mathrm{T}},\eta)^{\mathrm{T}} \mid {\nabla F(z^{(k)})}^{\mathrm{T}}(x - z^{(k)}) - \|\nabla F(z^{(k)})\|\,\eta \geq 0\},$$
$y^{(k+1)} := y^{(k)}$, $k := k+1$, and go to step *S 2*.

S 5 Replace the last objective cut by $c^{\mathrm{T}}x + \eta \leq c^{\mathrm{T}}x^{(k)} \Longrightarrow \mathscr{P}_{k+1}$.
If $F(x^{(k)}) > \alpha$, then set $y^{(k+1)} := x^{(k)}$,
else let $y^{(k+1)} := y^{(k)}$.
With $k := k+1$ go to step *S 2*.

□

An outer (feasibility) cut according to step *S 4* is illustrated in Fig. 1.8 whereas objective (central) cuts generated in step *S 5* are demonstrated in Fig. 1.9.

Remark 1.6. *The basic ideas of this algorithm are obviously related to the concept of Hesse's normal form of a linear equation: The equation $d^{\mathrm{T}}x = \rho$ is said to be in*

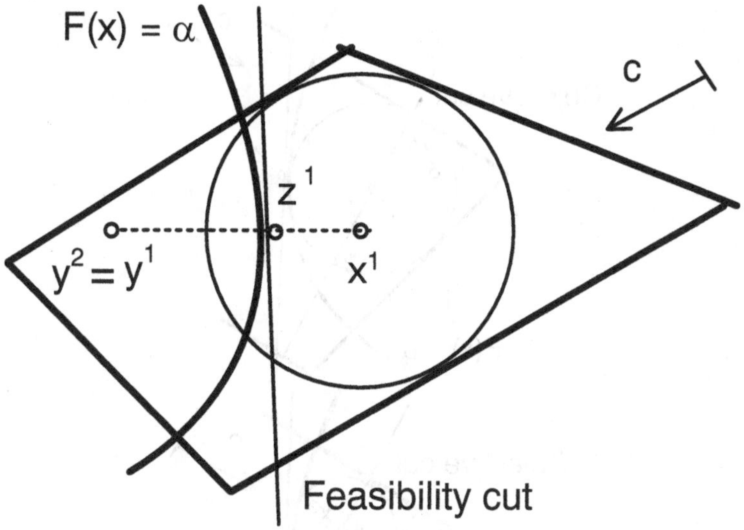

Fig. 1.8 Central cutting plane method: Outer cut.

normal form if $\|d\| = 1$. *Then, as is well known,* $\sigma = d^{\mathrm{T}}y - \rho$ *yields with* $|\sigma|$ *the Euclidean distance of* y *to the hyperplane* $\{x \mid d^{\mathrm{T}}x = \rho\}$, *with* $\sigma > 0$ *if and only if* $d^{\mathrm{T}}y > \rho$.
Hence, for an arbitrary equation $a^{\mathrm{T}}x = b$ *with* $a \neq 0$ *the equivalent equation* $\frac{a^{\mathrm{T}}}{\|a\|}x = \frac{b}{\|a\|}$ *is in normal form such that* $\eta \leq \frac{a^{\mathrm{T}}}{\|a\|}y - \frac{b}{\|a\|}$ *or equivalently* $\|a\|\eta \leq a^{\mathrm{T}}y - b$ *yields an upper bound for the distance* η *of any* $y \in \{y \mid a^{\mathrm{T}}y \geq b\}$ *to the hyperplane* $\{x \mid a^{\mathrm{T}}x = b\}$. *Now it is evident that solving an LP of the form* $\max\{\eta \mid {d^{(i)}}^{\mathrm{T}}x - \|d^{(i)}\|\eta \geq \rho_i,\ i \in I\}$ *as in step* S 2 *yields the center* $\hat{x}$ *and the radius* $\hat{\eta}$ *of the largest ball inscribed into the polyhedron* $\{x \mid {d^{(i)}}^{\mathrm{T}}x \geq \rho_i,\ i \in I\}$, *as was pointed out in Nemhauser–Widhelm [238].*

□

Therefore, with

$$\begin{aligned} J_k &:= \{j \leq k \mid \text{iteration } j \text{ generates a feasibility cut}\} \\ I_k &:= \{1, \cdots, k\} \setminus J^k \\ U_k &:= \min\{U, \min_{i \in I_k} c^{\mathrm{T}}x^{(i)}\}, \end{aligned}$$

in the k-th cycle of this algorithm we determine the center $x^{(k)}$ and the radius $\eta^{(k)}$ of the largest hypersphere inscribed into the polyhedron $\mathscr{P}_k$ defined by

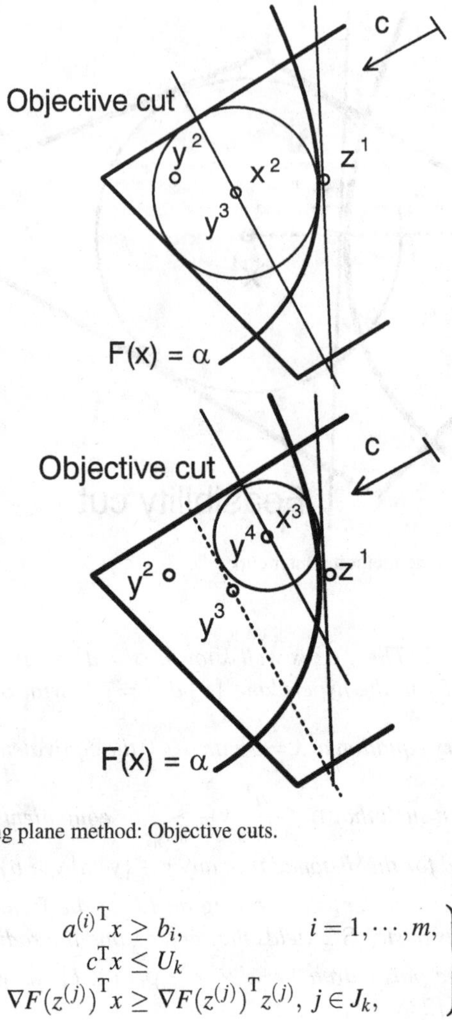

Fig. 1.9 Central cutting plane method: Objective cuts.

$$\left.\begin{array}{ll} {a^{(i)}}^{\mathrm{T}} x \geq b_i, & i = 1, \cdots, m, \\ c^{\mathrm{T}} x \leq U_k & \\ \nabla F(z^{(j)})^{\mathrm{T}} x \geq \nabla F(z^{(j)})^{\mathrm{T}} z^{(j)}, & j \in J_k, \end{array}\right\} \tag{1.78}$$

and, depending on $x^{(k)} \notin \mathscr{B}$ or $x^{(k)} \in \mathscr{B}$, we add a feasibility cut or else a central cut, respectively.

Proposition 1.31. *Under the assumptions for the central cutting plane method (Elzinga–Moore) holds* $\lim_{k \to \infty} \eta^{(k)} = 0$. *If* $U > \min_{x \in \mathscr{B}} c^{\mathrm{T}} x$, *then every convergent subsequence of* $\{x^{(k)} \mid k \in I_k\}$ *converges to a solution of (1.77).*

For the proof and for further details on the convergence behaviour of this algorithm we refer to Elzinga–Moore [86].

Exercises

1.13. Assume that the primal LP $\varphi(\zeta) := \min\{c^\mathrm{T}x \mid Ax = \zeta, x \geq 0\}$ is solvable for all $\zeta \in \mathbb{R}^m$. Show, that for an arbitrary $\zeta_0 \in \mathbb{R}^m$ any solution u_0 of the dual LP is a subgradient of $\varphi(\zeta_0)$, or equivalently, that $\arg\max\{\zeta_0^\mathrm{T}u \mid A^\mathrm{T}u \leq c\} \subset \partial\varphi(\zeta_0)$.

1.14. Assume an LP of the form $\min\{c^\mathrm{T}x \mid Ax \geq b, x \geq 0\}$ to be given. What can be said about vectors $\hat{x}$ and $\hat{u}$ satisfying the KKT conditions (1.70) for this linear program?

1.15. Consider the LP $\min\{c^\mathrm{T}x \mid Ax \geq 0\}$. Let $L(x,u)$ be the Lagrange function for this problem and assume $(\hat{x}, \hat{u})$ to be a saddle point satisfying (1.72) according to Prop. 1.28.. What does this mean for the given LP?

1.16. Given the quadratic program

$$\min\{x^2 + 4xy + y^2 \mid x^2 + y^2 - 1 \leq 0\},$$

find a solution using the KKT conditions. Is your solution unique? If the KKT conditions have nonoptimal solutions, why does this happen?

1.17. Find a solution $(\hat{x}, \hat{y}, \hat{z})$ of the NLP $\min\{x \mid x^2 + y^2 \leq 1, (y-2)^2 + z^2 \leq 1\}$. Can the KKT conditions be satisfied with this solution $(\hat{x}, \hat{y}, \hat{z})$? If not, why?

1.18. Show for problem $(A): \quad \min\{f(x) \mid g_i(x) \leq 0, i = 1, \cdots, m\}$ and the corresponding Lagrangian $L(x,u) = f(x) + \sum_{i=1}^m u_i g_i(x)$, that the solution $\hat{x}$ of problem $(P): \quad \min_x \sup_{u \geq 0} L(x,u)$ also solves (A); therefore, (P) is also called in NLP the primal problem, and the relation (1.72) (page 59) is denoted as *strong duality*.
Solve the NLP $\min\{c^\mathrm{T}x \mid x^\mathrm{T}x \leq 1\}$ with $c \in \mathbb{R}^n \setminus \{0\}$, and check the validity of strong duality for this problem.

1.19. Given the NLP $\min\{x + 5y \mid x^2 - 8x - y + 18 \leq 0; x, y \geq 0\}$. Choose $(\tilde{x}, \tilde{y}) = (4,3)$ as an interior point of the feasible set.

(a) With the first Polyhedron given as $\mathscr{P}_0 = \{(x,y)^T \mid x + y \leq 10, x \geq 0, y \geq 0\}$ start Veinott's cutting plane method and carry through 3 cycles; in each cycle compute the error estimate of Prop. 1.30..

(b) Determine the exact solution of this NLP.

Chapter 2
Single–stage SLP models

2.1 Introduction

In this chapter we consider stochastic programming problems which represent a single decision stage. The decision is to be made "here and now" and the models do not account for any corrective (recourse) actions which might be available after the realization of the random variables in the model becomes known. Such type of models typically involve, either in the constraints or in the objective function, or in both of them, random variables of the following form

$$\zeta(x,\xi) := T(\xi)x - h(\xi) \tag{2.1}$$

where $\xi : \Omega \to \mathbb{R}^r$ is a random vector on a probability space $(\Omega, \mathscr{F}, P)$. $T(\xi)$ denotes a random $s \times n$ matrix, $h(\xi) \in \mathbb{R}^s$ stands for a random vector, both depending on the random vector ξ. The support of ξ is defined as the smallest closed set $\Xi \subset \mathbb{R}^r$ having the property $\mathbb{P}(\xi \in \Xi) = 1$.

For being more specific, we assume that the dependence is defined in terms of affine linear relations as follows: for all $\xi \in \Xi$ we have

$$\begin{aligned} T(\xi) &= T + \sum_{j=1}^{r} T_j \xi_j, \\ h(\xi) &= h + \sum_{j=1}^{r} h_j \xi_j, \end{aligned} \tag{2.2}$$

where $T, T_j \in \mathbb{R}^{s\times n}$ are deterministic matrices and $h, h_j \in \mathbb{R}^s$ are deterministic vectors, $j = 1, \ldots, r$.

In this chapter the particular form (2.2) will not be used explicitly. All we need is the joint probability distribution of $(T(\xi), h(\xi))$ which will be presupposed as known throughout. As for stochastic programming in general, the basic assumption is that the probability distribution of $(T(\xi), h(\xi))$ does not depend on x. This means that our decision has no influence on the probability distribution of the random entries in the model data.

If in the constraints, $\zeta(x,\xi)$ frequently plays the role of a random slack variable in a random linear inequality. For instance, taking the inequality $T(\xi)x \geq h(\xi)$, this inequality can evidently also be written in the form $\zeta(x,\xi) \geq 0$.

For later reference, we write (2.1) also in a row–wise form as

$$\zeta_i(x;\xi) := t_i^{\mathrm{T}}(\xi)x - h_i(\xi), \;\; i = 1,\ldots,s, \tag{2.3}$$

where the components of the n–dimensional random vector $t_i(\xi)$ are the elements of the ith row of $T(\xi)$, $i = 1,\ldots,s$. Alternatively, (2.1), may be written in a column–wise fashion as

$$\zeta(x,\xi) = \sum_{j=1}^{n} T_j(\xi)x_j - h(\xi), \tag{2.4}$$

where the s–dimensional random vector $T_j(\xi)$ denotes the jth column of $T(\xi)$, $j = 1,\ldots,n$. Thus $\zeta(x,\xi)$ can be regarded as an affine linear combination of random vectors. Our assumption is that the joint probability distribution of these random vectors is known. The coefficients in the linear combination are the decision variables x_j, consequently the probability distribution of $\zeta(x,\xi)$ will depend on our decision. We control the probability distribution of $\zeta(x,\xi)$, by controlling its realizations, according to (2.4).

The question arises, what can be stated about the probability distribution of $\zeta(x,\xi)$? In particular, assuming that the joint probability distribution of $(T_j(\xi), j = 1,\ldots,n;\, h(\xi))$ belongs to a given parametric family of distributions, for which families will the affine linear combination $\zeta(x,\xi)$ belong to the same family? An example of a family, for which the answer is affirmative, is the class of multivariate normal distributions. This question will be further pursued in Section 2.2.3, in connection with separate probability constraints.

Note that a similar question also arises in mathematical statistics regarding linear statistical models. In that case $h(\xi)$ represents an error (noise) term, which is usually assumed as being stochastically independent of the random vectors $T_j(\xi)$. In mathematical statistics we are dealing with a random vector ζ with unknown distribution and the goal is to choose x in such a way, that the distribution of $\zeta(x,\xi)$ provides a good approximation to the distribution of ζ in a statistical sense. For achieving this, the x_j's are considered as random variables. The starting point is a joint sample according to the distribution of $(\zeta, T_j(\xi); j = 1,\ldots,n)$ and assuming the linear model (2.4), the aim is to construct unbiased estimators for the x_j's.

In stochastic programming we face a different situation. The primary entity is the given joint distribution of $(T_j(\xi), j = 1,\ldots,n; h(\xi))$ and the goal is to achieve a probability distribution of $\zeta(x,\xi)$ with advantageous properties, whereby x is considered as being deterministic. To make this precise, we will attach a quantitative meaning to the term "advantageous" and will arrive this way at a classification scheme for the different classes of SLP models as follows:

- First we define a function $\rho : \Upsilon \to \mathbb{R}^1$ for evaluating random vectors, where Υ is some linear space of s–dimensional random vectors defined on a probability space $(\Omega, \mathscr{F}, P)$. For instance, Υ will be frequently chosen as the linear space of random vectors with finite expected value. For each random vector $\vartheta \in \Upsilon$, $\rho(\vartheta)$ is interpreted as a quality measure in the corresponding modeling approach. Depending on the interpretation of $\rho(\vartheta)$ as either expressing opportunity or risk, "advantageous" will mean that higher or lower values of $\rho(\vartheta)$ are considered as preferable, respectively. In the latter case ρ will be called a risk measure. The probability distribution function of ϑ will be denoted by F_ϑ and Θ will denote the support of ϑ. In the special case $s = 1$, Υ is some linear space of random variables. In the sequel, the term random vector will always mean that $s > 1$ is permitted whereas the term random variable will indicate that $s = 1$ is assumed.
- Based on the chosen function ρ for evaluating random variables, decision vectors x will be evaluated as follows. We define the corresponding evaluation function $V : \mathbb{R}^n \to \mathbb{R}^1$ by substituting $\zeta(x, \xi)$ into ρ:

  $$V(x) := \rho(\zeta(x, \xi)) \tag{2.5}$$

  provided that $\zeta(x, \xi) \in \Upsilon$ holds for all x. $V(x)$ will be interpreted as a quality measure for x and will be employed for building SLP models. For indicating that the evaluation involves all components of the random vector simultaneously, we will call V a *joint evaluation function*.
- Alternatively, when dealing with constraints, it may make sense to assign quality measures to the components of $\zeta(x, \xi)$ separately. If ρ is defined for random variables and $\zeta_i(x, \xi) \in \Upsilon$ holds for all x and all i then $V_i(x) := \rho(\zeta_i(x, \xi))$ serves for evaluating x for the ith component of $\zeta(x, \xi)$, $i = 1, \ldots, s$. Concerning V_i, the term *separate evaluation function* will be employed, for pointing out the fact that the components of the random vector $\zeta(x, \xi)$ are evaluated separately. If $s = 1$ holds then $\zeta(x, \xi)$ is a random variable and both adjectives "separate" and "joint" apply. This ambiguity will have no substantial influence on the discussions concerning SLP models.

Having chosen ρ, the evaluation function V is uniquely defined. The different SLP model classes will correspond to different choices of the quality measure ρ for random vectors.

Υ will be one of the following linear spaces of s–dimensional random vectors:

$$\begin{aligned}
\mathscr{L}_s^0 &:= \{\text{the set of all random vectors on } (\Omega, \mathscr{F}, P)\},\\
\mathscr{L}_s^1 &:= \mathscr{L}_s^1(\Omega, \mathscr{F}, P) = \{\vartheta \mid \int_{\mathbb{R}^s} \|t\|_1 dF_\vartheta(t) < +\infty\},\\
\mathscr{L}_s^2 &:= \mathscr{L}_s^2(\Omega, \mathscr{F}, P) = \{\vartheta \mid \int_{\mathbb{R}^s} \|t\|_2^2 dF_\vartheta(t) < +\infty\},\\
\mathscr{L}_s^\infty &:= \mathscr{L}_s^\infty(\Omega, \mathscr{F}, P) = \{\vartheta \mid \exists C:\ \mathbb{P}(\|\vartheta\|_2 > C) = 0\},
\end{aligned} \tag{2.6}$$

where $\mathscr{L}_s^1$ is the space of s–dimensional random vectors with finite expected value, $\mathscr{L}_s^2$ stands for the space of random vectors with finite second moments, and $\mathscr{L}_s^\infty$ denotes the space of random vectors having a bounded support. $\|t\|_2 = \sqrt{\sum_{i=1}^s t_i^2}$ is the Euclidean norm and $\|t\|_1 = \sum_{i=1}^s |t_i|$ holds.

Note that up to this point we have viewed $\zeta(x,\xi) = T(\xi)x - h(\xi)$ as an affine linear combination of random vectors. Alternatively, we can also consider $\zeta(x,\xi)$ as a *deviation* between $T(\xi)x$ and $h(\xi)$. In mathematical statistics an interpretation could be fitting $T(\xi)x$ to $h(\xi)$ in a least squares sense. In this setting, $\zeta(x,\xi)$ would be an error term. Assuming some distributional properties of the error term and having a sample for $(T(\xi),h(\xi))$, the goal in mathematical statistics is to find a good fit. In stochastic programming we proceed analogously as before: quality measures for random variables will be introduced and stochastic programming models will be built by employing the corresponding evaluation function V. We interpret the quality measure in this case as *deviation measure*.

As mentioned above, SLP models will be built by employing evaluation functions V corresponding to some quality measure ρ. The different SLP model classes will be discussed in a framework of prototype models. For employing joint– and separate evaluation functions in the constraints, we consider the models

$$\left.\begin{array}{rl} \max & c^{\mathrm{T}}x \\ \text{s.t.} & V(x) \geq \kappa \\ & x \in \mathscr{B} \end{array}\right\} \qquad \left.\begin{array}{rl} \max & c^{\mathrm{T}}x \\ \text{s.t.} & V_i(x) \geq \kappa_i, \\ & i = 1,\ldots,s \\ & x \in \mathscr{B}, \end{array}\right\} \tag{2.7}$$

where κ and κ_i are prescribed, $i = 1,\ldots,s$, and $\mathscr{B}$ is a polyhedral set

$$\mathscr{B} = \{x \mid Ax \propto b,\ l \leq x \leq u\} \tag{2.8}$$

with A being an $m \times n$ matrix and x, b, l, and u having corresponding dimensions. The symbol $\propto$ means that any one of the relations $\leq$, $=$, and $\geq$ is permitted row–wise.

For models with the evaluation function being in the objective, we consider the prototype model

$$\left.\begin{array}{rl} \max & c^{\mathrm{T}}x + V(x) \\ \text{s.t.} & x \in \mathscr{B}. \end{array}\right\} \tag{2.9}$$

Alternatively, we will also employ prototype models with reversed direction of the inequalities in the constraints of (2.7) and with minimization instead of maximization in (2.9). To see the reason for this, let us assume first, that for some model class the evaluation function V is a concave function. In this case, both (2.7) and (2.9) are convex programming problems. Assume next that for some other SLP model class V turns out to be a nonlinear convex function. In this case our prototype models become non–convex optimization problems, whereas their counterparts with reversed inequality constraints and minimization in the objective will be convex programming problems. The point is that the chances for finding efficient algorithms are much better for convex optimization problems than for the non–convex case. This subject will be further pursued in Section 2.6.

From the modeling viewpoint, stochastic programming models can have a composite form, involving several different random vectors of the type (2.1). We have chosen to work with the above prototype models because they serve well for explaining the basic ideas which can then be applied to composite models in a straightforward way. For some model classes $c = 0$ will be required in (2.9). The reason is that, for those model classes, V has merely some generalized concavity property which might be destroyed by adding a linear term.

The objective function of (2.9) consists of a sum of two terms whereas in applications they are usually weighted with respect to each other. Weighting can also be interpreted in terms of duality. We take as an example the following weighted version of (2.9):

$$\left.\begin{array}{ll} \max & c^{\mathrm{T}}x + \lambda\, V(x) \\ \text{s.t.} & x \in \mathscr{B} \end{array}\right\} \tag{2.10}$$

with a positive weight λ. This can equivalently be written in the form

$$\left.\begin{array}{rl} v(\lambda) := \max & c^{\mathrm{T}}x + \lambda(V(x) - \kappa) \\ \text{s.t.} & x \in \mathscr{B} \end{array}\right\} \tag{2.11}$$

where $-\lambda\kappa$ is merely a shift in the optimal objective value. This problem is called a *Lagrangian relaxation* of the first optimization problem in (2.7). The corresponding Lagrange–dual–problem is then

$$\min\{\, v(\lambda) \mid \lambda \geq 0 \,\}. \tag{2.12}$$

For the duality relationships between (2.7), (2.11), and (2.12) see Bazaraa and Shetty [11].

For the sake of simplicity of presentation, we assume in the sequel that positive weighting factors (if any) are taken into account in the definition of c.

The simplest way for assigning a quality measure to $\zeta(x,\xi)$ is taking the expectation. To see how this works, let us discuss the application of the idea for including a system of random inequalities $\zeta(x,\xi) \geq 0$ into an SLP model. We choose separate evaluation for the components of $\zeta(x,\xi)$ and employ the quality measure

$$\rho_{\mathrm{E}}(\vartheta) := \mathbb{E}[\vartheta], \;\; \vartheta \in \mathscr{L}_1^1 \tag{2.13}$$

for the components. Assuming the existence of the expected values of $T(\xi)$ and $h(\xi)$ and setting $\kappa_i = 0$ for all i, this leads to the following formulation of (2.7):

$$\left.\begin{array}{ll} \max & c^{\mathrm{T}}x \\ \text{s.t.} & \bar{t}_i^{\mathrm{T}}x \geq \bar{h}_i, \;\; i = 1,\ldots,n \\ & x \in \mathscr{B} \end{array}\right\} \tag{2.14}$$

where $\bar{t}_i := \mathbb{E}_\xi[t_i(\xi)]$ and $\bar{h}_i := \mathbb{E}_\xi[h_i(\xi)]$ hold, with the components of $t_i(\xi)$ being the elements of the ith row of $T(\xi)$. The resulting deterministic LP problem is called *expected value problem.* Unfortunately, the expected value problem is frequently used as a substitute for the SLP problem. While in some (rare) situations this might be appropriate, in general it is a very crude approach: the whole probability distribution is collapsed into a one–point distribution. It should by no means be used as the single way for representing $\zeta(x,\xi)$ in the model. However, accompanied by a constraint or objective part involving some other quality measure, it can prove to be an important constituent of the SLP model. For examples of this kind see Section 2.7.3. In financial portfolio optimization, the most prominent and widely used model of the combined type is the model of Markowitz [217], see also Elton et al. [85].

For discussing the next idea, our starting point is again the system of random inequalities $\zeta(x,\xi) \geq 0$. We interpret this as prescribing the sign of $\zeta(x,\xi)$ and consider the inclusion of the system of random inequalities

$$T(\xi)x \geq h(\xi) \tag{2.15}$$

into the stochastic programming model. The difficulty is that, besides the decision vector x, the constraints also depend on the random vector ξ. One of the earliest proposals for overcoming this difficulty is due to Madansky [212], [213], who suggested a worst–case approach by prescribing the inequalities (2.15) for all $\xi \in \Xi$, with Ξ denoting the support of the random vector ξ. We assume that Ξ is a bounded set. This leads to the following formulation of (2.7):

$$\left.\begin{array}{ll} \max & c^{\mathrm{T}}x \\ \text{s.t.} & T(\xi)x \geq h(\xi), \;\; \xi \in \Xi \\ & x \in \mathscr{B}. \end{array}\right\} \tag{2.16}$$

Madansky termed the solution of this optimization problem as *fat solution.* The approach corresponds to the following choice of the quality measure ρ_{fat}: $\Upsilon = \mathscr{L}_s^\infty$ is the set of random vectors having a bounded support and

$$\rho_{\text{fat}}(\vartheta) := \min_{\widehat{\vartheta}\in\Theta} \min_{1\leq i\leq s} \widehat{\vartheta}_i, \;\; \vartheta \in \mathscr{L}_s^\infty, \tag{2.17}$$

where Θ is the support of ϑ. The formulation (2.16) corresponds to the model (2.7) with the inequality constraint chosen as $V(x) = \rho_{\text{fat}}(\zeta(x,\xi)) \geq 0$. The feasible domain $\mathscr{D}$ of (2.16) is the intersection of convex sets and thus it is obviously convex:

$$\mathscr{D} = \bigcap_{\xi\in\Xi} \{x \mid T(\xi)x \geq h(\xi), x \in \mathscr{B}\}.$$

In the special case of a finite discrete distribution, Ξ is a finite set and (2.16) reduces to a linear programming problem. In general, (2.16) may turn out in many cases as being infeasible, especially if Ξ contains infinitely many points.

Recently, after the new optimization area of semidefinite programming has emerged in the 1990s, it became numerically feasible to compute fat solutions also for bounded domains Ξ containing infinitely many points. The idea is that instead of considering Ξ as an index set, $\xi \in \Xi$ is explicitly handled as a constraint in (2.16) and ξ is considered as a deterministic variable. For instance, with ellipsoidal domains Ξ, (2.16) can be reformulated as an equivalent semidefinite programming problem, see Ben–Tal et al. [15] and the references therein. The cited paper also presents an extension of this approach to the class of semidefinite programming problems. Along with the extension, the approach has also been renamed as *robust optimization*. There are important application areas where working with fat solutions makes sense. As an example, let us mention structural design for mechanical structures, see Ben–Tal et al. [15]. Note that the term "robust optimization" is also used for other model classes; we will return to this point later.

Although in robust optimization, as defined above, Ξ is called the domain of uncertainty, the approach is only loosely connected to stochastic programming or to stochastic modeling in general. It can be considered as a kind of worst–case parametric programming approach. If, as in our case, Ξ is the support of a random variable ξ, the probability distribution of ξ does not play any role: the models will deliver identical results for all random variables having the same support. For these reasons, the topic of the above kind of robust optimization will not be pursued further in this book. Let us emphasize, however, that robust optimization is an important alternative modeling approach for dealing with uncertain data. For the interested reader we recommend the recent book of Ben–Tal et al. [16] and the references therein.

A straightforward idea for generalizing (2.16) is to consider x as a feasible solution, if it satisfies all random inequalities for restricted subsets of the support. A natural idea for imposing such a restriction is to consider subsets with prescribed probability levels. SLP models of this class have been introduced and first studied by Charnes and Cooper [41], Miller and Wagner [234] and by Prékopa [258].
The corresponding quality measure is

$$\rho_{\mathrm{P}}(\vartheta) := \mathbb{P}(\vartheta \geq 0), \;\; \vartheta \in \mathscr{L}_s^0,$$

defined on the set of all random vectors on $(\Omega, \mathscr{F}, P)$. The evaluation function $V(x)$ (see (2.5)) will be denoted for this model class by $G(x)$. This leads to the concept of probability functions, defined as follows:

$$G(x) := \mathbb{P}_{\xi}(\, T(\xi)x \geq h(\xi)\,). \tag{2.18}$$

Choosing constraints of the form $G(x) \geq \alpha$, with α being a high probability level (for instance, $\alpha = 0.99$), the prototype model (2.7) assumes the form

$$\left.\begin{array}{ll} \max & c^{\mathrm{T}}x \\ \text{s.t.} & \mathbb{P}_{\xi}(\, T(\xi)x \geq h(\xi)\,) \geq \alpha \\ & \qquad x \qquad\qquad\quad \in \mathscr{B}. \end{array}\right\} \tag{2.19}$$

By choosing $\alpha = 1$ in this model, we obtain a generalization of the concept of a fat solution, discussed on page 76. In this case $x \in \mathscr{B}$ is considered as feasible, if the random inequalities hold in an almost sure sense, meaning that they hold except for a subset of Ω having probability measure zero.

Taking the quality measure separately for the components of $\zeta(x,\xi)$, the constraints in (2.7) are $G_i(x) \geq \alpha_i$, with the probability functions $G_i(x) := \mathbb{P}_\xi(\, t_i^{\mathrm{T}}(\xi)x \geq h_i(\xi)\,)$. The probability levels α_i are specified separately for the individual rows.

Being in the objective, the probability function will be maximized.

Alternatively, we might be interested in constraints of the form $G(x) \leq \beta$, with β being small (for instance, $\beta = 0.01$). In this context, β frequently represents a ruin probability, meaning, for instance, the probability of financial ruin of a company, death of a patient, or crashing of a bridge. In such modeling situations, (2.9) would be formulated with minimizing G in the objective.

Constraints involving probability functions are called chance–constraints or probabilistic constraints. Depending on whether $G(x)$ or $G_i(x)$, $i = 1,\ldots,s$, is used, the constraints are called *joint* or *separate* constraints, respectively. From another point of view, a separate constraint is a special case of (2.18) with $T(\xi)$ consisting of a single row ($s = 1$). Models based on probability functions provide a natural way of building models in several application areas, see Prékopa [266]. Here we just point out two fields, where probabilities play an important part in planning anyhow: finance (ruin probability) and electrical power systems engineering (loss–of–load probability (LOLP)). Stochastic optimization problems involving probability functions will be discussed in detail in Section 2.2.

Let us consider a model involving a probability constraint of the form $G(x) = \mathbb{P}_\xi(\, T(\xi)x \geq h(\xi)\,) \geq \alpha$, with a high probability level α. For each fixed x we interpret the event, that some of the random inequalities do not hold, as loss. Such type of models have the following characteristic feature: On the one hand, they ensure that a loss may only occur with a small probability $(1-\alpha)$. On the other hand, losses may occur, and for the case when they occur, the models provide no control for the modeler on the size of the loss. In modeling situations, where considering the size of the loss makes sense at all, the second characteristic might be considered as a drawback. To distinguish between models based on probability constraints and models which account for the loss size, Klein Haneveld [188] calls the quality measure based on probability functions *qualitative* and quality measures accounting also for the loss size *quantitative*.

Let us discuss shortly situations where the size of the loss does not matter. As a hypothetical example let us imagine that a medical treatment is modeled and the random inequalities in (2.18) express the survival of the patient. Loss means in this case that the patient dies and the size of the loss is meaningless in the modeling context. As a more practical example let us consider mechanical truss optimization problems with a given topology. Such models contain several groups of constraints modeling the laws of mechanics. Under random loads these models may involve chance–constraints of the above type (see, for instance, Marti [224] and the references therein). The random inequalities in (2.18) express some mechanical requirements; if they do not hold, then the system crashes. The point is that if the system crashes, then the topology obviously changes and the whole model becomes invalid (the model crashes too). Therefore, it is pointless to include constraints accounting for the size of the loss.

For the case when taking into account the loss–size makes sense, several kinds of remedies have been suggested. It is usually assumed that penalty costs are available for the losses. Prékopa [266] proposes a combined model, involving both probabilistic constraints and recourse–constraints in a two stage recourse problem, with the expected penalty costs for the losses included as an additive term into the objective function. Dert [68] introduces besides the probabilistic constraint binary variables for indicating the occurrence of losses and uses a penalty term in the objective function for the expected penalty costs of losses.

For introducing the next model class we assume that negative values of $\zeta(x,\xi)$ represent losses and positive values correspond to gains. For the sake of simplicity of presentation we also assume that $\zeta(x,\xi)$ is a random variable ($s=1$ holds). The loss as a random variable can then be written as

$$\zeta^{-}(x,\xi) := (t^{\mathrm{T}}(\xi)x - h(\xi))^{-},$$

where $t(\xi)$ denotes the single row of $T(\xi)$, $h(\xi)$ is a random variable, and $z^{-} = \max\{0,-z\}$ denotes the negative part of $z \in \mathbb{R}$.

Using this, the probability constraint $G(x) \geq \alpha$ can be written in expected–value terms as

$$G(x) \geq \alpha \Longleftrightarrow \mathbb{E}_{\xi}[\chi(\zeta^{-}(x,\xi))] \leq 1-\alpha \tag{2.20}$$

with χ denoting the indicator function

$$\chi(z) = \begin{cases} 0 \text{ if } z \leq 0, \\ 1 \text{ if } z > 0. \end{cases}$$

In (2.20) the function χ enforces equality across different loss–sizes. Due to an idea of Klein Haneveld [188], χ is dropped and the following quality measure is introduced:

$$\rho^{-}_{\mathrm{sic}}(\vartheta) := \mathbb{E}[\vartheta^{-}], \ \ \vartheta \in \mathscr{L}^{1}_{1}.$$

This results in an evaluation function $H(x) := \mathbb{E}_{\xi}[\zeta^{-}(x,\xi)]$ which is simply the expected value of the random variable expressing losses. In models based on this evaluation function, constraints of the form $H(x) \leq \gamma$ will be employed, where γ is a prescribed maximal level of tolerable expected loss. Constraints based on $H(x)$ are called *integrated chance constraints*. If in the objective, $H(x)$ will be minimized. The prototype model with integrated chance constraint has the form

$$\left.\begin{array}{ll} \min c^{\mathrm{T}}x & \\ \text{s.t. } \mathbb{E}_{\xi}[\zeta^{-}(x,\xi)] & \leq \gamma \\ \qquad\qquad x & \in \mathscr{B}. \end{array}\right\} \tag{2.21}$$

For the integrated chance constraints which we have considered so far, only $\zeta^-(x,\xi)$ plays a role. It might be desirable to take into account the entire distribution of $\zeta(x,\xi)$. In fact, the following variant of integrated chance constraints takes into account also the expected gain $\zeta^+(x,\xi)$:

$$\mathbb{E}_\xi[\zeta^-(x,\xi)] \leq \alpha \mathbb{E}_\xi[|\zeta(x,\xi)|]$$

which can be derived from the quality measure

$$\rho^\alpha_{\text{sicm}}(\vartheta) := (1-\alpha)\mathbb{E}[\vartheta^-] - \alpha\mathbb{E}[\vartheta^+], \;\; \vartheta \in \mathscr{L}^1_1 \qquad (2.22)$$

and leads to a convex programming formulation for $\alpha \leq \frac{1}{2}$. Integrated chance constraints, including joint constraints for the case when $\zeta(x,\xi)$ is a random vector, will be presented in Section 2.4.1.

The remaining model types, which will be reviewed in the introduction, are only applicable in the case when $\zeta(x,\xi)$ is a random variable. Thus we have $\zeta(x,\xi) = t^{\mathrm{T}}(\xi)x - h(\xi)$, where the components of the n–dimensional random vector $t(\xi)$ are the elements of the single row of $T(\xi)$.

Motivated by reliability theory, Prékopa [260] has developed a model which is built by utilizing the conditional expectation of the loss size. The quality measure is chosen as

$$\rho_{\text{cexp}}(\vartheta) := \mathbb{E}[-\vartheta \mid \vartheta < 0], \;\; \vartheta \in \mathscr{L}^1_1.$$

Consequently, $\rho_{\text{cexp}}(\vartheta)$ is the conditional expectation of the loss, given that a loss occurs. With the corresponding evaluation function, constraints of the form

$$\mathbb{E}_\xi[-\zeta(x,\xi) \mid \zeta(x,\xi) < 0] \leq \gamma$$

are included into the model, where the prescribed γ is a maximal tolerable conditional expected loss size. This model will be the subject of Section 2.4.2.

In the following discussion it will be convenient to consider positive values of $\zeta(x,\xi)$ as losses and negative values as gains. A further idea to include the loss size and simultaneously also provide control on the probability of loss is utilizing quantiles. The first stochastic optimization model of this type has been proposed by Kataoka [178]. For a given $0 < \alpha < 1$, we utilize the following quality measure:

$$\rho^\alpha_{\text{VaR}}(\vartheta) := \nu(\vartheta,\alpha) := \min\{z \mid F_\vartheta(z) \geq \alpha\}, \;\; \vartheta \in \mathscr{L}^0_1, \qquad (2.23)$$

defined on the set of all random variables on $(\Omega, \mathscr{F}, P)$, and with F_ϑ standing for the probability distribution function of ϑ. In other words, for a given α, $\rho^\alpha_{\text{VaR}}(\vartheta)$ is the left endpoint of the closed interval of α–quantiles of ϑ. This leads to the following evaluation function

$$\nu(x,\alpha) := \min\{z \mid \Psi(x,z) \geq \alpha\},$$

where $\Psi(x,\cdot)$ denotes the probability distribution function of $\zeta(x,\xi)$ for each fixed x, and with α being a prescribed (high) probability level, for instance, $\alpha = 0.95$. This quality measure is widely used in the finance industry, it is called *Value at*

Risk (VaR) there. We will consider optimization problems involving $v(x,\alpha)$ in Section 2.3. In general, it is quite difficult to build numerically tractable optimization models which are based on VaR. The main difficulty is that $v(x,\alpha)$, as a function of x, is not convex in general.

An interesting recent approach for building SLP models is due to Rockafellar and Uryasev [282]. The idea is to combine VaR and the conditional expectation approach. The following quality measure is chosen:

$$\rho^{\alpha}_{\text{CVaR}}(\vartheta) := \min_{z} \left[z + \frac{1}{1-\alpha}\mathbb{E}[(\vartheta - z)^{+}]\right], \quad \vartheta \in \mathscr{L}^{1}_{1}.$$

The motivation for introducing this quality measure is twofold. On the one hand, utilizing a well–known fact from probability theory it can be shown that the solution set of the above minimization problem coincides with the set of α–quantiles of the distribution of ϑ. On the other hand, under the assumption that ϑ has a continuous distribution function, we have

$$\rho^{\alpha}_{\text{CVaR}}(\vartheta) = \mathbb{E}[\vartheta \mid \vartheta \geq v(\vartheta,\alpha)], \quad \vartheta \in \mathscr{L}^{1}_{1},$$

where $v(\vartheta,\alpha)$ is the value at risk (VaR) (see (2.23)). This means that $\rho^{\alpha}_{\text{CVaR}}(\vartheta)$ is the conditional expectation of the loss given that the loss exceeds VaR. The evaluation function

$$v_c(x,\alpha) := \rho^{\alpha}_{\text{CVaR}}(\zeta(x,\xi))$$

has nice convexity properties. Therefore, the prototype problems will involve inequality constraints of the form $v_c(x,\alpha) \leq \gamma$ and being in the objective, $v_c(x,\alpha)$ will be minimized. A further attractive feature is that, for finite discrete distributions, the optimization problems can be reduced to linear programming problems. A detailed discussion of this model class will be the subject of Section 2.4.3.

Finally we consider modeling approaches where $\zeta(x,\xi)$ is interpreted as a deviation between $t^{\mathrm{T}}(\xi)x$ and $h(\xi)$, with the quality measures penalizing this deviation. Admittedly, most quality measures which have been introduced so far, can also be interpreted from the purely mathematical viewpoint as measuring deviation. Nevertheless, we have chosen to discuss those quality measures as a separate class, which correspond to the following modeling attitude: both $t^{\mathrm{T}}(\xi)x$ and $h(\xi)$ represent important quantities in their own right, and the emphasis in modeling risk is on their deviation. Deviations are interpreted as risk and therefore the quality measure will be called a *risk measure* in this context. As a typical example let us mention portfolio optimization in finance, where $t^{\mathrm{T}}(\xi)x$ represents the random portfolio return and $h(\xi)$ models some benchmark return. For this approach see, for instance, Elton et al. [85] and also Section 2.7.3.

Our first example of a deviation measure is the risk measure

$$\rho_{\mathrm{Q}}(\vartheta) := \sqrt{\mathbb{E}[\vartheta^2]}, \quad \vartheta \in \mathscr{L}^{2}_{1}$$

defined on the linear space of random variables with finite second moment. The corresponding evaluation function is

$$Q(x) = \sqrt{\mathbb{E}_\xi[(t^{\mathrm{T}}(\xi)x - h(\xi))^2]}.$$

As a second example we take the mean absolute deviation, with the risk measure

$$\rho_{\mathrm{A}}(\vartheta) := \mathbb{E}[|\vartheta|], \ \vartheta \in \mathscr{L}_1^1$$

and the evaluation function

$$A(x) = \mathbb{E}_\xi[|t^{\mathrm{T}}(\xi)x - h(\xi)|].$$

Stochastic programming models, based on risk measures of this type, will be the subject of Section 2.5. Let us mention that stochastic optimization models in this class are by some authors also termed as *robust optimization problems.*

The basic question concerning the various quality measures is, how the stochastic optimization problems, based on these measures, behave from the numerical point of view. This will be the main subject of the present chapter.

From the point of view of efficient numerical solution, the most desirable property of a nonlinear optimization problem is that it should be a convex programming problem. Regarding the above–formulated prototype problems (2.7) and (2.9), in a strict sense these would count as convex programming problems under the assumption that V and V_i are concave functions.

For the subsequent discussion we will assume that in the objective function of (2.9) the additive linear term $c^{\mathrm{T}}x$ is missing, that is, we assume that $c = 0$ holds. The reason for this assumption is that we will work with functions V having some generalized concavity properties. For such functions the addition of a linear term may destroy the generalized concavity property. Examples for this phenomenon will be presented later on in this section.

We will employ the following generalization of the notion of a convex programming problem: we consider the above–mentioned problems as convex programming problems, if the feasible domain is convex and if $V(x)$ is a pseudo–concave function in (2.9). For general properties of optimization problems of this type see, for instance, Bazaraa and Shetty [11] and Avriel, Diewert, Schaible, and Zang [8].

We proceed with a short discussion concerning some generalizations of concave functions which will be utilized in this chapter.

Definition 2.11. *Let $f : C \to \mathbb{R}$ be a function defined over the convex set C.*

- *f is called quasi–concave, if the inequality*

$$f(\lambda x + (1-\lambda)y) \geq \min\{f(x), f(y)\}$$

  holds, for all $x \in C$, $y \in C$, and $\lambda \in [0,1]$.
- *f is called quasi–convex, if $-f$ is quasi–concave.*

Functions which are both quasi–convex and quasi–concave will be called *quasi–linear.* It is easy to see that f is quasi–concave if and only if the upper–level sets

$$\mathscr{U}_\gamma := \{x \mid f(x) \geq \gamma\} \tag{2.24}$$

are convex sets, for all $\gamma \in \mathbb{R}$. Thus, for ensuring the convexity of the feasible domain in (2.7), it will be sufficient to ensure that the function V is quasi–concave.

Definition 2.12. *Let $f : C \to \mathbb{R}$ be a continuously differentiable function defined over an open convex set C.*

- *f is called pseudo–concave, if the following implication*

$$\nabla^{\mathrm{T}} f(x)(y-x) \leq 0 \quad \Longrightarrow \quad f(y) \leq f(x)$$

  holds for all $x \in C$ and $y \in C$.
- *f is called pseudo–convex, if $-f$ pseudo–concave.*

The following facts are easy to check and are left as exercises for the reader: If f is a concave function, then it is quasi–concave and in the differentiable case it is also pseudo–concave. Pseudo–concave functions are also quasi–concave, for this assertion see e.g. Bazaraa and Shetty [11].

From our point of view, for maximization problems with quasi–concave restrictions (implying a convex feasible domain) and a pseudo–concave objective function, the most important properties are the following, see [11]:

- All local optimal solutions are global solutions.
- The Kuhn–Tucker optimality conditions are sufficient conditions of optimality.

Thus, in (2.9), V should be a pseudo–concave function. Note that requiring only quasi–concavity for V, results in general in non–convex optimization problems. Such problems may have local maxima which are not global.

A further remark concerns the quasi–concavity requirement for the constraint function V in (2.7). Although this way the convexity of the feasible domain is ensured, quasi–concavity is a rather weak property from the algorithmic point of view. One of the difficulties is that regularity conditions, which ensure the necessity of the Kuhn–Tucker conditions, are difficult to check in this case. From the algorithmic point of view it is much better, when besides the objective function, the constraint functions are pseudo–concave too. This implies, for instance, that the Slater–regularity can be utilized for enforcing the necessity of the Kuhn–Tucker conditions.

We will need the following fact concerning the pseudo–concavity of fractional functions:

Proposition 2.32. *Let f and g be continuously differentiable functions defined on $\mathbb{R}^n$ and let $C \subset \mathbb{R}^n$ be a convex set. We assume that $f(x) \geq 0$ and $g(x) > 0$ hold for all $x \in C$. If f is concave and g is convex then $h(x) := \dfrac{f(x)}{g(x)}$ is pseudo–concave on C.*

Proof: Let $x \in C, y \in C$, and assume that $\nabla^{\mathrm{T}} h(x)(y-x) \le 0$ holds. By straightforward computation this implies

$$g(x)\nabla^{\mathrm{T}} f(x)(y-x) - f(x)\nabla^{\mathrm{T}} g(x)(y-x) \le 0.$$

Utilizing the concavity of f, the convexity of g, the nonnegativity of f, and the positivity g, we get the inequality $g(x)f(y) - f(x)g(y) \le 0$ which immediately yields $h(y) - h(x) \le 0$. □

Concerning transformations of pseudo–concave functions, the following fact will also be needed later on:

Proposition 2.33. *Let C be an open convex set and let g be a continuously differentiable pseudo–concave or pseudo–convex function, defined on C. Let $f : \mathbb{R} \mapsto \mathbb{R}$ be a continuously differentiable, strictly monotonically increasing function, with $f'(x) \neq 0$ for all $x \in \mathbb{R}$. Then $h(x) := f(g(x))$ is pseudo–concave or pseudo–convex on C, respectively.*

Proof: For the gradient of h the relation $\nabla h(x) = f'(g(x))\nabla g(x)$ obviously holds. We assume that g is pseudo–concave, the proof for the pseudo–convex case runs analogously. Let $x \in C$, $y \in C$, and $\nabla^{\mathrm{T}} h(x)(y-x) \le 0$. Utilizing our assumptions, from this we get $\nabla^{\mathrm{T}} g(x)(y-x) \le 0$. The pseudo–concavity of g implies $g(y) \le g(x)$ and the monotonicity of f finally yields $h(y) \le h(x)$. □

Unfortunately, the sum of a linear and a pseudo–concave function is not necessarily pseudo–concave. As an example take $f_1(x) = -x$ and $f_2(x) = x + x^3$. It is easy to see that both functions are pseudo–concave, whereas their sum $f_1(x) + f_2(x) = x^3$ is not pseudo–concave. As a multivariate example let us take the function $f(x_1, x_2) = x_1 + x_1^3 + x_2 + x_2^3$ which is the sum of two pseudo–concave functions. The graph and the contour lines of this function are displayed in Figure 2.1. The function is clearly not quasi–concave, therefore it is not pseudo–concave, either.

A further important class of generalized concave functions consists of logarithmically concave (logconcave) functions.

Definition 2.13. *Let $f : C \to \mathbb{R}$ be a nonnegative function defined over the convex set C.*

- *f is called logarithmically concave or logconcave, if the inequality*

$$f(\lambda x + (1-\lambda)y) \ge [f(x)]^{\lambda}\,[f(y)]^{(1-\lambda)}$$

  holds, for all $x \in C$, $y \in C$, and $\lambda \in (0,1)$.
- *f is called logarithmically convex or logconvex, if the reverse inequality holds above.*

The definition immediately implies that for logconcave functions the set $\mathscr{C}^{+} := \{x \mid f(x) > 0, x \in C\}$ is convex. Observe, that the inequality in Definition 2.13. holds trivially, if either $x \notin \mathscr{C}^{+}$ or $y \notin \mathscr{C}^{+}$. This leads to the following simple alternative characterization of logconcave functions:

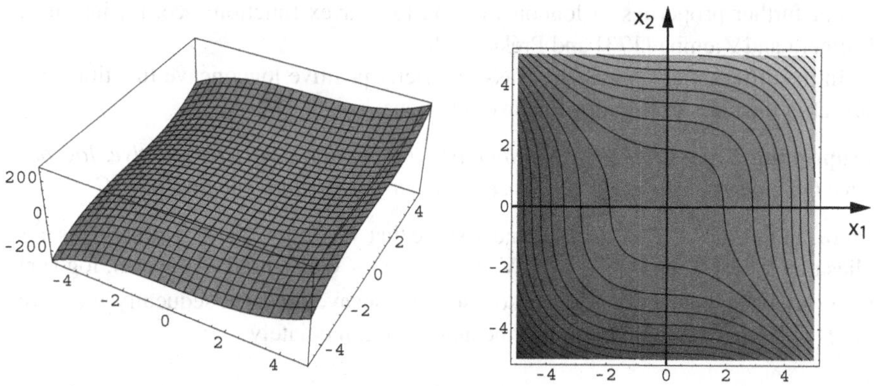

Fig. 2.1 The sum of two pseudo–concave functions needs not to be pseudo–concave. The picture shows the graph and the contour lines of the function $f(x_1,x_2) = x_1 + x_1^3 + x_2 + x_2^3$.

Proposition 2.34. *A nonnegative function f is logconcave over the convex set C, if and only if $\mathscr{C}^+ = \{x \mid f(x) > 0, x \in C\}$ is a convex set and* $\log f$ *is a concave function over $\mathscr{C}^+$.*

The next property involves products of logconcave functions. Let f_i, $i = 1,\ldots,r$, be logconcave functions on a convex set C and let $\mathscr{C}_i^+ := \{x \mid f_i(x) > 0, x \in C\}$ as before, for all i. Then the product $f(x) = \prod_{i=1}^{r} f_i(x)$ is also logconcave on C. In fact, let us observe that

$$C^+ := \{x \mid f(x) > 0, x \in C\} = \bigcap_{i=1}^{r} C_i^+$$

holds. Thus C^+ is a convex set and the assertion follows by considering $\log f$ on C^+.

A further fact concerning logconcave functions, which will be needed later on, is the following. Let f be a logconcave function on $\mathbb{R}^n$. Then $g(x) := f(x+y)$ is also logconcave on $\mathbb{R}^n$ for any fixed $y \in \mathbb{R}^n$. Moreover, $h(x,y) := f(x+y)$ is logconcave on $\mathbb{R}^{2n}$. In fact, for arbitrary $u, v \in \mathbb{R}^n$ and $\lambda \in (0,1)$ we have $g(\lambda u + (1-\lambda)v) = f(\lambda(u+y) + (1-\lambda)(v+y))$ from which the first assertion follows immediately. The second assertion follows also easily from the definition of logconcavity.

Considering logconvex functions, the definition implies the convexity of the set $\mathscr{C}^0 := \{x \mid f(x) = 0, x \in C\}$. Let $\mathrm{rint}C$ stand for the relative interior of C (see, for instance, Rockafellar [281]). It is easy to see, that $\mathrm{rint}C \cap \mathscr{C}^0 \neq \emptyset$ implies that $\mathrm{rint}C \subset \mathscr{C}^0$ holds. Thus, a logconvex function f for which $\mathrm{rint}C \cap \mathscr{C}^0 \neq \emptyset$ holds, can only have positive values at the (relative) boundary. Such functions are of no interest to us, therefore we will only consider positive logconvex functions. If $f(x) > 0$ for all $x \in C$, then f is logconvex, if and only if $\log f$ is convex. Finally let us remark that logconvex functions are also convex. This follows immediately from the inequality between the geometric and arithmetic means, see, for instance, Hardy et al. [132].

For further properties of logconcave and logconvex functions see, for instance, Kallberg and Ziemba [173] and Prékopa [266].

In the differentiable case, the class of strictly positive logconcave functions is a subset of the class of pseudo–concave functions:

Proposition 2.35. *Let f be a continuously differentiable, strictly positive, logconcave function over the open convex set C. Then f is pseudo–concave over C.*

Proof: Let $x \in C, y \in C, \lambda \in [0,1]$, and assume that $\nabla^{\mathrm{T}} f(x)(y-x) \le 0$ holds. This implies that $\nabla^{\mathrm{T}} \log f(x)(y-x) = \frac{1}{f(x)} \nabla^{\mathrm{T}} f(x)(y-x) \le 0$ also holds. However, $\log f(x)$ being a concave function, it is also pseudo–concave, and consequently we have $\log f(y) \le \log f(x)$, which implies the assertion immediately. □

Let us remark, that the notion of pseudo–concave functions can be extended to the non–differentiable case, see, for instance, [8]. We will not need this generalization in this book.

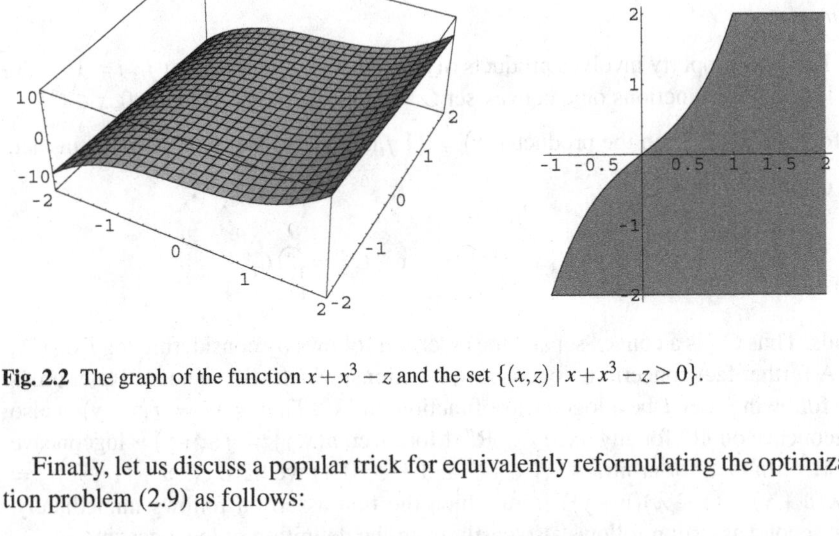

Fig. 2.2 The graph of the function $x + x^3 - z$ and the set $\{(x,z) \mid x + x^3 - z \ge 0\}$.

Finally, let us discuss a popular trick for equivalently reformulating the optimization problem (2.9) as follows:

$$\left.\begin{array}{lll} \max & c^{\mathrm{T}} x & +z \\ & V(x) & -z \ge 0 \\ \text{s.t.} & x & \in \mathscr{B}. \end{array}\right\} \tag{2.25}$$

This reformulation is used, for example, if we wish to apply cutting plane methods for solving (2.9). If V is a concave function, then (2.25) is obviously a convex programming problem. If, however, V is merely pseudo–concave, then this is in general not true. An example involving the pseudo–concave function $x + x^3$ is displayed in Figure 2.2; $x + x^3 - z$ is not quasi–concave and the feasible domain of the corresponding problem (2.25) is a non–convex set. Thus, the reformulated problem (2.25) is in general much harder to solve than the problem in the original formulation.

Requiring the stronger property of logconcavity (cf. Proposition 2.35.) does not help, either. Take e^x as an example. This function is obviously logconcave, whereas $e^x - z$ is a nonlinear convex function and the upper level set $\{(x,z) \mid e^x - z \geq 0\}$ is a non–convex set. Thus, $e^x - z$ is obviously not logconcave, in fact, it is not quasi–concave.

As already mentioned above, we will call our optimization problems (2.7) and (2.9) convex programming problems, if V is pseudo–concave in (2.9) and V is quasi–concave in (2.7), respectively. Whether or not our optimization problems are of the convex programming type, depends solely on (generalized) concavity properties of the function V.

Exercises

2.1. Let f be a concave function defined on an open convex set C. Show that

(a) f is quasi–concave;
(b) if f is differentiable then it is pseudo–concave;
(c) if $f(x) > 0$ holds for all $x \in C$ then f is logconcave.

2.2. For each of the following functions determine whether they are quasi–convex, quasi–concave, pseudo–convex, pseudo–concave, logconvex, or logconcave. With $C = \mathbb{R}$ let $f_1(x) = e^x$ and $f_2(x) = x^3$. With $C = \mathbb{R}^2$ let $f_3(x_1,x_2) = e^{-x_1^2-x_2^2}$ and let f_4 be the indicator function of a set $\mathscr{B}$, with $\mathscr{B}$ being a convex proper subset of $\mathbb{R}^2$. Formally, $f_4(x) = \begin{cases} 1, \text{ if } x \in \mathscr{B}, \\ 0, \text{ if } x \notin \mathscr{B}. \end{cases}$

2.3. Let f and g be positive logconvex functions defined on the convex set C. Show that their sum $h(x) = f(x) + g(x)$ is also logconvex over C.

2.2 Models involving probability functions

This section is devoted to pursuing the idea of using probability as a quality measure. We choose the following quality measure for evaluating random vectors

$$\rho_{\mathrm{P}}(\vartheta) := \mathbb{P}(\vartheta \geq 0), \;\; \vartheta \in \mathscr{L}_s^0, \tag{2.26}$$

which is defined on the set of all random vectors on $(\Omega, \mathscr{F}, P)$. The decision vector x will be evaluated by the corresponding evaluation function $G(x) := \rho_{\mathrm{P}}(\zeta(x,\xi)) := \mathbb{P}_\xi(\zeta(x,\xi) \geq 0)$. The function G will be called a *probability function.* In a detailed form we have

$$G(x) := \mathbb{P}_\xi(\,T(\xi)x - h(\xi) \geq 0\,). \tag{2.27}$$

Let $x \in \mathbb{R}^n$ be fixed arbitrarily and let $S(x) := \{z \in \mathbb{R}^r \mid T(z)x - h(z) \geq 0\}$. Due to our assumptions, $T(\cdot)$ and $h(\cdot)$ are affine linear functions (see (2.2) on page 71). Consequently, $S(x) \subset \mathbb{R}^r$ is a polyhedral set and

$$G(x) = \mathbb{P}(\,\xi \in S(x)\,) \tag{2.28}$$

holds.

The following prototype problems will be considered:

$$\left.\begin{array}{ll} \max\; c^{\mathrm{T}}x & \\ \text{s.t.}\;\; \mathbb{P}_\xi(\,T(\xi)x - h(\xi) \geq 0\,) & \geq \alpha \\ \qquad\qquad x & \in \mathcal{B} \end{array}\right\} \tag{2.29}$$

and

$$\left.\begin{array}{ll} \max\; \mathbb{P}_\xi(\,T(\xi)x - h(\xi) \geq 0\,) \\ \text{s.t.} \qquad\qquad x \in \mathcal{B}, \end{array}\right\} \tag{2.30}$$

where $\mathcal{B}$ is a polyhedral set given, for example, in the standard form

$$\mathcal{B} = \{\,x \mid Ax = b,\; l \leq x \leq u\,\}.$$

In this section we will assume throughout that $\mathcal{B} \neq \emptyset$ holds and that $\mathcal{B}$ is bounded.

Both optimization problems (2.29) and (2.30) are non–convex optimization problems in general. The emphasis in this section will be laid on identifying those subclasses, for which (2.29) and (2.30) belong to the class of convex optimization problems. We will throughout first consider the basic properties of the models above and will subsequently discuss the analogous results for the models with reversed direction of the inequality constraint and of optimization, respectively.

Notice that (2.30) is formulated without an additive linear term in the objective function. In the case, when the probability function is concave, the objective function in (2.30) would obviously remain concave with an additive linear term. However, in general, we will only be able to ensure some generalized concavity properties of probability functions, which are usually lost when adding a linear function to them.

As already mentioned above, the function G will be called a probability function. The constraint involving a probability function in (2.29) is called a *chance–constraint* or a *probabilistic constraint*. For constraints involving probability functions the following terminology will be used. In the case of $s = 1$ the constraint will be called *separate*, whereas in the case when $s > 1$ is permitted, the term *joint* constraint will be used. In this sense, *joint constraint* stands for the general case, which specializes to a *separate constraint* if $s = 1$ holds. The corresponding probability functions will be called joint and separate probability functions, respectively. This terminology has its roots in modeling. Let us consider a joint probability constraint

$$\mathbb{P}_\xi(t_i^{\mathrm{T}}(\xi)x \geq h_i(\xi),\, i = 1,\ldots,s) \geq \alpha,$$

where the components of $t_i(\xi)$ are the elements of the ith row of $T(\xi)$ and let us assume that $s > 1$ holds. In this constraint, the underlying event has the following interpretation: a system of random inequalities holds, meaning that all of the inequalities hold simultaneously (they hold *jointly*). Depending on the modeling situation, we may wish to consider separately for $i = 1,\ldots,s$ the events that the ith random inequality $t_i^{\mathrm{T}}(\xi)x \geq h_i(\xi)$ holds. In this case, the joint constraint above is split into s separate probability constraints, where the probability levels on the right–hand–side can now be chosen differently for different rows:

$$\mathbb{P}_\xi(t_i^{\mathrm{T}}(\xi)x \geq h_i(\xi)) \geq \alpha_i,\ i = 1,\ldots,s.$$

Let us make a further remark concerning terminology. In the literature, model (2.29) is called either *chance constrained* or alternatively, *probabilistic constrained* model. Both *chance* and *probabilistic* have a very general meaning, including virtually all aspects of randomness. None of them describes with sufficient accuracy the fact that we are dealing with constraints and objective functions which are defined via probabilities. In order to contrast models involving probability functions with other SLP models based on different quality measures, we use a terminology, which explicitly refers to probability. For this reason, we call G a probability function. This terminology has been coined by Uryasev, see, for instance, [331]. With our notations, a probability function in [331] is defined as a function of the following type:

$$\mathbb{P}_\xi(f(x,\xi) \geq 0),$$

where $f(x,\cdot)$ is Borel–measurable for all x. Our case fits this scheme by choosing $f(x,\xi) = T(\xi)x - h(\xi)$. In accordance with this, models like (2.29) and (2.30) will be generally called *SLP models with probability functions*.

Next we discuss the reformulation of the constraint $G(x) \geq \alpha$, as an equivalent constraint with reversed inequality. We have

$$\begin{aligned}\mathbb{P}_\xi(\zeta(x,\xi) \geq 0) \geq \alpha &\Longleftrightarrow \mathbb{P}_\xi([\min_{1\leq i\leq s} \zeta_i(x,\xi)] \geq 0) \geq \alpha \\ &\Longleftrightarrow \mathbb{P}_\xi([\min_{1\leq i\leq s} \zeta_i(x,\xi)] < 0) \leq 1-\alpha \\ &\Longleftrightarrow \mathbb{P}_\xi([\max_{1\leq i\leq s}(\zeta_i(x,\xi)^-)] > 0) \leq 1-\alpha,\end{aligned} \tag{2.31}$$

where for any real number z, $z^- := \max\{0,-z\}$ denotes the negative part of z. Note that, in comparison with the original probability function $G(x)$, the probability function on the left–hand–side of the equivalent reversed inequality is much more difficult to handle numerically. On the one hand, the underlying event in the probability function involves a strict inequality. On the other hand, for computing this probability function for a fixed x, the probability measure of the region $\mathbb{R}^r \setminus S(x)$ is to be computed, which is the complement of a polyhedral set and thus it is non–convex in general (cf. (2.28)). In the special case $s = 1$ the situation is much simpler: $S(x)$ is a half–space and thus $\mathbb{R}^r \setminus S(x)$ becomes an open half–space. (2.31) reduces to

$$\begin{aligned} \mathbb{P}_\xi(\zeta(x,\xi) \geq 0) \geq \alpha &\Longleftrightarrow \mathbb{P}_\xi(\zeta(x,\xi)^- > 0) \leq 1-\alpha \\ &\Longleftrightarrow \mathbb{P}_\xi(t^{\mathrm{T}}(\xi)x - h(\xi) < 0) \leq 1-\alpha, \end{aligned} \tag{2.32}$$

where the components of $t(\xi)$ are the elements of the single row of $T(\xi)$. This is the straightforward way for reversing a separate probability constraint. We still have a strict inequality which can be replaced by an inequality involving "$\leq$", if the probability distribution function of $\zeta(x,\xi)$ is continuous.

We will also need a reformulation of (2.31) in expectation terms:

$$\mathbb{P}_\xi(\zeta(x,\xi) \geq 0) \geq \alpha \iff \mathbb{E}_\xi[\chi(\max_{1\leq i\leq s} \zeta_i(x,\xi)^-)] \leq 1-\alpha, \tag{2.33}$$

where χ is the following indicator function

$$\chi(z) = \begin{cases} 0 \text{ if } z \leq 0, \\ 1 \text{ if } z > 0. \end{cases}$$

For the set of vectors which are feasible with respect to the probability constraint, we introduce the notation

$$\mathscr{B}(\alpha) = \{x \mid G(x) \geq \alpha\} \tag{2.34}$$

and for the sake of easy reference we formulate our prototype problems (2.29) and (2.30) also in terms of the probability function G as follows:

$$\left.\begin{array}{rl} \max & c^{\mathrm{T}}x \\ \text{s.t.} & G(x) \geq \alpha \\ & x \in \mathscr{B} \end{array}\right\} \tag{2.35}$$

and

$$\left.\begin{array}{rl} \max & G(x) \\ \text{s.t.} & x \in \mathscr{B}. \end{array}\right\} \tag{2.36}$$

Remark. Let us consider the case, when one of the rows of the matrix $(T(\xi),h(\xi))$ is constant almost surely, for instance, it is deterministic. Denoting by $t_i(\xi)$ the random vector with its components being the elements of the ith row of $T(\xi)$, we assume without loss of generality that $(t_1^{\mathrm{T}}(\xi),h_1(\xi)) = (t^{\mathrm{T}},h)$ a.s. holds, where $t \in \mathbb{R}^n$ and $h \in \mathbb{R}$ are deterministic. In this case

$$\mathscr{B}(\alpha) = \{x \mid \mathbb{P}_\xi(t_i(\xi)x \geq h_i(\xi),\, i=2,\ldots,s) \geq \alpha\} \cap \{x \mid t^{\mathrm{T}}x \geq h\}$$

holds. This implies, that $\mathscr{B}(\alpha) \cap \mathscr{B}$ remains unchanged if G and $\mathscr{B}$ are redefined as follows:

$$\begin{aligned} G(x) &:= \mathbb{P}_\xi(t_i(\xi)x \geq h_i(\xi),\, i=2,\ldots,s) \\ \mathscr{B} &:= \mathscr{B} \cap \{x \mid t^{\mathrm{T}}x \geq h\}. \end{aligned}$$

The meaning is the following: essentially deterministic inequalities within a probability constraint can be removed from this constraint, by appending them to the set of deterministic constraints. □

As already discussed in the introductory section 2.1, our optimization problems will be considered as convex programming problems, if G is pseudo–concave in (2.36), and if it is quasi–concave in (2.35). It may happen, however, that G is not a quasi–concave function but nevertheless (2.35) is a convex programming problem. The point is this. As we have discussed in the introduction to this chapter on page 83, a function is quasi–concave if and only if all upper level sets are convex. The domain $\mathscr{B}(\alpha)$ defined in (2.34) is clearly an upper level set corresponding to level α. The convexity of the feasible domain of (2.35) just means that this specific level set is convex. It will turn out that, for some model classes and probability distributions, $\mathscr{B}(\alpha)$ becomes convex for α large enough. In summary: whether or not (2.35) is a convex programming problem, may also depend on the prescribed probability level α.

2.2.1 Basic properties

The purpose of this section is to present some general results which hold without any assumptions concerning the probability distribution of ξ.

We consider the probability function

$$G(x) = \mathbb{P}_{\xi}(\, T(\xi)x \geq h(\xi)\,)$$

as well as the constraint involving this probability function

$$G(x) \geq \alpha. \tag{2.37}$$

This constraint requires, that for a feasible x the event

$$S(x) := \{\xi \mid T(\xi)x \geq h(\xi)\} \in \mathbb{B}^r$$

should belong to the set of events $\mathscr{G}_\alpha$ having probability measure of at least α

$$\mathscr{G}_\alpha = \{A \in \mathbb{B}^r \mid \mathbb{P}_{\xi}(A) \geq \alpha\}.$$

For the feasible set, determined by (2.37) and denoted by

$$\mathscr{B}(\alpha) = \{x \mid G(x) \geq \alpha\} = \{x \mid S(x) \in \mathscr{G}_\alpha\},$$

the following representation holds obviously:

$$\begin{aligned}\mathscr{B}(\alpha) &= \bigcup_{A\in\mathscr{G}_\alpha} \{x \mid S(x) = A\} \\ &= \bigcup_{A\in\mathscr{G}_\alpha} \{x \mid \forall\xi \in A : T(\xi)x \geq h(\xi)\} \\ &= \bigcup_{A\in\mathscr{G}_\alpha} \bigcap_{\xi\in A} \{x \mid T(\xi)x \geq h(\xi)\}.\end{aligned} \tag{2.38}$$

Both from the theoretical point of view concerning the existence of optimal solutions and from the standpoint of numerical solution it is an important question whether $\mathscr{B}(\alpha)$ is a closed set. The answer is affirmative:

Theorem 2.1. *The set $\mathscr{B}(\alpha)$ is closed.*

Proof: For a proof see Kall and Wallace [172], Proposition 1.7. □

Without any assumptions on the probability distribution of ξ, the sole available result concerning the convexity of $\mathscr{B}(\alpha)$ is the following:

Theorem 2.2. *Kall ([154]). $\mathscr{B}(\alpha)$ is convex for $\alpha = 0$ and $\alpha = 1$.*

Proof: For $\alpha = 0$ we clearly have $\mathscr{G}_\alpha = \mathbb{B}^r$ and consequently $\mathscr{B}(\alpha) = \mathbb{R}^n$ holds. For the case $\alpha = 1$ we first observe that $A \in \mathscr{G}_1$ and $B \in \mathscr{G}_1$ imply $A \cap B \in \mathscr{G}_1$ (consider the complement of $A \cap B$). Now let $x \in \mathscr{B}(1)$, $y \in \mathscr{B}(1)$, $\lambda \in [0,1]$, and $z = \lambda x + (1-\lambda)y$. Then we have $S(x) \in \mathscr{G}_1$ and $S(y) \in \mathscr{G}_1$ and consequently $S(x) \cap S(y) \in \mathscr{G}_1$. For arbitrary fixed $\xi \in \mathbb{R}^r$, the inequalities $T(\xi)x \geq h(\xi)$ and $T(\xi)y \geq h(\xi)$ obviously imply the inequality $T(\xi)z \geq h(\xi)$. Thus $S(x) \cap S(y) \subset S(z)$ holds, implying $S(z) \in \mathscr{G}_1$. □

In the case of $\alpha = 0$ the probability constraint is clearly redundant. If $\alpha = 1$, then the solution of (2.35) can be interpreted as a "fat solution", in a probabilistic sense.

Finally let us discuss the reverse inequality $G(x) \leq \beta$. We consider now

$$\mathscr{H}_\beta = \{A \in \mathbb{B}^r \mid \mathbb{P}_\xi(A) \leq \beta\}$$

and denoting the feasible set in this case also by $\mathscr{B}(\beta)$ we have

$$\mathscr{B}(\beta) = \{x \mid G(x) \leq \beta\} = \{x \mid S(x) \in \mathscr{H}_\beta\}.$$

Analogously as above, we get the following representation:

$$\mathscr{B}(\beta) = \bigcup_{A\in\mathscr{H}_\beta} \bigcap_{\xi\in A} \{x \mid T(\xi)x \geq h(\xi)\}. \tag{2.39}$$

Considering the analogous assertion to Theorem 2.2., $\mathscr{B}(1) = \mathbb{R}^n$ is obviously convex and the probability constraint is redundant. $\mathscr{B}(0)$ is in general not convex, though. To see this, let us consider the following example with $x \in \mathbb{R}^1$, $G(x) = \mathbb{P}(x \geq \xi_1, -x \geq \xi_2)$ where ξ has the singular distribution $\xi_1 \equiv -1, \xi_2 \equiv -1$. We have $\mathscr{B}(0) = (-\infty, -1) \cup (1, \infty)$ which is obviously not convex.

2.2.2 Finite discrete distribution

We consider the case, when ξ has a finite discrete distribution, given by a realization tableau

$$\begin{pmatrix} p_1 & \dots & p_N \\ \widehat{\xi}^1 & \dots & \widehat{\xi}^N \end{pmatrix} \tag{2.40}$$

with $p_i > 0\ \forall i$ and $\sum_{i=1}^{N} p_i = 1$.

The discussion will be focused on the model (2.35), formulated as follows

$$\left.\begin{array}{ll} \max c^{\mathrm{T}} x & \\ \text{s.t.} & x \in \mathscr{B}(\alpha) \cap \mathscr{B} \end{array}\right\} \tag{2.41}$$

with $\mathscr{B}(\alpha) = \{x \mid G(x) \geq \alpha\}$.

In the discretely distributed case the representation (2.38) on page 92 specializes as follows. Let $I = \{1,\dots,N\}$, then we have

$$\mathscr{B}(\alpha) = \bigcup_{\substack{J \subset I \\ \sum\limits_{j \in J} p_j \geq \alpha}} \bigcap_{j \in J} \{x \mid T(\widehat{\xi}^j)x \geq h(\widehat{\xi}^j)\}. \tag{2.42}$$

For the separate realizations of ξ let us introduce the notation

$$K_j = \{x \mid T(\widehat{\xi}^j)x \geq h(\widehat{\xi}^j)\},\ \ j = 1,\dots,N.$$

These sets are clearly convex polyhedral sets. Employing this notation, the representation above can be written in the form

$$\mathscr{B}(\alpha) = \bigcup_{\substack{J \subset I \\ \sum\limits_{j \in J} p_j \geq \alpha}} \bigcap_{j \in J} K_j. \tag{2.43}$$

Figure 2.3 shows the following example from Kall [154]:

$$\begin{aligned} K_1 &= \{x \in \mathbb{R}^2 \mid x_1 - x_2 \geq -2,\ x_2 \geq 3\}, \\ K_2 &= \{x \in \mathbb{R}^2 \mid x_1 - x_2 \geq 0,\ 2x_1 + 3x_2 \leq 25\}, \\ K_3 &= \{x \in \mathbb{R}^2 \mid x_1 + x_2 \leq 8,\ -x_1 + 3x_2 \geq 0\} \end{aligned}$$

with corresponding probabilities of realizations $p_1 = \frac{1}{4}$, $p_2 = \frac{1}{2}$, and $p_3 = \frac{1}{4}$. The probability level in the probability constraint is $\alpha = \frac{3}{4}$. The feasible domain is the shaded region in the figure, which is obviously non–convex. The following representation holds: $\mathscr{B}(\alpha) = [K_1 \cap K_2] \cup [K_2 \cap K_3]$.

A necessary condition for $\mathscr{B}(\alpha) \cap \mathscr{B} \neq \emptyset$ is the following. With the notation $I_0 = \{i,\ 1 \leq i \leq N \mid K_i \cap \mathscr{B} = \emptyset\}$, $\mathscr{B}(\alpha) \cap \mathscr{B} \neq \emptyset$ obviously implies that

$$\sum_{i \notin I_0} p_i \geq \alpha$$

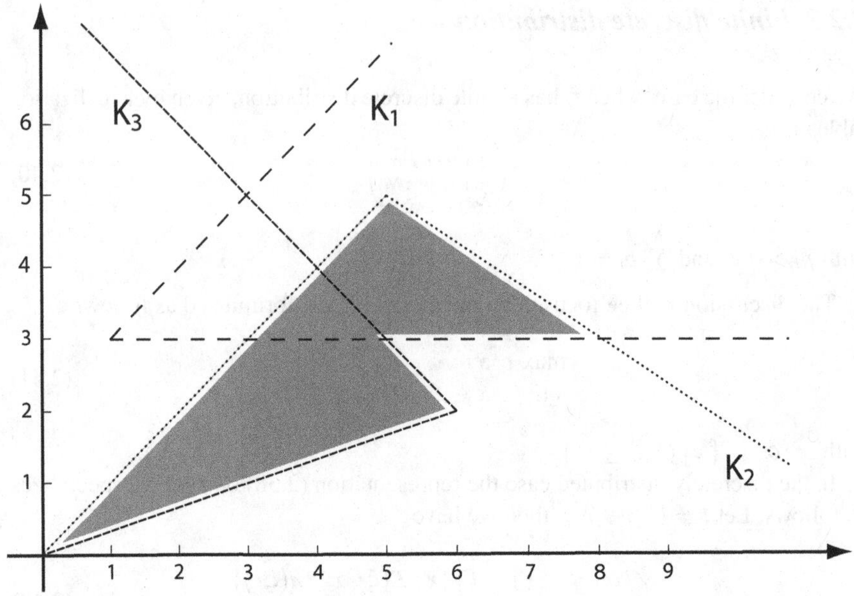

Fig. 2.3 Feasible domain of an SLP problem with joint probability constraint and finite discrete distribution.

must hold, otherwise each of the intersections in (2.43) would involve at least one $j \in I_0$, which would lead after intersecting $\mathscr{B}(\alpha)$ with $\mathscr{B}$ to a union of empty sets.

From (2.43) it is immediately clear, that our optimization problem (2.41) involves maximizing a linear function over a *union* of convex polyhedral sets. Thus, in general, the optimization problems do not belong to the class of convex optimization problems. This type of problems is called *disjunctive programming problem*, see, for instance, Nemhauser and Wolsey [241].

Utilizing the usual transformation of disjunctive programming, an equivalent mixed–integer formulation of (2.41) is the following (Raike [274]):

$$\left.\begin{array}{lll} \min & c^{\mathrm{T}}x & \\ \text{s.t.} & T(\widehat{\xi}^k)x + M\cdot(1-z_k)\mathbb{1} \geq h(\widehat{\xi}^k), & k=1,\ldots,N \\ & \sum_{i=1}^{N} p_i z_i \geq \alpha & \\ & z_k \in \{0,1\}, & k=1,\ldots,N \\ & x \in \mathscr{B}, & \end{array}\right\} \qquad (2.44)$$

where binary variables z_j have been introduced, M is a "big enough" constant and $\mathbb{1}^{\mathrm{T}} = (1,\dots,1)$. M is chosen in such a way, that $M \geq h(\widehat{\xi}^k) - T(\widehat{\xi}^k)x$ holds, $\forall x \in \mathscr{B}$, $k = 1,\dots,N$. Under our assumptions ($\mathscr{B} \neq \emptyset$, $\mathscr{B}$ bounded), such an M can be computed, for instance, by solving the following linear programming problems for $k \notin I_0$:

$$M_k = \max\{\gamma \mid \gamma + T(\widehat{\xi}^k)x \geq h(\widehat{\xi}^k), x \in \mathscr{B}\}$$

and setting $M = \max\limits_{k \notin I_0} M_k$.

For the case when only the right–hand–side is stochastic, further equivalent formulations as mixed–integer linear programming problems can be found in Prékopa [266].

There are some special cases, where the union in (2.42) amounts in a single convex polyhedral set.

Theorem 2.3. *Marti 1971 [222]. Let* $p_{i_0} = \min_{i \in I} p_i$. *Then* $\mathscr{B}(\alpha)$ *is convex for* $\alpha > 1 - p_{i_0}$.

Proof: For the proof see Kall [154]. □

Notice that $\alpha > 1 - p_{i_0}$ implies that $\mathscr{B}(\alpha) = \mathscr{B}(1)$ holds. Consequently, the constraint involving a probability function (2.37) can be replaced by the system of linear inequalities

$$T(\widehat{\xi}^i)x \geq h(\widehat{\xi}^i),\ i = 1,\dots,N. \tag{2.45}$$

Requiring that the inequalities should hold for all realizations, results in a "fat solution".

The result can be sharpened in a further special case:

Theorem 2.4. *Kall 1976 [154]. Let* $p_{i_0} = \min_{i \in I} p_i$ *and assume that* p_{i_0} *is uniquely determined. Let* $p_{i_1} = \min_{i \in I \setminus \{i_0\}} p_i$. *Then* $\mathscr{B}(\alpha)$ *is convex for* $\alpha > 1 - p_{i_1}$.

Proof: For the proof see Kall [154]. □

2.2.3 Separate probability functions

This section is devoted to discussing stochastic programming models which involve separate probability functions. The general prototype formulation of such problems has the same form as (2.29) and (2.30) with $\zeta(x,\xi)$ now being a random variable ($s = 1$). To emphasize one of the typical sources of such problems, we give a formulation for a random vector $\zeta(x,\xi)$ where the evaluation function has been applied component–wise:

$$\left.\begin{array}{ll}\max c^{\mathrm{T}}x & \\ \text{s.t. } \mathbb{P}_{\xi}(\, t_k^{\mathrm{T}}(\xi)x \geq h_k(\xi)\,) \geq \alpha_k, & k=1,\ldots,s \\ \quad x & \in \mathscr{B}\end{array}\right\} \tag{2.46}$$

and

$$\left.\begin{array}{ll}\max & \mathbb{P}_{\xi}(\, t^{\mathrm{T}}(\xi)x \geq h(\xi)\,) \\ \text{s.t.} & x \in \mathscr{B}\end{array}\right\} \tag{2.47}$$

where the components of the n–dimensional random vector $t_k(\xi)$ are the elements of the kth row of $T(\xi)$, $\forall k$; $t(\xi)$ is an n-dimensional random vector and $h_k(\xi)$, $h(\xi)$ are random variables $\forall k$. The term *separate* means, as we have discussed previously, that each of the probability functions appearing in the model formulations involves a single random inequality.

For the discussions regarding convexity of the feasible domain, it is clearly sufficient to consider a single separate probability function:

$$G(x) = \mathbb{P}_{\xi}(\, x \mid t(\xi)^{\mathrm{T}}x \geq h(\xi)\,).$$

For the sake of simplicity we introduce the notation $\eta := t(\xi)$ and replace the right–hand–side $h(\xi)$ by ξ, because only the probability distribution of $(t(\xi)^{\mathrm{T}}, h(\xi))$ counts anyway. Thus the probability function has the following form:

$$G(x) = \mathbb{P}_{\xi}(\, x \mid \eta^{\mathrm{T}}x - \xi \geq 0\,).$$

With our notation, the definition of $\zeta(x,\xi)$ on page 71 takes the form

$$\zeta(x,\eta,\xi) := \eta^{\mathrm{T}}x - \xi.$$

Note that $\zeta(x,\eta,\xi)$ is now a random variable.

The goal of this section is to identify subclasses of SLP models with separate probability functions, which lead to convex programming problems. We will also give equivalent formulations for these models in algebraic terms, which provide the basis for the numerical solution of the problems. It will turn out for this class of models that both type of constraints $G(x) \geq \alpha$ and $G(x) \leq \beta$ can lead, under appropriate assumptions, to convex optimization problems.

We will proceed as follows. Next we will discuss the special case when only the right–hand–side is stochastic. This will be followed by considering the case when (η,ξ) has a multivariate normal distribution. Next the results will be generalized to the class of stable distributions. Finally we discuss a distribution–free approach.

Considering other distributions, we mention that in the case when the components of (η,ξ) are independent and have exponential distributions, Biswal et al. [29] have presented an equivalent algebraic formulation as an NLP problem.

Only the right–hand–side is stochastic

We assume that $\eta \equiv t$ holds, with t being deterministic. In this case the probability function has the form

$$G(x) = \mathrm{P}_{\xi}(x \mid t^{\mathrm{T}}x \geq \xi).$$

For the case of reverse random inequalities $t^{\mathrm{T}}x \leq \xi$ we just consider the probability function corresponding to $(-t, -\xi)$. Denoting the probability distribution function of the random variable ξ by F_{ξ}, we have

$$G(x) = F_{\xi}(t^{\mathrm{T}}x).$$

The probability distribution function of a random variable being monotonically increasing, it is both quasi–convex and quasi–concave (it is quasi–linear). It is easy to see that substituting a linear function into a quasi–convex function results in a quasi–convex function, the same being true in the quasi–concave case. Consequently, $G(x)$ is both quasi–convex and quasi concave which immediately implies that both $\{x \mid G(x) \geq \alpha\}$ and $\{x \mid G(x) \leq \beta\}$ are convex sets. From the algorithmic point of view, however, it is desirable to obtain an explicit representation in terms of inequalities involving algebraic functions. This is easy to achieve in our case.

Considering first the constraint $G(x) \geq \alpha$, this is obviously equivalent to a linear constraint:

$$\mathrm{P}_{\xi}(x \mid t^{\mathrm{T}}x \geq \xi) \geq \alpha \iff F_{\xi}(t^{\mathrm{T}}x) \geq \alpha \iff t^{\mathrm{T}}x \geq Q_{\xi}^{-}(\alpha),$$

where $Q_{\xi}^{-}(\alpha)$ denotes the left end–point of the closed interval of α–quantiles of F_{ξ} (for properties of quantiles see, for instance, Cramér [47]).

Turning our attention to the reverse constraint $G(x) \leq \beta$ we observe that this can be written as $F_{\xi}(t^{\mathrm{T}}x) \leq \beta$. Assuming that F_{ξ} is continuous (for instance, ξ has a continuous distribution), we obtain again an equivalent linear inequality

$$\mathrm{P}_{\xi}(x \mid t^{\mathrm{T}}x \geq \xi) \leq \beta \iff F_{\xi}(t^{\mathrm{T}}x) \leq \beta \iff t^{\mathrm{T}}x \leq Q_{\xi}^{+}(\beta),$$

with $Q_{\xi}^{+}(\beta)$ denoting the right end–point of the interval of β–quantiles of F_{ξ}.

For arbitrary distributions, the equivalent reformulation should be set up with care. If F_{ξ} is continuous at the point $Q_{\xi}^{+}(\beta)$, then the above formulation holds. If, however, F_{ξ} is discontinuous at $Q_{\xi}^{+}(\beta)$, then the equivalent formulation is the following

$$\mathrm{P}_{\xi}(x \mid t^{\mathrm{T}}x \geq \xi) \leq \beta \iff F_{\xi}(t^{\mathrm{T}}x) \leq \beta \iff t^{\mathrm{T}}x < Q_{\xi}^{+}(\beta),$$

with a strict linear inequality implying the numerically unpleasant feature that the set $\{x \mid G(x) \leq \beta\}$ is an open half–space. This aspect reflects an asymmetry between the two setups $G(x) \geq \alpha$ and $G(x) \leq \beta$ of the constraints.

Having, for instance, a finite discrete distribution for ξ, the theoretically correct reformulation may consist of the strict inequality above. From the modeling point of view this is usually not a real problem: the unfavorable event (loss) can mostly be formulated as a strict inequality $\mathbb{P}_\xi(\, t^{\mathrm{T}}x < \xi\,)$ and thus we get

$$\begin{aligned}\mathbb{P}_\xi(\, t^{\mathrm{T}}x < \xi\,) \le \beta &\Longleftrightarrow 1 - \mathbb{P}_\xi(\, t^{\mathrm{T}}x \ge \xi\,) \le \beta \\ &\Longleftrightarrow F(t^{\mathrm{T}}x) \ge 1 - \beta \iff t^{\mathrm{T}}x \ge Q_\xi^-(1-\beta),\end{aligned}$$

that means, we obtain an equivalent linear constraint.

For discussing the situation concerning the objective function, we consider the problem (2.36) which in our case has the form

$$\left.\begin{array}{ll}\max F_\xi(t^{\mathrm{T}}x) \\ \text{s.t.} \;\; x \in \mathscr{B}.\end{array}\right\} \tag{2.48}$$

This is a linearly constrained nonlinear programming problem. Let us associate with (2.48) the following linear programming problem:

$$\left.\begin{array}{ll}\max t^{\mathrm{T}}x \\ \text{s.t.} \quad x \in \mathscr{B}.\end{array}\right\} \tag{2.49}$$

If F_ξ is strictly monotone, then (2.48) and (2.49) are clearly equivalent. In the general case, some care is needed. Provided that (2.49) has an optimal solution, this will be an optimal solution also for (2.48). Under our assumptions ($\mathscr{B} \neq \emptyset$, $\mathscr{B}$ bounded) this is always the case. For an unbounded polyhedral set $\mathscr{B}$ it may happen, however, that (2.49) has an unbounded objective over $\mathscr{B}$, whereas (2.48) has an optimal solution.

Analogous comments apply in the case when in (2.48) the objective is minimized.

Multivariate normal distribution

In this section we discuss the case, when $(\eta^{\mathrm{T}}, \xi)^{\mathrm{T}}$ has a joint multivariate normal distribution. For excluding the case already discussed in the previous section, we assume that η is stochastic, that means, that $\nexists d \in \mathbb{R}^n : \eta = d$ a.s.

Definition 2.14. *See, for example, Tong [328]. The r–dimensional random vector ζ has a multivariate normal distribution, if there exist an $(r \times s)$ matrix B and $\mu \in \mathbb{R}^r$, such that*

$$\zeta = B\tilde{\zeta} + \mu \tag{2.50}$$

holds, where $\tilde{\zeta}$ is an s–dimensional random vector with $\tilde{\zeta}_i$ being stochastically independent and having a standard normal distribution, $\forall i$.

Note that this definition allows for deterministic components of ζ: if the ith row of B is zero then we have $\zeta_i \equiv \mu_i$. From the definition immediately follows that

- $\mathbb{E}[\zeta] = \mu$ and
- $\Sigma = BB^{\mathrm{T}}$, where Σ denotes the covariance matrix of ζ

hold.

Σ is clearly a symmetric positive semidefinite matrix. The multivariate normal distribution is called *non–degenerate*, if Σ is positive definite. This is the case if and only if B has full row rank. Otherwise the distribution is called *degenerate* or *singular.*

The multivariate normal distribution is uniquely determined by the expected–value vector μ and the covariance matrix Σ, see, for instance, Tong [328]. We will use the notation $\zeta \sim \mathscr{N}(\mu, \Sigma)$, meaning that the random vector ζ has a normal distribution with expected value vector μ and covariance matrix Σ.

If the multivariate normal distribution is *non–degenerate*, then it is absolutely continuous w.r. to the Lebesgue–measure on $\mathbb{R}^r$, having the probability density function

$$f(y) = \frac{1}{(2\pi)^{\frac{n}{2}} |\Sigma|^{\frac{1}{2}}} e^{-\frac{1}{2}(y-\mu)^{\mathrm{T}}\Sigma^{-1}(y-\mu)} \tag{2.51}$$

where $|\Sigma|$ denotes the determinant of Σ.

Let R be the *correlation matrix* of ζ, defined as

$$R_{i,j} = \frac{\Sigma_{ij}}{\sigma_i \sigma_j}, \quad \forall i,j$$

where σ_i and σ_j denote the standard deviations of ζ_i and ζ_j, respectively. The non–degenerate multivariate normal distribution is called *standard multivariate normal distribution*, if the expected value vector is the zero–vector and the standard deviation of the components of ζ is 1. It is defined by the following density function

$$\varphi(y;R) = \frac{1}{(2\pi)^{\frac{n}{2}} |R|^{\frac{1}{2}}} e^{-\frac{1}{2}y^{\mathrm{T}}R^{-1}y}. \tag{2.52}$$

The corresponding distribution function will be denoted by $\Phi(y;R)$. In the univariate case we drop R in the notation; φ stands for the density function of the standard normal distribution, that means, we have

$$\varphi(y) = \frac{1}{\sqrt{2\pi}} e^{-\frac{x^2}{2}}$$

and the corresponding distribution function will be denoted by Φ.

Figure 2.4 shows the density– and distribution functions of the bivariate normal distribution with correlation $r = 0$. In Figure 2.5 these functions are displayed for the case $r = 0.9$.

Having a symmetric positive semidefinite matrix Σ and vector μ as primary data, a lower–triangular matrix B for relation (2.50) can be computed by the Cholesky–factorization for symmetric positive semidefinite matrices, see, for instance, Golub and Van Loan [127].

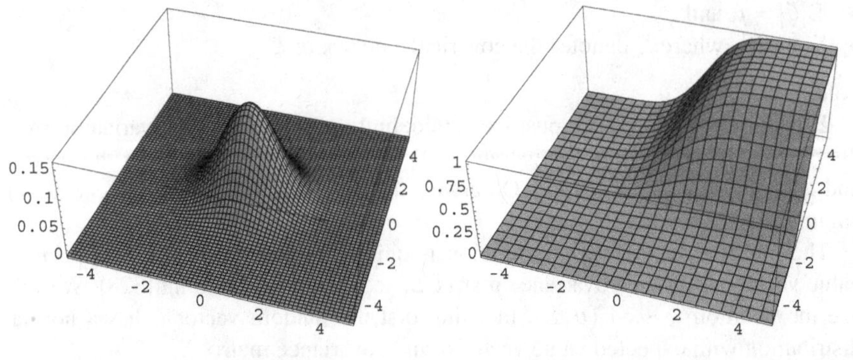

Fig. 2.4 The bivariate normal distribution function with correlation $r = 0$.

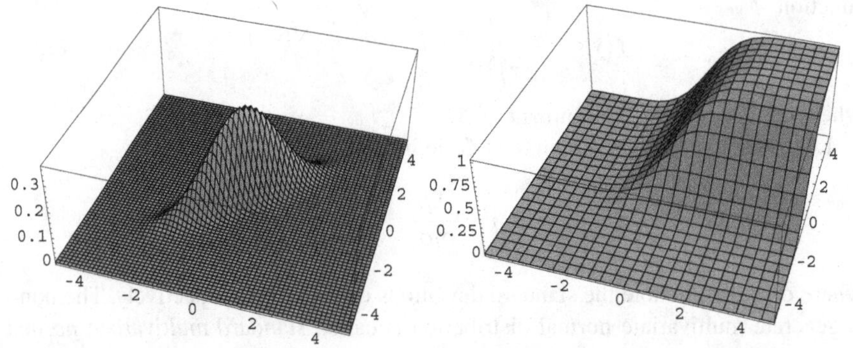

Fig. 2.5 The bivariate normal distribution function with correlation $r = 0.9$.

From the definition it follows immediately, that any affine linear transformation of a random vector with a multivariate normal distribution has again a multivariate normal distribution.

Assume now, that the $(n+1)$–dimensional random vector $\zeta^{\mathrm{T}} = (\eta^{\mathrm{T}}, \xi)^{\mathrm{T}}$ has a multivariate normal distribution:

$$\zeta = \begin{pmatrix} \eta \\ \xi \end{pmatrix} = \begin{pmatrix} D \\ d^{\mathrm{T}} \end{pmatrix} \cdot \tilde{\zeta} + \begin{pmatrix} \mu \\ \mu_{n+1} \end{pmatrix} \tag{2.53}$$

where D is an $(n \times s)$ matrix, $d \in \mathbb{R}^s$, $\mu \in \mathbb{R}^n$. We get

$$\begin{aligned} \zeta(x, \eta, \xi) &= \eta^{\mathrm{T}} x - \xi = (\eta^{\mathrm{T}}, \xi) \begin{pmatrix} x \\ -1 \end{pmatrix} \\ &= \tilde{\zeta}^{\mathrm{T}} \left(D^{\mathrm{T}} x - d \right) + \mu^{\mathrm{T}} x - \mu_{n+1}. \end{aligned} \tag{2.54}$$

It follows that $\zeta(x, \eta, \xi)$ is normally distributed with

$$\begin{aligned}\mathbb{E}[\zeta(x,\eta,\xi)] &= \mu^{\mathrm{T}}x - \mu_{n+1} \\ \mathbb{V}\mathrm{ar}[\zeta(x,\eta,\xi)] &= \|D^{\mathrm{T}}x - d\|^2 \\ &= x^{\mathrm{T}}DD^{\mathrm{T}}x - 2(Dd)^{\mathrm{T}}x + \|d\|^2,\end{aligned} \tag{2.55}$$

where $\|\cdot\|$ denotes the Euclidean norm. The first term on the right–hand–side is the variance of $\eta^{\mathrm{T}}x$ with DD^{T} being the covariance matrix of η. In the second term $(Dd)^{\mathrm{T}}x$ is the covariance between $\eta^{\mathrm{T}}x$ and ξ with Dd being the cross–covariance vector between η and ξ. The third term is the variance of ξ.

If $\mathbb{V}\mathrm{ar}[\zeta(x,\eta,\xi)] = 0$ then $\zeta(x,\eta,\xi) = \mathbb{E}[\zeta(x,\eta,\xi)]$, a.s., otherwise the standardized $\zeta(x,\eta,\xi)$ has a standard normal distribution.

In the case $\|D^{\mathrm{T}}x - d\| > 0$ we obtain via standardization

$$\begin{aligned}G(x) &= \mathbb{P}(\,\zeta(x,\eta,\xi) \ge 0) = 1 - \mathbb{P}(\,\zeta(x,\eta,\xi) \le 0) \\ &= 1 - \mathbb{P}\left(\frac{\zeta(x,\eta,\xi) - \mathbb{E}[\zeta(x,\eta,\xi)]}{\|D^{\mathrm{T}}x - d\|} \le \frac{-\mu^{\mathrm{T}}x + \mu_{n+1}}{\|D^{\mathrm{T}}x - d\|}\right) \\ &= 1 - \Phi\left(\frac{-\mu^{\mathrm{T}}x + \mu_{n+1}}{\|D^{\mathrm{T}}x - d\|}\right) = \Phi\left(\frac{\mu^{\mathrm{T}}x - \mu_{n+1}}{\|D^{\mathrm{T}}x - d\|}\right),\end{aligned} \tag{2.56}$$

where in the last step we utilized the symmetry of the standard normal distribution, that means, we made use of the relation $\Phi(x) = 1 - \Phi(-x)$, $\forall x \in \mathbb{R}$. Thus we get the following formula for $G(x)$:

$$G(x) = \begin{cases} 1, & \text{if } D^{\mathrm{T}}x - d = 0 \\ & \text{and } \mu^{\mathrm{T}}x - \mu_{n+1} \ge 0, \\ 0, & \text{if } D^{\mathrm{T}}x - d = 0 \\ & \text{and } \mu^{\mathrm{T}}x - \mu_{n+1} < 0, \\ \Phi\left(\dfrac{\mu^{\mathrm{T}}x - \mu_{n+1}}{\|D^{\mathrm{T}}x - d\|}\right), & \text{if } D^{\mathrm{T}}x - d \neq 0. \end{cases} \tag{2.57}$$

Regarding the constraint $G(x) \ge \alpha$, under the assumption $D^{\mathrm{T}}x - d \neq 0$ we get

$$\begin{aligned}G(x) \ge \alpha \iff\ & \Phi\left(\frac{\mu^{\mathrm{T}}x - \mu_{n+1}}{\|D^{\mathrm{T}}x - d\|}\right) \ge \alpha \\ \iff\ & \Phi^{-1}(\alpha)\,\|D^{\mathrm{T}}x - d\| - \mu^{\mathrm{T}}x \le -\mu_{n+1}.\end{aligned} \tag{2.58}$$

In the case when $D^{\mathrm{T}}x - d = 0$ holds, the last inequality reduces to the first case in (2.57), consequently the equivalence holds in all cases. Note that for $\alpha \ge \frac{1}{2}$ we have $\Phi^{-1}(\alpha) \ge 0$. The Euclidean norm being convex, $\|D^{\mathrm{T}}x - d\|$ is a convex function of x. Consequently, assuming that $\alpha \ge \frac{1}{2}$ holds, the function on the left–hand–side of the last inequality in (2.58) is a convex function. This implies that the set of feasible solutions w.r. to this constraint is a convex set. We have derived the following theorem:

Theorem 2.5. *Kataoka 1963 [178], Van de Panne and Popp 1963 [333]. Let the $(n+1)$–dimensional random vector $\zeta^{\mathrm{T}} = (\eta^{\mathrm{T}}, \xi)^{\mathrm{T}}$ have a multivariate normal distribution and let $\alpha \geq \frac{1}{2}$. Then the set $\mathscr{B}(\alpha) = \{x \mid G(x) \geq \alpha\}$ is convex.*

For the case, when $\alpha < \frac{1}{2}$ holds, we have the following assertion:

Theorem 2.6. *Kall 1976 [154]. Let $n > 1$ and assume that the $(n+1)$–dimensional random vector $\zeta^{\mathrm{T}} = (\eta^{\mathrm{T}}, \xi)^{\mathrm{T}}$ has a non–degenerate multivariate normal distribution. If $\alpha < \frac{1}{2}$ then either $\mathscr{B}(\alpha) = \mathbb{R}^n$ holds or otherwise $\mathscr{B}(\alpha)$ is a non–convex set.*

Proof: Let $\widehat{x} \in \mathbb{R}^n$ be such that $\widehat{x} \notin \mathscr{B}(\alpha)$ holds. We will show, that under our assumptions, there exist $x^{(1)} \in \mathscr{B}(\alpha)$ and $x^{(2)} \in \mathscr{B}(\alpha)$ such that $x^{(1)} \neq x^{(2)}$ and $\widehat{x} = \frac{1}{2}(x^{(1)} + x^{(2)})$ holds. From this our assertion follows immediately.
$n > 1$ implies that there exists $v \in \mathbb{R}^n$ such that $v \neq 0$ and $\mu^{\mathrm{T}} v = 0$ hold. Let us consider the constraint (2.58) along the line $x(\lambda) = \widehat{x} + \lambda v$, $\lambda \in \mathbb{R}$:

$$\Phi^{-1}(\alpha)\,\|D^{\mathrm{T}} x(\lambda) - d\| - \mu^{\mathrm{T}} \widehat{x} \leq -\mu_{n+1},$$

where we used that $\mu^{\mathrm{T}} x(\lambda) = \mu^{\mathrm{T}} \widehat{x}$, $\forall \lambda \in \mathbb{R}$ holds. We obviously have $\|D^{\mathrm{T}} x(\lambda) - d\| \geq \|D^{\mathrm{T}} x(\lambda)\| - \|d\|$, and an easy computation yields

$$\|D^{\mathrm{T}} x(\lambda)\|^2 = \lambda^2 v^{\mathrm{T}} D D^{\mathrm{T}} v + 2\lambda v^{\mathrm{T}} D D^{\mathrm{T}} \widehat{x} + \widehat{x}^{\mathrm{T}} D D^{\mathrm{T}} \widehat{x}.$$

Matrix D has full row rank and $v \neq 0$, therefore $\lim\limits_{\lambda \to \pm\infty} \|D^{\mathrm{T}} x(\lambda) - d\| = \infty$ holds. Taking into account $\Phi^{-1}(\alpha) < 0$, this implies that $\exists \lambda_0 \in \mathbb{R}$, such that both $x(\lambda_0) \in \mathscr{B}(\alpha)$ and $x(-\lambda_0) \in \mathscr{B}(\alpha)$. Obviously $x(\lambda_0) \neq x(-\lambda_0)$ and $\widehat{x} = \frac{1}{2}(x(\lambda_0) + x(-\lambda_0))$. □

For the probability function with reversed random inequalities, that means, for $\widehat{G}(x) := \mathbb{P}(\eta^{\mathrm{T}} x \leq \xi) = \mathbb{P}(\zeta(x, \eta, \xi) \leq 0)$ we get

$$\widehat{G}(x) = \begin{cases} 1, & \text{if } D^{\mathrm{T}} x - d = 0 \text{ and } \mu^{\mathrm{T}} x - \mu_{n+1} \leq 0 \\ 0, & \text{if } D^{\mathrm{T}} x - d = 0 \text{ and } \mu^{\mathrm{T}} x - \mu_{n+1} > 0 \\ \Phi\left(\dfrac{-\mu^{\mathrm{T}} x + \mu_{n+1}}{\|D^{\mathrm{T}} x - d\|}\right), & \text{if } D^{\mathrm{T}} x - d \neq 0. \end{cases} \tag{2.59}$$

This can either be derived by an analogous argumentation as above, or more directly as follows. Observe that if $\zeta(x, \eta, \xi)$ has a normal distribution, then $-\zeta(x, \eta, \xi)$ also has a normal distribution with the same variance and with reversed sign of the expected value. Thus (2.57) can be directly applied for $-\zeta(x, \eta, \xi)$, by writing $\widehat{G}$ as $\widehat{G}(x) = \mathbb{P}(-\zeta(x, \eta, \xi) \geq 0)$.

Utilizing the formulas (2.57) and (2.59), we obtain the following equivalent representations of probability constraints:

$$\begin{aligned} \mathbb{P}(\eta^{\mathrm{T}} x \geq \xi) \geq \alpha &\Longleftrightarrow \Phi^{-1}(\alpha)\,\|D^{\mathrm{T}} x - d\| - \mu^{\mathrm{T}} x \leq -\mu_{n+1} \\ \mathbb{P}(\eta^{\mathrm{T}} x \leq \xi) \geq \alpha &\Longleftrightarrow \Phi^{-1}(\alpha)\,\|D^{\mathrm{T}} x - d\| + \mu^{\mathrm{T}} x \leq \mu_{n+1} \end{aligned} \tag{2.60}$$

where for $\alpha \geq \frac{1}{2}$ the functions on the left–hand–side of the equivalent inequalities are convex, therefore the feasible domain determined by these inequalities is convex.

We turn our attention to the case with reverse inequalities in the constraints, that means, we deal with $G(x) \leq \beta$ and $\widehat{G}(x) \leq \beta$. In the case when the probability distribution is degenerate, the previously used technique for deriving the equivalent form leads to strict inequalities. Having $D^{\mathrm{T}}x - d = 0$, the formulas (2.57) and (2.59) imply a strict inequality (the second cases in these formulas apply). Assuming non–degeneracy of the probability distribution, we obtain the following equivalent representations by reversing the inequalities in (2.60):

$$\begin{aligned} \mathbb{P}(\eta^{\mathrm{T}}x \geq \xi) \leq \beta &\Longleftrightarrow \Phi^{-1}(\beta)\,\|D^{\mathrm{T}}x - d\| - \mu^{\mathrm{T}}x \geq -\mu_{n+1} \\ \mathbb{P}(\eta^{\mathrm{T}}x \leq \xi) \leq \beta &\Longleftrightarrow \Phi^{-1}(\beta)\,\|D^{\mathrm{T}}x - d\| + \mu^{\mathrm{T}}x \geq \mu_{n+1} \end{aligned} \tag{2.61}$$

where, provided that $\beta \leq \frac{1}{2}$ holds, the functions on the left–hand–side of the equivalent inequalities are concave, consequently the feasible domain determined by these inequalities is convex.

In the case when the probability distribution is degenerate, we observe a similar asymmetry as in the previous section on page 98 between the two formulations differing in the direction of the inequality ($G(x) \geq \alpha$ versus $G(x) \leq \beta$). The remedy is analogous: In practical modeling this difficulty can usually be overcome by working with strict inequalities in the model formulation. For instance, taking the constraint $\mathbb{P}_{\xi}(\zeta(x,\xi) < 0) \leq \beta$, this can be equivalently formulated as

$$\mathbb{P}_{\xi}(\zeta(x,\xi) \geq 0) \geq 1 - \beta$$

which results according to (2.60) in the linear constraint

$$\Phi^{-1}(1-\beta)\,\|D^{\mathrm{T}}x - d\| - \mu^{\mathrm{T}}x \leq -\mu_{n+1}$$

thus determining a convex feasible domain for $\beta \leq \frac{1}{2}$.

Next we turn our attention to models with probability functions in the objective and restrict our discussion to the case, when $\zeta^{\mathrm{T}} = (\eta^{\mathrm{T}}, \xi)^{\mathrm{T}}$ has a *non–degenerate* multivariate normal distribution. The distribution of ζ is non–degenerate, if and only if the matrix $\begin{pmatrix} D \\ d^{\mathrm{T}} \end{pmatrix}$ has full row rank, see Definition 2.14. and (2.53). Consequently, in the non–degenerate case $D^{\mathrm{T}}x - d \neq 0$ holds for all $x \in \mathbb{R}^n$. In particular, choosing $x = 0$ shows that $d \neq 0$ holds.

In the non–degenerate case we have, see (2.57):

$$G(x) = \Phi\left(\frac{\mu^{\mathrm{T}}x - \mu_{n+1}}{\|D^{\mathrm{T}}x - d\|}\right) \quad \forall x \in \mathbb{R}^n. \tag{2.62}$$

In a maximization problem the desired property of $G(x)$ would be pseudo–concavity. Unfortunately, $G(x)$ is not even quasi–concave. Quasi–concavity is namely equivalent with the convexity of all of the upper level sets (see page 83). This is implied by (2.60) for $\alpha \geq \frac{1}{2}$. For any $0 < \alpha < \frac{1}{2}$, however, the lower level set is convex according to (2.61). The upper level sets corresponding to the same α cannot be also convex, because this would mean that both the upper– and the lower level sets are half–spaces. This is not possible due to our non–degeneracy assumption $\|D^{\mathrm{T}}x - d\| \neq 0$ for all $x \in \mathbb{R}^n$. Consequently $G(x)$ is not quasi–concave. An analogous reasoning shows that $G(x)$ is not quasi–convex, either.

Introducing the notation

$$g(x) = \frac{\mu^{\mathrm{T}}x - \mu_{n+1}}{\|D^{\mathrm{T}}x - d\|}$$

we get $G(x) = \Phi(\,g(x)\,)$. Fortunately, by restricting $G(x)$ to certain half–spaces we have

Proposition 2.36. *If $\zeta^{\mathrm{T}} = (\eta^{\mathrm{T}}, \xi)^{\mathrm{T}}$ has a non–degenerate multivariate normal distribution, then both $g(x)$ and $G(x)$ are*

a) pseudo–concave on the half–space $\{\,x \mid \mu^{\mathrm{T}}x \geq \mu_{n+1}\,\}$ and
b) pseudo–convex on the half–space $\{\,x \mid \mu^{\mathrm{T}}x \leq \mu_{n+1}\,\}$.

Proof: Due to the non–degeneracy assumption $\|D^{\mathrm{T}}x - d\| > 0$, $\forall x \in \mathbb{R}^n$ holds. Due to Proposition 2.32. on page 83, the fractional function $g(x)$ is pseudo–concave on convex sets where the numerator is nonnegative, and pseudo–convex on convex sets where the numerator is non–positive. From this the result regarding $g(x)$ follows. Utilizing the fact that Φ is a strictly monotonically increasing, differentiable function, with $\Phi'(x) \neq 0\ \forall x \in \mathbb{R}$, the assertion concerning $G(x)$ follows from the already proved assertion regarding $g(x)$ and from Proposition 2.33. on page 84. □

Let us consider (2.36) on page 90, which in our case has the form

$$\left.\begin{array}{ll} \max\ \Phi\left(\dfrac{\mu^{\mathrm{T}}x - \mu_{n+1}}{\|D^{\mathrm{T}}x - d\|}\right) & \\ \text{s.t.} \qquad\qquad x \in \mathscr{B}. & \end{array}\right\} \tag{2.63}$$

According to Proposition 2.36., the objective function of this linearly constrained problem is pseudo–concave, if $x \in \mathscr{B}$ implies $\mu^{\mathrm{T}}x \geq \mu_{n+1}$. Thus, in this case, (2.63) is a convex programming problem. Taking into account the strict monotonicity of Φ, (2.63) is equivalent to the following linearly constrained convex programming problem

$$\left.\begin{array}{ll} \max\ \dfrac{\mu^{\mathrm{T}}x - \mu_{n+1}}{\|D^{\mathrm{T}}x - d\|} & \\ \text{s.t.} \qquad x \in \mathscr{B}. & \end{array}\right\} \tag{2.64}$$

This problem belongs to the class of fractional programming problems, see, for instance, Avriel, Diewert, Schaible, and Zang [8] and Schaible [297]. Proposition 2.36. implies that the objective function in (2.64) is pseudo–concave in the half–space $\{x \mid \mu^{\mathrm{T}}x \geq \mu_{n+1}\}$ and it is pseudo–convex in the half–space $\{x \mid \mu^{\mathrm{T}}x \leq \mu_{n+1}\}$.

Consequently, if $\mu^{\mathrm{T}}x \geq \mu_{n+1}\ \forall x \in \mathscr{B}$ holds, then (2.64) is a convex programming problem. This property can be enforced, for instance, by including a linear inequality of the form $\mu^{\mathrm{T}}x \geq \mu_{n+1}$ into the definition of $\mathscr{B}$. This might be well justified if a high probability is to be achieved by maximizing $\mathbb{P}(\eta^{\mathrm{T}}x \geq \xi)$. For achieving high probabilities it is necessary to have $\mathbb{E}[\eta^{\mathrm{T}}x] \geq \mathbb{E}[\xi]$, which is just the required inequality.

If the reverse inequality $\{x \mid \mu^{\mathrm{T}}x \leq \mu_{n+1}\}$ holds over $\mathscr{B}$, then our objective is pseudo–convex, (2.64) involves maximizing a pseudo–convex function, and thus it becomes much more difficult to solve numerically. In the general case, when none of the two inequalities involving expectations holds uniformly over $\mathscr{B}$, then (2.64) becomes a general non–convex optimization problem. In this case efficient solution methods are only available for rather low dimensions of x.

In the case when (2.63) and (2.64) are formulated as minimization problems, the above results can be adapted in a straightforward manner. If we take $\widehat{G}(x) = \mathbb{P}(\eta^{\mathrm{T}}x \leq \xi)$ instead of $G(x)$ then the above discussion applies with exchanged roles of the inequalities $\mu^{\mathrm{T}}x \geq \mu_{n+1}$ and $\mu^{\mathrm{T}}x \leq \mu_{n+1}$.

Finally we discuss the special case when ξ is deterministic. Note that the non–degeneracy assumption above implies that all components of η as well as ξ have non–degenerate univariate marginal distributions, that means, both the "technology matrix" and the right–hand–side are stochastic. We assume now that $\xi \equiv \mu_{n+1} := h$ holds with $h \in \mathbb{R}$ being deterministic. Considering (2.54), this means that $d = 0$ holds throughout. Non–degeneracy of the distribution in this case means that D has full row rank.

The explicit form of $\widehat{G}$ and the probability constraint can simply be obtained by setting $d = 0$ in (2.59) and in (2.61), respectively. Considering the problem of minimizing $\widehat{G}(x)$ results in:

$$\left.\begin{array}{ll} \min & \dfrac{-\mu^{\mathrm{T}}x + h}{\|D^{\mathrm{T}}x\|} \\[2ex] \text{s.t.} & x \in \mathscr{B} \end{array}\right\} \qquad (2.65)$$

which makes only sense under the assumption $0 \notin \mathscr{B}$. We have seen that problem (2.65) is a convex programming problem provided that $\mu^{\mathrm{T}}x \geq h,\ \forall x \in \mathscr{B}$ holds.

Figure 2.6 shows the graph and the contour lines of the function

$$f(x_1, x_2) = \frac{x_1 - x_2}{\sqrt{(x_1 + x_2)^2 + (x_1 - x_2)^2}}$$

which is the quotient of a linear and a convex function. In the contour plot darker regions represent lower values. Let $\varepsilon > 0$; for the figure we have chosen $\varepsilon = 0.1$. The function f is pseudo–concave for $\{x \in \mathbb{R}^2 \mid x_1 \geq x_2 + \varepsilon\}$ and pseudo–convex for $\{x \in \mathbb{R}^2 \mid x_1 \leq x_2 - \varepsilon\}$.

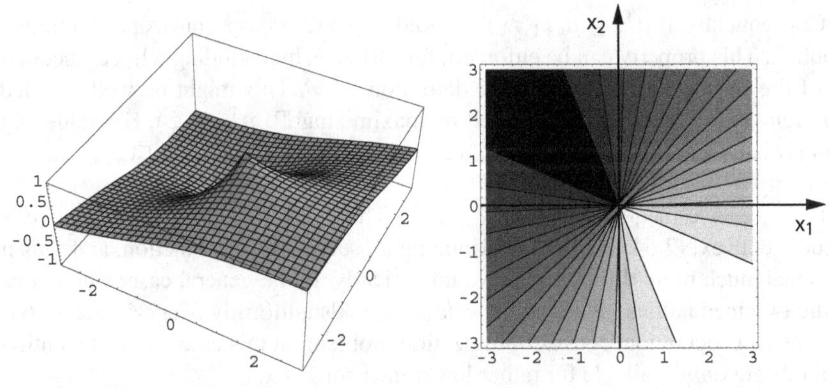

Fig. 2.6 Quotient of a linear and a nonlinear convex function.

Stable distributions

In the previous section, in the derivation of the explicit formula (2.58), it seems to be essential at a first glance, that both the expected value and the variance exist for $\zeta(x,\eta,\xi)$. A more careful analysis reveals, however, that quite other properties of the normal distribution are those, which matter.

Before carrying out this analysis, we discuss classifications of univariate distributions, which will be needed later on. We define a relation $\diamond$ between univariate distribution functions, see Feller [92]. Let F and H be two univariate distribution functions, then

$$F \diamond H \iff \exists a > 0,\, b:\; H(x) = F(ax+b)\; \forall x \in \mathbb{R}^1 \text{ holds} \tag{2.66}$$

or equivalently

$$F \diamond H \iff \exists a > 0,\, b:\; H\left(\frac{x-b}{a}\right) = F(x)\; \forall x \in \mathbb{R}^1. \tag{2.67}$$

This relation is obviously reflexive, symmetric, and transitive. Consequently we obtain a classification of all distribution functions. We may choose a representative from each class, and consider it as a *standard distribution* for that class. Let $\mathscr{D}$ be a class in this classification, and let H_0 be the standard distribution in $\mathscr{D}$. Then for any $F \in \mathscr{D}$ we have: $\exists a > 0,\, b$, such that $F(x) = H_0(\frac{x-b}{a})$, $\forall x \in \mathbb{R}^1$ holds. a is called the *scale–* and b the *location* parameter of F (w.r.t. to the standard distribution). The classes in this classification are also called *location–scale classes*.

Let ζ be a random variable with $F_\zeta \in \mathscr{D}$. This fact will be denoted as $\zeta \sim \mathscr{D}$. Then $\exists a, b \in \mathbb{R}$, $a > 0$ such that $F_\zeta(x) = H_0(\frac{x-b}{a})$ holds. This relation has the following interpretation: Let $\chi = \frac{\zeta - b}{a}$. Then we have

$$\mathbb{P}(\chi \le x) = \mathbb{P}(\zeta \le ax+b) = F_\zeta(ax+b) = H_0(x),$$

that means, χ has the standard distribution of $\mathscr{D}$. The transformation above is called *standardization* of ζ. This can also be expressed as follows: for any $\zeta \sim \mathscr{D}\ \exists a>0,\ b$, such that $\zeta = a\chi + b$ and $F_\chi = H_0$ holds. A final remark to this concept: let ζ be a random variable with $F_\zeta \in \mathscr{D}$, and let $p>0$, q be real numbers. Then obviously $p\zeta + q \sim \mathscr{D}$ holds.

We consider next the set of symmetric distributions. A distribution is called *symmetric* if for the distribution function F the following relation holds (see Feller [92]): $F(x) = 1 - F_-(-x)\ \forall x \in \mathbb{R}$, where $F_-(-x)$ stands for the left–sided limit of F at $-x$. If the density function f exists then the condition for symmetry can be written as $f(x) = f(-x)\ \forall x \in \mathbb{R}$. On the set of symmetric distributions the following equivalence relation establishes a classification:

$$F \diamond H \iff \exists a>0:\ H(x) = F(ax)\ \forall x \in \mathbb{R}^1.$$

The classes in this classification will be called *symmetric scale classes.*

If the random variable ζ has a symmetric distribution, this is clearly equivalent with ζ and $-\zeta$ having the same distribution function, that means, with $F_\zeta = F_{-\zeta}$. Let $\mathscr{S}$ be a class of symmetric distributions. Then $\zeta \sim \mathscr{S}$ implies $p\zeta \sim \mathscr{S}$ $\forall p \in \mathbb{R},\ p \neq 0$. For $p>0$ this is clear from the definition. If $p<0$ then we may write $p\zeta = (-p)(-\zeta)$. Now we have $F_\zeta = F_{-\zeta}$, and the assertion follows immediately.

If a location–scale class $\mathscr{D}$ contains a single symmetric distribution, then it obviously contains the whole symmetric scale class $\mathscr{S}$ of this distribution. In this case the standard distribution can be selected as a symmetric distribution, that means, $H_0 \in \mathscr{S}$. Let $\zeta \sim \mathscr{D}$ and $p,q \in \mathbb{R},\ p \neq 0$. Then, for such classes, $p\zeta + q \sim \mathscr{D}$ holds. For $p>0$ this is clear from the definition. Let us assume that $p<0$ holds. Standardization gives that $\exists a>0, b$ such that $\zeta = a\chi + b$ and $F_\chi = H_0$. Substitution results in $p\zeta + q = ap\chi + bp + q$. From this follows $ap\chi \sim \mathscr{S}$ and consequently $p\zeta + q \sim \mathscr{D}$.

Let us introduce the notion of a stable distribution next. For this concept see, for instance, Feller [92] and Uchaikin and Zolotarev [330].

A distribution function F, the corresponding probability distribution, and a random variable having this distribution are called *stable*, if for any real numbers $s_1>0$, m_1, $s_2>0$, and m_2 there exist real numbers $s>0$ and m, such that

$$F\left(\frac{x-m_1}{s_1}\right) * F\left(\frac{x-m_2}{s_2}\right) = F\left(\frac{x-m}{s}\right), \quad \forall x \in \mathbb{R}^1 \tag{2.68}$$

holds, where $*$ stands for the convolution operator. Let F be a stable distribution function and let $\mathscr{D}$ be its class in the above classification. From (2.68) immediately follows, that all $H \in \mathscr{D}$ are stable, that means, we may use the term *class of stable distributions.* In particular, the standard distribution $H_0 \in \mathscr{D}$ is also stable. Another easy consequence of (2.68) is the following: if $F \in \mathscr{D}$, $H \in \mathscr{D}$, and $\mathscr{D}$ is a stable class, then $F * H \in \mathscr{D}$ holds. Using the fact, that the distribution function of the sum of two stochastically independent random variables is the convolution of their distribution functions, we get the following: Let $\mathscr{D}$ be a stable class, $\zeta_i \sim \mathscr{D}$, $i=1,\ldots,s$, $\lambda_i \in \mathbb{R}\ \lambda_i > 0\ \forall i$. Assume that ζ_i, $i=1,\ldots,s$ are stochastically independent. Then the distribution function of $\sum_{i=1}^{s} \lambda_i \zeta_i$ also belongs to $\mathscr{D}$. This property is, however, not sufficient for our purposes: in (2.53) we deal with arbitrary linear combinations of independent random variables.

A distribution function F, the corresponding probability distribution, and a random variable having this distribution are called *strictly stable*, if for any real numbers $s_1 > 0$ and $s_2 > 0$ there exists a real number $s > 0$, such that

$$F\left(\frac{x}{s_1}\right) * F\left(\frac{x}{s_2}\right) = F\left(\frac{x}{s}\right), \ \forall x \in \mathbb{R}^1 \tag{2.69}$$

holds, where $*$ stands as before for the convolution operator. In the following we restrict our attention to symmetric distributions. Let F be a strictly stable distribution function and let $\mathscr{S}$ be its class in the classification of symmetric distributions. The analogous results hold, as for stable distributions. In particular, if $F \in \mathscr{S}$ and $H \in \mathscr{S}$, then $F * H \in \mathscr{S}$ follows. This implies for symmetric distributions the following: Let $\mathscr{S}$ be a strictly stable class of symmetric distributions, $\zeta_i \in \mathscr{S}$, $i = 1, \ldots, s$, $\lambda_i \in \mathbb{R}\ \forall i$, and not all λ_i's are zero. Assume that ζ_i, $i = 1, \ldots, s$ are stochastically independent. Then the distribution function of $\sum_{i=1}^{s} \lambda_i \zeta_i$ also belongs to $\mathscr{S}$.

As an example for a stable class of distributions let us shortly discuss the univariate normal distribution. The univariate normal distribution functions form a location–scale class, because they are of the form: $F(x) = \Phi\left(\frac{x-b}{a}\right)$, $0 < a \in \mathbb{R}$, $b \in \mathbb{R}$, where Φ is the distribution function of the standard normal distribution. This is a stable class. To see this, it is sufficient to check the stability of Φ. Considering the convolution (2.68)

$$\Phi\left(\frac{x-m_1}{\sigma_1}\right) * \Phi\left(\frac{x-m_2}{\sigma_2}\right) = \Phi\left(\frac{x-m}{s}\right), \ \forall x \in \mathbb{R}^1,$$

where the left–hand–side is the distribution function of the sum of two independent $\xi \sim \mathscr{N}(m_1, \sigma_1^2)$ and $\eta \sim \mathscr{N}(m_2, \sigma_2^2)$ random variables. We know that $\xi + \eta$ has a normal distribution. On the other hand, the expected value is additive w.r. to summation, and the variance is also additive provided that the random variables are stochastically independent. Therefore the above relation holds for $m = m_1 + m_2$ and $\sigma = \sqrt{\sigma_1^2 + \sigma_2^2}$. This argumentation also shows that the class of symmetric (centered) normal distribution functions $F(x) = \Phi\left(\frac{x}{a}\right)$, $0 < a \in \mathbb{R}$ form a strictly stable class of symmetric distributions.

Now we take the proposed second look at the derivation of the explicit form for G in Section 2.2.3.

1. The multivariate distribution of ζ was defined by the affine linear relations (2.53) for the realizations, in terms of the *i.i.d.* (independent and identically distributed) random variables $\tilde{\zeta}_i$, $i = 1 \ldots s$. In that particular case the distribution of $\tilde{\zeta}_i$ was

standard normal, $\forall i$, which, as discussed above, belongs to the strictly stable class of symmetric normal distributions.
2. Subsequently we have established in (2.54) an affine linear relation for $\zeta(x,\eta,\xi)$, in terms of $\tilde{\zeta}$.
3. Considering the linear part, this is a linear combination of random variables with distributions from a strictly stable class, therefore the linear combination belongs also to that class. Due to the additive deterministic term, $\zeta(x,\eta,\xi)$ belongs to the stable class of normal distributions. In addition, using the specific properties of the normal distribution, we were also able to compute the parameters of $\zeta(x,\eta,\xi)$, in terms of our decision variables x.
4. Finally, in (2.54), we have standardized $\zeta(x,\eta,\xi)$ in order to derive a formula for $G(x)$, involving the distribution function of the standard distribution in the location–scale class. Using this formula, the constraint $G(x) \geq \alpha$ has been reformulated as (2.58). By good luck, this resulted in a constraint of the convex programming type.

Another well–known stable univariate distribution is the Cauchy distribution, see, for instance, Feller [92]. For this distribution the expected value and consequently the variance do not exist. The density function of the Cauchy distribution $\mathscr{C}(m,t)$ is the following:

$$f(x) = \frac{1}{\pi} \frac{t}{t^2 + (x-m)^2}, \quad -\infty < x < \infty,$$

where m is a location parameter and $t > 0$ is a scale parameter. Taking $m = 0$ the resulting subclass of symmetric distributions is strictly stable. The distribution function of the standard Cauchy distribution $\mathscr{C}(0,1)$, defined by the density function with $t = 1$

$$\psi(x) = \frac{1}{\pi} \frac{1}{1+x^2}, \quad -\infty < x < \infty,$$

will be denoted by Ψ. The following fact is also well-known, see, for instance, Feller [92]: Let $\xi \sim \mathscr{C}(m_1,t_1)$ and $\eta \sim \mathscr{C}(m_2,t_2)$, and assume that ξ and η are stochastically independent. Then $\xi + \eta \sim \mathscr{C}(m_1+m_2, t_1+t_2)$ holds.

We will carry out the above procedure for a multivariate Cauchy distribution, see Marti [222].

Definition 2.15. *The r–dimensional random vector ζ has a non–degenerate multivariate Cauchy distribution, if there exist an $(r \times s)$ matrix B with full row rank and having at least one nonzero in each of its columns and $m \in \mathbb{R}^r$, such that*

$$\zeta = B\tilde{\zeta} + m, \tag{2.70}$$

where $\tilde{\zeta}$ is an s–dimensional random vector with its components being stochastically independent and $\tilde{\zeta}_i$ having a standard Cauchy distribution, $\forall i$.

Let us assume, that the $(n+1)$–dimensional random vector $\zeta^{\mathrm{T}} = (\eta^{\mathrm{T}}, \xi)^{\mathrm{T}}$ has a non–degenerate multivariate Cauchy distribution. In the same way, as in Section 2.2.3, we get:

$$\zeta(x,\eta,\xi) = \tilde{\zeta}^{\mathrm{T}}\left(D^{\mathrm{T}}x-d\right)+m^{\mathrm{T}}x-m_{n+1}. \tag{2.71}$$

Let us remark that $\|D^{\mathrm{T}}x-d\| \neq 0$ holds for all x, due to the assumption that the transformation matrix $\begin{pmatrix} D \\ d^{\mathrm{T}} \end{pmatrix}$ has full row rank (see also (2.53)). We conclude that $\zeta(x,\eta,\xi)$ has a Cauchy distribution, and proceed by computing its parameters. If $(D^{\mathrm{T}}x-d)_i \neq 0$ then $\tilde{\zeta}_i(D^{\mathrm{T}}x-d)_i \sim \mathscr{C}(0,|(D^{\mathrm{T}}x-d)_i|)$ holds. Consequently we have $\tilde{\zeta}^{\mathrm{T}}\left(D^{\mathrm{T}}x-d\right) \sim \mathscr{C}(0,\|D^{\mathrm{T}}x-d\|_1)$, where for $y \in \mathbb{R}^s$ $\|y\|_1 := \sum_{i=1}^{s}|y_i|$. Finally we get:

$$\zeta(x,\eta,\xi) \sim \mathscr{C}(m^{\mathrm{T}}x-m_{n+1},\|D^{\mathrm{T}}x-d\|_1).$$

Using standardization, as in (2.57) we get the following formula for $G(x)$:

$$G(x) = 1-\Psi\left(\frac{-m^{\mathrm{T}}x+m_{n+1}}{\|D^{\mathrm{T}}x-d\|_1}\right) = \Psi\left(\frac{m^{\mathrm{T}}x-m_{n+1}}{\|D^{\mathrm{T}}x-d\|_1}\right), \tag{2.72}$$

where we utilized the symmetry of the standard Cauchy distribution. Comparing this with the analogous formula (2.57) for the non–degenerate multivariate normal distribution, it can be observed that the sole difference is the different norm in the denominator.

We proceed now analogously as in (2.58) to arrive at:

$$G(x) \geq \alpha \iff \Psi^{-1}(\alpha)\,\|D^{\mathrm{T}}x-d\|_1 - m^{\mathrm{T}}x \leq -m_{n+1}. \tag{2.73}$$

The standard Cauchy distribution being symmetric, for $\alpha \geq \frac{1}{2}$, $\Psi^{-1}(\alpha) \geq 0$ holds. Because norms are convex functions, $\|D^{\mathrm{T}}x-d\|_1$ is a convex function of x. As for the normal distribution, we conclude that the function on the left–hand–side of the inequality is a concave function, and the set of x vectors, for which this inequality holds, is convex. We have derived the following theorem:

Theorem 2.7. *Marti 1971 [222]. Let the $(n+1)$–dimensional random vector $\zeta^{\mathrm{T}} = (\eta^{\mathrm{T}},\xi)^{\mathrm{T}}$ have a non–degenerate multivariate Cauchy distribution and let $\alpha \geq \frac{1}{2}$. Then the set $\mathscr{B}(\alpha) = \{x \mid G(x) \geq \alpha\}$ is convex.*

The alternative formulations of the probability constraints are analogous to those for the normal distribution. The difference is that, instead of the Euclidean norm, the $\|\cdot\|_1$–norm is to be substituted throughout. This seems to introduce, however, an additional difficulty: the $\|\cdot\|_1$–norm is a non–differentiable function of its argument. Under the assumption $\alpha \geq \frac{1}{2}$, a second look reveals, however, that by introducing additional variables the constraint (2.73) can be equivalently formulated as a set of linear constraints. In this respect, probability constraints are easier to deal with for the Cauchy distribution as for the normal distribution. For discussing the transformation let us formulate (2.73) in a detailed form:

$$\Psi^{-1}(\alpha)\sum_{i=1}^{s}|D_i^{\mathrm{T}}x-d_i| - m^{\mathrm{T}}x \leq -m_{n+1}, \tag{2.74}$$

where D_i is the ith column of D. This constraint is equivalent to the following system of linear constraints:

$$\begin{aligned} -m^{\mathrm{T}}x + \Psi^{-1}(\alpha) \sum_{i=1}^{s} y_i &\leq -m_{n+1} \\ D_i^{\mathrm{T}}x - \qquad y_k &\leq d_k, \ k=1,\ldots,s \\ D_i^{\mathrm{T}}x + \qquad y_k &\geq d_k, \ k=1,\ldots,s, \end{aligned} \tag{2.75}$$

in the following sense: Let $\bar{x}$ be a feasible solution of (2.74). Choosing $\bar{y}_k = |D_k^{\mathrm{T}}\bar{x} - d_k| \ \forall k$ implies that $(\bar{x}, \bar{y}_k, k=1,\ldots,s)$ is a feasible solution of (2.74). Vice versa, let $(\widehat{x}, \widehat{y}_k, k=1,\ldots,s)$ be feasible for (2.75). Then the inequality $|D_k^{\mathrm{T}}\widehat{x} - d_k| \leq y_k$ holds $\forall k$, which implies that $\widehat{x}$ is feasible for (2.74).

There is an important special case, as observed by Marti [222], in which the problem transforms into a deterministic LP problem, without introducing additional variables and constraints. Let us assume that $\mathscr{B} \subset \mathbb{R}_+^n$ holds which is the case, for instance, if the system of linear inequalities defining $\mathscr{B}$ includes $x \geq 0$. Assume further, that the components of (η, ξ) are stochastically independent and that they have Cauchy distributions $\eta_i \sim \mathscr{C}(m_i, t_i)$ $i=1,\ldots,n$ and $\xi \sim \mathscr{C}(m_{n+1}, t_{n+1})$. In this case the matrix $\begin{pmatrix} D \\ d^{\mathrm{T}} \end{pmatrix}$ is a diagonal $((n+1) \times (n+1))$ matrix, with the t_i's on its diagonal, see (2.2.3). Consequently we get $\|D^{\mathrm{T}}x - d\|_1 = \sum_{i=1}^{n} t_i x_i + t_{n+1}$ and (2.73) becomes a linear constraint.

A distribution–free approach

The sole assumption in this section is that the second moments of (η^{T}, ξ) exist. Let $(\mu^{\mathrm{T}}, \mu_{n+1}) = \mathbb{E}[(\eta^{\mathrm{T}}, \xi)]$ and Σ be the covariance matrix of (η^{T}, ξ). We assume that Σ is positive definite and take the Cholesky factorization $\Sigma = LL^{\mathrm{T}}$ with L being a lower triangular matrix (cf. the discussion on page 99). We consider L in the partitioned form

$$L = \begin{pmatrix} D \\ d^{\mathrm{T}} \end{pmatrix},$$

where D is an $(n \times n)$ matrix and $d \in \mathbb{R}^n$. For $\zeta(x, \eta, \xi) = \eta^{\mathrm{T}}x - \xi$ we get the same expression (2.55) as for the normal distribution

$$\begin{aligned} \mathbb{E}[\zeta(x, \eta, \xi)] &= \mu^{\mathrm{T}}x - \mu_{n+1} \\ \mathbb{V}\mathbf{ar}[\zeta(x, \eta, \xi)] &= \|D^{\mathrm{T}}x - d\|^2. \end{aligned}$$

The general idea is to employ upper bounds on the probability function $G(x) = \mathbb{P}(\eta^{\mathrm{T}}x - \xi \geq 0)$. Utilizing the Chebyshev–inequality we get

$$
\begin{aligned}
G(x) &= \mathbb{P}\big((\eta-\mu)^{\mathrm{T}} x - (\xi - \mu_{n+1}) \geq -\mu^{\mathrm{T}} x + \mu_{n+1} \big) \\
&\leq \mathbb{P}\big(|(\eta-\mu)^{\mathrm{T}} x - (\xi - \mu_{n+1})| \geq -\mu^{\mathrm{T}} x + \mu_{n+1} \big) \\
&\leq \frac{\mathbb{V}\mathrm{ar}(\eta^{\mathrm{T}} x - \xi)}{(-\mu^{\mathrm{T}} x + \mu_{n+1})^2} = \frac{\|D^{\mathrm{T}} x - d\|^2}{(-\mu^{\mathrm{T}} x + \mu_{n+1})^2}.
\end{aligned} \tag{2.76}
$$

We consider the probability constraint $G(x) \leq \beta$ with β small, for instance, $\beta = 0.01$. The idea is to require instead of this inequality the stronger inequality

$$
\frac{\|D^{\mathrm{T}} x - d\|^2}{(-\mu^{\mathrm{T}} x + \mu_{n+1})^2} \leq \beta. \tag{2.77}
$$

For having a nonempty solution set of this inequality, for small β values we may suppose that $-\mu^{\mathrm{T}} x + \mu_{n+1} > 0$ holds. This may be enforced by including a constraint $-\mu^{\mathrm{T}} x + \mu_{n+1} > \varepsilon$, with $\varepsilon > 0$, into the set of linear constraints of the problem. Assuming this, we can write (2.77) as follows

$$
\beta^{-\frac{1}{2}} \|D^{\mathrm{T}} x - d\| + \mu^{\mathrm{T}} x \leq \mu_{n+1} \tag{2.78}
$$

which defines a convex set.

For the case when (η, ξ) has a multivariate normal distribution, we have derived an equivalent formulation for $G(x) \leq \beta$ (first line in (2.59)). Slightly reformulated, this constraint is

$$
-\Phi^{-1}(\beta) \|D^{\mathrm{T}} x - d\| + \mu^{\mathrm{T}} x \leq \mu_{n+1} \tag{2.79}
$$

which is quite similar to (2.78). The sole difference is the different multiplier for the term $\|D^{\mathrm{T}} x - d\|$. Taking $\beta = 0.01$, for example, we have $\beta^{-\frac{1}{2}} = 10$ and $-\Phi^{-1}(\beta) \approx 2.32$. Thus, in the normally distributed case, requiring (2.78) instead of (2.79), a much stronger inequality results. Consequently, the feasible domain becomes much smaller in general. A prototype substitute problem takes the form

$$
\left.
\begin{array}{ll}
\min c^{\mathrm{T}} x & \\
\text{s.t. } \beta^{-\frac{1}{2}} \|D^{\mathrm{T}} x - d\| + \mu^{\mathrm{T}} x \leq \mu_{n+1} & \\
\qquad x \in \mathscr{B}. &
\end{array}
\right\} \tag{2.80}
$$

If for a given distribution, like the multivariate normal or the Cauchy distribution, an algebraic equivalent formulation exists, it makes no sense to use the stronger inequality (2.78). If, however, the distribution belongs to a class of distributions for which no equivalent algebraic formulation is known, or we have incomplete information regarding the distribution but have good estimates for the expected value and the covariance matrix, the substitute constraint (2.78) may provide a valuable modeling alternative. Notice that for *any distribution* with existing second moments, employing (2.78) in the model ensures that for the solution x^* the true inequality $G(x^*) \leq \beta$ holds also. In other words, employing (2.80) is a conservative approach, which might be quite acceptable if, for instance, β represents the ruin probability of a company. Nevertheless, it may happen that the optimal objective value in (2.80) becomes too high (too high costs, for instance), due to the narrower feasible domain in comparison with the feasible domain according to the true constraint $G(x) \leq \beta$.

Analogously, if $G(x)$ is to be minimized in an SLP model, one might consider a substitute model with the upper bound from (2.76) in the objective. Thus, instead of

$$\left.\begin{array}{ll} \min G(x) & \\ \text{s.t.} & x \in \mathscr{B} \end{array}\right\}$$

we may consider the substitute problem

$$\left.\begin{array}{l} \min \dfrac{\|D^{\mathrm{T}}x - d\|^2}{(-\mu^{\mathrm{T}}x + \mu_{n+1})^2} \\ \text{s.t. } x \in \mathscr{B}. \end{array}\right\}$$

Under the assumption that $-\mu^{\mathrm{T}}x + \mu_{n+1} > 0$ holds for all $x \in \mathscr{B}$, we get the equivalent formulation

$$\left.\begin{array}{l} \min \dfrac{\mu^{\mathrm{T}}x - \mu_{n+1}}{\|D^{\mathrm{T}}x - d\|} \\ \text{s.t. } x \in \mathscr{B}, \end{array}\right\} \tag{2.81}$$

where equivalence means that the set of optimal solution of the two problems coincide. According to Proposition 2.32. on page 83, the objective function in (2.81) is pseudo–convex over $\mathscr{B}$, thus (2.81) is a convex programming problem. A comparison with (2.64) shows that the substitute problem and the original problem are equivalent in the case of the non–degenerate multivariate normal distribution (notice that (2.64) corresponds to maximizing G). In the general case, the optimal objective value of the substitute problem (2.81) provides an upper bound on the optimal objective value of the original problem. Taking again the interpretation of $G(x)$ as ruin probability, for any optimal solution x^* of (2.81), the ruin probability $G(x^*)$ will not exceed the optimal objective value of (2.81). Concerning applicability of this approach, similar comments apply as for (2.80).

We would like to emphasize that, in general, both (2.80) and (2.81) are substitutes for the corresponding original problems, in general they are not equivalent to the true problems. Finally let us point out that this approach has first been suggested by Roy [292] and is utilized in the safety–first approaches to portfolio optimization, see Elton et al. [85].

2.2.4 The independent case

In this section we consider the joint probability function

$$G(x) = \mathbb{P}_{\xi}(T(\xi)x \geq h(\xi)) = \mathbb{P}_{\xi}(t_i^{\mathrm{T}}(\xi)x \geq h_i(\xi),\ i = 1, \ldots, s),$$

where the components of the n–dimensional random vector $t_i(\xi)$ are the elements of the ith row of the $(s \times n)$ random matrix $T(\xi)$. We will assume in this section throughout that $s > 1$ holds.

Our basic assumption is that the random vectors

$$(t_i^{\mathrm{T}}(\xi), h_i(\xi)), \; i = 1, \ldots, s$$

are stochastically independent. Models of this type have first been formulated and studied by Miller and Wagner [234].

The stochastic independence implies that the random vector $\zeta(x,\xi)$, with $\zeta_i(x,\xi) = t_i^{\mathrm{T}}(\xi)x - h_i(\xi)$, $i = 1, \ldots, s$, has stochastically independent components. Consequently, the probability function can be written in the independent case as follows:

$$\begin{aligned} G(x) = \mathbb{P}(\,\zeta(x,\xi) \geq 0\,) &= \prod_{i=1}^{s} \mathbb{P}(\,\zeta_i(x,\xi) \geq 0\,) \\ &= \prod_{i=1}^{s} \mathbb{P}(\,t_i^{\mathrm{T}}(\xi)x \geq h_i(\xi)\,). \end{aligned} \tag{2.82}$$

We observe, that the probability function $G(x)$ is the product of probability functions of the type, which have been studied in Section 2.2.3 on separate constraints; each term in the product involves a single random inequality.

Let us discuss the case first, when $t_i(\xi) \equiv t_i \; \forall i$ holds, that means, we assume that only the right–hand–side is stochastic. Setting $h(\xi) := \xi$, we have

$$\begin{aligned} G(x) &= \mathbb{P}(\,t_i^{\mathrm{T}} x \geq \xi_i, \; i = 1, \ldots, s\,) \\ &= F_{\xi_1, \ldots \xi_s}(t_1^{\mathrm{T}} x, \ldots, t_s^{\mathrm{T}} x) \\ &= \prod_{i=1}^{s} F_{\xi_i}(t_i^{\mathrm{T}} x). \end{aligned} \tag{2.83}$$

Distribution functions being monotonously increasing, the terms of the product are quasi–concave functions. This does not imply, however, the quasi–concavity of the product. Assuming positivity of the distribution functions, a natural idea is to transform the product into a sum, by a logarithmic transformation. The logarithm–function being strictly monotonically increasing, this would be suitable also from the optimization point of view. This way we get:

$$\log G(x) = \sum_{i=1}^{s} \log F_{\xi_i}(t_i^{\mathrm{T}} x).$$

$\log G(x)$ will be concave, if the univariate distribution functions F_{ξ_i} are logconcave. As already noted by Miller and Wagner [234], log–concavity of univariate distribution functions is a thoroughly studied subject in statistics, more closely in reliability theory. It has been found that many important distributions, including the normal distribution, have logconcave distribution functions. For a recent summary see, for instance, Sengupta and Nanda [303] and the references therein.

Let us assume that the distribution functions F_{ξ_i} are logconcave $\forall i$, in the sense of the general Definition 2.13. on page 84. $G(x)$, being the product of logconcave functions, is logconcave (see page 85). Consequently, the probability constraint

$$G(x) \geq \alpha$$

defines a convex set, $\forall \alpha \in [0,1]$. If the distribution functions are positive, the constraint can also be written as

$$\sum_{i=1}^{s} \log F_{\xi_i}(t_i^{\mathrm{T}} x) \geq \log \alpha \tag{2.84}$$

for all $\alpha \in (0,1]$.

If we drop the assumption of stochastic independence, but keep the supposition that only the right–hand–side is stochastic, then from (2.83) we see, that for the logconcavity of G it is sufficient, that the joint distribution function $F_{\xi_1,\dots\xi_s}$ is logconcave. This is true for several important distributions, and will be the subject of the subsequent Section 2.2.5.

Finally we discuss the situation under the stochastic independence assumption and random coefficients in the inequalities, see (2.82). We assume that the joint distributions of the rows are non–degenerate multivariate normal. For the separate terms of the product we can use the explicit form (2.57), derived in the section on separate probability constraints, thus resulting in:

$$G(x) = \prod_{i=1}^{s} \Phi\left(\frac{\mu^{(i)\mathrm{T}} x - \mu_{n+1}^{(i)}}{\|D^{(i)} x - d^{(i)}\|}\right), \tag{2.85}$$

where $\mu^{(i)}$, $D^{(i)}$, and $d^{(i)}$ are the parameters of the normal distribution corresponding to the ith row, $\forall i$. According to Proposition 2.36. on page 104, the terms of the product in (2.85) are pseudo–concave functions, at least on appropriate half–spaces. Unfortunately, this does not even imply that $G(x)$ is quasi–concave. To ensure the convexity of $\{x \mid G(x) \geq \alpha\}$ quite strong additional assumptions are needed. This topic will be further pursued in Section 2.2.6.

2.2.5 Joint constraints: random right–hand–side

In this section we consider a single probability constraint under the assumption that $T(\xi) \equiv T$ holds, that means, we assume that the technology matrix is deterministic. We also simplify the notation by setting $h(\xi) := \xi$. Consequently, the probability constraint has the following form:

$$G(x) := \mathbb{P}_{\xi}(Tx \geq \xi) \geq \alpha \tag{2.86}$$

where T is an $(s \times n)$ matrix and ξ is an s–dimensional random vector. Employing the probability distribution function F_ξ, $G(x)$ can be formulated as $G(x) = F_\xi(Tx)$. An alternative formulation for the probability constraint above is the following:

$$\begin{array}{rl} F_\xi(y) & \geq \alpha \\ y \; - Tx & = 0. \end{array} \tag{2.87}$$

From these representations it is clear, that for the convexity of the feasible domain

$$\mathscr{B}(\alpha) = \{x \mid G(x) \geq \alpha\}$$

it is sufficient, that the probability distribution function F_ξ is quasi–concave.

In the next subsection we will introduce the notion of generalized–concave probability measures. Via generalized–concavity properties of density functions this will lead to identifying several important classes of probability distributions for which F_ξ is quasi–concave. Subsequently we consider transformations which lead to generalized–concave probability functions. In the final subsection we consider SLP problems with joint probability functions in the objective.

Generalized–concave probability measures

We will assume in this section that the probability distribution P_ξ is absolutely continuous (w.r. to the Lebesgue–measure), that means, we assume that the probability measure is generated by a probability density function. We will discuss various conditions concerning the probability measure P_ξ induced by ξ, under which the probability distribution function F_ξ is quasi–concave.

We begin by discussing generalized means, see Hardy, Littlewood, and Pólya [132].

Let $a \geq 0$, $b \geq 0$, and $\lambda \in [0,1]$. The generalized means $\mathscr{M}_\gamma^\lambda(a,b)$ are defined as follows: for $ab = 0$ let $\mathscr{M}_\gamma^\lambda(a,b) = 0$, for all $\gamma \in \mathbb{R} \cup \{-\infty\} \cup \{\infty\}$. Otherwise, that is, if $ab > 0$ holds, we define

$$\mathscr{M}_\gamma^\lambda(a,b) = \begin{cases} [\lambda a^\gamma + (1-\lambda) b^\gamma]^{\frac{1}{\gamma}}, & \text{if } -\infty < \gamma < \infty \\ & \text{and } \gamma \neq 0 \\ a^\lambda b^{1-\lambda}, & \text{if } \gamma = 0 \\ \min\{a,b\}, & \text{if } \gamma = -\infty \\ \max\{a,b\}, & \text{if } \gamma = \infty. \end{cases} \tag{2.88}$$

The following monotonicity property of these generalized means will be used, see [132]:

$$\gamma_1 < \gamma_2 \implies \mathscr{M}_{\gamma_1}^\lambda(a,b) \leq \mathscr{M}_{\gamma_2}^\lambda(a,b), \;\; \forall a,b \geq 0$$

with the inequality being strict, unless $a = b$ or $ab = 0$. Based on these generalized means we define:

Definition 2.16. *A nonnegative function $f : \mathbb{R}^n \to \mathbb{R}_+$ will be called γ–concave, if for any $x, y \in \mathbb{R}^n$ and $\lambda \in [0,1]$ the following inequality holds:*

$$f(\lambda x + (1-\lambda)y) \geq \mathcal{M}_\gamma^\lambda(f(x), f(y)).$$

Let us note that in the literature this kind of generalized concave functions, as well as the generalized concave measures introduced later in this section, are usually called α–concave, see for instance, Dancs and Uhrin [48] and Norkin and Roenko [245]. Because α is used for probability levels in this chapter, we use the term γ–concave, instead.

Let f be a γ–concave function and $\mathscr{C}^+ := \{x \mid f(x) > 0\}$. The γ–concavity immediately implies that $\mathscr{C}^+$ is a convex set. As already discussed for the logconcave case (c.f. Proposition 2.34. on page 85), this observation leads to the following alternative characterization: the nonnegative function f is γ–concave, if and only if $\mathscr{C}^+$ is a convex set and the inequality in Definition 2.16. holds for all $x, y \in \mathscr{C}^+$.

For various γ values, γ–concavity can be interpreted over $\mathscr{C}^+$ as follows (see the definition of the generalized means):

- $\gamma = +\infty$: f is constant;
- $0 < \gamma < +\infty$: f^γ is a concave function, note that $\gamma = 1$ corresponds to ordinary concavity;
- $\gamma = 0$: f is logconcave, that means, $\log f$ is concave;
- $-\infty < \gamma < 0$: f^γ is a convex function;
- $\gamma = -\infty$: f is quasi–concave.

Notice that we have stated the properties only over $\mathscr{C}^+$. To see the reason, let us discuss the case $\gamma = 1$. A nonnegative function f is 1–concave, if it is concave over the convex set $\mathscr{C}^+$, where it is positive. If f is defined over $\mathbb{R}^n$, this does not mean that f is a concave function there. The following nonnegative function $g : \mathbb{R} \to \mathbb{R}_+$

$$g(x) = \begin{cases} 1 - x^2 & \text{if } x \in [-1,1] \\ 0 & \text{if } x \in (-\infty,-1) \text{ or } x \in (1,\infty) \end{cases}$$

is obviously 1–concave but it is not concave. Considering the well–known properties of concave functions, some caution is needed when 1–concave functions are dealt with. For instance, let $g : \mathbb{R} \to \mathbb{R}_+$ and $h : \mathbb{R} \to \mathbb{R}_+$ be both 1–concave functions, with $\mathscr{C}_g^+ := \{x \mid g(x) > 0\}$ and $\mathscr{C}_h^+ := \{x \mid h(x) > 0\}$. Then for $g + h$ we have $\mathscr{C}_{g+h}^+ := \{x \mid g(x) + h(x) > 0\} = \mathscr{C}_g^+ \cup \mathscr{C}_h^+$, which is a non–convex set in general. Thus, the sum of 1–concave functions is not necessarily 1–concave.

The monotonicity property of the generalized means implies: if f is γ_2–concave, then it is γ_1–concave, for all $\gamma_1 < \gamma_2$. In particular, if f is γ–concave for any $\gamma \in [-\infty, \infty]$ then f is quasi–concave. For the implications concerning the various types of generalized concavity see Figure 2.8 on page 121.

Although pseudo–concavity does not fit into the class of γ–concave functions, logconcave functions, which are continuously differentiable over their domain of positivity, are also pseudo-concave there, see Proposition 2.35. on page 86. Consequently, for $\gamma \geq 0$ the γ–concave functions, having the above smoothness property, are also pseudo–concave over their positivity domain.

We wish to extend the notion of γ–concavity to probability measures. For this we have to specify first, how a linear combination of sets should be defined. Let A and B two subsets of $\mathbb{R}^r$ and let $\lambda \in \mathbb{R}$. We employ the following definitions:

$$\begin{aligned} A+B &= \{x \mid \exists y \in A \text{ and} \exists z \in B, \text{ such that } x = y+z\}, \\ \lambda A &= \{x \mid \exists y \in A \text{ such that } x = \lambda y\}. \end{aligned} \tag{2.89}$$

Figure 2.7 shows the convex combination of two sets. For the properties of these operations on sets see, for instance, Rockafellar [281].

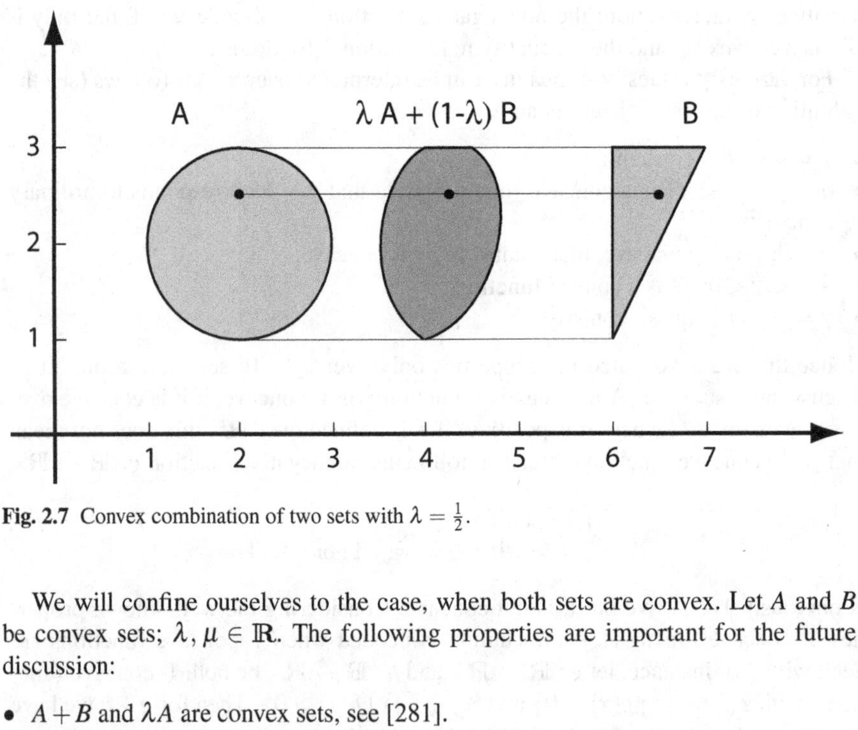

Fig. 2.7 Convex combination of two sets with $\lambda = \frac{1}{2}$.

We will confine ourselves to the case, when both sets are convex. Let A and B be convex sets; $\lambda, \mu \in \mathbb{R}$. The following properties are important for the future discussion:

- $A+B$ and λA are convex sets, see [281].
- Let $\lambda \geq 0$ and $\mu \geq 0$. Then $(\lambda+\mu)A = \lambda A + \mu A$ (without the convexity of A only $(\lambda+\mu)A \subset \lambda A + \mu A$ holds). See [281].
- If either A or B is open, then $A+B$ is open.
- If both A and B are closed, and at least one of them is bounded, then $A+B$ is closed. The sum of two unbounded closed convex sets need not to be closed, see [281]. If both A and B are closed then $A+B$ is Borel–measurable, see, for instance, [88].

- If A is convex, then it is obviously Lebesgue–measurable, because the boundary has Lebesgue–measure 0.
- If A is convex, then it is not necessarily Borel–measurable. To see this, let us construct a convex set in $\mathbb{R}^2$ as follows: Let us take a non–Borel–measurable set $\mathscr{K}$ on the interval $[0,2\pi)$ (for the existence of such a set see, for instance, Billingsley [21]) and let us map this set onto the boundary of the open unit disc in $\mathbb{R}^2$ by the mapping $\Psi : \mathscr{K} \to \mathbb{R}^2,\ x \to (\cos x, \sin x)$. The union of the open unit disk and the image of $\mathscr{K}$ under Ψ is obviously convex, and, as a union of a Borel–measurable set (the open unit disc), and a non–Borel–measurable set, it cannot be Borel–measurable.
- The sum of two Borel–measurable sets is not necessarily Borel–measurable, see Erdős and Stone [88].

As a next step, we will define generalized concavity properties of probability measures, in analogy with Definition 2.16. Considering the list of properties above, one must be careful in working with convex combinations of Borel–sets. Therefore we formulate the definition as follows:

Definition 2.17. *The probability measure $\mathbb{P}$ on the Borel–sets $\mathbb{B}^r$ is called γ–concave, if for any convex, measurable sets A and B and any $\lambda \in [0,1]$, for which $\lambda A + (1-\lambda)B$ is Borel–measurable, the following inequality holds:*

$$\mathbb{P}(\lambda A + (1-\lambda)B) \geq \mathscr{M}_\gamma^\lambda(\mathbb{P}(A), \mathbb{P}(B)).$$

A γ–concave probability measure with $\gamma = -\infty$ will be called a *quasi–concave*. In this case the defining inequality takes the form

$$\mathbb{P}(\lambda A + (1-\lambda)B) \geq \min\{\mathbb{P}(A), \mathbb{P}(B)\}.$$

For $\gamma = 0$ we have a *logconcave* probability measure, with the defining inequality

$$\mathbb{P}(\lambda A + (1-\lambda)B) \geq \mathbb{P}(A)^\lambda\, \mathbb{P}(B)^{1-\lambda}.$$

Let ξ be a random variable and $\mathbb{P}_\xi$ the induced measure on $\mathbb{B}^r$. We denote by F_ξ the probability distribution function of ξ. For any convex, closed set A in $\mathbb{R}^r$ let us introduce the function $\Gamma_A(y) = \mathbb{P}_\xi(A + \{y\})$. Then the following proposition holds.

Proposition 2.37. *If $\mathbb{P}_\xi$ is a γ–concave measure, then Γ_A is a γ–concave function.*

Proof: Let $x, y \in \mathbb{R}^r$, $\lambda \in [0,1]$. Then we have

$$\begin{aligned}
\Gamma_A(\lambda x + (1-\lambda)y) &= \mathbb{P}_\xi(A + \{\lambda x + (1-\lambda)y\}) \\
&= \mathbb{P}_\xi([\lambda + (1-\lambda)]A + \{\lambda x + (1-\lambda)y\}) \\
&= \mathbb{P}_\xi(\lambda[A + \{x\}] + (1-\lambda)[A + \{y\}]) \\
&\geq \mathscr{M}_\gamma^\lambda(\mathbb{P}(A + \{x\}), \mathbb{P}(A + \{y\})) \\
&= \mathscr{M}_\gamma^\lambda(\Gamma_A(x), \Gamma_A(y)).
\end{aligned}$$

□

Let us assume that $\mathbb{P}_\xi$ is γ–concave. Taking $A = \mathbb{R}^r_-$ we get from Proposition 2.37., that F_ξ is γ–concave. Consequently, $\{x \mid \mathbb{P}_\xi(Tx \geq \xi) \geq \alpha\}$ is a convex set, $\forall \alpha \in [0,1]$ (see (2.86)).

Let us consider

$$H(y) := \mathbb{P}_\xi\{y \mid \xi \geq y\}.$$

Choosing now $A = \mathbb{R}^r_+$, Proposition 2.37. implies, that H is also γ–concave. Consequently, $\{x \mid \mathbb{P}_\xi(Tx \leq \xi) \geq \alpha\}$ is also a convex set, $\forall \alpha \in [0,1]$.

The above considerations imply, that for showing that the distribution function F is γ–concave, it is sufficient to prove the γ–concavity of the probability distribution $\mathbb{P}_\xi$.

The following fundamental theorem links, for continuous distributions, the γ–concavity of the probability density function with the γ–concavity of $\mathbb{P}_\xi$.

Theorem 2.8. *Let f be a γ–concave probability density function for the probability distribution of the r–dimensional random variable ξ. Let $-\frac{1}{r} \leq \gamma \leq \infty$. Then $\mathbb{P}_\xi$ is an $\dfrac{\gamma}{1+r\gamma}$–concave probability measure.*

Proof: Let $\lambda \in [0,1]$ and assume that the convex sets A, B, and $\lambda A + (1-\lambda)B$ are Borel–measurable. The γ-concavity of f implies:

$$\begin{aligned}
\mathbb{P}_\xi(\lambda A + (1-\lambda)B) &= \int\limits_{\lambda A+(1-\lambda)B} f(z)dz \\
&\geq \int\limits_{\lambda A+(1-\lambda)B} \Big(\sup_{\lambda x+(1-\lambda)y=z} \mathscr{M}^\lambda_\gamma(f(x), f(y)) \Big) dz.
\end{aligned}$$

Now we apply an integral–inequality, see Prékopa [266] (for $\gamma = 0$ it is called Prékopa's inequality):

$$\begin{aligned}
\mathbb{P}_\xi(\lambda A + (1-\lambda)B) &\geq \int\limits_{\lambda A+(1-\lambda)B} \Big(\sup_{\lambda x+(1-\lambda)y=z} \mathscr{M}^\lambda_\gamma(f(x), f(y)) \Big) dz \\
&\geq \mathscr{M}^\lambda_{\frac{\gamma}{1+r\gamma}} \Big(\int\limits_A f(x)dx, \int\limits_B f(y)dy \Big) \\
&= \mathscr{M}^\lambda_{\frac{\gamma}{1+r\gamma}} \big(\mathbb{P}_\xi(A), \mathbb{P}_\xi(B) \big)
\end{aligned}$$

which completes the proof. □

For some ranges of γ–values we summarize the assertion of the theorem, together with the implications from Theorem 2.37., see also Figure 2.8. For this let $C := \{x \mid f(x) > 0\}$ and let us assume that C is a convex set.

- f is constant over $C \Longrightarrow F^{\frac{1}{r}}$ and $H^{\frac{1}{r}}$ are concave, consequently both F and H are logconcave and therefore also quasi–concave.
- f is logconcave $\Longrightarrow F$ and H are logconcave and therefore quasi–concave, too.
- $f^{-\frac{1}{r}}$ is convex $\Longrightarrow F$ and H are quasi–concave.

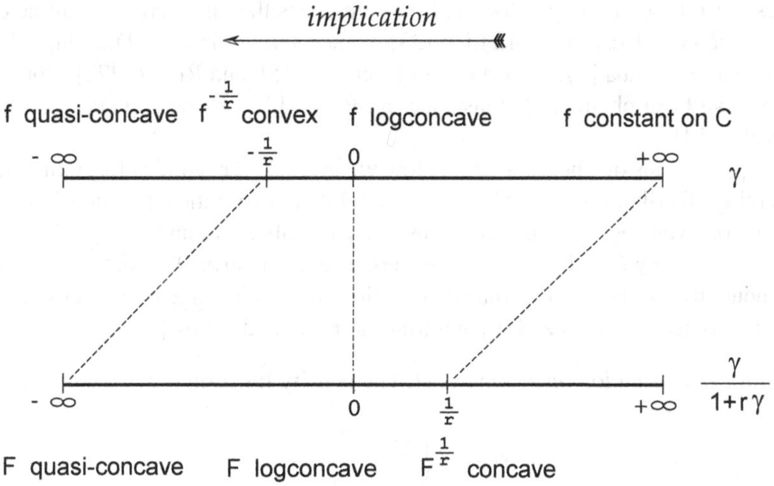

Fig. 2.8 γ–concave density functions versus $\frac{\gamma}{1+r\gamma}$–concave distribution functions.

Logconcave functions have several nice properties. We will need the following fact:

Theorem 2.9. *Prékopa [259]. Let $f : \mathbb{R}^{n+m} \to \mathbb{R}_+$ be a logconcave function. Then*

$$g(x) := \int_{\mathbb{R}^m} f(x,y)dy$$

is a logconcave function on $\mathbb{R}^n$.

Proof: See Prékopa [266]. □

If f is a logconcave density function then this theorem implies that all marginal density functions are logconcave, too.

If f and g are two logconcave density functions on $\mathbb{R}^n$ then their convolution is also logconcave. In fact, the logconcavity of f implies that $h(x,y) := f(x-y)$ is logconcave in $\mathbb{R}^{2n}$ (see the remark on page 85). Thus $f(x-y)g(y)$ is logconcave in $\mathbb{R}^{2n}$. Applying Theorem 2.9. yields the result.

For $\gamma = 0$ Theorem 2.8. has first been established by Prékopa in 1971 [258], by Leindler 1972 [203], and in its final form by Prékopa 1973 [259]. Dinghas 1957 [72] proved the theorem for $\gamma > 0$. Borell proved the theorem in full generality in 1975 [31].

The breakthrough in the field of generalized concave measures and their application in stochastic programming has been achieved by Prékopa, who developed the theory of logarithmic concave probability measures. These fundamental results have inspired several authors: papers with alternative proofs have appeared, the theory has been extended to quasi–concave measures, and applications in stochastic programming, statistics, and economics have been studied. For a comprehensive discussion of these results see Prékopa [266] and the references therein. Here we confine ourselves to refer to Brascamp and Lieb [33], Dancs and Uhrin [48], Das Gupta [54], Kallberg and Ziemba [173], Norkin and Roenko [245], and Rinott [277]. Converse results have been obtained, for instance, by Borell [31], Brascamp and Lieb [33], and Kall [154].

As applications of Theorem 2.8., below we give some examples for multivariate probability distributions, for which the probability distribution function is quasi–concave or even logconcave. The probability distribution– and density functions will be denoted by F and f, respectively. For a square matrix D, its determinant will be denoted by $|D|$. For multivariate distributions and their usage in statistics see, for instance, Johnson and Kotz [149] and Mardia, Kent, and Bibby [216].

- *Uniform distribution on a convex set.* The density function is

$$f(x) = \begin{cases} \frac{1}{\lambda(C)} & \text{if } x \in C \\ 0 & \text{otherwise,} \end{cases}$$

  where $C \subset \mathbb{R}^s$ is a bounded convex set with a positive Lebesgue–measure $\lambda(C)$. f is obviously logconcave thus F is logconcave, too.
- *Non–degenerate normal distribution.* The density function of this distribution is positive on $\mathbb{R}^r$ and is given in (2.51) on page 99. Taking logarithm and neglecting the additive constant results in

$$-\frac{1}{2}(y-\mu)^{\mathrm{T}}\Sigma^{-1}(y-\mu).$$

  This is a concave function, because with Σ, Σ^{-1} is also positive definite, see for instance, Horn and Johnson [141]. Thus f is logconcave implying the logconcavity of F. Figure 2.9 shows the standard bivariate normal distribution function and its logarithm.

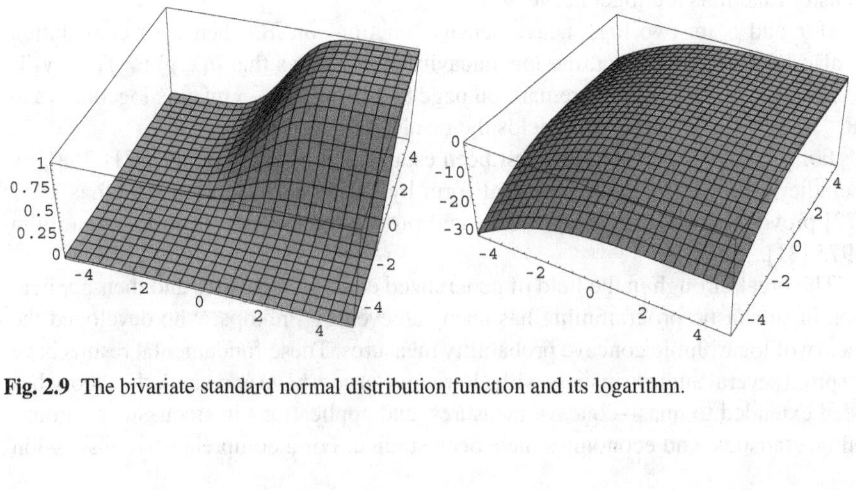

Fig. 2.9 The bivariate standard normal distribution function and its logarithm.

- *Dirichlet distribution.* This is the joint distribution of the random variables $\eta_j = \xi_j \left[\sum_{i=0}^{r} \xi_i\right]^{-1}$, $j = 1,\ldots,r$, where ξ_j, $j = 0,\ldots,r$ are independent random variables, ξ_j having χ^2–distribution with $\nu_j > 0$ degrees of freedom. The density function of this distribution is

$$f(x) = \begin{cases} \dfrac{\Gamma(\sum\limits_{j=0}^{s} \theta_j)}{\prod\limits_{j=0}^{s} \Gamma(\theta_j)}(1 - \sum\limits_{j=1}^{s} x_j)^{\theta_0 - 1} \prod\limits_{j=1}^{s} x_j^{\theta_j - 1}, & \text{if } x > 0,\ \sum\limits_{j=1}^{r} x_j < 1, \\ 0, & \text{otherwise}, \end{cases}$$

where θ_j, $j = 0,\ldots,r$ are the parameters of the distribution; $\theta_j = \frac{1}{2}\nu_j\ \forall j$. On the convex set $\mathscr{C}^+ := \{x \mid x > 0,\ \sum_{j=1}^{s} x_j < 1\}$ the density function is positive and it is zero if $x \notin \mathscr{C}^+$. Therefore, see Proposition 2.34. on page 85, for checking logconcavity, it is sufficient to consider $\log f(x)$ over $\mathscr{C}^+$. Apart of an additive constant, we have for $x \in \mathscr{C}^+$:

$$\log f(x) = (\theta_0 - 1)\log(1 - \sum_{j=1}^{s} x_j) + \sum_{j=1}^{s} (\theta_j - 1)\log x_j.$$

If $\theta_j \geq 1\ \forall j$ then this is a linear combination, with nonnegative coefficients, of concave functions, therefore $\log f(x)$ is concave on $\mathscr{C}^+$. Let us remark that the concavity of the first term in the right–hand–side follows from the fact, that substitution of an affine–linear function into a concave function preserves concavity. We have got: provided that $\theta_j \geq 1$ for $j = 0,\ldots,s$ holds, f is logarithmic concave implying the logconcavity of F.
- *Wishart distribution.* This is the joint distribution of the elements of the sample covariance matrix for a multivariate normal population. Let us consider a sample with sample–size $N > s$ from a population consisting of s–dimensional random vectors having a multivariate normal distribution with covariance matrix C. The density function for this distribution is the following:

$$f(X) = \begin{cases} \gamma |X|^{\frac{1}{2}(N-s-2)} e^{-\frac{1}{2}\mathrm{Tr}(C^{-1}X)}, & \text{if } X \text{ is positive definite} \\ 0, & \text{otherwise}, \end{cases}$$

where X is an $(s \times s)$ symmetric matrix and

$$\gamma = |C|^{-\frac{N-1}{2}} \left[2^{\frac{(N-1)s}{2}} \pi^{\frac{s(s-1)}{4}} \prod_{j=1}^{s} \Gamma\left(\frac{N-j}{2}\right)\right]^{-1}$$

holds. For an $(s \times s)$ matrix D, $\mathrm{Tr} D := \sum_{j=1}^{s} D_{jj}$ denotes the *trace* of D. We wish to check whether f is logconcave. For this we first observe that the set

$\mathscr{C}^+ := \{X \mid X$ is symmetric positive definite$\}$ is obviously a convex subset of the linear space of symmetric $(s \times s)$ matrices. Therefore it is sufficient to consider $\log f$ on C^+:

$$\log f(X) = \log \gamma + \frac{1}{2}(N-s-2)\log|X| - \frac{1}{2}\mathrm{Tr}C^{-1}X.$$

The third term is obviously linear in X. According to an inequality of Fan (see, for instance, Beckenbach and Bellman [13]), the function $|X|$ is a logconcave function of X. Therefore, if $N \geq s-2$ then f is logconcave and so F is logconcave, too.

- *t–distribution (Student–distribution).* We consider the joint distribution of $\eta_j = \xi_j \left(\frac{\zeta}{\sqrt{\nu}}\right)^{-1}$, $j = 1, \ldots, r$, where $(\xi_1, \ldots, \xi_r)$ has a joint standardized non–degenerate multivariate normal distribution with correlation matrix R. ζ has a χ–distribution with ν degrees of freedom. The density function for this distribution is positive on $\mathbb{R}^s$ and has the analytical form

$$f(x) = \frac{\Gamma(\frac{1}{2}(\nu+s))}{(\pi\nu)^{\frac{s}{2}}\Gamma(\frac{\nu}{2})|R|^{\frac{1}{2}}}(1+\frac{1}{\nu}x^{\mathrm{T}}R^{-1}x)^{-\frac{1}{2}(\nu+s)},$$

where the parameters are R, a symmetric positive definite matrix, and the positive integer ν, interpreted as degrees of freedom. $f^{-\frac{1}{s}}$ is, apart of a positive multiplicative constant, as follows:

$$g(x) := (1+\frac{1}{\nu}x^{\mathrm{T}}R^{-1}x)^{\frac{1}{2}(1+\frac{\nu}{s})},$$

which is a convex function on $\mathbb{R}^s$. To see this, let us remark first that

$$h(x) := (1+\frac{1}{\nu}x^{\mathrm{T}}R^{-1}x)^{\frac{1}{2}} = \left((x^{\mathrm{T}},1)\begin{pmatrix} R^{-1} & 0 \\ 0^{\mathrm{T}} & 1 \end{pmatrix}\begin{pmatrix} x \\ 1 \end{pmatrix}\right)^{\frac{1}{2}}$$

is convex because the positive definite matrix above induces a norm in $\mathbb{R}^{s+1}$. We have $g = h^{1+\frac{\nu}{s}}$, therefore the convexity of g follows from the fact, that substituting a convex function into a monotonically increasing convex function results in a convex composite function. Thus $f^{-\frac{1}{s}}$ is convex, implying that F is quasi–concave.

- *Univariate gamma distribution.* The density function of this distribution is

$$f(x) = \begin{cases} \frac{\lambda^{\vartheta}}{\Gamma(\vartheta)}x^{\vartheta-1}e^{-\lambda x}, & \text{if } x > 0 \\ 0, & \text{otherwise}, \end{cases}$$

where $\lambda > 0$ and $\vartheta > 0$ are parameters. This distribution will be denoted by $\mathscr{G}(\lambda, \vartheta)$. If $\lambda = 1$, then the distribution is called a standard gamma distribution. Assuming $x > 0$ and taking logarithm we observe that f is logconcave, provided that $\vartheta \geq 1$ holds. If an s–dimensional random vector η has stochastically independent components $\eta_i \sim \mathcal{G}(\lambda_i, \vartheta_i)$ and $\vartheta_i \geq 1\ \forall i$ holds, then η has a logconcave distribution. This follows by considering the density function of η, which is the product of the one–dimensional density functions of the components. The univariate densities being logconcave, their product is logconcave, too.

Generalized–concave distribution functions

So far we have discussed one way for ensuring generalized concavity of the distribution function F_ξ. The method, applicable for continuous distributions, has been the following: the generalized concavity of the probability density function has been studied, which implied via Theorem 2.8. the generalized concavity of F_ξ. For several important multivariate distributions it turned out that F_ξ is pseudo–concave, or that they even have the more important logconcavity property.

Another possibility has been discussed in Section 2.2.4. Under the assumption that the components of ξ are stochastically independent, the joint distribution function is the product of the one–dimensional marginal distribution functions, that means,

$$F_\xi(y) = \prod_{i=1}^{s} F_{\xi_i}(y_i).$$

If the marginal distribution functions F_{ξ_i} are logconcave then F_ξ will be logconcave, too.

In the sequel we explore further ways for ensuring generalized concavity properties of the probability distribution function. The idea is to apply transformations to random vectors having generalized–concave distributions, in order to obtain distributions for which the probability distribution function again has some generalized concavity properties.

The subsequent theorems and the insight behind their application in stochastic programming have been first found by Prékopa for the logconcave case. Their extension to the γ–concave case is straightforward.

The following general theorem gives a sufficient condition for generalized concavity of composite functions. See, for instance, Prékopa [266] and for an extension Tamm [324].

We consider the following probability function:

$$M(x) = \mathbb{P}_\xi\{\, g_i(x,\xi) \geq 0,\ i = 1,\ldots,s \,\} = \mathbb{P}_\xi\{\, g(x,\xi) \geq 0 \,\}, \tag{2.90}$$

where $g_i : \mathbb{R}^n \times \mathbb{R}^r \to \mathbb{R}$, $i = 1,\ldots,s$; $g^{\mathrm{T}}(x,\xi) = (g_1^{\mathrm{T}}(x,\xi),\ldots,g_s^{\mathrm{T}}(x,\xi))$.

Theorem 2.10. *Let g_i, $i = 1,\ldots,m$ be quasi–concave functions, that means, let $g_i(\cdot,\cdot)$ be jointly quasi–concave in both arguments. For the sake of simplicity we also assume that g is continuous. Assume further, that ξ has a γ–concave probability distribution. Then $M(x)$ is a γ–concave function.*

Proof: Let $\mathscr{H}(x) := \{z \mid g(x,z) \geq 0\} \subset \mathbb{R}^s$. Due to our assumptions these sets are convex and closed $\forall x$ and we have $M(x) = \mathbb{P}_\xi(\mathscr{H}(x))$. Let $\lambda \in (0,1)$, $x, y \in \mathbb{R}^n$. The basic ingredient of the proof is the following inclusion, which can be proved in a straightforward way:

$$\mathscr{H}(\lambda x + (1-\lambda)y) \supset \lambda \mathscr{H}(x) + (1-\lambda)\mathscr{H}(y).$$

Using this and the γ–concavity of the probability measure, we immediately get:

$$\begin{aligned} M(\lambda x + (1-\lambda)y) &= \mathbb{P}_\xi(\mathscr{H}(\lambda x + (1-\lambda)y)) \\ &\geq \mathbb{P}_\xi(\lambda \mathscr{H}(x) + (1-\lambda)\mathscr{H}(y)) \\ &\geq \mathscr{M}_\gamma^\lambda(\mathbb{P}_\xi(\mathscr{H}(x)), \mathbb{P}_\xi(\mathscr{H}(y))) \\ &= \mathscr{M}_\gamma^\lambda(M(x), M(y)). \end{aligned}$$

□

As an application of this theorem we will show, how it can be applied to prove logconcavity of the log–normal distribution function.

- *Log–normal distribution.* Let the random variables $\xi_1, \ldots, \xi_s$ have a joint non–degenerate multivariate normal distribution. The joint distribution of the random variables $\eta_i = e^{\xi_i}$, $i = 1, \ldots, s$ is called a multivariate log–normal distribution. The density function of this distribution is not logconcave, see Prékopa [266]. For the joint distribution function F_η we have:

$$\begin{aligned} F_\eta(x_1, \ldots, x_s) &= \mathbb{P}_\eta(\eta_1 \leq x_1, \ldots, \eta_s \leq x_s) \\ &= \mathbb{P}_\xi(x_1 - e^{\xi_1} \geq 0, \ldots, x_s - e^{\xi_s} \geq 0). \end{aligned}$$

In the preceding section we have seen that the probability measure of a non–degenerate multivariate normal distribution is logconcave. Theorem 2.10. can be applied with $\gamma = 0$ thus showing that F_η is a logconcave function.

Let us consider next the effect of linear transformations of random variables having γ–concave distributions. The following theorem holds:

Theorem 2.11. *Let ξ be an s–dimensional random vector, D an $(r \times s)$ matrix, and $\zeta = D\xi + d$. If ξ has a γ–concave distribution then the distribution of ζ is also γ–concave.*

Proof: Let $\lambda \in (0,1)$ and let A, B, $C_\lambda := \lambda A + (1-\lambda)B$ be Borel–measurable convex sets in $\mathbb{R}^r$. Then their inverse images in $\mathbb{R}^s$ under the affine linear transformation defined by D and d, that means,

$$\begin{aligned} \bar{A} &:= \{x \mid Dx + d \in A\}, \\ \bar{B} &:= \{x \mid Dx + d \in B\}, \text{ and} \\ \bar{C}_\lambda &:= \{x \mid Dx + d \in C_\lambda\} \end{aligned}$$

are Borel–measurable convex sets in $\mathbb{R}^s$. It is easy to see that

$$\bar{C}_\lambda \supset \lambda \bar{A} + (1-\lambda)\bar{B}$$

holds. Using this we get

$$\begin{aligned} \mathbb{P}_\zeta(\lambda A + (1-\lambda)B) &= \mathbb{P}_\xi(\bar{C}_\lambda) \\ &\geq \mathbb{P}_\xi(\lambda \bar{A} + (1-\lambda)\bar{B}) \\ &\geq \mathscr{M}_\gamma^\lambda(\mathbb{P}_\xi(\bar{A}), \mathbb{P}_\xi(\bar{B})) \\ &= \mathscr{M}_\gamma^\lambda(\mathbb{P}_\zeta(A), \mathbb{P}_\zeta(B)). \end{aligned}$$

□

This theorem can be utilized to study generalized concavity properties of distributions, which are derived in a similar way, as the multivariate normal distribution. We take s stochastically independent continuous random variables, each of them having a γ–concave density function. The joint density function is then the product of the density functions of the components. If this joint density function is γ–concave, then via Theorem 2.11., $\zeta = D\xi + d$ will have a γ–concave distribution. Especially, if the components of ξ have logconcave densities ($\gamma = 0$), then the joint density function of ξ will be logconcave (see page 85).

- *The multivariate normal distribution.* We consider the multivariate normal distribution, see Definition 2.14. on Page 98. In Section 2.2.5 we have proved, by applying Theorem 2.8., that the non–degenerate multivariate normal distribution is logconcave. Without the non–degeneracy assumption we can proceed as follows. Recall (Definition 2.14. on page 98) that the r–dimensional random vector ζ has a multivariate normal distribution, if $\zeta = B\xi + \mu$ holds, where B is an $(r \times s)$ matrix, $\mu \in \mathbb{R}^r$ holds, and the components of ξ are independent and have a standard normal distribution. The joint probability distribution of ξ is then obviously non–degenerate multivariate normal. Thus, Theorem 2.8. implies that ξ has a logconcave probability distribution. Consequently, the application of Theorem 2.11. yields the logconcavity of the probability distribution of ζ and thus the logconcavity of the multivariate normal distribution in the general case.
- *A multivariate gamma distribution.* In the preceding section on Page 124 we have seen that the univariate gamma distribution has a logconcave density function, therefore our technique can be used in this case, too. Prékopa and Szántai [270] have defined a multivariate gamma distribution as follows. Let ξ be a $s = 2^r - 1$ dimensional random vector with stochastically independent components. The components are assumed to have standard gamma distributions, see Page 124. Let D be an $(r \times 2^r - 1)$ matrix with nonzero columns and components equal to 0 or 1. The distribution of $\zeta := D\xi$ is called a multivariate gamma distribution. If for the parameter $\vartheta \geq 1$ holds, then Theorem 2.11. implies that the distribution of ζ is logconcave. If $\vartheta < 1$ then the distribution of ζ is not

necessarily logconcave, but the joint distribution function F_ζ is still a logconcave function, see [270].

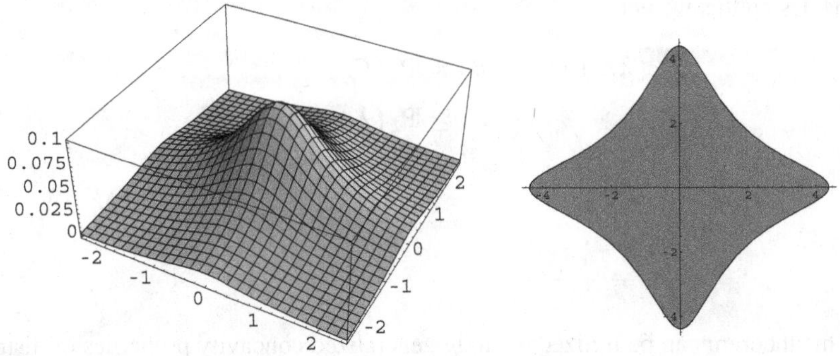

Fig. 2.10 The graph of the bivariate standard Cauchy–distribution and the upper level set corresponding to level 0.005.

In Section 2.2.3 we have considered a multivariate Cauchy distribution, which is derived on the basis of an affine linear transformation as discussed above, see Definition 2.15. on page 109. A natural idea is trying to apply Theorem 2.11. for deriving some γ–concavity property of the multivariate Cauchy distribution. Notice, however, that the density function of the univariate Cauchy distribution is not logconcave. Moreover, as it can easily be seen, the product of the density functions of standard univariate Cauchy distributions is not even quasi–concave, see Figure 2.10. Therefore, see Figure 2.8, the joint density function of ξ is not γ–concave, for any γ. Consequently, our technique does not go through in this case. Notice, however, that there are other generalizations of the Cauchy distribution to the multivariate case, where the distribution is quasi–concave, see Prékopa [268].

Finally let us comment on the case when ξ has a finite discrete distribution. Prékopa [266] gave a definition of logconcavity of such distributions and studied their properties. In Dentcheva et al. [65], the authors extend this notion to r–concave discrete distributions, where r–concavity corresponds to γ–concavity (see Section 2.2.5) in the continuous case, and is appropriately modified for the discretely distributed case. The authors also report on algorithmically relevant applications by providing bounds on the optimal objective value of SLP problems with probabilistic constraints.

Maximizing joint probability functions

For the case when the probability function is in the objective, we formulate the prototype problem

$$\begin{array}{ll} \max & G(x) \\ \text{s.t.} & x \in \mathscr{B}, \end{array} \tag{2.91}$$

where G is the probability function $G(x) = \mathbb{P}_\xi(Tx \geq \xi)$ and $\mathscr{B}$ is a polyhedral set determined by linear inequalities and/or equalities, see (2.8) on page 74.

If G is logconcave and differentiable then it is also pseudo–concave, see Proposition 2.35. on page 86. This is the case, for instance, for the non–degenerate multivariate normal distribution or for the log–normal distribution (see pages 122 and 126). Consequently, for logconcave distributions, (2.91) is a linearly constrained convex optimization problem. Some other distributions only have the quasi–concavity property, like the multivariate t–distribution (see page 124). In such cases (2.91) has a quasi–concave objective function and the problem may have local maxima which are not global solutions; the problem becomes much more difficult to solve numerically.

Note that (2.91) has been formulated as a maximization problem. Assuming $s > 1$, that is, assuming that ξ is a random vector, this is the only way for arriving at convex programming problems. Reversing the random inequality does not help in this respect: with G the function $\hat{G} : \hat{G}(x) = \mathbb{P}_\xi(Tx \leq \xi)$ is also logconcave, see the discussion on page 120. For reversing the random inequality in the multivariate case see also (2.31) on page 89. Thus, for ξ having a logconcave distribution and assuming $s > 1$, the counterpart of (2.91) involving minimization is a much more difficult problem numerically than (2.91).

2.2.6 Joint constraints: random technology matrix

In this section we consider the probability function in full generality

$$G(x) = \mathbb{P}_\xi(\, x \mid T(\xi)x \geq h(\xi)\,),$$

where the $(s \times n)$ technology matrix $T(\xi)$ is also allowed to be stochastic. In Section 2.2.3 on separate probability constraints we have assumed that $s = 1$ holds. We have seen that the feasible domain is convex under various further assumptions concerning the probability distribution and the probability level α. If $s > 1$, then the convexity of the feasible domain can only be ensured under quite strong assumptions. We will discuss the case, when the joint distribution of the random entries is multivariate normal.

The matrix of random entries $(T(\xi), h(\xi))$ will be considered both column–wise and row–wise, therefore we introduce the notation:

$$(T(\xi), h(\xi)) = \left(\zeta^{(1)}, \ldots, \zeta^{(n)}, \zeta^{(n+1)}\right) = \begin{pmatrix} \eta^{(1)\mathrm{T}} \\ \vdots \\ \eta^{(s)\mathrm{T}} \end{pmatrix}.$$

Here the s–dimensional random vector $\zeta^{(j)}$ denotes the jth column of $T(\xi)$ for $j \leq n$, and the right–hand–side $h(\xi)$ for $j = n+1$.

Let $\zeta^{\mathrm{T}} = (\zeta^{(1)\mathrm{T}}, \ldots, \zeta^{(n+1)\mathrm{T}})$ be the random vector consisting of all random entries in columns major order. The $n+1$–dimensional random vector $\eta^{(i)}$ stands for the ith row, $1 \leq i \leq s$, that is, $\eta^{(i)\mathrm{T}} = (T_{i1}(\xi), \ldots, T_{in}(\xi), h_i(\xi))$. Let $\eta^{\mathrm{T}} = (\eta^{(1)\mathrm{T}}, \ldots, \eta^{(s)\mathrm{T}})$ be the random vector consisting of all random entries in rows major order.

For any vector $x \in \mathbb{R}^n$ let $\widehat{x} \in \mathbb{R}^{n+1}$ be $\widehat{x}^{\mathrm{T}} = (x_1, \ldots, x_n, -1)$. For simplicity of notation in this section we drop the explicit reference to ξ in $\zeta(x, \xi)$. We have the following alternative representations

$$\begin{aligned} \zeta(x) &:= \sum_{i=1}^{n} \zeta^{(i)} x_i - \zeta^{(n+1)} \\ &= \left(\eta^{(1)\mathrm{T}}\widehat{x}, \ldots, \eta^{(s)\mathrm{T}}\widehat{x}\right)^{\mathrm{T}} \end{aligned} \tag{2.92}$$

and

$$G(x) = \mathbb{P}(\zeta(x) \geq 0).$$

Please note that we distinguish between the random vector ζ and $\zeta(x)$ defined in (2.92).

We assume that ζ has a multivariate normal distribution. This implies a multivariate normal distribution for η, as well as for the marginal distributions of $\zeta^{(j)}$ and $\eta^{(i)}$, $\forall i, j$, and for the distribution of $\zeta(x)$ (see Section 2.2.3 and [328]).

Let $\mu(x)$ be the expected–value vector and $\Sigma(x)$ be the covariance matrix of $\zeta(x)$. We proceed with computing these moments in terms of the moments of ζ and η. To this we introduce some further notation:

- M is the $(s \times (n+1))$ matrix of expected values of $(T(\xi), h(\xi))$;
- $C^{(i,j)}$ is the $(s \times s)$ covariance matrix of $\zeta^{(i)}$ if $i = j$, otherwise the cross–covariance matrix of $\zeta^{(i)}$ and $\zeta^{(j)}$, $i = 1, \ldots, n+1$, $j = 1, \ldots, n+1$;
- $\bar{C}^{(i,j)}$ is the $((n+1) \times (n+1))$ covariance matrix of $\eta^{(i)}$ if $i = j$, otherwise the cross–covariance matrix of $\eta^{(i)}$ and $\eta^{(j)}$, $i = 1, \ldots, s$, $j = 1, \ldots, s$.

For the expected value of $\zeta(x)$ we immediately get

$$\mu(x) = \mathbb{E}[\zeta(x)] = \sum_{j=1}^{n} \mathbb{E}[\zeta^{(j)}] x_i - \mathbb{E}[\zeta^{(n+1)}] = M\widehat{x}.$$

For the covariance matrix of $\zeta(x)$ we obtain two alternative forms corresponding to the column–wise and row–wise representations, respectively. We proceed with the column–wise form. For computing the covariance matrix, we note that $\zeta(x)$ is defined by an affine linear transformation:

$$\zeta(x) = \begin{pmatrix} x_1 & \dots x_n & & -1 & & \\ & \ddots & \dots & \ddots & & \ddots & \\ & & x_1 \dots & & x_n & & -1 \end{pmatrix} \begin{pmatrix} \zeta^{(1)} \\ \vdots \\ \zeta^{(n)} \\ \zeta^{(n+1)} \end{pmatrix}.$$

A straightforward computation gives for the covariance matrix of $\zeta(x)$:

$$\begin{aligned} \Sigma(x) &= \mathbb{C}\mathrm{ov}[\zeta(x), \zeta(x)] \\ &= \sum_{i,j=1}^{n} x_i x_j C^{(i,j)} + \sum_{j=1}^{n} x_j C^{(n+1,j)} + C^{(n+1,n+1)} \\ &= \sum_{i,j=1}^{n+1} \widehat{x}_i \widehat{x}_j C^{(i,j)}. \end{aligned} \tag{2.93}$$

In the alternative representation we observe:

$$\zeta(x) = \begin{pmatrix} \widehat{x}^{\mathrm{T}} & & \\ & \ddots & \\ & & \widehat{x}^{\mathrm{T}} \end{pmatrix} \begin{pmatrix} \eta^{(1)} \\ \vdots \\ \eta^{(s)} \end{pmatrix},$$

which immediately leads to

$$\begin{aligned} \Sigma(x) &= \mathbb{C}\mathrm{ov}[\zeta(x), \zeta(x)] \\ &= \begin{pmatrix} \widehat{x}^{\mathrm{T}} \bar{C}^{(1,1)} \widehat{x} \dots \widehat{x}^{\mathrm{T}} \bar{C}^{(1,s)} \widehat{x} \\ \vdots \quad \ddots \quad \vdots \\ \widehat{x}^{\mathrm{T}} \bar{C}^{(s,1)} \widehat{x} \dots \widehat{x}^{\mathrm{T}} \bar{C}^{(s,s)} \widehat{x} \end{pmatrix}. \end{aligned} \tag{2.94}$$

Next we observe that $\mathbb{P}(\zeta(x) \geq 0) = \mathbb{P}(-\zeta(x) \leq 0)$, where $-\zeta(x)$ is also normally distributed with the same covariance matrix as $\zeta(x)$ and expected value vector $-\mu(x)$. We will consider the case when all covariance matrices $C^{(i,j)}$ are multiples of a fixed symmetric positive semidefinite matrix. Therefore it is sufficient to prove convexity for one of the sets

$$\begin{aligned} \mathscr{B}(\alpha) &:= \{x \mid \mathbb{P}(\zeta(x) \geq 0) \geq \alpha\}, \\ \mathscr{A}(\alpha) &:= \{x \mid \mathbb{P}(\zeta(x) \leq 0) \geq \alpha\}, \end{aligned}$$

the convexity of the other one follows immediately.

Theorem 2.12. *Prékopa [261]. Let us assume that ζ has a joint multivariate normal distribution and that*

1. *either there exists an $((n+1) \times (n+1))$ matrix S and a symmetric positive semidefinite matrix C, such that $C^{(i,j)} = S_{ij} C$ holds, $\forall i, j$,*

2. *or there exists an* $(s \times s)$ *matrix* $\bar{S}$ *and a symmetric positive semidefinite matrix* $\bar{C}$*, such that* $\bar{C}^{(i,j)} = \bar{S}_{ij}\bar{C}$ *holds,* $\forall i, j$.

In both cases, if $\alpha \geq \frac{1}{2}$ *then* $\mathscr{A}(\alpha)$ *is a convex set.*

Proof: We begin with proving the first assertion of the theorem. We will assume that S is nonsingular; for the general case see Prékopa [266]. For the covariance matrix of $\zeta(x)$ we have (cf. (2.93))

$$\Sigma(x) = \sum_{i,j=1}^{n+1} \widehat{x}_i\widehat{x}_j S_{i,j} C = C \cdot \widehat{x}^{\mathrm{T}} S \widehat{x}. \tag{2.95}$$

In particular, for the variance we get

$$\mathbb{V}\mathbf{ar}[\zeta_i(x)] = C_{ii}\widehat{x}^{\mathrm{T}} S \widehat{x}, \ \forall i. \tag{2.96}$$

We may assume that $C_{ii} > 0$ holds $\forall i$.

In fact, $C_{ii} = 0$ implies that $\mathbb{V}\mathbf{ar}[\zeta_i(x)] = 0$, $\forall x$. Consequently, the coefficients and right–hand–side in the ith row of the system of random inequalities are a.s. constant. Therefore (see the Remark on page 90) the ith inequality can be moved to the set of deterministic constraints in the corresponding optimization problem.

From relation (2.96) immediately follows, that S is a symmetric positive semidefinite matrix. We have assumed that S is nonsingular, therefore S is positive definite.

Another implication of (2.95) is, that the correlation matrix R of $\zeta(x)$ does not depend on x. In fact, $R_{ij} := \mathbb{C}\mathbf{orr}[\zeta_i(x), \zeta_j(x)] = \dfrac{C_{ij}}{\sqrt{C_{ii}C_{jj}}}$ holds.

By standardizing $\zeta(x)$ (see page 99) we get:

$$\mathbb{P}(\,\zeta(x) \leq 0\,) = \Phi\left(-\frac{\mu_1(x)}{\sqrt{C_{1,1}\widehat{x}^{\mathrm{T}} S \widehat{x}}}, \ldots, -\frac{\mu_{n+1}(x)}{\sqrt{C_{n+1,n+1}\widehat{x}^{\mathrm{T}} S \widehat{x}}}; R\right),$$

where $\widehat{x}^{\mathrm{T}} S \widehat{x} > 0$ holds, due to our assumption concerning S and the fact that $\widehat{x} \neq 0$ $\forall x \in \mathbb{R}^n$.

Let

$$h^{\mathrm{T}}(x) = \left(-\frac{\mu_1(x)}{\sqrt{C_{1,1}}}, \ldots, -\frac{\mu_{n+1}(x)}{\sqrt{C_{n+1,n+1}}}\right),$$

$$\|\widehat{z}\|_S = \left(\widehat{z}^{\mathrm{T}} S \widehat{z}\right)^{\frac{1}{2}}, \ \forall \widehat{z} \in \mathbb{R}^{n+1},$$

where $\|\cdot\|_S$ is clearly a norm in $\mathbb{R}^{n+1}$. With this notation we have

$$\mathbb{P}(\,\zeta(x) \leq 0\,) = \Phi\left(\frac{1}{\|\widehat{z}\|_S} h(x); R\right).$$

$\Phi(z;R)$ is a multivariate distribution function, consequently it is monotonically increasing in each of its arguments. This implies that $\Phi(z_i) \geq \Phi(z;R)\ \forall i$ holds. Under our assumption $\alpha \geq \frac{1}{2}$, we deduce that $h(x) \geq 0$ holds $\forall x \in \mathscr{A}(\alpha)$.

Let $x \in \mathscr{A}(\alpha), y \in \mathscr{A}(\alpha), \lambda \in (0,1)$ and let $\widehat{x}$ and $\widehat{y}$ be the corresponding $(n+1)$–dimensional vectors with their last coordinate being equal to -1, cf. page 130.

With the notation $x_\lambda = \lambda x + (1-\lambda)y$ and $\widehat{x}_\lambda = \lambda\widehat{x} + (1-\lambda)\widehat{y}$, using the triangle inequality for norms we get:

$$\begin{aligned} \mathbb{P}(\zeta(x_\lambda) \leq 0) &= \Phi\left(\tfrac{1}{\|\widehat{x}_\lambda\|_S} h(x_\lambda); R\right) \\ &\geq \Phi\left(\tfrac{1}{\lambda\|\widehat{x}\|_S + (1-\lambda)\|\widehat{y}\|_S}(\lambda h(x) + (1-\lambda)h(y); R\right). \end{aligned} \tag{2.97}$$

We will make use of the following trivial fact: for $A,B,C,D \in \mathbb{R}$, $C>0$, and $D>0$ we have

$$\frac{A+B}{C+D} = \kappa\frac{A}{C} + (1-\kappa)\frac{B}{D}$$

with $\kappa = \frac{C}{C+D}$; $0 < \kappa < 1$. Applying this componentwise in (2.97) with the setting $A = \lambda h(x)$, $B = (1-\lambda)h(y)$, $C = \lambda\|\widehat{x}\|_S$, and $D = (1-\lambda)\|\widehat{y}\|_S$, and utilizing the logconcavity of $\Phi(z;R)$ we get:

$$\begin{aligned} \mathbb{P}(\zeta(x_\lambda) \leq 0) &\geq \Phi\left(\kappa\tfrac{1}{\|\widehat{x}\|_S}h(x) + (1-\kappa)\tfrac{1}{\|\widehat{y}\|_S}h(y); R\right) \\ &\geq \Phi\left(\tfrac{1}{\|\widehat{x}\|_S}h(x); R\right)^{\kappa} \Phi\left(\tfrac{1}{\|\widehat{y}\|_S}h(y); R\right)^{1-\kappa} \\ &= \mathbb{P}(\zeta(x) \leq 0)^{\kappa}\, \mathbb{P}(\zeta(y) \leq 0)^{1-\kappa} \\ &\geq \alpha^{\kappa}\alpha^{1-\kappa} = \alpha. \end{aligned} \tag{2.98}$$

The proof of the second assertion runs along analogous lines. For the covariance matrix of $\zeta(x)$ we now have (see (2.93))

$$\Sigma(x) = \bar{S} \cdot \widehat{x}^{\mathrm{T}}\bar{C}\widehat{x}. \tag{2.99}$$

For the variance we get

$$\mathbb{V}\mathbf{ar}[\zeta_i(x)] = \bar{S}_{ii}\widehat{x}^{\mathrm{T}}\bar{C}\widehat{x},\ \ \forall i. \tag{2.100}$$

Arguing similarly as for the first assertion, we conclude that $\bar{S}_{ii} > 0\ \forall i$ may be assumed. If $\bar{C}$ is positive definite, then the rest of the proof runs analogously to the proof of the first assertion. For the general case see Prékopa [266]. □

Let us remark, that the second assertion of the theorem has originally been proved in [261] under the assumption of the stochastic independence of the rows of $(T(\xi), h(\xi))$; the general case has been proved by Burkauskas [39].

2.2.7 Summary on the convex programming subclasses

SLP models with probability functions are non–convex in general but in the preceding sections we have found important subclasses consisting of convex programming problems. From the practical modeling point of view it is important to know, whether a particular model instance involving probability functions is a convex programming problem. Having namely a convex programming problem there are good chances for finding efficient solution algorithms, or in many cases general–purpose software can be used for solving the problem.

Therefore, for the sake of easy reference, in this section we summarize those model classes which consist of convex programming problems. For further such model classes see Prékopa [266]. If a particular model instance does not belong to any one of these model classes then most probably it is a non–convex optimization problem. This is not certain in general, of course; further research is needed which may lead to the discovery of new convex programming classes of SLP problems with probability functions.

For direct reference we repeat some of the notation and introduce some new one:

$$\begin{aligned}
G(x) &= \mathbb{P}_{\xi}(\, T(\xi)x \geq h(\xi)\,) \\
\hat{G}(x) &= \mathbb{P}_{\xi}(\, T(\xi)x \leq h(\xi)\,) \\
\mathscr{B}(\alpha) &= \{x \mid G(x) \geq \alpha\} \\
\widehat{\mathscr{B}}(\alpha) &= \{x \mid \widehat{G}(x) \geq \alpha\} \\
\mathscr{D}(\beta) &= \{x \mid G(x) \leq \beta\} \\
\widehat{\mathscr{D}}(\beta) &= \{x \mid \widehat{G}(x) \leq \beta\},
\end{aligned}$$

where $T(\xi)$ denotes a random $s \times n$ matrix, $h(\xi) \in \mathbb{R}^s$ stands for a random vector. The components of the n–dimensional random vector $t_i(\xi)$ are the elements of the ith row of $T(\xi)$, $\forall i$ and $T_j(\xi)$ stands for the jth column of $T(\xi)$, $\forall j$. If $s = 1$ (separate probability function) holds, we use the notation $t(\xi) = T(\xi)$; $\mu = \mathbb{E}[t(\xi)]$, $\mu_{t+1} = \mathbb{E}[h(\xi)]$.

A. General cases: convex models are identified by choosing specific probability levels. If $\alpha = 1$ or $\alpha = 0$ or $\beta = 1$ then $\mathscr{B}(\alpha)$, $\widehat{\mathscr{B}}(\alpha)$, $\mathscr{D}(\beta)$, and $\widehat{\mathscr{D}}(\beta)$ are all convex sets. (Proposition 2.2. on page 92 and the discussion on page 92).

B. ξ has a finite discrete distribution: convex models are identified by choosing specific probability levels. If α is high enough (as precisely formulated in the assumptions of Proposition 2.3. on page 95 and Proposition 2.4. on page 95) then $\mathscr{B}(\alpha)$ and $\widehat{\mathscr{B}}(\alpha)$ are convex. In general, however, $\mathscr{B}(\alpha)$, $\widehat{\mathscr{B}}(\alpha)$, $\mathscr{D}(\beta)$, and $\widehat{\mathscr{D}}(\beta)$ are non–convex sets. Equivalent linear mixed–integer programming reformulations are available, see (2.44) on page 94.

C. Separate probability functions, $s = 1$: convex cases are identified by choosing specific probability distributions and probability levels.

1. If only the right–hand–side is stochastic then $\mathscr{B}(\alpha)$, $\widehat{\mathscr{B}}(\alpha)$, $\mathscr{D}(\beta)$, and $\widehat{\mathscr{D}}(\beta)$ are half–spaces, determined by linear inequalities (Section 2.2.3) although for $\mathscr{D}(\beta)$, and $\widehat{\mathscr{D}}(\beta)$ some care is needed if ξ does not have a continuous distribution (page 98). (2.91) can be formulated as a deterministic linear program if $\mathscr{B}$ is bounded, otherwise some caution is needed, see (2.49) on page 98. These results hold for arbitrary values $0 < \alpha < 1$ and $0 < \beta < 1$.
2. If $(t(\xi), h(\xi))$ has a multivariate normal distribution and $\alpha \geq \frac{1}{2}$ and $\beta \leq \frac{1}{2}$ hold, then $\mathscr{B}(\alpha)$, $\widehat{\mathscr{B}}(\alpha)$, $\mathscr{D}(\beta)$, and $\widehat{\mathscr{D}}(\beta)$ are convex sets, determined by convex nonlinear constraints, see Section 2.2.3, (2.60), and (2.61) on page 103. Some care is needed concerning $\mathscr{D}(\beta)$ and $\widehat{\mathscr{D}}(\beta)$, see (2.61) on page 103, in the case when the distribution is degenerate, see page 103. If the distribution is non–degenerate then $G(x)$ is pseudo–concave on $\mathscr{B}$, if $\mu^{\mathrm{T}} x \geq \mu_{n+1}$ holds for all $x \in \mathscr{B}$. It is pseudo–convex on $\mathscr{B}$, if $\mu^{\mathrm{T}} x \leq \mu_{n+1}$ holds for all $x \in \mathscr{B}$. Similar assertions hold for $\widehat{G}(x)$ with exchanged roles of the inequalities for the expected values (Proposition 2.36.) on page 104. Thus (2.91) is a convex programming problem if for all $x \in \mathscr{B}$, $\mu^{\mathrm{T}} x \geq \mu_{n+1}$ holds. The corresponding minimization problem is a convex programming problem provided that for all $x \in \mathscr{B}$ the reverse strict inequalities $\mu^{\mathrm{T}} x \leq \mu_{n+1}$ hold.
3. If $(t(\xi), h(\xi))$ has a multivariate Cauchy distribution, similar remarks apply as in the normally distributed case, see Section 2.2.3. This section outlines also a technique for carrying out the analysis for distributions belonging to the class of stable distributions.

D. Stochastically independent random variables, $s > 1$: convex cases are identified by choosing specific probability distributions. If only $h(\xi)$ is stochastic, $(h_1(\xi), \ldots, h_s(\xi))$ are stochastically independent, and each $h_i(\xi)$ has a log–normal distribution function, then $G(x)$ is a logconcave function and $\mathscr{B}(\alpha)$ is convex (Section 2.2.4).

E. Only the right–hand–side is stochastic: convex cases are identified by choosing specific probability distributions. In the case of $s = 1$ this has been discussed above in item B.1 and under the assumption of stochastic independence the discussion can be found under item D. In the general case $G(x)$ and $\widehat{G}(x)$ are logconcave for the following multivariate distributions: uniform (page 122), non–degenerate normal (page 122), Dirichlet (page 123), Wishart (page 123), log–normal (page 126), and gamma (page 127). The probability functions G and $\widehat{G}(x)$ are quasi–concave for the multivariate t–distribution (page 124). Consequently, $\mathscr{B}(\alpha)$ and $\widehat{\mathscr{B}}(\alpha)$ are convex. Having $G(x)$ or $\widehat{G}(x)$ in the objective function, (2.91) is a convex programming problem for the logconcave distributions listed above. Regarding the case with reverse inequality constraints and the same distributions, $\mathscr{D}(\beta)$, and $\widehat{\mathscr{D}}(\beta)$ are non–convex sets in general and the minimization variant of (2.91) is a non–convex optimization problem.

F. Random technology matrix: for the case $s = 1$ the discussion can be found under items C.2 and C.3. For $s > 1$, $\mathscr{B}(\alpha)$ and $\widehat{\mathscr{B}}(\alpha)$ are convex under the following assumptions: $(T_1(\xi), \ldots, T_n(\xi), h(\xi))$ have a joint multivariate normal distribution and the covariance matrices of the columns as well as the cross–covariance matrices are constant multiples of a fixed covariance matrix, then $\mathscr{B}(\alpha)$ and $\widehat{\mathscr{B}}(\alpha)$ are convex sets. This holds also under the analogous assumption concerning the rows. For both facts see Proposition 2.12. on page 131.

Exercises

2.4. Show the following assertion concerning probabilities, which has been utilized in the proof of Theorem 2.2. on page 92:

$$\mathbb{P}(A) = \mathbb{P}(B) = 1 \;\Rightarrow\; \mathbb{P}(A \cap B) = 1.$$

2.5. Consider the following chance–constrained problem:

$$\left.\begin{array}{lll} \min & 2x_1 + x_2 & \\ \text{s.t.} & x_1 & \geq 1 \\ & \mathbb{P}(x_1 + x_2 \geq \xi) \geq 0.9, & \end{array}\right\}$$

with $\xi \in \mathscr{U}(1,2)$, meaning that the random variable ξ is uniformly distributed over the interval $[1,2]$.

(a) Formulate the equivalent LP problem.
(b) Solve this LP graphically and solve the original problem by employing SLP–IOR; compare the results.

2.6. In the proof of Proposition 2.36. on page 104 we have utilized the fact that $h(x) := \|D^{\mathrm{T}}x - d\|$ is a convex function. Prove that this holds for any norm $\|\cdot\|$.

2.7. Consider the following pair of chance–constrained problems:

$$\left.\begin{array}{ll} \min & 2x_1 + x_2 \\ \text{s.t.} & x_1 + x_2 \leq 8 \\ & \mathbb{P}(x_1 + x_2 \geq \xi_1) \geq 0.95 \\ & \mathbb{P}(x_1 \geq \xi_2) \geq 0.95 \\ & x_1, \; x_2 \geq 0 \end{array}\right\} (S) \qquad \left.\begin{array}{ll} \min & 2x_1 + x_2 \\ \text{s.t.} & x_1 + x_2 \leq 8 \\ & \mathbb{P}\begin{pmatrix} x_1 + x_2 \geq \xi_1 \\ x_1 \geq \xi_2 \end{pmatrix} \geq 0.95 \\ & x_1, \; x_2 \geq 0, \end{array}\right\} (J)$$

where the probability distribution of $\xi = (\xi_1, \xi_2)^{\mathrm{T}}$ is a normal distribution with parameters $\mathbb{E}[\xi] = (2,1)^{\mathrm{T}}$, standard deviations $\sigma[\xi] = (0.5, 0.5)^{\mathrm{T}}$ and correlation

$\sigma_{1,2}(\xi_1,\xi_2) = 0.2$. Problems (S) and (J) are formulated on the basis of the same data–set, (S) with separate chance constraints and (J) with a joint chance constraint. In (S) the marginal distributions of (ξ_1,ξ_2) are chosen for the probability distributions of ξ_1 and ξ_2. Notice that for both of the separate constraints the same probability level is prescribed as for the joint constraint.

(a) Solve both problems by utilizing SLP–IOR and compare the optimal objective values z_S^* and z_J^*.
(b) It will turn out that $z_S^* < z_J^*$ holds. Show that this is not just by chance: if formulating two chance–constrained problems on the same data–set in the above way, the optimal (minimal) objective value for the problem with separate constraints never exceeds the optimal objective value of the optimization problem with joint constraints.

2.3 Quantile functions, Value at Risk

One way for including simultaneously the loss size and the probability of loss into an SLP model leads via quantiles. Recall that for a random variable ϑ with distribution function F_ϑ and for $0 < \alpha < 1$, $z \in \mathbb{R}$ is an α–quantile, if both inequalities

$$\mathbb{P}(\vartheta \le z) \ge \alpha \quad \text{and} \quad \mathbb{P}(\vartheta \ge z) \ge 1-\alpha$$

hold. The set of α–quantiles is a non–empty closed interval for $0 < \alpha < 1$, see, for instance, Cramér [47]. We assume that $0 < \alpha < 1$ holds and assign the following quality measure to random variables:

$$\rho_{\mathrm{VaR}}^{\alpha}(\vartheta) := \nu(\vartheta,\alpha) := \min\{z \mid F_\vartheta(z) \ge \alpha\}, \;\; \vartheta \in \mathscr{L}_1^0, \tag{2.101}$$

defined on the set of all random variables over Ω. According to this definition, for a given α, $\nu(\vartheta,\alpha)$ is the left endpoint of the closed interval of α–quantiles of ϑ.

Similarly as in Section 2.2.3 on separate probability functions, for the sake of simplicity of notation, we consider the random variable

$$\zeta(x,\eta,\xi) := \eta^{\mathrm{T}}x - \xi. \tag{2.102}$$

We interpret positive values of $\zeta(x,\eta,\xi)$ as loss and negative values as gain. The evaluation function corresponding to the risk measure (2.101) will be the following:

$$v(x,\alpha) := \min\{z \mid \Psi(x,z) \ge \alpha\}, \tag{2.103}$$

where $\Psi(x,\cdot)$ denotes the probability distribution function of $\zeta(x,\eta,\xi)$. We will call $v(x,\alpha)$ a *quantile function*. α will typically have a large value, for instance, $\alpha = 0.95$. The interpretation of $v(x,\alpha)$ is in this case a minimal loss level, corresponding to the decision vector x, with the following property: the probability of the event that the loss will not exceed $v(x,\alpha)$ is at least α. In financial applications $v(x,\alpha)$ is called Value at Risk (VaR), see Elton et al. [85], and the references therein. We will adopt this terminology for our more general setting.

We consider minimizing the sum of a linear function and VaR, under linear constraints:

$$\left.\begin{array}{ll} \min & c^{\mathrm{T}}x + v(x,\alpha) \\ \text{s.t.} & x \in \mathscr{B}. \end{array}\right\} \tag{2.104}$$

By using the definition of $v(x,\alpha)$ and introducing an additional variable z, the following equivalent formulation results:

$$\left.\begin{array}{ll} \min & c^{\mathrm{T}}x + z \\ \text{s.t.} & \Psi(x,z) \geq \alpha \\ & x \quad\quad \in \mathscr{B}. \end{array}\right\} \tag{2.105}$$

The equivalence with (2.104) is immediate by noting that for each fixed $x \in \mathscr{B}$ in (2.105), it is sufficient to take into account the minimal z in the constraint, this minimal z is however $v(x,\alpha)$. Substituting the definition of Ψ finally leads to the formulation

$$\left.\begin{array}{ll} \min & c^{\mathrm{T}}x + z \\ \text{s.t.} & \mathbb{P}(\eta^{\mathrm{T}}x - \xi \leq z) \geq \alpha \\ & x \quad\quad\quad\quad\quad \in \mathscr{B}. \end{array}\right\} \tag{2.106}$$

This model clearly belongs to the class of SLP models with separate probability functions, see (2.46) with $s = 1$, in Section 2.2.3. The probability function in the model above is a special case of the general form with the "technology vector" containing a deterministic component

$$\hat{G}(x,z) = \mathbb{P}\left((\eta^{\mathrm{T}}, -1) \begin{pmatrix} x \\ z \end{pmatrix} - \xi \leq 0 \right).$$

It is an interesting fact, that the first SLP model for minimizing VaR has been formulated by Kataoka [178] in the form (2.106) already in 1963.

Being a special case of SLP models with separate probability functions, the whole machinery developed in Section 2.2.3 applies. We will illustrate this by discussing the case of the multivariate normal distribution. Let

$$\zeta(x,z,\eta,\xi) := \eta^{\mathrm{T}}x - \xi - z$$

and assume that (η,ξ) has a multivariate normal distribution (see page 100). For a fixed (x,z), the z–term can be interpreted as merely modifying the expected value of ξ, therefore for $\widehat{G}(x,z)$ the explicit form (2.59) on page 102 applies with μ_{n+1} replaced by $\mu_{n+1} + z$. Consequently, see (2.60) on page 102, (2.106) can be written as

$$\left.\begin{array}{ll} \min & c^{\mathrm{T}}x + z \\ \text{s.t.} & \Phi^{-1}(\alpha)\|D^{\mathrm{T}}x - d\| + \mu^{\mathrm{T}}x - z \leq \mu_{n+1} \\ & x \quad\quad\quad\quad\quad\quad\quad\quad\quad\quad\quad \in \mathscr{B}. \end{array}\right\} \tag{2.107}$$

At the optimal solution the nonlinear constraint is clearly active. This observation leads, by eliminating z, to the following linearly constrained alternative formulation:

$$\left.\begin{array}{ll} \min & c^{\mathrm{T}}x + \Phi^{-1}(\alpha)\|D^{\mathrm{T}}x - d\| + \mu^{\mathrm{T}}x - \mu_{n+1} \\ \text{s.t.} & x \qquad\qquad\qquad\qquad\qquad\qquad \in \mathcal{B}. \end{array}\right\} \tag{2.108}$$

Assuming that $\alpha \geq \frac{1}{2}$ holds, due to the convexity of the Euclidean norm both models (2.107) and (2.108) are convex programming problems.

Except of those cases, discussed in Section 2.2.3, which can be formulated as convex programming problems, the model (2.106) is in general a non–convex optimization problem.

Turning now our attention to SLP problems with VaR–constraints, we consider problems of the following form:

$$\left.\begin{array}{ll} \min\limits_{x} & c^{\mathrm{T}}x \\ \text{s.t.} & \min\limits_{z}\{z \mid \Psi(x,z) \geq \alpha\} \leq \kappa \\ & \qquad\qquad x \qquad\qquad \in \mathcal{B}. \end{array}\right\} \tag{2.109}$$

Observe that the minimum in the minimization problem involved in the first constraint is attained. Therefore, for a fixed x this constraint holds, if and only if there exists a $z \in \mathbb{R}$ such that it holds for that z. Thus the optimization problem (2.109) can be equivalently formulated as follows:

$$\left.\begin{array}{ll} \min\limits_{x,z} & c^{\mathrm{T}}x \\ \text{s.t.} & \Psi(x,z) \geq \alpha \\ & \qquad\quad z \leq \kappa \\ & \quad x \qquad \in \mathcal{B}. \end{array}\right\} \tag{2.110}$$

Finally, substituting the definition of Ψ results in

$$\left.\begin{array}{ll} \min\limits_{x,z} & c^{\mathrm{T}}x \\ \text{s.t.} & \mathbb{P}(\eta^{\mathrm{T}}x - z \leq \xi) \geq \alpha \\ & \qquad\quad z \qquad\quad \leq \kappa \\ & \quad x \qquad\qquad\quad \in \mathcal{B}. \end{array}\right\} \tag{2.111}$$

Thus, also in this case, we have obtained an equivalent problem which belongs to the class of SLP problems with separate probability functions, see Section 2.2.3. Therefore, analogous comments and formulations apply, as for the SLP problem in which VaR is minimized, see (2.106).

For further stochastic programming problems based on quantile functions see Kibzun and Kan [182].

2.4 Models based on expectation

The simplest way of including expectations into an SLP model is based on choosing the quality measure

$$\rho_{\mathrm{E}}(\vartheta) := \mathbb{E}[\vartheta], \;\; \vartheta \in \mathscr{L}_1^1,$$

defined on the linear space of random variables with finite expected value. We consider the random variable

$$\zeta(x,\xi) = t^{\mathrm{T}}(\xi)x - h(\xi),$$

where $t(\xi)$ is an n–dimensional random vector and $h(\xi)$ is a random variable. Under the assumption that the expected value of $(T(\xi), h(\xi))$ exists, we obtain the following deterministic linear–affine evaluation function for x:

$$\mathbb{E}[\zeta(x,\xi)] = \bar{t}^{\mathrm{T}}x - \bar{h}$$

with $\bar{t} = \mathbb{E}[t(\xi)]$ and $\bar{h} = \mathbb{E}[h(\xi)]$. In the case when $\zeta(x,\xi)$ is a random vector, this holds componentwise. Consequently, the prototype models (2.7) and (2.9) become linear programming problems. These LP's are called *expected value problems*, corresponding to the SLP problem.

On the one hand, having an equivalent linear programming problem is an attractive feature from the numerical point of view. On the other hand, however, replacing the probability distribution by a one–point distribution leads to a very crude approximation of the original distribution in general. In some modeling situations it may happen that the solution $\bar{x}$ of the expected value problem also solves a corresponding SLP problem. However, this is usually an indication of a modeling or data error: the corresponding SLP model is not "truly stochastic". Unfortunately, the expected value problem is frequently used by modelers as a substitute for the SLP problem, without further considerations. When doing this, extreme care is needed, since the solution obtained this way may turn out to be quite risky when evaluated by an alternative evaluation function. Taking the expected value problem should by no means be used as the single way representing $\zeta(x,\xi)$ in the model. Accompanied with other constraints or objective functions, based on alternative quality measures, utilizing $\rho_{\mathrm{E}}(\vartheta)$ may lead to important and meaningful model formulations. As an example we refer to the portfolio optimization model of Markowitz [217] which has been applied with tremendous success in finance.

The picture radically changes if the expectation is taken separately for the positive– or negative part of $\zeta(x,\xi)$, or if conditional expectations are utilized. In this section we will discuss several important model classes based on these ideas.

We shall need some basic facts from probability theory concerning expectations. Let ϑ be a random variable and assume that $\mathbb{E}[\,\vartheta\,]$ exists. Recall from probability theory, that this assumption means the finiteness of the integral $\int_{-\infty}^{\infty} |t|\,dF_{\vartheta}(t)$, where F_{ϑ} denotes the probability distribution of ϑ.

The following well–known integral representations will be used in this section, for which, for the sake of completeness, we also present a proof. Introducing the notation $u^+ := \max\{0,u\}$ and $u^- := \max\{0,-u\}$ for all $u \in \mathbb{R}$, we have

Proposition 2.38. *Assume that $\mathbb{E}[\vartheta]$ exists. Then for all $z \in \mathbb{R}$ both $\mathbb{E}[(\vartheta - z)^+]$ and $\mathbb{E}[(\vartheta - z)^-]$ exist and we have:*

$$\begin{aligned} \mathbb{E}[(\vartheta - z)^+] &= \int_z^\infty (1 - F_\vartheta(t))dt \\ \mathbb{E}[(\vartheta - z)^-] &= \int_{-\infty}^z F_\vartheta(t)dt. \end{aligned} \tag{2.112}$$

Proof: The existence of $\mathbb{E}[\vartheta]$ obviously implies the existence of the expected values on the left–hand–side in (2.112). Using integration by parts we get for $z < y$

$$\begin{aligned} \int_z^y (t-z)dF_\vartheta(t) &= (t-z)F_\vartheta(t)]_z^y - \int_z^y F_\vartheta(t)dt \\ &= -(y-z)(1-F_\vartheta(y)) + \int_z^y (1-F_\vartheta(t))dt \end{aligned}$$

and consequently

$$\begin{aligned} \mathbb{E}[(\vartheta - z)^+] &= \int_z^\infty (t-z)dF_\vartheta(t) \\ &= \lim_{y\to\infty} \int_z^y (t-z)dF_\vartheta(t) = \int_z^\infty (1-F_\vartheta(t))dt, \end{aligned}$$

where we have used the fact that the existence of the expected value of ϑ implies that $\lim_{y\to\infty} y(1-F_\vartheta(y)) = 0$ holds. For the second relation we get similarly via integration by parts:

$$\begin{aligned} \mathbb{E}[(\vartheta - z)^-] &= \int_{-\infty}^z (z-t)dF_\vartheta(t) \\ &= (z-t)F_\vartheta(t)]_{-\infty}^z + \int_{-\infty}^z F_\vartheta(t)dt = \int_{-\infty}^z F_\vartheta(t)dt, \end{aligned}$$

where we used that $\lim_{x\to-\infty} xF_\vartheta(x) = 0$ holds, due to the existence of the expected value of ϑ. □

2.4.1 Integrated chance constraints

Similarly as in Section 2.2 concerning probability functions, also in this section we will distinguish two cases: first we discuss the case when $\zeta(x,\xi) := T(\xi)x - h(\xi)$ is a random variable ($s = 1$ holds, see (2.1) on page 71). Afterwards we consider the general case when $\zeta(x,\xi) := T(\xi)x - h(\xi)$ is a random vector, that means, $s \geq 1$ holds. We will assume throughout that the expected values of $T(\xi)$ and $h(\xi)$ exist.

Separate integrated probability functions

We consider the random variable

$$\zeta(x,\xi) := t(\xi)^{\mathrm{T}}x - h(\xi),$$

where $t(\xi)$ is an n–dimensional random vector and $h(\xi)$ is a random variable. Depending on whether positive or negative values of $\zeta(x,\xi)$ are considered as losses, the loss as a random variable can be written as

$$\zeta^{+}(x,\xi) := [t(\xi)^{\mathrm{T}}x - h(\xi)]^{+}$$

or

$$\zeta^{-}(x,\xi) := [t(\xi)^{\mathrm{T}}x - h(\xi)]^{-},$$

respectively. Here we have made use of the notation $z^{+} = \max\{0,z\}$ and $z^{-} = \max\{0,-z\}$, $z \in \mathbb{R}$. z^{+} will be called the positive part and z^{-} the negative part of the real number z.

For being in accordance with the literature, let us assume that losses are modeled as negative values of $\zeta(x,\xi)$. Using the notation above, the probability constraint corresponding to the random linear inequality $\zeta(x,\xi) \geq 0$ can obviously be written in expectation terms (see (2.33) on page 90) as follows

$$\mathbb{P}_{\xi}(\zeta(x,\xi) \geq 0) \geq \alpha \iff \mathbb{E}_{\xi}[\chi(\zeta^{-}(x,\xi))] \leq 1 - \alpha \tag{2.113}$$

with the indicator function

$$\chi(z) := \begin{cases} 0 \text{ if } z \leq 0, \\ 1 \text{ if } z > 0. \end{cases}$$

In the second inequality in (2.113) the application of the function χ results in assigning the constant value 1 to the loss irrespectively of its size. This can heuristically be viewed as the source of the generally non–convex behavior of probability functions, see the nice examples in Klein Haneveld [188] and Klein Haneveld and Van der Vlerk [191]. This observation leads to integrated chance constraints by dropping χ in (2.113) and by prescribing an upper bound for $\mathbb{E}_{\xi}[\zeta^{-}(x,\xi)]$. More specifically, we choose two risk measures for random variables as

$$\begin{aligned} \rho^+_{\text{sic}}(\vartheta) &:= \mathbb{E}[\vartheta^+], \\ \rho^-_{\text{sic}}(\vartheta) &:= \mathbb{E}[\vartheta^-], \ \vartheta \in \mathscr{L}^1_1, \end{aligned} \tag{2.114}$$

defined on the linear space of random variables with finite expected value. The corresponding evaluation functions K and H will be

$$\begin{aligned} K(x) &:= \rho^+_{\text{sic}}(\zeta(x,\xi)) = \mathbb{E}_\xi[\zeta^+(x,\xi)] \quad \text{and} \\ H(x) &:= \rho^-_{\text{sic}}(\zeta(x,\xi)) = \mathbb{E}_\xi[\zeta^-(x,\xi)], \end{aligned}$$

respectively. The functions $K(x)$ and $H(x)$ will be called *separate integrated probability functions*. Assuming, for instance, that losses correspond to negative values of $\zeta(x,\xi)$, a *separate integrated chance constraint* has the form

$$\mathbb{E}_\xi[(\zeta(x,\xi))^-] \le \gamma, \tag{2.115}$$

where γ is a prescribed maximal tolerable expected loss. The following relation provides an explanation of the term "integrated": due to Proposition 2.38. on page 141 we have

$$\mathbb{E}_\xi[(\zeta(x,\xi))^-] = \int_{-\infty}^{0} \mathbb{P}(\zeta(x,\xi) \le z)dz = \int_{-\infty}^{0} \mathbb{P}(\zeta(x,\xi) < z)dz.$$

The second equality holds because the set of jump-points of the distribution function $\Psi(x,\cdot) := \mathbb{P}(\zeta(x,\xi) \le z)$ is countable and therefore it has (Lebesgue) measure 0.

Let us define the positive– and negative–part functions φ^+ and φ^- according to $\varphi^+(z) := z^+$ and $\varphi^-(z) := z^-$ for $z \in \mathbb{R}$, respectively. Both of these functions are obviously convex. From the optimization point of view the most attractive property of separate integrated probability functions is formulated in the subsequent proposition:

Proposition 2.39. *Both $H(x)$ and $K(x)$, and consequently $\mathbb{E}_\xi[|\zeta(x,\xi)|] = H(x) + K(x)$ are convex functions on $\mathbb{R}^n$.*

Proof: The assertion follows easily from the convexity of the functions $\varphi^+(\cdot)$ and $\varphi^-(\cdot)$. We prove the assertion for $K(x)$; the proof for $H(x)$ is analogous. We have $K(x) = \mathbb{E}_\xi[\varphi^+(\zeta(x,\xi))]$. Because $\zeta(x,\xi)$ is linear in x and φ^+ is a convex function, $\varphi^+(\zeta(x,\xi))$ is convex for each fixed ξ. Taking the expected value preserves convexity. For a formal proof let $x,y \in \mathbb{R}^n$ and $0 \le \lambda \le 1$. We have

$$\begin{aligned} K(\lambda x + (1-\lambda)y) &= \mathbb{E}_\xi[\varphi^+(\zeta(\lambda x + (1-\lambda)y, \xi))] \\ &= \mathbb{E}_\xi[\varphi^+(\lambda \zeta(x,\xi) + (1-\lambda)\zeta(y,\xi))] \\ &\le \mathbb{E}_\xi[\lambda \varphi^+(\zeta(x,\xi)) + (1-\lambda)\varphi^+(\zeta(y,\xi))] \\ &= \lambda K(x) + (1-\lambda)K(y). \end{aligned}$$

□

This result implies that $K(x)$ and $H(x)$ are convex, in particular also for finite discrete distributions. This is in sharp contrast with probability functions, where (generalized) concavity holds only under various assumptions, excluding finite discrete distributions in general.

With $\zeta^+(x,\xi)$ representing losses, the following prototype models will be considered:

$$\left.\begin{array}{ll} \min c^{\mathrm{T}}x & \\ \text{s.t. } \mathbb{E}_\xi[\zeta^+(x,\xi)] \le \gamma & \\ \qquad\qquad x \in \mathscr{B} & \end{array}\right\} \tag{2.116}$$

and

$$\left.\begin{array}{l} \min c^{\mathrm{T}}x + \mathbb{E}_\xi[\zeta^+(x,\xi)] \\ \text{s.t.} \quad x \in \mathscr{B}, \end{array}\right\} \tag{2.117}$$

where $\gamma > 0$ is a prescribed maximally tolerable loss level. Due to Proposition 2.39., both problems are convex programming problems. Convex functions being continuous (see, for instance, Rockafellar [281]), the feasible set of (2.116) is obviously closed.

Note that there is no way of building convex programming models of the above type with reversed inequality constraints in (2.116) or with maximization in (2.117) which are based on separate integrated probability functions. Because both $K(x)$ and $H(x)$ are convex, it is immaterial whether the loss is represented by $\zeta^+(x,\xi)$ or by $\zeta^-(x,\xi)$.

Next we assume that ξ has a finite discrete distribution with N realizations and corresponding probabilities given in the tableau

$$\begin{pmatrix} p_1 & \dots & p_N \\ \widehat{\xi}^1 & \dots & \widehat{\xi}^N \end{pmatrix} \tag{2.118}$$

with $p_i > 0\ \forall i$ and $\sum_{i=1}^{N} p_i = 1$. We introduce the notation $T^k = T(\widehat{\xi}^k)$, $h^k = h(\widehat{\xi}^k)$, $k = 1,\dots,N$, and $\mathscr{N} = \{1,\dots,N\}$. The ith row of T^k will be denoted by t_i^k and if $s = 1$ then the single row of T^k will be denoted by t^k. For notational convenience, both t_i^k and t^k will be considered as row–vectors ($(1 \times n)$ matrices).

Problems (2.116) and (2.117) can in this case be formulated as follows:

$$\left.\begin{array}{l} \min c^{\mathrm{T}}x \\ \text{s.t. } \displaystyle\sum_{k=1}^{N} p_k(t^k x - h^k)^+ \le \gamma \\ \qquad\qquad x \in \mathscr{B} \end{array}\right\} \tag{2.119}$$

and

$$\left.\begin{array}{ll}\min\ c^{\mathrm{T}}x+\sum\limits_{k=1}^{N}p_k(t^kx-h^k)^+ \\ \text{s.t.} \quad x\in\mathscr{B}.\end{array}\right\} \tag{2.120}$$

These nonlinear programming problems can be equivalently formulated as linear programming problems by introducing the auxiliary variables y^k, $k=1,\ldots,N$:

$$\left.\begin{array}{lrl}\min\ c^{\mathrm{T}}x & & \\ \text{s.t.}\ \sum\limits_{k=1}^{N}p_ky^k\le\gamma & & \\ t^kx & -y^k & \le h^k,\ k=1,\ldots,N \\ & y^k & \ge 0,\ k=1,\ldots,N \\ x & & \in\mathscr{B}\end{array}\right\} \tag{2.121}$$

and

$$\left.\begin{array}{lrl}\min\ c^{\mathrm{T}}x+\sum\limits_{k=1}^{N}p_ky^k & & \\ \text{s.t.}\ t^kx & -y^k & \le h^k,\ k=1,\ldots,N \\ & y^k & \ge 0,\ k=1,\ldots,N \\ x & & \in\mathscr{B}.\end{array}\right\} \tag{2.122}$$

The equivalence of (2.119) and (2.121) as well as the equivalence of (2.120) and (2.122) follows easily from the following fact: if $\bar{x},\bar{y}^k,k=1,\ldots,N$ is a feasible solution of either (2.121) or (2.122), then the following inequality holds:

$$\sum_{k=1}^{N}p_k(t^k\bar{x}-h^k)^+\le\sum_{k=1}^{N}p_k\bar{y}^k.$$

Let $\mathscr{S}(\gamma)=\{x\mid\mathbb{E}_\xi[\zeta^+(x,\xi)]\le\gamma\}$ be the set of feasible solutions corresponding to the integrated chance constraint. The following representation holds, which plays an important role in the dual decomposition algorithm (see Section 4.4.3).

Theorem 2.13. *Klein Haneveld and Van der Vlerk [191]. For $\gamma\ge 0$, $\mathscr{S}(\gamma)$ is a polyhedral set. In fact the following representation holds:*

$$\{x\mid K(x)\le\gamma\}=\bigcap_{\mathscr{K}\subset\mathscr{N}}\{x\mid\sum_{k\in\mathscr{K}}p_k(t^kx-h^k)\le\gamma\}. \tag{2.123}$$

Proof: We have

$$K(x) = \mathbb{E}_\xi[\zeta^+(x,\xi)] = \sum_{k=1}^{N} p_k \zeta^+(x,\widehat{\xi}^k)$$
$$= \sum_{k:\,\zeta(x,\widehat{\xi}^k)>0} p_k \zeta(x,\widehat{\xi}^k)$$
$$= \max_{\mathscr{K}\subset\mathscr{N}} \sum_{k\in\mathscr{K}} p_k \zeta(x,\widehat{\xi}^k).$$

Using this representation we get

$$\mathscr{S}(\gamma) = \{x \mid K(x) \le \gamma\} = \bigcap_{\mathscr{K}\subset\mathscr{N}} \{x \mid \sum_{k\in\mathscr{K}} p_k \zeta(x,\widehat{\xi}^k) \le \gamma\}$$

from which the result immediately follows. □

In (2.123), for $\mathscr{K} = \emptyset \subset \mathscr{N}$ the sum over the empty index set is interpreted as having the value 0 thus the corresponding inequality holds for any x. Consequently, $\mathscr{S}(\gamma)$ is represented by a system of $2^N - 1$ proper linear inequalities. Models (2.116) and (2.117) deliver identical solutions for random variables for which $\zeta^+(x,\xi)$ is the same almost surely. This is not the case with the following variant of integrated chance constraints:

$$\mathbb{E}_\xi[\zeta^+(x,\xi)] \le \alpha \mathbb{E}_\xi[|\zeta(x,\xi)|] \tag{2.124}$$

with α being prescribed. Because $K(x) \ge 0$ and $K(x) \le \mathbb{E}_\xi[|\zeta(x,\xi)|]$ obviously hold for all x, it is sufficient to consider α–values with $\alpha \in [0,1]$. Using the relations $z = z^+ - z^-$ and $|z| = z^+ + z^-$, $z \in \mathbb{R}$, the above inequality can be equivalently written as

$$(1-2\alpha)\mathbb{E}_\xi[\zeta^+(x,\xi)] + \alpha\mathbb{E}_\xi[\zeta(x,\xi)] \le 0 \tag{2.125}$$

or as

$$(1-2\alpha)\mathbb{E}_\xi[\zeta^+(x,\xi)] + \alpha(\bar{t}x - \bar{h}) \le 0 \tag{2.126}$$

with $\bar{t} = \mathbb{E}_\xi[t(\xi)]$ and $\bar{h} = \mathbb{E}_\xi[h(\xi)]$.

This motivates the choice of the following quality measure for evaluating random variables in our framework for constructing SLP models:

$$\rho^\alpha_{\text{sic}}(\vartheta) := (1-\alpha)\mathbb{E}[\vartheta^+] - \alpha\mathbb{E}[\vartheta^-] = \alpha\mathbb{E}[\vartheta] + (1-2\alpha)\mathbb{E}[\vartheta^+],\ \ \vartheta \in \mathscr{L}^1_1.$$

We obtain the evaluation function as usual by substituting $\vartheta = \zeta(x,\xi)$:

$$K_\alpha(x) := \alpha(\bar{t}x - \bar{h}) + (1-2\alpha)\mathbb{E}_\xi[\zeta^+(x,\xi)].$$

Proposition 2.39. implies that $K_\alpha(x)$ is convex for $\alpha < \frac{1}{2}$ and it is concave for $\alpha > \frac{1}{2}$. For $\alpha = \frac{1}{2}$ the function is clearly linear–affine.

Choosing α such that $\alpha \in [0, \frac{1}{2}]$ holds, the parameter α will be interpreted as a risk–aversion parameter. Decreasing α means increasing risk–aversion. The prototype models will have the form

$$\left.\begin{array}{ll}\min c^{\mathrm{T}}x & \\ \text{s.t. } \alpha(\bar{t}x-\bar{h})+(1-2\alpha)\mathbb{E}_{\xi}[\zeta^{+}(x,\xi)] & \leq 0 \\ \qquad x & \in \mathscr{B}\end{array}\right\} \tag{2.127}$$

and

$$\left.\begin{array}{l}\min c^{\mathrm{T}}x+\alpha(\bar{t}x-\bar{h})+(1-2\alpha)\mathbb{E}_{\xi}[\zeta^{+}(x,\xi)] \\ \text{s.t.} \quad x \in \mathscr{B}\end{array}\right\} \tag{2.128}$$

with $\alpha \in [0, \frac{1}{2}]$ prescribed. Both problems are clearly convex programming problems.

Interpreting $\zeta^{+}(x,\xi)$ as gain (and, consequently, $\zeta^{-}(x,\xi)$ as loss), we choose the parameter α such that $\alpha \in [\frac{1}{2}, 1]$ holds. By utilizing K_α with $\alpha \in [\frac{1}{2}, 1]$, the corresponding optimization problems are analogous to the two models above, with reversed inequality constraint in (2.127) and with changing "min" to "max" in (2.128).

If the probability distribution of ξ is finite discrete, problems (2.127) and (2.128) can be equivalently formulated as linear programming problems. This can be done analogously as above for (2.116) and (2.117), we leave the details as an exercise for the reader.

For the case of a finite discrete distribution the feasible set is polyhedral; an analogous representation holds as in Theorem 2.13.. We formulate it for the case $\alpha \in [0, \frac{1}{2}]$, the variant with $\alpha \in [\frac{1}{2}, 1]$ can be obtained from this in a straightforward way. Let $\widehat{\mathscr{S}}(\alpha) = \{x \mid K_\alpha(x) \leq 0\}$ be the set of feasible solutions corresponding to this type of integrated chance constraint.

Theorem 2.14. *Klein Haneveld and Van der Vlerk [191]. For $\gamma \geq 0$, $\mathscr{S}(\gamma)$ is a polyhedral set and the following representation holds:*

$$\{x \mid K_\alpha(x) \leq 0\} = \bigcap_{\mathscr{K} \subset \mathscr{N}} \{x \mid (1-2\alpha) \sum_{k \in \mathscr{K}} p_k(t^k x - h^k) + \alpha(\bar{t}x - \bar{h}) \leq 0\}.$$

Proof: The proof runs along the same lines as the proof for Theorem 2.13.. □

Joint integrated probability functions

Let $s > 1$ and $\zeta(x,\xi) = T(\xi)x - h(\xi)$. Analogously as before, define

$$\zeta^{+}(x,\xi) := [T(\xi)x - h(\xi)]^{+}$$

and

$$\zeta^{-}(x,\xi) := [T(\xi)x - h(\xi)]^{-},$$

where on the right–hand–side the positive– and negative parts of the vectors are defined in a componentwise fashion. For a heuristic introduction of joint integrated chance constraints we proceed analogously as in the case $s = 1$. We assume that losses are represented by negative values of $\zeta(x,\xi)$. The probability constraint in expected value terms looks as follows (see (2.33) on page 90):

$$\mathbb{P}_\xi(\zeta(x,\xi)\geq 0)\geq\alpha \iff \mathbb{E}_\xi[\chi(\max_{1\leq i\leq s}\zeta_i^-(x,\xi))]\leq 1-\alpha.$$

Analogously to the special case $s=1$, dropping χ results in the *joint integrated chance constraint* (cf. (2.115)):

$$\mathbb{E}_\xi[\max_{1\leq i\leq s}\zeta_i^-(x,\xi)]\leq\gamma$$

with prescribed maximal tolerable loss γ. We proceed by defining the quality measures for random variables by

$$\begin{aligned}\rho^+_{\mathrm{jic}}(\vartheta) &:= \mathbb{E}[\max_{1\leq i\leq s}\vartheta_i^+] \quad\text{and}\\ \rho^-_{\mathrm{jic}}(\vartheta) &:= \mathbb{E}[\max_{1\leq i\leq s}\vartheta_i^-],\ \vartheta\in\mathcal{L}_s^1.\end{aligned}$$

This results in the evaluation functions

$$\begin{aligned}K_J(x) &:= \rho^+_{\mathrm{jic}}(\zeta(x,\xi)) = \mathbb{E}_\xi[\max_{1\leq i\leq s}\zeta_i^+(x,\xi)] \quad\text{and}\\ H_J(x) &:= \rho^-_{\mathrm{jic}}(\zeta(x,\xi)) = \mathbb{E}_\xi[\max_{1\leq i\leq s}\zeta_i^-(x,\xi)],\end{aligned}$$

respectively. The functions $K_J(x)$ and $H_J(x)$ will be called *joint integrated probability functions*.

The attractive property of convexity remains preserved by the generalization:

Proposition 2.40. *Both $H_J(x)$ and $K_J(x)$ are convex functions on $\mathbb{R}^n$.*

Proof: The proof is similar to the proof for Proposition 2.39.. For any fixed ξ, $\zeta_i(\cdot,\xi)^+$ is a convex function for the same reasons as in the case $s=1$, see the proof of Proposition 2.39.. $\max\limits_{1\leq i\leq s}\zeta_i^+(\cdot,\xi)$ is the point–wise maximum of convex functions, consequently this function is also convex for each fixed ξ (see, for instance, Rockafellar [281]). Taking the expected value w.r. to ξ preserves convexity, see the proof of Proposition 2.39., therefore K_J is a convex function. The proof for H_J is analogous. □

Let us emphasize that the convexity property holds also for a random technology matrix and without any restriction on the probability distribution of ξ, beyond the existence of the expected value.

The prototype SLP problems are formulated as follows:

$$\left.\begin{array}{ll}\min & c^{\mathrm{T}}x\\ \text{s.t.} & \mathbb{E}_\xi[\max\limits_{1\leq i\leq s}\zeta_i^+(x,\xi)]\leq\gamma\\ & x\in\mathcal{B}\end{array}\right\}\qquad(2.129)$$

and

$$\left.\begin{array}{ll}\min & c^{\mathrm{T}}x + \mathbb{E}_{\xi}[\max\limits_{1\le i\le s} \zeta_i^{+}(x,\xi)] \\ \text{s.t.} & x \in \mathscr{B}\end{array}\right\} \tag{2.130}$$

with the prescribed maximal loss level $\gamma \geq 0$. Both of them are obviously convex programming problems and due to the convexity of K_J, the feasible set is closed also for (2.129).

Reversing the inequality in the integrated chance constraint in (2.129) and changing min to max in (2.130) leads in general in both cases to non–convex optimization problems.

There is no change in the behavior of the optimization problems if we utilize H_J instead of K_J in the problem formulations.

Assume next that ξ has a finite discrete distribution, specified in (2.118). Then (2.129) and (2.130) take the form

$$\left.\begin{array}{lll}\min & c^{\mathrm{T}}x & \\ \text{s.t.} & \sum\limits_{k=1}^{N} p_k \max\limits_{1\le i\le s}(t_i^k x - h^k)^{+} & \le \gamma \\ & x & \in \mathscr{B}\end{array}\right\} \tag{2.131}$$

and

$$\left.\begin{array}{ll}\min & c^{\mathrm{T}}x + \sum\limits_{k=1}^{N} p_k \max\limits_{1\le i\le s}(t_i^k x - h^k)^{+} \\ \text{s.t.} & x \in \mathscr{B}.\end{array}\right\} \tag{2.132}$$

We introduce auxiliary variables y_i^k and z^k, $i = 1,\dots,s$, $k = 1,\dots,N$ and formulate equivalent linear programming problems to (2.131) and (2.132) as follows (see Klein Haneveld and Van der Vlerk [191]):

$$\left.\begin{array}{lll}\min & c^{\mathrm{T}}x & \\ \text{s.t.} & \sum\limits_{k=1}^{N} p_k z^k & \le \gamma \\ & t_i^k x - y_i^k & \le h_i^k,\ k = 1,\dots,N,\ i = 1,\dots,s \\ & -y_i^k + z^k & \ge 0,\ k = 1,\dots,N,\ i = 1,\dots,s \\ & y_i^k & \ge 0,\ k = 1,\dots,N,\ i = 1,\dots,s \\ & z^k & \ge 0,\ k = 1,\dots,N \\ & x & \in \mathscr{B}\end{array}\right\} \tag{2.133}$$

and

$$\left.\begin{array}{lll} \min\ c^{\mathrm{T}}x + \sum_{k=1}^{N} p_k z^k & & \\ \text{s.t. } t_i^k x - y_i^k & \le h_i^k, & k=1,\dots,N,\ i=1,\dots,s \\ \qquad -y_i^k + z^k & \ge 0, & k=1,\dots,N,\ i=1,\dots,s \\ \qquad y_i^k & \ge 0, & k=1,\dots,N,\ i=1,\dots,s \\ \qquad z^k & \ge 0, & k=1,\dots,N \\ x & \in \mathscr{B}. & \end{array}\right\} \quad (2.134)$$

The equivalence can readily be proved, based on the following fact: if $\bar{x}$, $\bar{y}_i^k$, $\bar{z}^k$, $k=1,\dots,N$, $i=1,\dots,s$ is a feasible solution of either (2.133) or (2.134) then the following inequality holds:

$$\sum_{k=1}^{N} p_k \max_{1\le i\le s}(t_i^k \bar{x} - h^k)^+ \le \sum_{k=1}^{N} p_k \bar{z}^k.$$

Let $\mathscr{S}_J(\gamma) = \{x \mid \mathbb{E}_\xi[\max_{1\le i\le s} \zeta_i^+(x,\xi)] \le \gamma\}$. For ξ having a finite discrete distribution, Theorem 2.13. has the following generalization:

Theorem 2.15. *Klein Haneveld and Van der Vlerk [191]. For $\gamma \ge 0$, $\mathscr{S}_J(\gamma)$ is a polyhedral set and the following representation holds:*

$$\{x \mid K_J(x) \le \gamma\} = \bigcap_{\mathscr{K}\subset\mathscr{N}} \bigcap_{l\in\mathscr{I}^{\mathscr{K}}} \{x \mid \sum_{k\in\mathscr{K}} p_k(t_{l_k}^k x - h_{l_k}^k) \le \gamma\}, \qquad (2.135)$$

where $\mathscr{I} = \{1,\dots,s\}$, $\mathscr{I}^{\mathscr{K}} := \{l := (l_k, k\in\mathscr{K}) \mid l_k \in \mathscr{I}$ for all $k\in\mathscr{K}\}$, and $t_{l_k}^k$ is the l_k^{th} row of T^k.

Proof: We have

$$\begin{aligned} K_J(x) = \mathbb{E}_\xi[\max_{1\le i\le s} \zeta_i^+(x,\xi)] &= \sum_{k=1}^{N} p_k \max_{1\le i\le s} \zeta_i^+(x,\widehat{\xi}^k) \\ &= \sum_{k\in\mathscr{N}^+} p_k \max_{1\le i\le s} \zeta_i(x,\widehat{\xi}^k) \end{aligned}$$

where $\mathscr{N}^+ := \{k\in\mathscr{N} \mid \max_{1\le i\le s} \zeta_i(x,\widehat{\xi}^k) > 0\}$. Thus we get for the constraint:

$$\begin{aligned} K_J(x) \le \gamma &\Longleftrightarrow \sum_{k\in\mathscr{N}^+} p_k \max_{1\le i\le s} \zeta_i(x,\widehat{\xi}^k) \le \gamma \\ &\Longleftrightarrow \max_{\mathscr{K}\subset\mathscr{N}} \sum_{k\in\mathscr{K}} p_k \max_{1\le i\le s} \zeta_i(x,\widehat{\xi}^k) \le \gamma \\ &\Longleftrightarrow \max_{\mathscr{K}\subset\mathscr{N}} \max_{l\in\mathscr{I}^{\mathscr{K}}} \sum_{k\in\mathscr{K}} p_k \zeta_{l_k}(x,\widehat{\xi}^k) \le \gamma. \end{aligned}$$

Substituting the definition of $\zeta_i(x,\widehat{\xi}^k)$ and noting that the number of linear inequalities which determine $\mathscr{S}_J(\gamma)$ is obviously finite yields the result. □

For counting the inequalities in (2.135) let us observe first that the number of inequalities for a fixed index set $\mathscr{K}$ (the cardinality of $\mathscr{I}^{|\mathscr{K}|}$) is $s^{|\mathscr{K}|}$. Adding up for all subsets of $\mathscr{N}$ (except of $\emptyset$) results in

$$\sum_{k=1}^{N} \binom{N}{k} s^k = (s+1)^N - 1.$$

The models in this section are due to Klein Haneveld [188] and have been subsequently investigated by Klein Haneveld and Van der Vlerk [191]. For further properties of integrated chance constraints see these references.

2.4.2 A model involving conditional expectation

We consider negative values of the random variable $\zeta(x,\xi)$ as losses and will discuss constraints which are based on the conditional expectation of the loss given that a loss occurs. This corresponds to the quality measure for random variables

$$\rho_{\text{cexp}}(\vartheta) := \mathbb{E}[-\vartheta \mid \vartheta < 0], \;\; \vartheta \in \mathscr{L}_1^1, \tag{2.136}$$

if $\mathbb{P}(\vartheta < 0) > 0$ and $\rho_{\text{cexp}}(\vartheta) := 0$ otherwise. Assuming that ϑ has a continuous distribution, we have the following close relation between ρ_{cexp} and the quality measure ρ_{sic}^- which lead to integrated chance constraints (see (2.114) on page 143): $\rho_{\text{sic}}^-(\vartheta) = \rho_{\text{cexp}}(\vartheta)\mathbb{P}(\vartheta < 0)$. This follows immediately from

$$\mathbb{E}[\vartheta^-] = -\int_{-\infty}^{0} t dF_\vartheta(t) = \mathbb{E}[-\vartheta \mid \vartheta < 0] \cdot \mathbb{P}(\vartheta < 0).$$

Constraints of the form

$$\mathbb{E}_\xi[-\zeta(x,\xi) \mid \zeta(x,\xi) < 0] \le \gamma \tag{2.137}$$

will be considered, with γ being a prescribed upper bound for the conditional loss size. In general, constraints of this type result in non–convex optimization problems. In the special case when only the right–hand–side is stochastic, the feasible set corresponding to the constraint (2.137) is convex for a broad class of univariate distributions. We choose

$$\zeta(x,\xi) = t^{\mathrm{T}}x - \xi,$$

where $t \in \mathbb{R}^n$ is a deterministic vector and ξ is a random variable.

The following result will be utilized:

Proposition 2.41. *Assume that ξ has a continuous distribution with a logconcave density function. Assume furthermore that the expected value of ξ exists. Then*

$$l(t) := \mathbb{E}[\xi - t \mid \xi - t > 0]$$

is a monotonically decreasing function of t.

Proof: This is a well–known fact in reliability theory where $l(t)$ is called mean residual life. For a proof see, for instance, Prékopa [266]. □

We assume that for ξ the assumptions of the theorem hold. Then (2.137) takes the form

$$l(t^{\mathrm{T}}x) \leq \gamma \iff t^{\mathrm{T}}x \geq l^{-1}(\gamma),$$

where l^{-1} is to be understood as a generalized inverse defined as follows $l^{-1}(z) := \inf\{z \mid l(z) \leq \gamma\}$. Consequently, the constraint (2.137) can be reformulated as a deterministic linear constraint.

2.4.3 Conditional Value at Risk

We assume that positive values of $\zeta(x,\xi)$ represent losses. For motivating the quality (risk) measure which will be introduced, let us start with computing a conditional expected value. Let ϑ be a random variable with finite expected value, F_ϑ its distribution function, $0 < \alpha < 1$, and ν_α an α–quantile of the distribution of ϑ (see Section 2.3 for the definition of quantiles). Note that due to $\alpha < 1$, $\mathbb{P}(\vartheta < \nu_\alpha) < 1$ holds and consequently we have $\mathbb{P}(\vartheta \geq \nu_\alpha) > 0$. Introducing the notation $\pi_\alpha = \mathbb{P}(\vartheta \geq \nu_\alpha)$ we get

$$\begin{aligned}
\mathbb{E}[\vartheta \mid \vartheta \geq \nu_\alpha] &= \frac{1}{\mathbb{P}(\vartheta \geq \nu_\alpha)} \int_{\nu_\alpha}^{\infty} t\, dF_\vartheta(t) \\
&= \frac{1}{\pi_\alpha}\left[\int_{\nu_\alpha}^{\infty} t\, dF_\vartheta(t) - \nu_\alpha \int_{\nu_\alpha}^{\infty} dF_\vartheta(t) + \nu_\alpha \pi_\alpha\right] \\
&= \frac{1}{\pi_\alpha}\left[\int_{\nu_\alpha}^{\infty} (t-\nu_\alpha)\, dF_\vartheta(t) + \nu_\alpha \pi_\alpha\right] \\
&= \nu_\alpha + \frac{1}{\pi_\alpha}\int_{-\infty}^{\infty} (t-\nu_\alpha)^+\, dF_\vartheta(t) \\
&= \nu_\alpha + \frac{1}{\mathbb{P}(\vartheta \geq \nu_\alpha)} \mathbb{E}[(\vartheta - \nu_\alpha)^+].
\end{aligned} \tag{2.138}$$

If F_ϑ is continuous, we have $F_\vartheta(\nu_\alpha) = \alpha$ and $\mathbb{P}(\vartheta \geq \nu_\alpha) = 1-\alpha$. Consequently, in this case the above relation takes the form

$$\mathbb{E}[\,\vartheta \mid \vartheta \geq \nu_\alpha\,] = \nu_\alpha + \frac{1}{1-\alpha}\mathbb{E}[(\vartheta - \nu_\alpha)^+]. \tag{2.139}$$

On the other hand, due to a well–known fact in probability theory, the following optimization problem has a solution for any $0 < \alpha < 1$ and the solution set is the interval of α–quantiles:

$$\min_z \left(\alpha\mathbb{E}[(\vartheta - z)^+] + (1-\alpha)\mathbb{E}[(\vartheta - z)^-]\right). \tag{2.140}$$

Using $z = z^+ - z^-$, we have

$$\alpha(\vartheta - z)^+ + (1-\alpha)(\vartheta - z)^- = (1-\alpha)\left[z - \vartheta + \frac{1}{1-\alpha}(\vartheta - z)^+\right].$$

Taking expectation, this leads to the equivalent formulation of the objective function of the unconstrained minimization problem (2.140) as

$$(1-\alpha)(z + \frac{1}{(1-\alpha)}\mathbb{E}[(\vartheta - z)^+]) - (1-\alpha)\mathbb{E}[\vartheta]$$

which results in the following equivalent formulation of (2.140):

$$\min_z \left(z + \frac{1}{(1-\alpha)}\mathbb{E}[(\vartheta - z)^+]\right), \tag{2.141}$$

see Rockafellar and Uryasev [282]. Utilizing (2.38.) and introducing the notation $u_c(z)$ for the objective function of this unconstrained optimization problem, we have

$$u_c(z) := z + \frac{1}{(1-\alpha)}\mathbb{E}[(\vartheta - z)^+] = z + \frac{1}{(1-\alpha)}\int_z^\infty (1 - F_\vartheta(t))dt. \tag{2.142}$$

The function $u_c(\cdot)$ is obviously convex. In fact, for each fixed ϑ, $(\vartheta - z)^+$ is obtained from the convex function $(\cdot)^+$ by substituting a linear function, therefore it is convex. Taking the expected value clearly preserves convexity, see, for instance, the proof of Proposition 2.39. on page 143. Thus (2.141) is a convex programming problem. As mentioned above, the set of solutions of (2.141) consists of the set of α–quantiles of the distribution of ϑ. This is easy to see under the assumption that F_ϑ is continuous. In fact, due to the integral representation in (2.142) it follows immediately that $u_c(z)$ is continuously differentiable. Taking into account that (2.141) is a convex programming problem, the set of optimal solutions is determined by the equation $\frac{du_c(z)}{dz} = 0$ which can be written as

$$1 - \frac{1}{(1-\alpha)}(1 - F_\vartheta(z)) = 0 \iff F_\vartheta(z) = \alpha,$$

which obviously has as solution set the interval of α–quantiles. Based on the fact that for $u_c(z)$, being a convex function, the left– and right–sided derivatives exist for all $z \in \mathbb{R}$, a proof for the general case can be found in Rockafellar and Uryasev [283]. An elementary proof is given by Pflug [254].

The solution set of problem (2.141) being the interval of α–quantiles, it follows that in particular the Value at Risk $\nu_\alpha := \nu(\vartheta, \alpha)$ (for the definition see (2.101) on page 137) is an optimal solution of (2.140). Taking into account (2.139) it follows that for continuous F_ϑ the optimal objective value in (2.141) is $\mathbb{E}[\,\vartheta \mid \vartheta \geq \nu(\vartheta, \alpha)]$. Consequently, in this case, the optimal objective value of (2.141) is the conditional expected value of the loss, given that the loss is greater than or equal to VaR. This motivates introducing the following risk measure for random variables:

$$\rho^{\alpha}_{\text{CVaR}}(\vartheta) := \nu_c(\vartheta, \alpha) := \min_z [z + \frac{1}{1-\alpha} \mathbb{E}[(\vartheta - z)^+]], \; \vartheta \in \mathscr{L}^1_1, \tag{2.143}$$

defined on the linear space of random variables with finite expected value. For the case when F_ϑ is continuous, we have

$$\nu_c(\vartheta, \alpha) = \mathbb{E}[\,\vartheta \mid \vartheta \geq \nu(\vartheta, \alpha)], \tag{2.144}$$

where $\nu(\vartheta, \alpha)$ is the Value at Risk corresponding to ϑ and α, see (2.101) on page 137.

Because the Value at Risk $\nu(\vartheta, \alpha)$ is an optimal solution of (2.140), substituting it for z in (2.140) immediately leads to the inequality

$$\nu_c(\vartheta, \alpha) \geq \nu(\vartheta, \alpha).$$

The risk measure $\nu_c(\vartheta, \alpha)$ has been introduced by Rockafellar and Uryasev [282] for a financial application where the authors call it Conditional Value–at–Risk (CVaR). We will use this terminology also in our context. For the corresponding evaluation function for x we consider the random variable

$$\zeta(x, \xi) := t(\xi)^{\mathrm{T}} x - h(\xi),$$

where $t(\xi)$ is an n–dimensional random vector and $h(\xi)$ is a random variable. The evaluation function, denoted by $\nu_c(x, \alpha)$, is

$$\nu_c(x, \alpha) := \min_z \left[z + \frac{1}{1-\alpha} \mathbb{E}[(\zeta(x, \xi) - z)^+] \right].$$

Introducing the notation

$$w^{\alpha}_c(x, z) := z + \frac{1}{1-\alpha} \mathbb{E}[(\zeta(x, \xi) - z)^+]$$

we have the shorthand form

$$\nu_c(x, \alpha) = \min_z w^{\alpha}_c(x, z).$$

Let $\Psi(x,\cdot)$ denote the probability distribution function of $\zeta(x,\xi)$. For later reference we formulate the specialization of the general findings above for the case $\vartheta = \zeta(x,\xi)$ as a separate proposition:

Proposition 2.42. *Let $x \in \mathbb{R}^n$ be arbitrary and assume $0 < \alpha < 1$. For the unconstrained optimization problem*

$$\min_z \left[z + \frac{1}{1-\alpha} \mathbb{E}[(\zeta(x,\xi) - z)^+] \right]$$

the following assertions hold: this is a convex optimization problem; the optimal solution exists and is attained; the set of optimal solutions coincides with the set of α–quantiles of $\Psi(x,\cdot)$.

Proof: The proof follows readily from the general case. □

Let us consider $w_c^\alpha(x,z)$ as a function in the joint variables (x,z).

Proposition 2.43. *w_c^α is a convex function.*

Proof: $\zeta(x,\xi) - z$ being a linear–affine function of (x,z) and $(\cdot)^+$ being a convex function implies that the composite function $(\zeta(x,\xi)-z)^+$ is jointly convex in (x,z) for each fixed ξ. Proceeding analogously as in the proof of Proposition 2.39. on page 143 it is easy to see that taking the expected value preserves convexity. □

Next we formulate the corresponding optimization problems. SLP models involving CVaR in the objective can be formulated as follows:

$$\left.\begin{array}{ll} \min\limits_x c^{\mathrm{T}}x + \min\limits_z \left[z + \dfrac{1}{1-\alpha} \mathbb{E}[(\zeta(x,\xi) - z)^+] \right] & \\ \text{s.t.} \quad x \in \mathscr{B}. & \end{array}\right\} \tag{2.145}$$

This can obviously be written in the equivalent form

$$\left.\begin{array}{ll} \min\limits_{(x,z)} c^{\mathrm{T}}x + z + \frac{1}{1-\alpha} \mathbb{E}[(\zeta(x,\xi) - z)^+] & \\ \text{s.t.} \quad x \in \mathscr{B}. & \end{array}\right\} \tag{2.146}$$

The equivalence is due to the fact, that for each fixed $x \in \mathscr{B}$ in (2.146) it is sufficient to take into account the corresponding z for which the sum of the second and third terms in the objective is minimal with fixed x (this minimum is attained for any x, see the discussion above).

Proposition 2.43. immediately implies that (2.146) is a convex programming problem for arbitrary probability distribution of ξ. Let (x^*,z^*) be an optimal solution of (2.146). Then z^* is an α-quantile of $\Psi(x^*,\cdot)$ and we have

$$\begin{aligned} v(x^*,\alpha) &\le z^* \\ v_c(x^*,\alpha) &= z^* + \tfrac{1}{1-\alpha} \mathbb{E}[(\zeta(x^*,\xi) - z^*)^+], \end{aligned}$$

where $v(x^*,\alpha)$ is the Value at Risk corresponding to x^* (for VaR see (2.103) on page 137).

Let us turn our attention to the particular case when ξ has a finite discrete distribution with N realizations and corresponding probabilities given as

$$\begin{pmatrix} p_1 & \dots & p_N \\ \widehat{\xi}^1 & \dots & \widehat{\xi}^N \end{pmatrix} \tag{2.147}$$

with $p_i > 0\ \forall i$ and $\sum_{i=1}^{N} p_i = 1$. Let us introduce the notation $t^k := t(\widehat{\xi}^k)$, $h^k := h(\widehat{\xi}^k)$, $k = 1,\dots,N$. The optimization problem (2.146) specializes as follows:

$$\left.\begin{array}{ll} \min\limits_{(x,z)} & c^{\mathrm{T}}x + z + \frac{1}{1-\alpha}\sum\limits_{k=1}^{N} p_k(\zeta(x,\widehat{\xi}^k) - z)^+ \\ \text{s.t.} & x \in \mathscr{B}. \end{array}\right\} \tag{2.148}$$

Using a well–known idea in optimization, this nonlinear programming problem can be transformed into a linear programming problem by introducing additional variables y_k for representing $(\zeta(x,\widehat{\xi}^k) - z)^+$ for all $k = 1,\dots,N$. The equivalent linear programming problem is the following:

$$\left.\begin{array}{lrl} \min\limits_{(x,z,y)} & \multicolumn{2}{l}{c^{\mathrm{T}}x + z + \frac{1}{1-\alpha}\sum\limits_{k=1}^{N} p_k y_k} \\ \text{s.t.} & \zeta(x,\widehat{\xi}^k) - z - y_k & \le 0,\ k = 1,\dots,N \\ & y_k & \ge 0,\ k = 1,\dots,N \\ & x & \in \mathscr{B}. \end{array}\right\} \tag{2.149}$$

The equivalence can be seen as follows. If $(\bar{x},\bar{z})$ is a feasible solution of (2.148) then taking $\bar{y}_k = (\zeta(\bar{x},\widehat{\xi}^k) - \bar{z})^+$ for all k, the resulting $(\bar{x},\bar{z},\bar{y}_k,\ k = 1,\dots,N)$ is obviously feasible in (2.149) with equal objective values. Vice versa, let $(\hat{x},\hat{z},\hat{y}_k,\ k = 1,\dots,N)$ be a feasible solution of (2.149). Then $(\hat{x},\hat{z})$ is evidently feasible in (2.148) and due to the first constraint in (2.149), the corresponding objective value in (2.148) does not exceed the objective value in (2.149). This proves the equivalence. Substituting for $\zeta(x,\widehat{\xi}^k)$ results in the final form of the equivalent LP problem:

$$\left.\begin{array}{lrl} \min\limits_{(x,z,y)} & \multicolumn{2}{l}{c^{\mathrm{T}}x + z + \frac{1}{1-\alpha}\sum\limits_{k=1}^{N} p_k y_k} \\ \text{s.t.} & t^k x - z - y_k & \le h^k,\ k = 1,\dots,N \\ & y_k & \ge 0,\ k = 1,\dots,N \\ & x & \in \mathscr{B}. \end{array}\right\} \tag{2.150}$$

Let us turn our attention to the optimization problems with CVaR constraints

$$\left.\begin{array}{ll}\min & c^{\mathrm{T}}x \\ \text{s.t.} & v_c(x,\alpha) \le \gamma \\ & x \in \mathscr{B},\end{array}\right\} \tag{2.151}$$

where γ is a prescribed threshold. Substituting for $v_c(x,\alpha)$ results in

$$\left.\begin{array}{ll}\min\limits_{x} & c^{\mathrm{T}}x \\ \text{s.t.} & \min\limits_{z} w_c^{\alpha}(x,z) \le \gamma \\ & x \in \mathscr{B}.\end{array}\right\} \tag{2.152}$$

Due to Proposition 2.42. the minimum in the first inequality is attained for any $x \in \mathscr{B}$. Therefore, for any fixed x, the first inequality holds if and only if there exists a $z \in \mathbb{R}$ for which $w_c^{\alpha}(x,z) \le \gamma$ holds. Substituting for $w_c^{\alpha}(x,z)$, the following equivalent formulation results:

$$\left.\begin{array}{ll}\min\limits_{(x,z)} & c^{\mathrm{T}}x \\ \text{s.t.} & z + \frac{1}{1-\alpha}\mathbb{E}[(\zeta(x,\xi)-z)^{+}] \le \gamma \\ & x \in \mathscr{B}.\end{array}\right\} \tag{2.153}$$

This is a nonlinear optimization problem involving a nonlinear constraint. From Proposition 2.43. immediately follows that this problem belongs to the class of convex optimization problems, for an arbitrary probability distribution of ξ.

Let us consider the case of a finite discrete distribution of ξ, as specified in (2.147). This leads to the specialized form

$$\left.\begin{array}{ll}\min\limits_{(x,z)} & c^{\mathrm{T}}x \\ \text{s.t.} & z + \frac{1}{1-\alpha}\sum\limits_{k=1}^{N} p_k(\zeta(x,\widehat{\xi}^k)-z)^{+} \le \gamma \\ & x \in \mathscr{B}.\end{array}\right\} \tag{2.154}$$

Using the same transformation as for deriving (2.149), we get the equivalent formulation as

$$\left.\begin{array}{llll}\min\limits_{(x,z,y)} & c^{\mathrm{T}}x & & \\ \text{s.t.} & & z + \frac{1}{1-\alpha}\sum\limits_{k=1}^{N} p_k y_k & \le \gamma \\ & \zeta(x,\widehat{\xi}^k) - z & -y_k & \le 0,\ k=1,\ldots,N \\ & & y_k & \ge 0,\ k=1,\ldots,N \\ & x & & \in \mathscr{B}.\end{array}\right\} \tag{2.155}$$

The final equivalent form is obtained by substituting for $\zeta(x,\widehat{\xi}^k)$:

$$
\left.
\begin{array}{llll}
\min\limits_{(x,z,y)} & c^{\mathrm{T}}x & & \\
\text{s.t.} & z+\frac{1}{1-\alpha}\sum\limits_{k=1}^{N} p_k y_k & \le \gamma & \\
& t^k x \ -z \qquad -y_k & \le h^k, & k=1,\dots,N \\
& y_k & \ge 0, & k=1,\dots,N \\
& x & \in \mathscr{B}. &
\end{array}
\right\}
\qquad (2.156)
$$

Finally let us discuss the interpretation of $v_c(x,\alpha)$, $0<\alpha<1$. From our general discussions at the beginning of this section it follows readily, that under the assumption that the probability distribution function $\Psi(x,\cdot)$ of the random variable $\zeta(x,\xi)$ is continuous, we have the relation

$$
v_c(x,\alpha) = \mathbb{E}[\,\zeta(x,\xi)\mid \zeta(x,\xi)\ge v(x,\alpha)],
$$

where $v(x,\alpha)$ is the Value at Risk corresponding to $\zeta(x,\xi)$ and α. If ξ has for example a finite discrete distribution then this relation does not hold anymore in general.

For the following discussion let us consider again a random variable ϑ and assume that the expected value exists. In this terms the above relation has been formulated in (2.144) under the assumption that the distribution function F_ϑ of ϑ is continuous. It has the form

$$
\rho^{\alpha}_{\mathrm{CVaR}}(\vartheta) = \mathbb{E}[\,\vartheta \mid \vartheta \ge \nu(\vartheta,\alpha)],
$$

where $\nu(\vartheta,\alpha)$ is the VaR corresponding to ϑ and α, see (2.101) on page 137.

For general distributions an interpretation has been given by Rockafellar and Uryasev in [283]. The conditional expectation relation above holds in general, if the original distribution function F_ϑ is replaced by the upper–tail distribution function F^{α}_{ϑ} defined as follows:

$$
F^{\alpha}_{\vartheta}(y) = \begin{cases} 0 & \text{if } y<\nu(\vartheta,\alpha) \\ \frac{F_\vartheta(y)-\alpha}{1-\alpha} & \text{if } y\ge\nu(\vartheta,\alpha). \end{cases}
$$

Another interpretation, representing α–CVaR as a mean over α of α–VaR, has been found by Acerbi [3]. For further properties of CVaR see Rockafellar and Uryasev [283] and Acerbi and Tasche [4]. In the latter paper several related risk measures and their interrelations are also presented.

Exercises

2.8. With integrated chance constraints of the second kind the SLP–problems have been formulated on page 147 as (2.127) and (2.128), with $\alpha \in [0,\frac{1}{2}]$ prescribed.

Formulate the equivalent linear programming problems for the case when ξ has a finite discrete distribution.

2.9. Consider the following pair of SLP problems:

$$\left.\begin{array}{ll}\min\limits_{x} & x_1 + 2x_2 \\ \text{s.t.} & \mathbb{E}[\zeta^+(x,\xi)] \le 0.8 \\ & x_1 + x_2 \ge 3 \\ & x_1, \quad x_2 \ge 0\end{array}\right\} \quad (ICC)$$

and

$$\left.\begin{array}{ll}\min\limits_{x,z} & x_1 + 2x_2 \\ \text{s.t.} & \rho^{\alpha}_{\text{CVaR}}(\zeta(x,\xi)) \le 0.8 \\ & x_1 + x_2 \ge 3 \\ & x_1, \quad x_2 \ge 0\end{array}\right\} \quad (CVaR)$$

with $\zeta(x,\xi) := \xi_1 x_1 + \xi_2 x_2 - 0.5$, $\alpha = 0.95$ and (ξ_1, ξ_2) having a finite discrete distribution with three realizations, given by the realizations tableau:

p_k	0.2	0.5	0.3
ξ_1^k	1	-0.5	0.5
ξ_2^k	2	0.5	-0.5

Notice that both problems are set up on the same data–set; *(ICC)* is formulated with an integrated chance constraint whereas *(CVaR)* is an SLP model with a CVaR–constraint.

(a) Solve both problems by utilizing SLP–IOR and compare the optimal objective values z^*_{ICC} and z^*_{CVaR}.
(b) You will see that $z^*_{ICC} < z^*_{CVaR}$ holds. This is a typical result when setting up and solving both type of problems based on the same data–set. Give an explanation for this phenomenon.

2.5 Models built with deviation measures

In this section we deal exclusively with quality measures expressing risk. Similarly as in Section 2.2.3, for the sake of simplicity we employ the notation $\eta := t(\xi)$ and replace the right–hand–side $h(\xi)$ by ξ. Thus we consider the random variable

$$\zeta(x,\eta,\xi) := \eta^{\mathrm{T}} x - \xi,$$

where η denotes an n–dimensional random vector and ξ is a random variable. We will assume in this section that the expected value of (η^{T}, ξ) exists and will use the notation $\mu := \mathbb{E}[\eta] \in \mathbb{R}^n$ and $\mu_{n+1} := \mathbb{E}[\xi] \in \mathbb{R}$.

2.5.1 Quadratic deviation

The risk measure is chosen as

$$\rho_{\mathrm{Q}}(\vartheta) := \sqrt{\mathbb{E}[\vartheta^2]} = \sqrt{\mathbb{V}\mathbf{ar}[\vartheta] + (\mathbb{E}[\vartheta])^2},\ \ \vartheta \in \mathscr{L}_1^2,$$

defined on the linear space of random variables with finite variance. We assume that the second moments for the random vector (η^{T}, ξ) exist and the distribution is non–degenerate, meaning that the covariance matrix of this random vector is positive definite. The corresponding evaluation function will be denoted by $Q(x)$ and has the form

$$Q(x) := \sqrt{\mathbb{E}[(\eta^{\mathrm{T}}x - \xi)^2]} = \sqrt{\mathbb{V}\mathbf{ar}[\eta^{\mathrm{T}}x - \xi] + (\mu^{\mathrm{T}}x - \mu_{n+1})^2}. \tag{2.157}$$

It is interpreted as measuring the deviation between the random variables $\eta^{\mathrm{T}}x$ and ξ.

Proposition 2.44. *Q is a convex function.*

Proof: An elegant proof of this assertion can be obtained by combining Propositions 2.47. and 2.50. in Section 2.7. Here we present a direct elementary proof. Let us consider the functions $q : \mathbb{R}^{n+1} \to \mathbb{R}$ and $\hat{q} : \mathbb{R}^{n+1} \to \mathbb{R}$ defined as $q(x, x_{n+1}) = \mathbb{E}[(\eta^{\mathrm{T}}x + \xi x_{n+1})^2]$ and $\hat{q}(x, x_{n+1}) = \sqrt{q(x, x_{n+1})}$, respectively. We will prove that $\hat{q}$ is a convex function. Due to the relation $Q(x) = \hat{q}(x, -1)$, the assertion follows from this.

We consider $q(x, x_{n+1})$ first. This function is obviously nonnegative, $q(x, x_{n+1}) \geq 0$ holds for all $x \in \mathbb{R}^n$, $x_{n+1} \in \mathbb{R}$. The function is quadratic

$$\begin{aligned} \mathbb{E}[(\eta^{\mathrm{T}}x + \xi x_{n+1})^2] &= \mathbb{E}\left[(x^{\mathrm{T}}, x_{n+1}) \begin{pmatrix} \eta \\ \xi \end{pmatrix} (\eta^{\mathrm{T}}, \xi) \begin{pmatrix} x \\ x_{n+1} \end{pmatrix}\right] \\ &= (x^{\mathrm{T}}, x_{n+1}) \begin{pmatrix} \mathbb{E}[\eta\eta^{\mathrm{T}}] & \mathbb{E}[\xi\eta] \\ \mathbb{E}[\xi\eta^{\mathrm{T}}] & \mathbb{E}[\xi^2] \end{pmatrix} \begin{pmatrix} x \\ x_{n+1} \end{pmatrix} \end{aligned} \tag{2.158}$$

therefore, because of the nonnegativity of q, the symmetric matrix in the second line in (2.158) is positive semidefinite. Thus q is a convex function. In general, the square root of a convex function need not to be convex (take z and $\sqrt{z}$ for $z \geq 0$). In our case $\hat{q}(x, x_{n+1}) = \sqrt{q(x, x_{n+1})}$ is the square root of a positive semidefinite quadratic form, therefore it is convex. To see this let D be an $(n \times n)$ symmetric positive semidefinite matrix, we shall prove that $d(x) := \sqrt{x^{\mathrm{T}}Dx}$ is a convex function. For this function $d(\lambda x) = \lambda d(x)$ holds obviously for all $\lambda \geq 0$ and $x \in \mathbb{R}^n$. Therefore, for proving the convexity of d, it is sufficient to prove that $d(x+y) \le d(x)+d(y)$ (subadditivity) holds for all $x,y \in \mathbb{R}^n$ (see Proposition 2.46. on page 174). We have

$$d^2(x+y) = (x+y)^{\mathrm{T}} D(x+y) = x^{\mathrm{T}} D(x+y) + y^{\mathrm{T}} D(x+y). \tag{2.159}$$

By applying the Cauchy–Schwarz inequality to the first term on the right–hand–side we get

$$x^{\mathrm{T}} D(x+y) = [x^{\mathrm{T}} D^{\frac{1}{2}}][D^{\frac{1}{2}}(x+y)] \le \sqrt{x^{\mathrm{T}} Dx}\sqrt{(x+y)^{\mathrm{T}} D(x+y)},$$

where $D^{1/2}$ denotes the symmetric square root of the positive semidefinite matrix D. The latter is defined as follows: take the spectral decomposition $D = T\Lambda T^{\mathrm{T}}$ of D, where the columns of T consist of an orthonormal system of eigenvectors of D and the diagonal elements of the diagonal matrix Λ are the corresponding eigenvalues. Taking $D^{1/2} := T\Lambda^{1/2}T^{\mathrm{T}}$ we obviously have $D = D^{1/2}D^{1/2}$. Performing analogously with the second term in (2.159) and substituting into (2.159) yields the result. □

Applying (2.158) for $\hat{\eta}$ and $\hat{\xi}$, defined as $\hat{\eta} := \eta - \mu$ and $\hat{\xi} := \xi - \mu_{n+1}$, and setting $x_{n+1} = -1$ from (2.158) it follows that

$$\mathbb{V}\mathrm{ar}[\eta^{\mathrm{T}} x - \xi] = x^{\mathrm{T}} V x - 2d^{\mathrm{T}} x + v$$

holds, where $V := \mathbb{E}[\hat{\eta}\hat{\eta}^{\mathrm{T}}] = \mathbb{C}\mathrm{ov}[\eta,\eta]$ is the covariance matrix of η, $d := \mathbb{E}[\hat{\eta}\hat{\xi}] = \mathbb{C}\mathrm{ov}[\eta,\xi]$ is the cross–covariance vector between η and ξ, and $v := \mathbb{E}[\hat{\xi}^2] = \mathbb{V}\mathrm{ar}[\xi]$ is the variance of ξ. Note that V is a positive semidefinite matrix. Thus we have have derived the formula

$$Q(x) = \sqrt{x^{\mathrm{T}} V x - 2d^{\mathrm{T}} x + v + (\mu^{\mathrm{T}} x - \mu_{n+1})^2}.$$

We obtain the following convex optimization problems

$$\left.\begin{array}{ll} \min c^{\mathrm{T}} x & \\ \text{s.t.} \ \sqrt{x^{\mathrm{T}} V x - 2d^{\mathrm{T}} x + v + (\mu^{\mathrm{T}} x - \mu_{n+1})^2} & \le \sqrt{\kappa} \\ \qquad x & \in \mathscr{B} \end{array}\right\} \tag{2.160}$$

and

$$\left.\begin{array}{ll} \min & \sqrt{x^{\mathrm{T}} V x - 2d^{\mathrm{T}} x + v + (\mu^{\mathrm{T}} x - \mu_{n+1})^2} \\ \text{s.t.} & x \in \mathscr{B}. \end{array}\right\} \tag{2.161}$$

Due to the definition of Q, the expression under the square root is nonnegative for all $x \in \mathbb{R}^n$, and the positive square root function is strictly monotonically increasing, consequently we have the equivalent formulation

$$\left.\begin{array}{ll} \min c^{\mathrm{T}} x & \\ \text{s.t.} \ x^{\mathrm{T}} V x + (\mu^{\mathrm{T}} x - \mu_{n+1})^2 - 2d^{\mathrm{T}} x & \le \kappa - v \\ \qquad x & \in \mathscr{B} \end{array}\right\} \tag{2.162}$$

and

$$\left.\begin{array}{l}\min\; x^{\mathrm{T}}Vx+(\mu^{\mathrm{T}}x-\mu_{n+1})^2-2d^{\mathrm{T}}x\\ \text{s.t.}\;\; x\in\mathscr{B}.\end{array}\right\}\tag{2.163}$$

The matrix V is positive semidefinite, therefore both problems are convex optimization problems. Note that (2.163) is a convex quadratic optimization problem.

A widely used variant of the above risk measure for random variables is the standard deviation:

$$\rho_{\text{Std}}(\vartheta):=\sigma(\vartheta):=\sqrt{\mathbb{E}[(\vartheta-\mathbb{E}[\vartheta])^2]},\;\;\vartheta\in\mathscr{L}_1^2.$$

The evaluation function becomes

$$Q_d(x)=\sqrt{x^{\mathrm{T}}Vx-2d^{\mathrm{T}}x+v}\tag{2.164}$$

leading to the convex optimization problems

$$\left.\begin{array}{lll}\min & c^{\mathrm{T}}x & \\ \text{s.t.} & x^{\mathrm{T}}Vx-2d^{\mathrm{T}}x & \le\kappa-v\\ & x & \in\mathscr{B}\end{array}\right\}\tag{2.165}$$

and

$$\left.\begin{array}{l}\min\; x^{\mathrm{T}}Vx-2d^{\mathrm{T}}x\\ \text{s.t.}\;\; x\in\mathscr{B}.\end{array}\right\}\tag{2.166}$$

An important special case is $\xi\equiv 0$. The evaluation function becomes

$$\sigma(x):=\sqrt{\mathbb{E}[(\eta^{\mathrm{T}}x-\mu^{\mathrm{T}}x)^2]}=\sqrt{x^{\mathrm{T}}Vx},\tag{2.167}$$

where V is covariance matrix of η. Because of their practical importance we formulate also the resulting optimization problems. For obvious reasons, these can equivalently be formulated in terms of $\sigma^2(x)$ as follows:

$$\left.\begin{array}{lll}\min & c^{\mathrm{T}}x & \\ \text{s.t.} & x^{\mathrm{T}}Vx & \le\kappa\\ & x & \in\mathscr{B}\end{array}\right\}\tag{2.168}$$

and

$$\left.\begin{array}{lll}\min & x^{\mathrm{T}}Vx & \\ \text{s.t.} & x & \in\mathscr{B}.\end{array}\right\}\tag{2.169}$$

Optimization problems of this type are widely used in financial portfolio optimization, see Markowitz [217] and Elton et al. [85].

2.5.2 Absolute deviation

Let the risk measure be

$$\rho_A(\vartheta) := \mathbb{E}[\,|\vartheta|\,],\ \ \vartheta \in \mathscr{L}_1^1, \tag{2.170}$$

defined on the linear space of random variables with finite expected value. Assuming that the expected value of (η^{T}, ξ) exists we get the corresponding evaluation function

$$A(x) := \mathbb{E}[\,|\eta^{\mathrm{T}}x - \xi\,|\,], \tag{2.171}$$

which is interpreted as measuring deviation between the random variables $\eta^{\mathrm{T}}x$ and ξ. Let $\zeta(x, \eta, \xi) := \eta^{\mathrm{T}}x - \xi$. We have:

Proposition 2.45. *$A(\cdot)$ is a convex function.*

Proof: The absolute–value function being convex and $\zeta(\cdot, \eta, \xi)$ being linear, the composite function $|\zeta(\cdot, \eta, \xi)|$ is convex for any fixed realization of (η, ξ). Taking expected value preserves convexity. The full proof runs analogously as the proof of Proposition 2.39. on page 143. □

Thus the optimization problems

$$\left.\begin{array}{ll} \min & c^{\mathrm{T}}x \\ \text{s.t.} & \mathbb{E}[\,|\eta^{\mathrm{T}}x - \xi\,|\,] \le \kappa \\ & x \in \mathscr{B} \end{array}\right\} \tag{2.172}$$

and

$$\left.\begin{array}{ll} \min & \mathbb{E}[\,|\eta^{\mathrm{T}}x - \xi\,|\,] \\ \text{s.t.} & x \in \mathscr{B} \end{array}\right\} \tag{2.173}$$

are convex optimization problems for arbitrary random variables with finite expected value.

The model (2.173) is closely related to simple recourse problems. To see this let $\eta \equiv t$ with t being an n–dimensional deterministic vector and let us formulate this problem equivalently as follows. We introduce the nonnegative random variables y and z and make use of the relations $|u| = u^+ + u^-$ and $u = u^+ - u^-$ which hold for any real number u. This results in the following equivalent simple recourse formulation of (2.173)

$$\left.\begin{array}{lllll} \min & \mathbb{E}[y+z] & & & \\ \text{s.t.} & t^{\mathrm{T}}x - \xi & -y & +z & = 0 \\ & & y & & \ge 0 \\ & & & z & \ge 0 \\ & x & & & \in \mathscr{B}, \end{array}\right\} \tag{2.174}$$

where the constraints involving random variables are interpreted as usual: they should hold in the almost sure sense. For proving the equivalence let $\hat{x}$, along with the random variables $\hat{y}$ and $\hat{z}$ be a feasible solution of (2.174). Then $\hat{x}$ is obviously a feasible solution of (2.173) and for the corresponding objective function values we have

$$\mathbb{E}[\hat{y}+\hat{z}] \geq \mathbb{E}[|\hat{y}-\hat{z}|] = \mathbb{E}[|t^{\mathrm{T}}\hat{x}-\xi|].$$

In the reverse direction, when $\bar{x}$ is a feasible solution of (2.173) then setting $\bar{y} = (t^{\mathrm{T}}\bar{x}-\xi)^{+}$ and $\bar{z} = (t^{\mathrm{T}}\bar{x}-\xi)^{-}$ we get a feasible solution to (2.174) and the objective values are equal. This proves the equivalence.

Let us consider the case of a finite discrete distribution next. Assume that (η^{T},ξ) has N distinct realizations with corresponding probabilities given in the table:

$$\begin{pmatrix} p_1 & \dots & p_N \\ \widehat{\eta}^1 & \dots & \widehat{\eta}^N \\ \widehat{\xi}^1 & \dots & \widehat{\xi}^N \end{pmatrix} \tag{2.175}$$

with $p_i > 0\ \forall i$ and $\sum\limits_{i=1}^{N} p_i = 1$. Let $t^k := (\widehat{\eta}^k)^{\mathrm{T}}$, $h^k := \widehat{\xi}^k$, $k = 1,\dots,N$. In this case our optimization problems have the form

$$\left.\begin{array}{ll} \min c^{\mathrm{T}}x & \\ \text{s.t.}\ \sum\limits_{k=1}^{N} p_k|t^k x - h^k| & \leq \kappa \\ \qquad\qquad x & \in \mathscr{B} \end{array}\right\} \tag{2.176}$$

and

$$\left.\begin{array}{ll} \min \sum\limits_{k=1}^{N} p_k|t^k x - h^k| & \\ \text{s.t.}\qquad\quad x & \in \mathscr{B}. \end{array}\right\} \tag{2.177}$$

Both of these problems are nonlinear programming problems. We utilize a transformation, analogous to the transformation which leads to the formulation (2.174). Introducing this time deterministic auxiliary variables y_k and z_k, $k = 1,\dots,N$, we obtain the equivalent deterministic linear programming formulations

$$\left.\begin{array}{lllll} \min c^{\mathrm{T}}x & & & & \\ \text{s.t.}\ \sum\limits_{k=1}^{N} p_k(& y_k & +z_k) & \leq \kappa & \\ t^k x & -y_k & +z_k & = h^k, & k=1,\dots,N \\ & y_k & & \geq 0, & k=1,\dots,N \\ & & z_k & \geq 0, & k=1,\dots,N \\ x & & & \in \mathscr{B} & \end{array}\right\} \tag{2.178}$$

and

$$\left.\begin{array}{lll} \min \sum_{k=1}^{N} p_k(y_k+z_k) & & \\ \text{s.t. } \ t^k x - y_k + z_k & = h^k, & k=1,\ldots,N \\ \qquad\qquad y_k & \geq 0, & k=1,\ldots,N \\ \qquad\qquad\qquad z_k & \geq 0, & k=1,\ldots,N \\ \qquad x & \in \mathscr{B}. & \end{array}\right\} \tag{2.179}$$

The proof of the equivalence of (2.177) and (2.179) runs along the same lines as the proof for the equivalence of (2.173) and (2.174). For the equivalence of (2.176) and (2.178) it is sufficient to remark, that for any feasible solution $(\hat{x}, \hat{y}_k, \hat{z}_k, k=1,\ldots,N)$ of (2.178), $\hat{x}$ is feasible in (2.176), due to the following inequality:

$$\sum_{k=1}^{N} p_k|t^k x - h^k| = \sum_{k=1}^{N} p_k|y_k - z_k| \leq \sum_{k=1}^{N} p_k(y_k+z_k) \leq \kappa.$$

Analogously to the quadratic measure, we consider the variant which measures absolute deviations from the expected value and is called *mean absolute deviation* (MAD):

$$\rho_{\text{MAD}}(\vartheta) := \mathbb{E}[|\vartheta - \mathbb{E}[\vartheta]|], \ \vartheta \in \mathscr{L}_1^1.$$

The evaluation function becomes

$$A_d(x) = |(\eta - \mu)^{\mathrm{T}} x - (\xi - \mu_{n+1})|$$

leading to convex optimization problems, which are analogous to (2.172) and (2.173). The linear programming formulations for the case of a finite discrete distribution coincide with (2.178) and (2.179) when we set $t^k := (\widehat{\eta}^k - \mu)^{\mathrm{T}}$ and $h^k := \widehat{\xi}^k - \mu_{n+1}$.

An important special case in practice (for instance, in portfolio optimization in finance) is the case $\xi \equiv 0$ thus leading to the deviation measure

$$A_d(x) = |(\eta - \mu)^{\mathrm{T}} x| = |\eta^{\mathrm{T}} x - \mu^{\mathrm{T}} x|. \tag{2.180}$$

For the discretely distributed case we formulate the particular form of the optimization problems explicitly. Let $t^k := \widehat{\eta}^k$ (note that in (2.178) and (2.179) we have had $t^k = (\widehat{\eta}^k)^{\mathrm{T}}$). (2.176) and (2.177) have the form now

$$\left.\begin{array}{ll} \min c^{\mathrm{T}} x & \\ \text{s.t. } \sum_{k=1}^{N} p_k|(t^k - \mu)^{\mathrm{T}} x| \leq \kappa & \\ \qquad\qquad\qquad x \in \mathscr{B} & \end{array}\right\} \tag{2.181}$$

and

$$\left.\begin{array}{ll} \min \sum\limits_{k=1}^{N} p_k |(t^k - \mu)^{\mathrm{T}} x| & \\ \text{s.t.} & x \in \mathscr{B}, \end{array}\right\} \tag{2.182}$$

respectively. The equivalent linear programming formulations can easily be obtained from (2.178) and (2.179), by substituting $t^k x$ with $(t^k - \mu)^{\mathrm{T}} x$ there.

Models of this type have first been proposed in the framework of portfolio optimization in finance by Konno and Yamazaki [194]. In this paper the authors propose a variant for the equivalent linear program (2.179) (with the substitution described above), by introducing fewer auxiliary variables on the cost of a larger amount of constraints, as follows:

$$\left.\begin{array}{ll} \min \sum\limits_{k=1}^{N} p_k y_k & \\ \text{s.t.} & (t^k - \mu)^{\mathrm{T}} x + y_k \geq 0, \; k = 1, \ldots, N \\ & (t^k - \mu)^{\mathrm{T}} x - y_k \leq 0, \; k = 1, \ldots, N \\ & x \in \mathscr{B}. \end{array}\right\} \tag{2.183}$$

The equivalence with (2.179) can easily be seen, for instance, by considering separately the cases $(t^k - \mu)^{\mathrm{T}} x \geq 0$ and $(t^k - \mu)^{\mathrm{T}} x < 0$.

Let us assume next that η has a non–degenerate multivariate normal distribution and let $x \in \mathbb{R}^n$ be fixed. Then the random variable $\eta^{\mathrm{T}} x$, being a linear transformation of a random vector with a non–degenerate normal distribution, is normally distributed (see Section 2.2.3). We obviously have $\hat{\mu} := \mathbb{E}[\eta^{\mathrm{T}} x] = \mu^{\mathrm{T}} x$ and $\hat{\sigma}^2 := \mathbb{V}\mathbf{ar}[\eta^{\mathrm{T}} x] = x^{\mathrm{T}} V x$. An easy computation gives:

$$\begin{aligned} \mathbb{E}[|\eta^{\mathrm{T}} x - \mu^{\mathrm{T}} x|] &= \hat{\sigma} \mathbb{E}\left[\left|\frac{\eta^{\mathrm{T}} x - \hat{\mu}^{\mathrm{T}} x}{\hat{\sigma}}\right|\right] \\ &= \frac{\hat{\sigma}}{\sqrt{2\pi}} \int_{-\infty}^{\infty} |z| e^{-\frac{z^2}{2}} dz = \sqrt{\frac{2}{\pi}} \hat{\sigma} \int_{0}^{\infty} z e^{-\frac{z^2}{2}} dz \\ &= \sqrt{\frac{2}{\pi}} \hat{\sigma} = \sqrt{\frac{2}{\pi}} \sqrt{x^{\mathrm{T}} V x}. \end{aligned}$$

This implies that for a non–degenerate normal distribution the models with absolute deviation and those with quadratic deviation are equivalent. Note, however, that due to the scaling factor $\sqrt{2\pi}$ above, in the model (2.168) with a quadratic constraint, a scaling in the parameter κ has to be accounted for.

From the statistical point of view the natural measure for absolute deviations would be the absolute deviation from the median, instead of the expected value. The difficulty is that we are dealing with linear combinations of random variables $\eta^{\mathrm{T}} x = \sum_{i=1}^{n} \eta_i x_i$. The median of $\eta^{\mathrm{T}} x$ is in general by no means equal to the linear combination of the medians of the components of η. This makes it extremely difficult to build numerically tractable median–based optimization problems of the deviation type.

2.5.3 Quadratic semi–deviation

In both of the previous sections we employed risk measures which penalized deviations in both directions. The quadratic risk measure $Q(x) = \sqrt{\mathbb{E}[(\eta^{\mathrm{T}}x - \xi)^2]}$ (2.157) evaluates upper– and lower deviations of $\eta^{\mathrm{T}}x$ with respect to the target random variable ξ in the same manner. This observation holds also for the standard deviation $\sigma(x) = \sqrt{\mathbb{E}[(\eta^{\mathrm{T}}x - \mu^{\mathrm{T}}x)^2]}$ (2.167) with respect to the deterministic target $\mu^{\mathrm{T}}x$, and for the absolute–deviation counterparts $A(x)$ (2.171), and $A_d(x)$ (2.180). All of these risk measures model risk as deviation from a target, irrespectively of the direction of this deviation.

In many modeling situations, however, the direction of deviation matters. In such cases one of them is favorable (gain) and the other is disadvantageous (loss).

We introduce the following risk measures for random variables:

$$\begin{aligned} \rho_{\mathrm{Q}}^{+}(\vartheta) &:= \sqrt{\mathbb{E}[(\vartheta^{+})^2]}, \\ \rho_{\mathrm{Q}}^{-}(\vartheta) &:= \sqrt{\mathbb{E}[(\vartheta^{-})^2]}, \quad \vartheta \in \mathscr{L}_1^2, \end{aligned}$$

both of them being defined on the linear space of random variables with finite variance and with $z^{-} = \max\{0, -z\}$, $z^{+} = \max\{0, z\}$ standing for the negative– and positive part for a real number z, respectively.

Let us assume that the second moments for (η^{T}, ξ) exist. The corresponding evaluation functions, denoted by $Q^{-}(x)$ and $Q^{+}(x)$, respectively, are defined as

$$\begin{aligned} Q^{+}(x) &:= \sqrt{\mathbb{E}[((\eta^{\mathrm{T}}x - \xi)^{+})^2]} \\ Q^{-}(x) &:= \sqrt{\mathbb{E}[((\eta^{\mathrm{T}}x - \xi)^{-})^2]}. \end{aligned} \qquad (2.184)$$

These measures are interpreted as measuring the upper/lower deviation between $\eta^{\mathrm{T}}x$ and ξ. Both $Q^{+}(x)$ and $Q^{-}(x)$ are convex functions; this will be proved in a general framework in Section 2.7.2, see Propositions 2.47. and 2.50. there.

Let us assume that negative values of the random variable $\zeta(x, \eta, \xi) := \eta^{\mathrm{T}}x - \xi$ represent losses. Then the following prototype optimization problems result

$$\left.\begin{array}{lrl} \min & c^{\mathrm{T}}x & \\ \text{s.t.} & \sqrt{\mathbb{E}[((\eta^{\mathrm{T}}x - \xi)^{-})^2]} & \leq \sqrt{\kappa} \\ & x & \in \mathscr{B} \end{array}\right\} \qquad (2.185)$$

and

$$\left.\begin{array}{ll} \min & \sqrt{\mathbb{E}[((\eta^{\mathrm{T}}x-\xi)^{-})^{2}]} \\ \text{s.t.} & x \in \mathscr{B}, \end{array}\right\} \tag{2.186}$$

both of which are convex optimization problems. They can be equivalently written, due to the fact that the function $\sqrt{z}$ is strictly monotonically increasing, as

$$\left.\begin{array}{ll} \min & c^{\mathrm{T}}x \\ \text{s.t.} & \mathbb{E}[((\eta^{\mathrm{T}}x-\xi)^{-})^{2}] \le \kappa \\ & x \qquad \in \mathscr{B} \end{array}\right\} \tag{2.187}$$

and

$$\left.\begin{array}{ll} \min & \mathbb{E}[((\eta^{\mathrm{T}}x-\xi)^{-})^{2}] \\ \text{s.t.} & x \in \mathscr{B}. \end{array}\right\} \tag{2.188}$$

Let us discuss the case when (η, ξ) has a discrete distribution specified in (2.175) on page 164. In this case our problems assume the form:

$$\left.\begin{array}{ll} \min & c^{\mathrm{T}}x \\ \text{s.t.} & \sum_{k=1}^{N} p_k((t^k x - h^k)^{-})^2 \le \kappa, \ k = 1,\dots,N \\ & x \qquad \in \mathscr{B} \end{array}\right\} \tag{2.189}$$

and

$$\left.\begin{array}{ll} \min & \sum_{k=1}^{N} p_k((t^k x - h^k)^{-})^2 \\ \text{s.t.} & x \qquad \in \mathscr{B}. \end{array}\right\} \tag{2.190}$$

By introducing auxiliary variables y_k, $k = 1,\dots,N$, these problems can be written equivalently as follows:

$$\left.\begin{array}{ll} \min & c^{\mathrm{T}}x \\ \text{s.t.} & \sum_{k=1}^{N} p_k y_k^2 \le \kappa, \ k = 1,\dots,N \\ & t^k x + y_k \ge h^k, \ k = 1,\dots,N \\ & \qquad y_k \ge 0, \ k = 1,\dots,N \\ & x \qquad \in \mathscr{B} \end{array}\right\} \tag{2.191}$$

and

$$\left.\begin{array}{ll} \min & \sum_{k=1}^{N} p_k y_k^2 \\ \text{s.t.} & t^k x + y_k \ge h^k, \ k = 1,\dots,N \\ & \qquad y_k \ge 0, \ k = 1,\dots,N \\ & x \qquad \in \mathscr{B}. \end{array}\right\} \tag{2.192}$$

Problem (2.192) is a convex quadratic programming problem whereas (2.191) is a quadratically constrained convex optimization problem.

For proving the equivalence of (2.189) and (2.191), as well as of (2.190) and (2.192) let us make the following observation. If $(\hat{x}, \hat{y}_k, k = 1, \ldots, N)$ is a feasible solution of either (2.191) or (2.192), then the constraints imply the inequality $(t^k\hat{x} - h^k)^- \leq y_k$ for all k. Consequently, in both cases

$$\sum_{k=1}^{N} p_k((t^k\hat{x} - h^k)^-)^2 \leq \sum_{k=1}^{N} p_k\hat{y}_k^2$$

holds. From this the equivalence follows in a straightforward way; the detailed proof is left as an easy exercise to the reader.

Analogously as in both previous sections we discuss the variants measuring deviations from the expected value:

$$\begin{aligned}
\rho^+_{\text{Std}}(\vartheta) &:= \sigma^+(x) := \sqrt{\mathbb{E}[((\vartheta - \mathbb{E}[\vartheta])^+)^2]}, \\
\rho^-_{\text{Std}}(\vartheta) &:= \sigma^-(x) := \sqrt{\mathbb{E}[((\vartheta - \mathbb{E}[\vartheta])^-)^2]}, \ \ \vartheta \in \mathscr{L}^2_1.
\end{aligned}$$

These are called *upper standard semi–deviation* and *lower standard semi–deviation*, respectively. The evaluation functions are obtained by performing the substitution $(\eta - \mu)^{\mathrm{T}}x - (\xi - \mu_{n+1})$ for ϑ, which leads to convex optimization problems analogous to (2.187) and (2.173). In the case of a finite discrete distribution, the linear programming formulations coincide with (2.191) and (2.190) provided that the definitions $t^k := (\widehat{\eta}^k - \mu)^{\mathrm{T}}$ and $h^k := \widehat{\xi}^k - \mu_{n+1}$ are used.

We discuss the important special case where $\xi \equiv 0$ holds separately. The valuation functions take the form:

$$\left.\begin{aligned}
\sigma^+(x) &:= \sqrt{\mathbb{E}[((\eta^{\mathrm{T}}x - \mu^{\mathrm{T}}x)^+)^2]} \\
\sigma^-(x) &:= \sqrt{\mathbb{E}[((\eta^{\mathrm{T}}x - \mu^{\mathrm{T}}x)^-)^2]},
\end{aligned}\right. \tag{2.193}$$

which are interpreted as measuring the upper/lower deviation between $\eta^{\mathrm{T}}x$ and its expected value $\mu^{\mathrm{T}}x$. Because of its importance in practice we formulate the optimization problems for the case when ξ has a finite discrete distribution explicitly. With t^k now considered as a column vector, (2.189) and (2.190) have the form:

$$\left.\begin{aligned}
&\min\ c^{\mathrm{T}}x \\
&\text{s.t.}\ \sum_{k=1}^{N} p_k(((t^k - \mu)^{\mathrm{T}}x)^-)^2 \leq \kappa,\ k = 1, \ldots, N \\
&\qquad\qquad\qquad x \qquad \in \mathscr{B}
\end{aligned}\right\} \tag{2.194}$$

and

$$\left.\begin{aligned}
&\min \sum_{k=1}^{N} p_k(((t^k - \mu)^{\mathrm{T}}x)^-)^2 \\
&\text{s.t.} \qquad\qquad x \qquad \in \mathscr{B}
\end{aligned}\right\} \tag{2.195}$$

whereas (2.191) and (2.192) assume the form

$$\left.\begin{array}{rrl} \min c^{\mathrm{T}}x & & \\ \text{s.t.} & \sum_{k=1}^{N} p_k y_k^2 & \leq \kappa,\ k=1,\ldots,N \\ (t^k-\mu)^{\mathrm{T}}x & +y_k & \geq 0,\ k=1,\ldots,N \\ & y_k & \geq 0,\ k=1,\ldots,N \\ x & & \in \mathscr{B} \end{array}\right\} \tag{2.196}$$

and

$$\left.\begin{array}{rl} \min \sum_{k=1}^{N} p_k y_k^2 & \\ \text{s.t.}\ \ (t^k-\mu)^{\mathrm{T}}x + y_k & \geq 0,\ k=1,\ldots,N \\ y_k & \geq 0,\ k=1,\ldots,N \\ x & \in \mathscr{B}. \end{array}\right\} \tag{2.197}$$

The importance of introducing this type of risk measures has first been recognized by Markowitz [217] who also applied them in financial portfolio optimization.

2.5.4 Absolute semi–deviation

Similarly to the way for constructing a semi–deviation variant for quadratic deviation, we get the following semi–deviation measures:

$$\rho^{+}_{\text{sic}}(\vartheta) := \mathbb{E}[\vartheta^{+}],$$
$$\rho^{-}_{\text{sic}}(\vartheta) := \mathbb{E}[\vartheta^{-}],\ \vartheta \in \mathscr{L}^1_1,$$

defined on the space of random variables with finite expected value and with $z^- = \max\{0,-z\}$, $z^+ = \max\{0,z\}$ for any real number z. Note that we do not obtain new risk measures: These risk measures have been already discussed in Section 2.4.1, in connection with integrated chance constraints, see (2.114) on page 143. Now we consider them again, this time in relation with the absolute–deviation risk measure ρ_{A} defined in (2.170). Using the relations $z = z^+ - z^-$ and $|z| = z^+ + z^-$ we obtain that $z^- = \frac{1}{2}(|z| - z)$ and $z^+ = \frac{1}{2}(|z| + z)$ hold. Thus we have

$$\begin{aligned} \rho^{+}_{\text{sic}}(\vartheta) &= \tfrac{1}{2}\left(\mathbb{E}[|\vartheta|] + \mathbb{E}[\vartheta]\right) = \tfrac{1}{2}\left(\rho_{\text{A}}(\vartheta) + \mathbb{E}[\vartheta]\right) \\ \rho^{-}_{\text{sic}}(\vartheta) &= \tfrac{1}{2}\left(\mathbb{E}[|\vartheta|] - \mathbb{E}[\vartheta]\right) = \tfrac{1}{2}\left(\rho_{\text{A}}(\vartheta) - \mathbb{E}[\vartheta]\right). \end{aligned} \tag{2.198}$$

The evaluation function $A(x)$ (2.171) defined on page 163 has now the semi–deviation counterparts:

$$
\begin{aligned}
K(x) &= \mathbb{E}[(\eta^{\mathrm{T}}x-\xi)^{+}] = \tfrac{1}{2}\left(\mathbb{E}[\,|\eta^{\mathrm{T}}x-\xi|\,]+(\mu^{\mathrm{T}}x-\mu_{n+1})\right) \\
H(x) &= \mathbb{E}[(\eta^{\mathrm{T}}x-\xi)^{-}] = \tfrac{1}{2}\left(\mathbb{E}[\,|\eta^{\mathrm{T}}x-\xi|\,]-(\mu^{\mathrm{T}}x-\mu_{n+1})\right),
\end{aligned} \tag{2.199}
$$

which are the separate integrated probability functions defined in Section 2.4.1 on page 143. The following relations hold:

$$
\begin{aligned}
K(x) &= \tfrac{1}{2}\left(A(x)+(\mu^{\mathrm{T}}x-\mu_{n+1})\right) \\
H(x) &= \tfrac{1}{2}\left(A(x)-(\mu^{\mathrm{T}}x-\mu_{n+1})\right).
\end{aligned} \tag{2.200}
$$

According to Proposition 2.45. on page 163 $A(\cdot)$ is a convex function, consequently both $K(x)$ and $H(x)$ are convex functions, too.

Turning our attention to the case when the lower/upper absolute deviation is measured with respect to the expected value, we obviously have (see (2.198))

$$
\begin{aligned}
\rho^{+}_{\mathrm{MAD}}(\vartheta) &:= \mathbb{E}[(\vartheta-\mathbb{E}[\vartheta])^{+}] = \tfrac{1}{2}\rho_{\mathrm{MAD}}(\vartheta) \\
\rho^{-}_{\mathrm{MAD}}(\vartheta) &:= \mathbb{E}[(\vartheta-\mathbb{E}[\vartheta])^{-}] = \tfrac{1}{2}\rho_{\mathrm{MAD}}(\vartheta).
\end{aligned}
$$

This implies that the optimization model for minimizing the corresponding evaluation function will deliver the same results as its mean–absolute–deviation counterpart. With the valuation function in the constraint, the only difference with respect to (2.181) will be the right–hand–side of this constraint: with the semi–deviation measure this will be 2κ.

Exercises

2.10. Give a direct, detailed proof for the fact that $A(x) := \mathbb{E}[\,|\eta^{\mathrm{T}}x-\xi|\,]$ is a convex function.

2.11. Complete the proof of the equivalence of (2.189) and (2.191), as well as of (2.190) and (2.192). The main ingredient of the proof is sketched in the paragraph next to (2.192).

2.6 Modeling risk and opportunity

The different SLP model classes in the previous sections have been identified as follows: a quality measure ρ has been chosen first which characterizes the model class. Based on the selected quality measure, the corresponding evaluation function $V(x) := \rho(\zeta(x,\xi))$ was utilized in building SLP models belonging to the class of models.

In this section we will take a look on some modeling issues concerning SLP models. For the sake of simplicity we will consider the following pair of prototype problems

$$\begin{cases} \min V(x) \\ \text{s.t.} \quad x \in \mathscr{B} \end{cases} \qquad \begin{cases} \max V(x) \\ \text{s.t.} \quad x \in \mathscr{B} \end{cases} \tag{2.201}$$

with the evaluation functions V in the objective. Analogous reasoning applies for SLP models involving constraints with V.

Before proceeding let us emphasize that the two problems in (2.201) are substantially different from the numerical point of view. Assuming, for instance, that V is a nonlinear convex function, this implies that the minimization problem is in general much more easier to solve numerically than its maximization counterpart. Applying the usual trick for transforming the maximization problem into a minimization problem involving $-V$ in the objective, does not help in this respect, of course.

Let us point out next that, from the modeling viewpoint, the mere definition and mathematical properties of a quality measure ρ do not a priori imply a selection between the two possible models in (2.201). To see this, consider the standard deviation $\rho_{\text{Std}}(\vartheta) := \sigma(\vartheta) := \mathbb{E}[(\vartheta - \mathbb{E}[\vartheta])^2]^{\frac{1}{2}}$ as a quality measure, discussed in Section 2.5.1. Notice that the implied evaluation function V is a convex function. With this evaluation function, the SLP model (2.166) on page 162 corresponds to the minimization formulation in (2.201). The usage of this model presupposes the following modeling attitude: the modeler interprets any deviation from the expected value as risk, quantifies the deviations by choosing the standard deviation as quality measure, and seeks to minimize this quality measure. In this modeling context, the quality measure ρ_{Std} can be interpreted as a risk measure. Note that assuming a symmetric distribution, a large standard deviation indicates that ϑ exhibits large deviations both in the upward and downward direction with respect to the expected value. Consider now a gambler. For she/he the upward deviations represent an opportunity for winning, therefore larger standard deviations will be preferred to smaller values. This modeler would choose the maximization problem in (2.201). Consequently, for such a modeler the interpretation of the same quality measure ρ_{Std} is clearly an opportunity measure. The modeler faces a non–convex optimization problem.

In the previous example the same quality measure served simultaneously as risk– and opportunity–measure, the sole difference was the way, how it has been used for building SLP models. Both for the risk–averse modeler and for the gambler the standard deviation is not the best way for building an SLP model. To see this, and to further explore the ways for modeling risk and opportunity, let us assume that

- negative values of $\zeta(x,\xi) - \mathbb{E}[\zeta(x,\xi)]$ are interpreted as something unpleasant, like costs, loss in wealth, or loss in health;
- positive values of $\zeta(x,\xi) - \mathbb{E}[\zeta(x,\xi)]$ quantify something desirable, like monetary gains or stability of an engineering structure;
- $\zeta(x,\xi) - \mathbb{E}[\zeta(x,\xi)] = 0$ expresses neutrality in the risk–opportunity aspect.

Instead of the standard deviation, in this situation it makes sense to choose the lower– and upper standard semi–deviations (see Section 2.5.3) as quality measures. The risk–averse modeler would choose the lower semi–deviation $\rho_{\text{Q}}^{-}(\vartheta) := \mathbb{E}[(\vartheta^{-})^2]^{\frac{1}{2}}$, interpreted as a risk measure. The corresponding optimization problem is the minimization problem in (2.201). A modeler who does not care for risk would choose the upper semi–deviation $\rho_{\mathrm{Q}}^{+}(\vartheta) := \mathbb{E}[(\vartheta^{+})^{2}]^{\frac{1}{2}}$ with the corresponding maximization problem in (2.201). The corresponding evaluation functions are convex functions for both the lower– and for the upper semi–deviation, see the discussion on page 167. Therefore, again, the risk–averse modeler faces a convex optimization problem whereas the modeler neglecting risk has a non–convex optimization problem to solve. The idea of combining the two quality measures, for instance as $\rho_{\mathrm{Q}}^{-} - \lambda \rho_{\mathrm{Q}}^{+}$ with $\lambda > 0$, and minimizing the resulting evaluation function, still results in a non–convex optimization problem.

Another possibility for employing a suitable quality measure is to work with separate integrated probability functions, see Section 2.4.1. The risk–averse modeler would choose $\rho_{\mathrm{sic}}^{-}(\vartheta) := \mathbb{E}[\vartheta^{-}]$ with the corresponding minimization problem whereas her/his risk–seeking counterpart would employ $\rho_{\mathrm{sic}}^{+}(\vartheta) := \mathbb{E}[\vartheta^{+}]$ and the maximization problem. Both corresponding evaluation functions are convex, see Proposition 2.39., therefore analogous comments apply as for the semi–deviations. There is, however, an essential difference: now it makes sense to combine the two quality measures. This leads to the quality measure $\rho_{\mathrm{sic}}^{\alpha}$ discussed on page 146 with $\alpha \in [0,1]$. For $\alpha \in [0,\frac{1}{2})$ it serves as a risk measure with a convex evaluation function whereas for $\alpha \in (\frac{1}{2},1]$ it can be interpreted as quantifying opportunity with a corresponding concave evaluation function.

Finally let us discuss the usage of probability functions in modeling. Concerning separate probability functions, we have seen in Section 2.2.3, that, for certain special cases convex programming problems arise. This is true both for the risk–averse and for the risk–seeking attitude. For joint probability constraints the situation is different, see Section 2.2.5. Convex programming problems can only be obtained when interpreting the quality measure as a measure of opportunity, that means, the evaluation function is to be maximized.

2.7 Risk measures

We consider random variables of the form

$$\zeta(x,\eta,\xi) := \eta^{\mathrm{T}} x - \xi,$$

where (η^{T},ξ) is an $n+1$–dimensional random vector defined on a probability space $(\Omega,\mathscr{F},\mathbb{P})$; η denotes an n–dimensional random vector and ξ is a random variable. Whenever the expected value of (η^{T},ξ) exists we will employ the notation $\mu := \mathbb{E}[\eta] \in \mathbb{R}^{n}$ and $\mu_{n+1} := \mathbb{E}[\xi] \in \mathbb{R}$.

In the previous sections we used a two–step scheme in presenting the various stochastic programming model classes. In a first step we have specified a function $\rho : \Upsilon \to \mathbb{R}$ for evaluating random variables with Υ being some linear space of random variables, defined on a probability space $(\Omega,\mathscr{F},P)$. We have called ρ a quality measure concerning random variables. In a second step, provided that $\zeta(x,\eta,\xi) \in \Upsilon$ holds for all x, we have substituted $\zeta(x,\eta,\xi)$ into ρ thus getting the evaluation function V, $V(x) := \rho(\zeta(x,\eta,\xi))$. V has been subsequently used for building SLP models. Assuming that ρ quantifies risk, V has been built into SLP models as follows: If in the objective, then $V(x)$ was minimized and if in a constraint then constraints of the type $V(x) \le \kappa$ were employed. This modeling attitude justifies the usage of the term "risk measure" for ρ. For optimization models involving V in the above outlined fashion, the (generalized) convexity of V is clearly an advantageous property. It leads to optimization problems for which we have good chances for finding an efficient numerical solution procedure.

For fixed (η,ξ), $\zeta(\cdot,\eta,\xi)$ is a linear–affine function, thus there is a close relation between structural properties of ρ and (generalized) convexity properties of V. The purpose of this section is to discuss properties of various risk measures and their impact on the evaluation function.

Let $(\Omega,\mathscr{F},P)$ be a probability space and ϑ be a random variable on it. The distribution function of ϑ will be denoted by F_ϑ and Θ denotes the support of ϑ. Recall, that Υ has been chosen as one of the linear spaces listed in (2.6) on page 73.

A function $g : X \to \mathbb{R}$, defined on a linear space X, is called *positively homogeneous*, if for any $\lambda \ge 0$ and $x \in X$, the relation $g(\lambda x) = \lambda g(x)$ holds. g is called *subadditive*, if for any $x, y \in X$ the inequality $g(x+y) \le g(x) + g(y)$ holds. For later reference the following simple facts are formulated as an assertion:

Proposition 2.46. *Let $g : X \to \mathbb{R}$ be a function defined on a linear space X. Then*

a) if g is both positively homogeneous and subadditive then it is convex.
b) Suppose that g is positively homogeneous and convex. This implies subadditivity.

Proof: In fact, let $x, y \in X$ and $\lambda \in (0,1)$ then we have

$$g(\lambda x + (1-\lambda)y) \le g(\lambda x) + g((1-\lambda)y) = \lambda g(x) + (1-\lambda)g(y),$$

where the inequality follows from subadditivity and the equality from positive homogeneity. This proves *a)*. Suppose that g is positively homogeneous and convex and let $x, y \in X$ then

$$g(x+y) = g(2[\frac{1}{2}x + \frac{1}{2}y]) = 2g(\frac{1}{2}x + \frac{1}{2}y) \le g(x) + g(y)$$

from which *b)* follows. □

The next proposition establishes a relation between properties of ρ and properties of the corresponding evaluation function V.

Proposition 2.47. *Let Υ be a linear space of random variables and $\rho : \Upsilon \to \mathbb{R}$ a real-valued function on Υ. Assume that $\eta^{\mathrm{T}} x - \xi \in \Upsilon$ holds for all x and let $V(x) := \rho(\eta^{\mathrm{T}} x - \xi)$. Then we have:*

a) If ρ is convex then V is convex too.
b) If $\xi \equiv 0$ and ρ is subadditive then V is also subadditive.
c) If $\xi \equiv 0$ and ρ is positively homogeneous then V is also positively homogeneous.

Proof:
a) Let $x, y \in \mathbb{R}^n$ and $\lambda \in [0,1]$ then we have

$$\begin{aligned} V(\lambda x + (1-\lambda)y) &= \rho(\eta^{\mathrm{T}}(\lambda x + (1-\lambda)y) - \xi) \\ &= \rho(\lambda(\eta^{\mathrm{T}}x + \xi) + (1-\lambda)(\eta^{\mathrm{T}}y + \xi)) \\ &\le \lambda\rho(\eta^{\mathrm{T}}x + \xi) + (1-\lambda)\rho(\eta^{\mathrm{T}}y + \xi) \\ &= \lambda V(x) + (1-\lambda)V(y). \end{aligned}$$

Assertions *b)* and *c)* follow similarly. □

Notice that in the above assertions the stated properties of V hold for any probability distribution of $\vartheta \in \Upsilon$. Thus we obtain convex SLP problems under the sole assumption $\vartheta \in \Upsilon$. By proving the convexity of a specific risk measure ρ, we obtain alternative proofs of convexity of the corresponding SLP problems discussed in the previous sections.

2.7.1 Risk measures in finance

In financial theory and praxis, more closely in portfolio optimization, an increasing effort in research is devoted to identify those properties of risk measures, which are distinguishing features. The general aim of the research is twofold. On the one hand, the goal is to develop an axiomatically founded risk theory in finance. On the other hand, the aim is to provide guidelines for practitioners for choosing an appropriate risk measure in their daily work and to support the construction of appropriate standards for risk management in the finance industry. Several different definitions and systems of axioms have been proposed in the financial literature. Below we simply list some of the current definitions without discussing their intuitive background and implications, these being application–specific. We assume throughout that positive values of the random variables $\vartheta \in \Upsilon$ represent losses.

Kijima and Ohnisi [186] propose the following definition: ρ is a risk measure, if the following properties hold for any $\vartheta, \vartheta_1, \vartheta_2 \in \Upsilon$ and $\lambda, C \in \mathbb{R}$, $\lambda \ge 0$:

$$\begin{array}{lll} (K1) & \rho(\vartheta_1 + \vartheta_2) \le \rho(\vartheta_1) + \rho(\vartheta_2) & \text{(subadditivity)} \\ (K2) & \rho(\lambda\vartheta) = \lambda\rho(\vartheta) & \text{(positive homogeneity)} \\ (K3) & \rho(\vartheta) \ge 0 & \text{(nonnegativity)} \\ (K4) & \rho(\vartheta + C) = \rho(\vartheta) \ \text{for}\ C \ge 0 & \text{(shift invariance)} \end{array} \tag{2.202}$$

The important issue of axiomatic foundation of risk measures has first been addressed in the seminal paper of Artzner, Delbaen, Eber, and Heath [7]. The authors propose the axioms below and explore their implications. We formulate the axioms for random variables representing losses, whereas in the original paper the interpretation is future value.

$$\begin{array}{lll} (A1) & \rho(\vartheta_1+\vartheta_2)\le\rho(\vartheta_1)+\rho(\vartheta_2) & \text{(subadditivity)} \\ (A2) & \rho(\lambda\vartheta)=\lambda\rho(\vartheta) & \text{(positive homogeneity)} \\ (A3) & \text{If } \vartheta_1\le\vartheta_2 \text{ then } \rho(\vartheta_1)\le\rho(\vartheta_2) & \text{(monotonicity)} \\ (A4) & \rho(\vartheta+C)=\rho(\vartheta)+C & \text{(translation invariance)} \end{array} \tag{2.203}$$

The authors call a function ρ for which the above axioms hold a *coherent risk measure*. Concerning SLP models in general, in an intuitive sense the axiom *A4* looks rather unusual. The reason for including it in this form is that the authors consider capital requirement problems, see [7]. For distinguishing between the different requirements concerning translation in (2.202) and (2.203), we use the terms “shift invariance” and “translation invariance”, respectively.

In the system of axioms of Föllmer and Schied [94], [95], subadditivity and positive homogeneity is replaced by the weaker requirement of convexity

$$\begin{array}{lll} (F1) & \rho(\vartheta) \text{ is a convex function} & \text{(convexity)} \\ (F2) & \text{If } \vartheta_1\le\vartheta_2 \text{ then } \rho(\vartheta_1)\le\rho(\vartheta_2) & \text{(monotonicity)} \\ (F3) & \rho(\vartheta+C)=\rho(\vartheta)+C & \text{(translation invariance)} \end{array} \tag{2.204}$$

leading to *convex risk measures.* Coherent risk measures are obviously convex; a convex risk measure is coherent, if it is positively homogeneous (see Proposition 2.46.).

Rockafellar, Uryasev, and Zabarankin [284] introduce the notion of *deviation measure* for $\Upsilon=\mathscr{L}_1^1$. Their axioms are

$$\begin{array}{lll} (D1) & \rho(\vartheta_1+\vartheta_2)\le\rho(\vartheta_1)+\rho(\vartheta_2) & \text{(subadditivity)} \\ (D2) & \rho(\lambda\vartheta)=\lambda\rho(\vartheta) & \text{(positive homogeneity)} \\ (D3) & \rho(\vartheta)>0 \text{ for } \vartheta \text{ non–constant,} & \\ & \rho(\vartheta)=0 \text{ otherwise} & \text{(nonnegativity)} \\ (D4) & \rho(\vartheta+C)=\rho(\vartheta) & \text{(shift invariance)} \end{array} \tag{2.205}$$

The authors also define an associated risk measure called *expectation–bounded risk measure*, see [284], and explore the implications in portfolio theory. Notice that for a deviation measure the axioms (2.202) hold; the axioms for a deviation measure can be considered as a refinement (restriction) of (2.202).

Due to the different prescription for the case of a translation, the set of risk measures obeying (2.202) or (2.205) and the risk measures for which either (2.203) or (2.204) hold, are disjunct sets.

We feel that there is not much chance that a general definition of a risk measure can be given, which would be acceptable also beyond the field of finance. From our general stochastic programming point of view, the convexity of a risk–measure is surely a desirable property. Proposition 2.47. implies, namely, that the SLP models, which are built on the basis of such a measure, are convex optimization problems. From this viewpoint, a risk measure can be considered as more valuable, when beyond serving as a diagnostic metric, it can also be built into efficiently solvable optimization models which involve, for instance, minimizing risk. Without exception, all of the above definitions correspond to risk measures of this type.

2.7.2 Properties of risk measures

This section is devoted to discussing convexity properties of risk measures which have been utilized for building stochastic programming models. Unless explicitly referring to the axioms (2.204) of convex risk measures, under convexity we will simply mean convexity of the risk–measure–function ρ. We will use the following notation: the functions $\varphi^+ : \mathbb{R} \to \mathbb{R}_+$ and $\varphi^- : \mathbb{R} \to \mathbb{R}_+$ are defined as $\varphi^+(z) := z^+$ and $\varphi^-(z) = z^-$, respectively, where $z^+ = \max\{0, z\}$ and $z^- = \max\{0, -z\}$ are the positive– and negative part of the real number z. Let further φ^A denote the absolute–value function $\varphi^A(z) := |z|$ for all $z \in \mathbb{R}$. The relation $\varphi^A = \varphi^+ + \varphi^-$ obviously holds. Note that φ^+, φ^-, and φ^A are positively homogeneous and subadditive functions, therefore they are convex.

Proposition 2.48. *The following risk measures are positively homogeneous and subadditive. Moreover, they are also monotonously increasing and translation invariant. Consequently, they are convex risk measures in the sense of axioms (2.204) and being positively homogeneous they are also coherent according to axioms (2.203).*

(A) $\rho_{\mathrm{E}}(\vartheta) := \mathbb{E}[\vartheta], \ \vartheta \in \mathscr{L}_1^1$, *(Section 2.4);*

(B) $\rho_{\mathrm{fat}}(\vartheta) := \max\limits_{\widehat{\vartheta} \in \Theta} \widehat{\vartheta}, \ \vartheta \in \mathscr{L}_1^\infty$ *where Θ is the support of ϑ (Section 2.1, page 76); note that we have changed minimum to maximum for getting a risk measure. ρ_{fat} is called the maximum loss risk measure.*

(C) $\rho^{\alpha}_{\mathrm{CVaR}}(\vartheta) := v_c(\vartheta, \alpha) := \min\limits_{z} [z + \frac{1}{1-\alpha}\mathbb{E}[(\vartheta - z)^+]], \ \vartheta \in \mathscr{L}_1^1, 0 < \alpha < 1,$ *(Section 2.4.3).*

Proof:

(A): The assertion holds trivially because ρ_{E} is a linear function.

(B): Let $\lambda \geq 0$ be a real number and $\vartheta \in \mathscr{L}_1^\infty$ with support Θ. Then $\lambda\Theta$ is a closed set, therefore it is the support of $\lambda\vartheta$. For the definition of the operation $\lambda\Theta$ see (2.89) on page 118. Thus we have

$$\rho_{\mathrm{fat}}(\lambda\vartheta) = \max_{\widehat{\vartheta} \in \lambda\Theta} \widehat{\vartheta} = \max_{\bar{\vartheta} \in \Theta} \lambda\bar{\vartheta} = \lambda\rho_{\mathrm{fat}}(\vartheta).$$

For proving subadditivity let $\vartheta_1, \vartheta_2 \in \mathscr{L}_1^\infty$ with supports Θ_1 and Θ_2, respectively. Let $\Theta := \Theta_1 + \Theta_2$ where the sum of the two sets is defined according to (2.89) on page 118. From the discussion on that page it follows that Θ is a closed set. Consequently, Θ contains the support of $\vartheta_1 + \vartheta_2$. Thus we have

$$\begin{aligned}\rho_{\mathrm{fat}}(\vartheta_1 + \vartheta_2) &= \max_{\widehat{\vartheta}_1 + \widehat{\vartheta}_2 \in supp\{\vartheta_1 + \vartheta_2\}} [\widehat{\vartheta}_1 + \widehat{\vartheta}_2] \\ &\leq \max_{\widehat{\vartheta}_1 + \widehat{\vartheta}_2 \in \Theta} [\widehat{\vartheta}_1 + \widehat{\vartheta}_2] \leq \max_{\bar{\vartheta}_1 \in \Theta_1} \bar{\vartheta}_1 + \max_{\bar{\vartheta}_2 \in \Theta_2} \bar{\vartheta}_2.\end{aligned}$$

For any real number C, the support of $\vartheta + C$ is $\Theta + \{C\}$, from which the translation invariance immediately follows. If $\vartheta_1(\omega) \le \vartheta_2(\omega)$ holds for all $\omega \in \Omega$ then we obviously have $\rho_{\text{fat}}(\vartheta_1) \le \rho_{\text{fat}}(\vartheta_2)$.

(C): Let $\lambda \ge 0$ and $\vartheta \in \mathcal{L}_1$. If $\lambda = 0$, then we have

$$\rho^{\alpha}_{\text{CVaR}}(\lambda\vartheta) = \rho^{\alpha}_{\text{CVaR}}(0) = \min_z [z + \frac{1}{1-\alpha}\mathbb{E}[(-z)^+] = 0,$$

where the last equality follows from

$$z + \frac{1}{1-\alpha}\mathbb{E}[(-z)^+] = \begin{cases} z & \text{if } z \ge 0 \\ (1 - \frac{1}{1-\alpha})z & \text{if } z < 0. \end{cases}$$

Assuming $\lambda > 0$ we have

$$\begin{aligned} \rho^{\alpha}_{\text{CVaR}}(\lambda\vartheta) &= \min_z [z + \frac{1}{1-\alpha}\mathbb{E}[(\lambda\vartheta - z)^+]] \\ &= \lambda \min_z [\frac{z}{\lambda} + \frac{1}{1-\alpha}\mathbb{E}[(\vartheta - \frac{z}{\lambda})^+]] \\ &= \lambda \min_y [y + \frac{1}{1-\alpha}\mathbb{E}[(\vartheta - y)^+]] = \lambda \rho^{\alpha}_{\text{CVaR}}(\vartheta). \end{aligned}$$

For proving subadditivity we utilize the fact (see Section 2.4.3) that the minimum in the definition is attained. Let $\vartheta_1, \vartheta_2 \in \mathcal{L}_1$ and z_1, z_2 be corresponding solutions of the minimization problem in the definition. For proving subadditivity it is sufficient to prove convexity (see Proposition 2.46.). Let $0 < \lambda < 1$, z_λ be the minimum for $\lambda\vartheta_1 + (1-\lambda)\vartheta_2$, and $\bar{z}_\lambda = \lambda z_1 + (1-\lambda)z_2$. Utilizing the convexity of φ^+ we get

$$\begin{aligned} \rho^{\alpha}_{\text{CVaR}}(\vartheta_1 + \vartheta_2) &= \min_z [z + \frac{1}{1-\alpha}\mathbb{E}[\varphi^+(\lambda\vartheta_1 + (1-\lambda)\vartheta_2 - z)]] \\ &= z_\lambda + \tfrac{1}{1-\alpha}\mathbb{E}[\varphi^+(\lambda\vartheta_1 + (1-\lambda)\vartheta_2 - z_\lambda)] \\ &\le \bar{z}_\lambda + \tfrac{1}{1-\alpha}\mathbb{E}[\varphi^+(\lambda\vartheta_1 + (1-\lambda)\vartheta_2 - \bar{z}_\lambda)] \\ &\le \rho^{\alpha}_{\text{CVaR}}(\vartheta_1) + \rho^{\alpha}_{\text{CVaR}}(\vartheta_2). \end{aligned}$$

Due to the fact that φ^+ is a monotonically increasing function, the monotonicity of $\rho^{\alpha}_{\text{CVaR}}$ follows immediately. Let $C \in \mathbb{R}$ then we have

$$\begin{aligned} \rho^{\alpha}_{\text{CVaR}}(\vartheta + C) &= \min_z [z + \frac{1}{1-\alpha}\mathbb{E}[(\vartheta - (z - C))^+]] \\ &= C + \min_z [z - C + \frac{1}{1-\alpha}\mathbb{E}[(\vartheta - (z - C))^+]] \\ &= \rho^{\alpha}_{\text{CVaR}}(\vartheta) + C \end{aligned}$$

thus the translation invariance follows. □

For the next group of risk measures translation– or shift–invariance does not hold, but we have:

Proposition 2.49. *The risk measures listed below are positively homogeneous and subadditive and they are also monotonous.*

(D1) $\rho^+_{\text{sic}}(\vartheta) := \mathbb{E}[\vartheta^+]$, $\vartheta \in \mathscr{L}^1_1$, *(Section 2.4.1);*

(D2) $\rho^-_{\text{sic}}(\vartheta) := \mathbb{E}[\vartheta^-]$, $\vartheta \in \mathscr{L}^1_1$, *(Section 2.4.1);*

(E) $\rho^\alpha_{\text{sic}}(\vartheta) := \alpha\mathbb{E}[\vartheta] + (1-2\alpha)\mathbb{E}[\vartheta^+] = \alpha\rho_{\text{E}}(\vartheta) + (1-2\alpha)\rho^+_{\text{sic}}(\vartheta)$, $\vartheta \in \mathscr{L}^1_1$, $0 \le \alpha \le \frac{1}{2}$, *(Section 2.4.1).*

Proof:
(D1): We have $\rho^+_{\text{sic}}(\vartheta) = \mathbb{E}[\varphi^+(\vartheta)]$. The function φ^+ being positively homogeneous and subadditive, as well as monotonously increasing, the assertion follows immediately. In fact, for proving subadditivity let $\vartheta_1, \vartheta_2 \in \mathscr{L}_1$ then we have

$$\rho^+_{\text{sic}}(\vartheta_1 + \vartheta_2) = \mathbb{E}[\varphi^+(\vartheta_1 + \vartheta_2)] \le \mathbb{E}[\varphi^+(\vartheta_1) + \varphi^+(\vartheta_2)] = \rho^+_{\text{sic}}(\vartheta_1) + \rho^+_{\text{sic}}(\vartheta_2).$$

The proof for positive homogeneity is analogous. ρ^+_{sic} turns out to be a monotonically increasing function.
(D2): In this case $\rho^-_{\text{sic}}(\vartheta) = \mathbb{E}[\varphi^-(\vartheta)]$ holds. The positive homogeneity and subadditivity of φ^- implies these properties for ρ^-_{sic}. φ^- being monotonously decreasing, ρ^-_{sic} is monotonically decreasing, too.
(E): This follows immediately from the linearity of the first term and from *(D1).* □

In the next group neither translation–invariance nor monotonicity holds. Nevertheless, we have

Proposition 2.50. *The following risk measures are positively homogeneous and subadditive.*

(F) $\rho_{\text{Q}}(\vartheta) := \sqrt{\mathbb{E}[\vartheta^2]}$, $\vartheta \in \mathscr{L}^2_1$, *(Section 2.5.1);*

(G) $\rho_{\text{A}}(\vartheta) := \mathbb{E}[|\vartheta|]$, $\vartheta \in \mathscr{L}^1_1$, *(Section 2.5.2);*

(H) $\rho^+_{\text{Q}}(\vartheta) := \sqrt{\mathbb{E}[(\vartheta^+)^2]}$ *and*
$\rho^-_{\text{Q}}(\vartheta) := \sqrt{\mathbb{E}[(\vartheta^-)^2]}$, $\vartheta \in \mathscr{L}^1_1$, *(Section 2.5.3).*

Proof: The positive homogeneity is trivial for all cases therefore we confine ourselves to proving subadditivity.
(F): Let $\vartheta_1, \vartheta_2 \in \mathscr{L}_2$ then the Minkowski–inequality immediately yields

$$\begin{aligned}\rho_{\text{Q}}(\vartheta_1 + \vartheta_2) &= \left(\mathbb{E}[(\vartheta_1 + \vartheta_2)^2]\right)^{\frac{1}{2}} = \left(\mathbb{E}[|\vartheta_1 + \vartheta_2|^2]\right)^{\frac{1}{2}} \\ &\le \left(\mathbb{E}[|\vartheta_1|^2]\right)^{\frac{1}{2}} + \left(\mathbb{E}[|\vartheta_2|^2]\right)^{\frac{1}{2}} = \rho_{\text{Q}}(\vartheta_1) + \rho_{\text{Q}}(\vartheta_2).\end{aligned}$$

(G): We have $\rho_{\text{A}}(\vartheta) = \mathbb{E}[\varphi^A(\vartheta)]$ and the assertion follows from the subadditivity of φ^A.

(H): We prove the assertion for ρ_{Q}^+, the proof for ρ_{Q}^- is analogous. We utilize the subadditivity of φ^+ and again the Minkowski-inequality:

$$\begin{aligned}\rho_{\text{Q}}^+(\vartheta_1 + \vartheta_2) &= \left(\mathbb{E}[((\vartheta_1 + \vartheta_2)^+)^2]\right)^{\frac{1}{2}} \\ &\le \left(\mathbb{E}[(\vartheta_1^+ + \vartheta_2^+)^2]\right)^{\frac{1}{2}} \\ &\le \left(\mathbb{E}[(\vartheta_1^+)^2]\right)^{\frac{1}{2}} + \left(\mathbb{E}[(\vartheta_2^+)^2]\right)^{\frac{1}{2}} = \rho_{\text{Q}}^+(\vartheta_1) + \rho_{\text{Q}}^+(\vartheta_2).\end{aligned}$$

□

Finally we turn our attention to the deviation measures in Section 2.5.

Proposition 2.51. *The following risk measures are deviation measures according to the axioms (2.205).*

(I) $\rho_{\text{Std}}(\vartheta) := \sigma(\vartheta) := \sqrt{\mathbb{E}[(\vartheta - \mathbb{E}[\vartheta])^2]},\ \vartheta \in \mathscr{L}_1^2;$

(J) $\rho_{\text{MAD}}(\vartheta) := \mathbb{E}[|\vartheta - \mathbb{E}[\vartheta]|],\ \vartheta \in \mathscr{L}_1^1;$

(K) $\rho_{\text{Std}}^+(\vartheta) := \sigma^+(x) := \sqrt{\mathbb{E}[((\vartheta - \mathbb{E}[\vartheta])^+)^2]}$ *and*
$\rho_{\text{Std}}^-(\vartheta) := \sigma^-(x) := \sqrt{\mathbb{E}[((\vartheta - \mathbb{E}[\vartheta])^-)^2]},\ \vartheta \in \mathscr{L}_1^1;$

(L) $\rho_{\text{MAD}}^+(\vartheta) := \mathbb{E}[(\vartheta - \mathbb{E}[\vartheta])^+] = \frac{1}{2}\rho_{\text{MAD}}(\vartheta)$ *and*
$\rho_{\text{MAD}}^-(\vartheta) := \mathbb{E}[(\vartheta - \mathbb{E}[\vartheta])^-] = \frac{1}{2}\rho_{\text{MAD}}(\vartheta),\ \vartheta \in \mathscr{L}_1^1.$

Proof: Note that each one of these risk measures results from an already considered risk measure by substituting ϑ by $\vartheta - \mathbb{E}[\vartheta]$. Therefore it is clear that each one is positively homogeneous and subadditive. All of them are nonnegative and can only be zero if ϑ is constant. Finally the shift–property *D4* holds trivially. □

Recall, that due to Proposition 2.46., the positive homogeneity and subadditivity of the risk functions considered so far implies that all of them are convex.

The risk measures listed below have been used for building SLP models but have not yet been considered:

(M) $\rho_{\text{P}}(\vartheta) := \mathbb{P}(\vartheta \ge 0),\ \vartheta \in \mathscr{V}$, (Section 2.2);

(N) $\rho_{\text{cexp}}(\vartheta) := \mathbb{E}[-\vartheta \mid \vartheta < 0],\ \vartheta \in \mathscr{L}_1^1$, (Section 2.4.2);

(O) $\rho_{\text{VaR}}^{\alpha}(\vartheta) := \nu(\vartheta, \alpha) := \min\{z \mid F_\vartheta(z) \ge \alpha\},\ \vartheta \in \mathscr{V}, 0 < \alpha < 1$, (Section 2.3).

They are non–convex in general. Despite this fact, we have seen in the previous sections that under some assumptions concerning the probability distribution and parameter values, using these quality measures resulted in convex or in generalized convex optimization problems. The point is the following: having a convex risk measure ρ, this leads automatically to convex evaluation functions (see Proposition 2.47.) and thus to convex optimization problems. In other words, the convexity of ρ is a sufficient condition for getting convex optimization problems. The convexity of ρ is by no means also necessary for this, as the convex optimization models, built on the basis of the above risk measures, and presented in the previous sections demonstrate.

At last let us consider risk measures for random vectors, introduced in Section 2.4.1 as

$$\text{(P)} \quad \rho^{+}_{\text{jic}}(\vartheta) := \mathbb{E}[\max_{1 \le i \le s} \vartheta_i^{+}] \text{ and } \rho^{-}_{\text{jic}}(\vartheta) := \mathbb{E}[\max_{1 \le i \le s} \vartheta_i^{-}], \quad \vartheta \in \mathscr{L}_1^1,$$

where ϑ is now an s–dimensional random vector. These risk measures have the properties:

Proposition 2.52. *Both ρ^{+}_{jic} and ρ^{-}_{jic} are positively homogeneous and subadditive. Moreover, both of them are monotonous.*

Proof: The positive homogeneity is obvious. We prove the subadditivity for ρ^{+}_{jic}, the proof for ρ^{-}_{jic} is analogous. Let $\vartheta^{(1)}, \vartheta^{(2)} \in \mathscr{L}_1^1$ then we have

$$\begin{aligned}
\rho^{+}_{\text{jic}}(\vartheta^{(1)} + \vartheta^{(2)}) &= \mathbb{E}[\max_{1 \le i \le s} (\vartheta_i^{(1)} + \vartheta_i^{(2)})^{+}] \\
&\le \mathbb{E}\left[\max_{1 \le i \le s} [(\vartheta_i^{(1)})^{+} + (\vartheta_i^{(2)})^{+}]\right] \\
&\le \mathbb{E}\left[\max_{1 \le i \le s} [(\vartheta_i^{(1)})^{+}]\right] + \mathbb{E}\left[\max_{1 \le i \le s} [(\vartheta_i^{(2)})^{+}]\right] \\
&= \rho^{+}_{\text{jic}}(\vartheta^{(1)}) + \rho^{+}_{\text{jic}}(\vartheta^{(2)})
\end{aligned}$$

where for the first inequality we used the subadditivity of φ^{+} and the second inequality follows from the properties of the max operator. From the properties of φ^{+} and φ^{-} it is also clear that ρ^{+}_{jic} is monotonically increasing whereas ρ^{-}_{jic} is monotonically decreasing. □

2.7.3 Portfolio optimization models

For illustrating the use of various risk measures in practice, we present some portfolio optimization models. We consider a one–period financial portfolio optimization problem with n risky assets. Let $\eta^{\mathrm{T}} = (\eta_1, \ldots, \eta_n)$ be the vector of random returns of the assets and $r_i := \mathbb{E}[\eta_i]$, $i = 1, \ldots, n$ be the expected returns. The asset–weights in the portfolio will be denoted by $x_1, \ldots, x_n$, thus $\eta^{\mathrm{T}} x$ represents the random portfolio return. Since for risk measures we have interpreted positive values of random variables as losses, we take $\zeta(x, \eta) := -\eta^{\mathrm{T}} x$. With μ_p standing in this section for a prescribed minimal expected portfolio return level, we consider optimization problems of the following form:

$$\left.\begin{aligned} \psi(\mu_p) = \min\ & \rho(-\eta^{\mathrm{T}}x) \\ \text{s.t.}\ & r^{\mathrm{T}}x \geq \mu_p \\ & \mathbb{1}^{\mathrm{T}}x = 1 \\ & x \in \mathscr{B}, \end{aligned}\right\} \tag{2.206}$$

where $\mathscr{B}$ is a polyhedral set determined by additional linear constraints, ρ is a risk measure, and $\mathbb{1}^{\mathrm{T}} = (1,\ldots,1)$ holds. The interpretation is the following: we are looking for a portfolio with minimum risk, under prescribing a minimum acceptable level μ_p of expected portfolio return. This formulation of the portfolio selection problem is called a *risk–reward* model, with $\rho(-\eta^{\mathrm{T}}x)$ standing for risk and $r^{\mathrm{T}}x$ representing reward.

Some well–known particular cases, differing in the choice of the risk measure are the following:

- $\rho = \rho_{\mathrm{Std}}$ corresponds to the classical minimum–variance model of Markowitz [217];
- $\rho = \rho_{\mathrm{Std}}^{-}$ leads to the mean–semivariance model of Markowitz [217];
- $\rho = \rho_{\mathrm{MAD}}$ gives the mean–absolute–deviation model of Konno and Yamazaki [194];
- $\rho = \rho_{\mathrm{CVaR}}^{\alpha}$ corresponds to the mean–CVaR model of Rockafellar and Uryasev [282];
- $\rho = \rho_{\mathrm{VaR}}^{\alpha}$ results in the mean–VaR model widely used in the finance industry, see, for instance, Jorion [151].

Note that all of these risk measures belong to the class of deviation measures. Although problem (2.206) is also useful in its own right, in finance this problem is considered as a parametric optimization problem with parameter μ_p. The optimal objective value $\psi(\mu_p)$, as a function of μ_p, plays an important role. Its graph in $\mathbb{R}^2$ is called the *efficient frontier*, corresponding to the risk ρ and return μ_p. Traditionally, the efficient frontier is represented graphically with the horizontal axis corresponding to risk and the vertical one corresponding to return.

The reason behind considering the efficiency curve is the following: we actually face a bi–objective optimization problem, where we would like to maximize the expected return and at the same time minimize risk. In all cases listed above, $\psi(\mu_p)$ is strictly monotonically increasing in μ_p, on the interval where the constraint $r^{\mathrm{T}}x \geq \mu_p$ is active at the optimal solution. Consequently, on the ψ–interval corresponding to this interval, ψ^{-1} exists and is strictly monotonically increasing. Thus it makes sense to consider the following alternative representation of the efficiency curve:

$$\left.\begin{aligned} \mu(\psi_p) = \max\ & r^{\mathrm{T}}x \\ \text{s.t.}\ & \rho(-\eta^{\mathrm{T}}x) \leq \psi_p \\ & \mathbb{1}^{\mathrm{T}}x = 1 \\ & x \in \mathscr{B}, \end{aligned}\right\} \tag{2.207}$$

where now $\psi_p > 0$ plays the role of a parameter. The interpretation of this problem is the following: we maximize expected return under the condition that the maximum acceptable risk is ψ_p. The portfolio optimization model in the above form is called a *reward–risk* model.

Due to the multi–objective character of the problem setting, it is not surprising that a third characterization of the efficient frontier is via the optimization problem

$$\left.\begin{array}{ll} \max & r^{\mathrm{T}}x - \nu\rho(-\eta^{\mathrm{T}}x) \\ \text{s.t.} & \mathbb{1}^{\mathrm{T}}x = 1 \\ & x \in \mathscr{B}, \end{array}\right\} \tag{2.208}$$

where in this case $\nu \geq 0$ is acting as a (risk–aversion) parameter for the efficiency curve. The evaluation function for the risk is accounted for by an additive term with a negative sign.

For details on the relationship between these three problems see, for instance, Palmquist, Uryasev, and Krokhmal [249].

Let us finally remark that taking $\zeta(x,\eta,\xi) := \eta^{\mathrm{T}}x - \xi$ instead of $\zeta(x,\eta) := \eta^{\mathrm{T}}x$ also leads to an important class of portfolio optimization problems. In this case ξ may represent, for instance, the random return of a benchmark which can be, for instance, an index like the Dow Jones Industrial Average.

2.7.4 Optimizing performance

In this section we consider random variables of the form

$$\zeta(x,\eta,\xi) := \eta^{\mathrm{T}}x - \xi$$

with positive values representing gains and negative values representing losses.

Gains will be measured via the expected value

$$f(x) := \mathbb{E}[\eta^{\mathrm{T}}x - \xi] = \mu^{\mathrm{T}}x - \mu_{n+1}$$

(termed in this context as *reward*), where we have employed the notation $\mu := \mathbb{E}[\eta] \in \mathbb{R}^n$ and $\mu_{n+1} := \mathbb{E}[\xi] \in \mathbb{R}$. Losses will be measured via the evaluation function corresponding to a risk measure by

$$g(x) := \rho(-\eta^{\mathrm{T}}x + \xi),$$

termed as *risk*. Notice that we have substituted $\vartheta = -\zeta(x,\eta,\xi)$ into $\rho(\cdot)$ in order to being in accordance with the convention in Section 2.7.1, where positive values of ϑ represented losses.

As a performance measure we choose the reward–to–risk ratio

$$\frac{f(x)}{g(x)} = \frac{\mathbb{E}[\eta^{\mathrm{T}}x - \xi]}{\rho(-\eta^{\mathrm{T}}x + \xi)} = \frac{\mu^{\mathrm{T}}x - \mu_{n+1}}{\rho(-\eta^{\mathrm{T}}x + \xi)}$$

representing reward per unit risk.

Note that the selection of an appropriate performance measure heavily depends on the specific application. Considering, for example, financial portfolio optimization, there is a wide variety of performance evaluation measures in use, see e.g. Cogneau and Hubner [44]. In the financial portfolio optimization context (see Section 2.7.3), our performance measure corresponds to the classical Sharpe–ratio when choosing the standard deviation ρ_{Std} as the risk measure. The Sharpe–ratio is one of the most widely used performance measure in the field of financial portfolio management.

The performance optimization problem is formulated as

$$\left.\begin{array}{ll} \max\limits_{x} & \dfrac{f(x)}{g(x)} \\ \text{s.t.} & x \in \mathscr{B} \end{array}\right\} \equiv \left.\begin{array}{ll} \max\limits_{x} & \dfrac{\mu^{\mathrm{T}}x - \mu_{n+1}}{\rho(-\eta^{\mathrm{T}}x + \xi)} \\ \text{s.t.} & x \in \mathscr{B}, \end{array}\right\} \tag{2.209}$$

where $\mathscr{B}\{\, x \mid Ax = b, x \geq 0 \,\}$ is a polyhedral set. Considering the financial portfolio optimization case, the solutions of (2.209) are called *tangential portfolios*. Concerning (2.209) we make the following assumptions:

A1. The set $\mathscr{B}$ of feasible solutions is nonempty and bounded.
A2. $g(x) > 0$ holds for all $x \in \mathscr{B}$.
A3. $\exists \hat{x} \in \mathscr{B}$ for which $f(\hat{x}) > 0$ holds.
A4. ρ is a positively homogeneous risk measure and $g(x)$ is continuous.

Assumptions *A1* and *A4* imply that for (2.209) optimal solutions exist. Notice also that *A2* and *A3* immediately imply that at optimal solutions x^* the inequality $\dfrac{f(x^*)}{g(x^*)} > 0$ must hold, that is, $f(x^*) > 0$ holds.

From the optimization point of view problem (2.209) belongs to the class of fractional programming problems. Assuming that $f(x) \geq 0$ holds for all $x \in \mathscr{B}$ and that ρ is a convex risk measure, Proposition 2.32. on page 83 implies that the objective function of (2.209) is pseudo–concave, thus (2.209) has favorable properties from the numerical point of view.

It is a well–known fact in fractional programming that under appropriate assumptions (2.209) can be equivalently formulated as a convex programming problem, see e.g. Avriel et al. [8] and Schaible [298], [297]. Under the positive homogeneity assumption *A4* concerning ρ, the equivalent convex programming problem can be further simplified, see Stoyanov, Rachev and Fabozzi [314].

For deriving the equivalent convex programming problem let us consider the following problem first:

$$\left.\begin{array}{rl} \max\limits_{y,t} & \mu^{\mathrm{T}} y \quad -\mu_{n+1} t \\ \text{s.t.} & \rho(-\eta^{\mathrm{T}} y + \xi t) = 1 \\ & Ay \qquad -bt = 0 \\ & y \qquad\qquad \geq 0 \\ & \qquad\qquad t \geq 0. \end{array}\right\} \tag{2.210}$$

We observe that for any (y,t) for which the last three constraints hold we have: if $t = 0$ then $y = 0$ follows. In fact, if $t = 0$ and $y \neq 0$ then $Ay = 0$, $y \geq 0$ has a nontrivial solution which contradicts the boundedness of $\mathscr{B}$ (cf. assumption *A1*).

We will show that under our assumptions (2.209) and (2.210) are equivalent. In fact, let x be a feasible solution of (2.209). Then, due to the positive homogeneity of ρ, with

$$t := \frac{1}{g(x)} = \frac{1}{\rho(-\eta^{\mathrm{T}} x + \xi)} > 0, \qquad y := tx,$$

(y,t) is a feasible solution of (2.210) and the corresponding objective function values of the two problems are equal.
Conversely, assume that (y,t) is a feasible solution of (2.210). Then $t > 0$ must hold, because otherwise, as we have seen above, $y = 0$ follows. But for $(y,t) = (0,0)$ the first constraint in (2.210) cannot hold since for the positively homogeneous ρ we have $\rho(0) = 0$. Thus, for any feasible solution $t > 0$ holds. Then $x := \frac{1}{t} y$ is obviously feasible in (2.209) and the positive homogeneity of ρ implies that the corresponding objective function values are equal.

Consequently, (2.209) and (2.210) are equivalent. In particular, (2.210) has an optimal solution (y^*, t^*) and the optimal objective values are equal. Since the optimal objective value of (2.209) is positive, we have

$$\mu^{\mathrm{T}} y^* - \mu_{n+1} t^* > 0.$$

Next we consider the following relaxation of (2.210)

$$\left.\begin{array}{rl} \max\limits_{y,t} & \mu^{\mathrm{T}} y \quad -\mu_{n+1} t \\ \text{s.t.} & \rho(-\eta^{\mathrm{T}} y + \xi t) \leq 1 \\ & Ay \qquad -bt = 0 \\ & y \qquad\qquad \geq 0 \\ & \qquad\qquad t \geq 0, \end{array}\right\} \tag{2.211}$$

where the first equality constraint has been relaxed. We are going to prove that under our assumptions the equivalence of (2.210) and (2.211) follows.

Since (2.211) is a relaxation of (2.210), the optimal objective value of (2.211) must be positive provided that an optimal solution exists. Although $y = 0$,

$t = 0$ is a feasible solution of (2.211), it cannot be optimal since the corresponding objective value is 0. In general, concerning optimality it is sufficient to consider feasible solutions of (2.211) with positive objective function values.

Let (y,t) be a feasible solution with $\mu^{\mathrm{T}} y - \mu_{n+1} t > 0$. Assume that for such a solution the first inequality constraint in (2.211) is inactive, that is, $\rho(-\eta^{\mathrm{T}} y + \xi t) < 1$ holds. We take

$$\gamma := \frac{1}{\rho(-\eta^{\mathrm{T}} y + \xi t)} > 1, \qquad \bar{y} := \gamma y, \qquad \bar{t} := \gamma t.$$

Then, due to the positive homogeneity of ρ, $(\bar{y},\bar{t})$ is obviously a feasible solution of (2.211) with the first inequality constraint being active. For the corresponding objective function value we get:

$$\mu^{\mathrm{T}} \bar{y} - \mu_{n+1} \bar{t} \;=\; \gamma(\mu^{\mathrm{T}} y - \mu_{n+1} t) \;>\; \mu^{\mathrm{T}} y - \mu_{n+1} t.$$

Consequently, with any feasible (y,t) for which $\mu^{\mathrm{T}} y - \mu_{n+1} t > 0$ holds and for which the first inequality constraint in (2.211) is inactive, we can associate a feasible solution $(\bar{y},\bar{t})$ having the following properties: it has a higher objective function value than (y,t) and for this feasible solution the first constraint is active in (2.211). This implies that problems (2.210) and (2.211) are equivalent, the optimal solution of (2.211) exists and the first constraint in (2.211) is active at the optimum.

We have shown that the following proposition holds:

Proposition 2.53. *Let us assume that A1–A4 hold. Then the fractional programming problem (2.209) is equivalent to (2.211).*

If ρ is a coherent risk measure, then (2.211) is clearly a convex programming problem, serving as an equivalent formulation of the original fractional programming problem (2.209).

For the equivalent reformulation the assumption $g(x) > 0, \forall x \in \mathscr{B}$ is an essential one. From the modeling point of view, enforcing this could be done by adding the inequality $g(x) \geq \varepsilon$ to the set of constraints of (2.209), with some $\varepsilon > 0$. The difficulty: if ρ is a convex risk measure, this is a reverse convex constraint, transforming (2.209) and its reformulation (2.211) into non–convex optimization problems. Therefore, we consider the following alternative formulation of (2.209):

$$\left.\begin{array}{ll} \min\limits_{x} & \dfrac{g(x)}{f(x)} \\[2ex] \text{s.t.} & x \in \mathscr{B} \end{array}\right\} \quad \equiv \quad \left\{\begin{array}{ll} \min\limits_{x} & \dfrac{\rho(-\eta^{\mathrm{T}} x + \xi)}{\mu^{\mathrm{T}} x - \mu_{n+1}} \\[2ex] \text{s.t.} & x \in \mathscr{B}. \end{array}\right. \tag{2.212}$$

If $f(x) > 0$ and $g(x) > 0$ hold for $\forall x \in \mathscr{B}$ then this problem is clearly equivalent to (2.209). Therefore we keep assumptions *A1, A2, A4* and replace *A3* with the stronger requirement

A3'. $f(x) > 0$ holds for all $x \in \mathscr{B}$.

If g is a convex function then (2.212) involves the minimization of a pseudo–convex objective function under linear constraints. For deriving the equivalent convex program we proceed analogously as for the case of (2.209), cf. Stoyanov, Rachev and Fabozzi [314]. We consider the problem

$$\left.\begin{aligned} &\min_{y,t} \rho(-\eta^{\mathrm{T}} y + \xi t) \\ &\text{s.t.}\ \ \mu^{\mathrm{T}} y \quad - \mu_{n+1} t = 1 \\ &\qquad\quad Ay \qquad\quad -bt = 0 \\ &\qquad\quad\ \ y \qquad\qquad\quad \geq 0 \\ &\qquad\qquad\qquad\qquad\ \ t \geq 0. \end{aligned}\right\} \tag{2.213}$$

If x is a feasible solution of (2.212) then

$$t := \frac{1}{f(x)} = \frac{1}{\mu^{\mathrm{T}} x - \mu_{n+1}} > 0, \qquad y := tx,$$

is feasible for (2.213), due to the positive homogeneity of ρ and the corresponding objective values are equal. Conversely, let (y,t) be a feasible solution of (2.213) then $t > 0$ must hold and $x := \frac{1}{t} y$ is feasible for (2.212) with the same objective value. Thus, (2.212) and (2.213) are equivalent.

By relaxing the first equality constraint in (2.213) we get the following problem:

$$\left.\begin{aligned} &\min_{y,t} \rho(-\eta^{\mathrm{T}} y + \xi t) \\ &\text{s.t.}\ \ \mu^{\mathrm{T}} y \quad - \mu_{n+1} t \geq 1 \\ &\qquad\quad Ay \qquad\quad -bt = 0 \\ &\qquad\quad\ \ y \qquad\qquad\quad \geq 0 \\ &\qquad\qquad\qquad\qquad\ \ t \geq 0. \end{aligned}\right\} \tag{2.214}$$

It is easy to see that the relaxed constraint must be active at any optimal solution of (2.214). In fact, let us assume that for a feasible solution (y,t) the inequality $\mu^{\mathrm{T}} y - \mu_{n+1} t > 1$ holds. Taking

$$\kappa := \frac{1}{\mu^{\mathrm{T}} y - \mu_{n+1} t} < 1, \qquad \bar{y} := \kappa y, \qquad \bar{t} := \kappa t,$$

$(\bar{y},\bar{t})$ is obviously feasible for (2.214) and the first inequality constraint becomes active. Utilizing the positive homogeneity of ρ, for the objective function values we get

$$\rho(-\eta^{\mathrm{T}} \bar{y} + \xi \bar{t}) \;=\; \kappa \rho(-\eta^{\mathrm{T}} y + \xi t) \;<\; \rho(-\eta^{\mathrm{T}} y + \xi t),$$

implying that the first inequality constraint in (2.214) must be active at optimum. Consequently, the optimization problems (2.213) and (2.214) are equivalent.

The results above can be summarized in

Proposition 2.54. *Let us assume that A1, A2, A3' and A4 hold. Then the fractional programming problem (2.212) is equivalent to (2.214).*

If ρ is convex then (2.214) is clearly a convex programming problem.

Concerning (2.212), the requirement $f(x) > 0$ for all $x \in \mathscr{B}$ can be enforced by adding a linear constraint of the form $\mu^{\mathrm{T}}x - \mu_{n+1} \geq \varepsilon$ to the defining set of linear relations of $\mathscr{B}$, with a suitably chosen $\varepsilon > 0$.

As we have seen, under our assumptions the optimization problems (2.212), (2.213), (2.214) are equivalent formulations of the original fractional programming problem (2.209). Notice that for proving the equivalence we merely needed the positive homogeneity of ρ. Assuming additionally that ρ is subadditive, the equivalent problem (2.214) becomes a convex programming problem.

Exercises

2.12. The portfolio optimization model of Young [352] corresponds to the choice of ρ_{fat} as a risk measure in our general portfolio optimization model (2.206). The idea is to select a portfolio which minimizes the maximum loss. We consider the case of a finite discrete distribution of the asset returns η: $(p_i, \hat{\eta}^i)$, $p_i = \mathbb{P}[\eta = \hat{\eta}^i]$, $i = 1, \dots, N$. Let $r := \mathbb{E}[\eta]$. The model formulation is:

$$\left.\begin{array}{rl} \max\limits_{x} \min\limits_{1 \leq i \leq N} (\hat{\eta}^i)^{\mathrm{T}} x & \\ r^{\mathrm{T}} x \geq \mu_p & \\ \mathbb{1}^{\mathrm{T}} x = 1 & \\ x \geq 0 & \end{array}\right\}$$

with $\mathbb{1} = (1, \dots, 1)^{\mathrm{T}}$.

(a) Explain why in the above model the maximum loss is minimized.
(b) Give for this model an equivalent linear programming formulation.

2.13. A portfolio optimization problem is given as follows. There are two risky assets, with random returns $\eta = (\eta_1, \eta_2)$, where (η_1, η_2) has a finite discrete distribution with three realizations, given by the scenario tableau:

p_k	0.3	0.5	0.2
η_1^k	-0.003	0.02	0.01
η_2^k	0.06	-0.006	0.02

The minimum expected return is $\mu_p = 0.018$. As the risk measure to be minimized choose CVaR. With the random portfolio return $\zeta(x, \eta) = \eta_1 x_1 + \eta_2 x_2$ this means that the objective function in (2.206) will be $\rho^{\alpha}_{\text{CVaR}}(-\zeta(x, \eta))$ where we choose

$\alpha = 0.99$. Set up the portfolio optimization problem and solve it by employing SLP–IOR.

2.14. Prove the following fact: if x^* is an optimal solution of (2.208) with $\nu > 0$ then

(a) x^* is an optimal solution of (2.206) with $\mu_p = r^{\mathrm{T}}x^*$ and
(b) it is an optimal solution of (2.207) with $\psi_p = \rho(-\eta^{\mathrm{T}}x^*)$.

Chapter 3
SLP models with recourse

For various SLP models with recourse, we present in this chapter properties which are relevant for the particular solution methods developed for various model types, to be discussed later on.

3.1 The general multi-stage SLP

As briefly sketched in Section 1.1 an SLP with recourse is a dynamic decision model with $T \geq 2$ stages, as illustrated in Fig. 3.1,

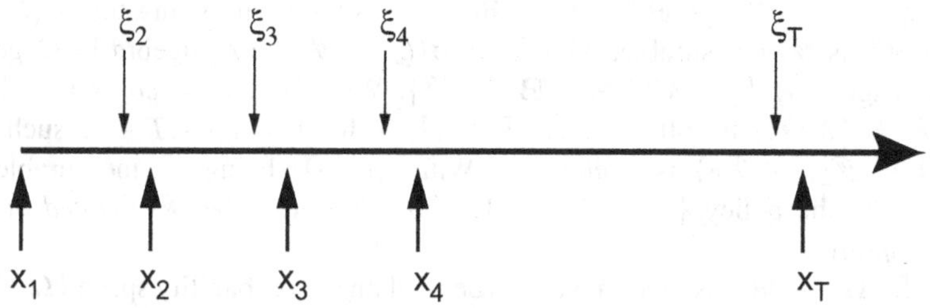

Fig. 3.1 Dynamic decision structure.

where for feasibility sets, emerging stagewise during the horizon $\mathscr{T} = \{1, 2, \cdots, T\}$,

$$\mathscr{B}_t(x_1, \cdots, x_{t-1}; \xi_2, \cdots, \xi_t), t \in \mathscr{T},$$

we take successively

- a first stage decision $x_1 \in \mathscr{B}_1 \subset \mathbb{R}^{n_1}$; then, after observing the realization of a random variable (or vector) ξ_2,
- a second stage decision $x_2(x_1; \xi_2) \in \mathscr{B}_2(x_1; \xi_2) \subset \mathbb{R}^{n_2}$; then after observing the realization of a further random variable (or vector) ξ_3,

- a third stage decision $x_3(x_1,x_2;\xi_2,\xi_3) \in \mathscr{B}_3(x_1,x_2;\xi_2,\xi_3) \subset \mathbb{R}^{n_3}$; and so on until, after observing the realization of ξ_T, finally
- a T-th stage decision $x_T(x_1,\cdots,x_{T-1};\xi_2,\cdots,\xi_T) \in \mathscr{B}_T(x_1,\cdots,x_{T-1};\xi_2,\cdots,\xi_T) \subset \mathbb{R}^{n_T}$.

Here the feasibility set $\mathscr{B}_t(x_1,\cdots,x_{t-1};\xi_2,\cdots,\xi_t)$ for x_t is given by (random) linear constraints, depending on the previous decisions $x_1,\cdots,x_{t-1}$ and the observations of $\xi_2,\cdots,\xi_t$.

For each stage t the decision $x_t(x_1,\cdots,x_{t-1};\xi_2,\cdots,\xi_t)$ involves the t-th stage objective value $c_t^{\mathrm{T}}(\xi_2,\cdots,\xi_t)x_t(x_1,\cdots,x_{t-1};\xi_2,\cdots,\xi_t)$, and the goal is to minimize the expected value of the sum of these T objectives.

More precisely, with any set $\Omega \neq \emptyset$, some σ-algebra $\mathscr{G}$ of subsets of Ω and a probability measure $P : \mathscr{G} \to [0,1]$, the general model may be stated as follows: Given the probability space $(\Omega,\mathscr{G},P)$, random vectors $\xi_t : \Omega \longrightarrow \mathbb{R}^{r_t}$, and the probability distribution $\mathbb{P}_\xi$ induced by $\xi = (\xi_2^{\mathrm{T}},\cdots,\xi_T^{\mathrm{T}})^{\mathrm{T}} : \Omega \longrightarrow \mathbb{R}^R$, $R = r_2+\cdots+r_T$, on the Borel σ-field of $\mathbb{R}^R$, with $\zeta_t = (\xi_2^{\mathrm{T}},\cdots,\xi_t^{\mathrm{T}})^{\mathrm{T}}$ being the state variable at stage t, the multi-stage stochastic linear program (MSLP) reads as

$$\left.\begin{aligned}
&\min\{c_1^{\mathrm{T}}x_1 + \mathbb{E}\sum_{t=2}^{T} c_t^{\mathrm{T}}(\zeta_t)x_t(\zeta_t)\} \\
&A_{11}x_1 \qquad\qquad\qquad\qquad\quad = b_1 \\
&A_{t1}(\zeta_t)x_1 + \sum_{\tau=2}^{t} A_{t\tau}(\zeta_t)x_\tau(\zeta_\tau) = b_t(\zeta_t) \text{ a.s., } t = 2,\cdots,T, \\
&\qquad\qquad x_1 \geq 0,\ x_t(\zeta_t) \geq 0 \qquad \text{a.s., } t = 2,\cdots,T,
\end{aligned}\right\} \tag{3.1}$$

where $x_t : \mathbb{R}^{r_2+\cdots+r_t} \longrightarrow \mathbb{R}^{n_t}$ is to be Borel measurable, implying that $x_t(\zeta_t(\cdot)) : \Omega \longrightarrow \mathbb{R}^{n_t}$ is $\mathscr{F}_t$-measurable, with $\mathscr{F}_t = \sigma(\zeta_t) \subset \mathscr{G}$, the σ-algebra in Ω generated at stage t by $\{\zeta_t^{-1}[M] \mid M \in \mathbb{B}^{r_2+\cdots+r_t}\}$. With $\zeta_1 \equiv \xi_1 = const$ and therefore $\mathscr{F}_1 = \{\emptyset,\Omega\}$, it follows that $\mathscr{F}_t \subset \mathscr{F}_{t+1}$ for $t = 1,\cdots,T-1$, such that $\mathscr{F} = \{\mathscr{F}_1,\mathscr{F}_2,\cdots,\mathscr{F}_T\}$ is a *filtration*. With $x_t(\zeta_t(\cdot))$ being $\mathscr{F}_t$-measurable for $t = 1,\cdots,T$, the policy $\{x_t(\zeta_t(\cdot)); t = 1,\cdots,T\}$ is said to be $\mathscr{F}$*-adapted* or else *nonanticipative*.

The $\xi_t : \Omega \longrightarrow \mathbb{R}^{r_t}$ as random vectors defined on the probability space $\{\Omega,\mathscr{G},P\}$ are obviously $\mathscr{G}$-measurable. According to the definition in (2.6) on page 73, we say that $\xi_t \in \mathscr{L}_{r_t}^2 := \mathscr{L}_{r_t}^2(\Omega,\mathscr{G},\mathbb{R}^{r_t})$ if, in addition, the ξ_t are square integrable, i.e. if $\displaystyle\int_\Omega \|\xi_t(\omega)\|^2 P(d\omega)$ exists. In particular, for any arbitrary $\mathscr{F}_t$-simple function $\gamma_t(\omega) := \sum_{i=1}^K g_i \cdot \chi_{M_i}(\omega)$ with $g_i \in \mathbb{R}^{r_t}$, $\chi_{M_i}(\omega) = 1$ if $\omega \in M_i$ and $\chi_{M_i}(\omega) = 0$ otherwise, $M_i \in \mathscr{F}_t$, $M_i \cap M_j = \emptyset$ for $i \neq j$, and $\cup_{i=1}^K M_i = \Omega$, it obviously follows that $\gamma_t \in \mathscr{L}_{r_t}^2(\Omega,\mathbb{R}^{r_t})$.

Assumption 3.1. *Let*

- $\xi_t \in \mathscr{L}_{r_t}^2 := \mathscr{L}_{r_t}^2(\Omega,\mathscr{G},\mathbb{R}^{r_t})\ \forall t$,
- $A_{t\tau}(\cdot), b_t(\cdot), c_t(\cdot)$ *be linear affine in* ζ_t *(and therefore* $\mathscr{F}_t$*-measurable), where* $A_{t\tau}(\cdot)$ *is a* $m_t \times n_\tau$*-matrix.*

Due to this assumption, also the elements of $A_{t\tau}(\cdot), b_t(\cdot), c_t(\cdot)$ are square-integrable with respect to P. Hence, requiring that $\xi_t \in \mathscr{L}^2_{r_t}\ \forall t$ holds, Schwarz's inequality (see e.g. Zaanen [353]) implies in particular that $\mathbb{E}[c_t^{\mathrm{T}}(\zeta_t)x_t(\zeta_t)]$, $t = 2, \cdots, T$, exist, such that problem (3.1) is well defined.
Sometimes the following reformulation of (3.1) may be convenient: Given

- a probability space $(\Omega, \mathscr{G}, P)$;
- $\mathscr{F}_t, t = 1, \cdots, T$, being σ-algebras such that $\mathscr{F}_t \subset \mathscr{G}\ \forall t$ and $\mathscr{F}_t \subset \mathscr{F}_{t+1}$ for $t = 1, \cdots, T-1$ (i.e. $\{\mathscr{F}_t \mid t = 1, \cdots, T\}$ being a filtration);
- $\mathscr{F} := \{\mathscr{F}_1, \cdots, \mathscr{F}_T\}$, where possibly, but not necessarily, $\mathscr{F}_T = \mathscr{G}$;
- X_t a linear subspace of $\mathscr{L}^2_{n_t}$ (with respect to $(\Omega, \mathscr{G}, P)$), including the set of $\mathscr{F}_t$-simple functions;
- M_t the set of $\mathscr{F}_t$-measurable functions $\Omega \longrightarrow \mathbb{R}^{n_t}$ and hence, $X_t \cap M_t$ being a closed linear subspace of X_t;

then problem (3.1) may be restated as

$$\left.\begin{array}{l} \min \mathbb{E}\left\{\sum_{t=1}^{T} c_t^{\mathrm{T}} x_t\right\} \\ \left.\begin{array}{r} \sum_{\tau=1}^{t} A_{t\tau} x_\tau = b_t \text{ a.s.} \\ x_t \geq 0 \text{ a.s.} \\ x_t \in X_t \cap M_t \end{array}\right\} t = 1, \cdots, T, \end{array}\right\} \tag{3.2}$$

with $A_{t\tau}, b_t, c_t$ assumed to be $\mathscr{F}_t$-measurable for $1 \leq \tau \leq t$, $t = 1, \cdots, T$, and to have finite second moments, as implied by Assumption 3.1. (remember: $\mathscr{F}_1 = \{\emptyset, \Omega\}$, such that A_{11}, b_1, c_1 are constant).

Following S.E. Wright [347] various aggregated problems may be derived from (3.2) by using coarser information structures, chosen as subfiltrations $\widehat{\mathscr{F}} = \{\widehat{\mathscr{F}_t}\}$, $\widehat{\mathscr{F}_t} \subset \widehat{\mathscr{F}_{t+1}}$, such that $\widehat{\mathscr{F}_t} \subseteq \mathscr{F}_t$, $\forall t$, instead of the original filtration $\mathscr{F} = \{\mathscr{F}_t\}$, $\mathscr{F}_t \subset \mathscr{F}_{t+1}$, $t = 1, \cdots, T-1$.
Denoting problem (3.2) as $\mathscr{P}(\mathscr{F}, \mathscr{F})$, we then may consider

- the *decision-aggregated* problem $\mathscr{P}(\widehat{\mathscr{F}}, \mathscr{F})$,

$$\left.\begin{array}{l} \min \mathbb{E}\left\{\sum_{t=1}^{T} c_t x_t\right\} \\ \left.\begin{array}{r} \sum_{\tau=1}^{t} A_{t\tau} x_\tau = b_t \text{ a.s.} \\ x_t \geq 0 \text{ a.s.} \\ x_t \in X_t \cap \widehat{M}_t \end{array}\right\} t = 1, \cdots, T, \end{array}\right\} \tag{3.3}$$

  where $\widehat{M}_t$ is the set of $\widehat{\mathscr{F}_t}$-measurable functions $\Omega \longrightarrow \mathbb{R}^{n_t}$, thus requiring that $x = (x_1^{\mathrm{T}}, \cdots, x_T^{\mathrm{T}})^{\mathrm{T}}$ is $\widehat{\mathscr{F}}$-adapted;

- the *constraint-aggregated* problem $\mathscr{P}(\mathscr{F},\widehat{\mathscr{F}})$,

$$\left.\begin{array}{l} \min \mathbb{E}\left\{\sum_{t=1}^{T} c_t x_t\right\} \\ \left.\begin{array}{rl} \mathbb{E}\left\{\sum_{\tau=1}^{t} A_{t\tau}x_\tau \,\Big|\, \widehat{\mathscr{F}}_t\right\} = \mathbb{E}\left\{b_t \,\Big|\, \widehat{\mathscr{F}}_t\right\} & \text{a.s.} \\ x_t \geq 0 & \text{a.s.} \\ x_t \in X_t \cap M_t & \end{array}\right\} t = 1,\cdots,T, \end{array}\right\} \tag{3.4}$$

i.e. x is $\mathscr{F}$-adapted as in (3.2), and the constraints are stated in conditional expectation given $\widehat{\mathscr{F}}_t$;
- and the *fully aggregated* problem $\mathscr{P}(\widehat{\mathscr{F}},\widehat{\mathscr{F}})$ defined as:

$$\left.\begin{array}{ll} \min \mathbb{E}\left\{\sum_{t=1}^{T} \mathbb{E}[c_t \mid \widehat{\mathscr{F}}_t]x_t\right\} & \\ \mathbb{E}\left\{\sum_{\tau=1}^{t} A_{t\tau}x_\tau \,\Big|\, \widehat{\mathscr{F}}_t\right\} = \mathbb{E}\left\{b_t \,\Big|\, \widehat{\mathscr{F}}_t\right\} & \text{a.s. } \forall t \\ x_t \geq 0 & \text{a.s. } \forall t \\ x_t \in X_t \cap \widehat{M}_t & \forall t\,. \end{array}\right\} \tag{3.5}$$

Observe that by Assumption 3.1. the expected values

$$\mathbb{E}\left\{\sum_{\tau=1}^{t} A_{t\tau}x_\tau\right\} \quad \text{and} \quad \mathbb{E}\{b_t\}$$

exist and hence, the conditional expectations in (3.4) and (3.5),

$$\mathbb{E}\left\{\sum_{\tau=1}^{t} A_{t\tau}x_\tau \,\Big|\, \widehat{\mathscr{F}}_t\right\} \quad \text{and} \quad \mathbb{E}\left\{b_t \,\Big|\, \widehat{\mathscr{F}}_t\right\},$$

are a.s. uniquely determined and $\widehat{\mathscr{F}}_t$-measurable due to the Radon-Nikodym theorem (see e.g. Halmos [131]).

Denoting for the above problems $\mathscr{P}(\mathscr{F},\mathscr{F})$, $\mathscr{P}(\widehat{\mathscr{F}},\mathscr{F})$, $\mathscr{P}(\mathscr{F},\widehat{\mathscr{F}})$, $\mathscr{P}(\widehat{\mathscr{F}},\widehat{\mathscr{F}})$

- their feasible sets by $\mathscr{B}(\mathscr{F},\mathscr{F})$, $\mathscr{B}(\widehat{\mathscr{F}},\mathscr{F})$, $\mathscr{B}(\mathscr{F},\widehat{\mathscr{F}})$ and $\mathscr{B}(\widehat{\mathscr{F}},\widehat{\mathscr{F}})$, and
- their optimal values by $\inf(\mathscr{P}(\mathscr{F},\mathscr{F}))$, $\inf(\mathscr{P}(\widehat{\mathscr{F}},\mathscr{F}))$, $\inf(\mathscr{P}(\mathscr{F},\widehat{\mathscr{F}}))$ and $\inf(\mathscr{P}(\widehat{\mathscr{F}},\widehat{\mathscr{F}}))$,

respectively, and with the usual convention that $\inf\{\varphi(x) \mid x \in \mathscr{B}\} = \infty$ if $\mathscr{B} = \emptyset$, the following relations between the above problems are mentioned in S.E. Wright [347]:

Proposition 3.1. *For the feasible sets of the above problems hold the inclusions*

$$\begin{array}{ll} \mathscr{B}(\mathscr{F},\mathscr{F}) \supseteq \mathscr{B}(\widehat{\mathscr{F}},\mathscr{F}) & \mathscr{B}(\mathscr{F},\mathscr{F}) \subseteq \mathscr{B}(\mathscr{F},\widehat{\mathscr{F}}) \\ \mathscr{B}(\widehat{\mathscr{F}},\mathscr{F}) \subseteq \mathscr{B}(\widehat{\mathscr{F}},\widehat{\mathscr{F}}) & \mathscr{B}(\mathscr{F},\widehat{\mathscr{F}}) \supseteq \mathscr{B}(\widehat{\mathscr{F}},\widehat{\mathscr{F}}), \end{array}$$

implying for the corresponding optimal values the inequalities

$$\begin{array}{l} \inf(\mathscr{P}(\mathscr{F},\widehat{\mathscr{F}})) \le \inf(\mathscr{P}(\mathscr{F},\mathscr{F})) \le \inf(\mathscr{P}(\widehat{\mathscr{F}},\mathscr{F})) \\ \inf(\mathscr{P}(\mathscr{F},\widehat{\mathscr{F}})) \le \inf(\mathscr{P}(\widehat{\mathscr{F}},\widehat{\mathscr{F}})) \le \inf(\mathscr{P}(\widehat{\mathscr{F}},\mathscr{F})). \end{array}$$

Proof: The above inclusions result from the following observations:

$\underline{\mathscr{B}(\mathscr{F},\mathscr{F}) \supseteq \mathscr{B}(\widehat{\mathscr{F}},\mathscr{F})}$: Any $\{x_t\} \in \mathscr{B}(\widehat{\mathscr{F}},\mathscr{F})$ satisfies the constraints of (3.3) and hence in particular the conditions $x_t \in X_t \cap \widehat{M}_t \ \forall t$. Since $\widehat{\mathscr{F}_t} \subseteq \mathscr{F}_t \ \forall t$, we then have $x_t \in X_t \cap M_t \ \forall t$, such that $\{x_t\} \in \mathscr{B}(\mathscr{F},\mathscr{F})$.

$\underline{\mathscr{B}(\mathscr{F},\mathscr{F}) \subseteq \mathscr{B}(\mathscr{F},\widehat{\mathscr{F}})}$: Any $\{x_t\} \in \mathscr{B}(\mathscr{F},\mathscr{F})$ is $\mathscr{F}$-adapted and satisfies all other constraints in (3.2), in particular the random vectors $\sum_{\tau=1}^{t} A_{t\tau}x_\tau$ and b_t, measurable w.r.t. $\mathscr{F}_t$, coincide almost surely, such that for any sub–σ–algebras $\widehat{\mathscr{F}_t} \subseteq \mathscr{F}_t$ their conditional expectations $\mathbb{E}\left\{ \sum_{\tau=1}^{t} A_{t\tau}x_\tau \,\middle|\, \widehat{\mathscr{F}_t} \right\}$ and $\mathbb{E}\left\{ b_t \,\middle|\, \widehat{\mathscr{F}_t} \right\}$, being a.s. uniquely determined and $\widehat{\mathscr{F}_t}$-measurable as mentioned above, coincide a.s. as well. Hence we have $\{x_t\} \in \mathscr{B}(\mathscr{F},\widehat{\mathscr{F}})$.

The two remaining inclusions,

$\underline{\mathscr{B}(\widehat{\mathscr{F}},\mathscr{F}) \subseteq \mathscr{B}(\widehat{\mathscr{F}},\widehat{\mathscr{F}})}$ and $\underline{\mathscr{B}(\mathscr{F},\widehat{\mathscr{F}}) \supseteq \mathscr{B}(\widehat{\mathscr{F}},\widehat{\mathscr{F}})}$,

as well as the inequalities for the optimal values, are now obvious. □

Remark 3.1. *Concerning the fully aggregated problem (3.5) we have the following facts:*

- *If $\mathscr{F}$ is infinite, i.e. at least one of the σ-algebras $\mathscr{F}_t = \sigma(\zeta_t)$, $t = 1,\cdots,T$, is not finitely generated (equivalently, at least one random vector ζ_t has not a finite discrete distribution), and $\widehat{\mathscr{F}}$ is finite, then $\mathscr{P}(\widehat{\mathscr{F}},\widehat{\mathscr{F}})$ with finitely many constraints and variables is clearly simpler to deal with than $\mathscr{P}(\mathscr{F},\mathscr{F})$;*
- *for a sequence $\{\widehat{\mathscr{F}^\nu}\}$ of (finite) filtrations with successive refinements, i.e. $\widehat{\mathscr{F}_t^\nu} \subseteq \widehat{\mathscr{F}_t^{\nu+1}} \ \forall t$, under appropriate assumptions, e.g. for a corresponding sequence of measures P_ν on $\widehat{\mathscr{F}_T^\nu}$ converging weakly to P (see Billingsley [20]), we may expect convergence of the optimal values of (3.5) to that one of (3.2);*
- *according to Prop. 3.1., in general there is no definite relationship between the optimal values of (3.5) and of (3.2), as remarked for instance by Wright [347] (p. 900); however there are special problem classes—in particular in the two-stage case—and particular assumptions for the multi-stage case implying that*

$\inf(\mathscr{P}(\widehat{\mathscr{F}},\widehat{\mathscr{F}}))$ yields a lower bound for $\inf(\mathscr{P}(\mathscr{F},\mathscr{F}))$, which can be used in designing solution methods, as we shall see later. □

First we shall deal with two-stage SLP's. Under various assumptions on the model structure and the underlying probability distributions, we shall reveal properties of the recourse function and its expectation which turn out to be useful when designing solution methods. Unfortunately, not all of these results can be generalized to corresponding statements for multi-stage SLP's in general.

3.2 The two-stage SLP: Properties and solution appraoches

In the previous section, for the T-stage SLP we had the following general probabilistic setup: On some probability space $(\Omega,\mathscr{G},P)$ a sequence of random vectors $\xi_t : \Omega \longrightarrow \mathbb{R}^{r_t}$, $t = 2,\cdots,T$, was defined, such that $\xi = (\xi_2^{\mathrm{T}},\cdots,\xi_T^{\mathrm{T}})^{\mathrm{T}}$ induced the probability distribution $\mathbb{P}_\xi$ on the Borel σ-field of $\mathbb{R}^{r_2+\cdots+r_T}$. Then the random vectors $\zeta_t = (\xi_2^{\mathrm{T}},\cdots,\xi_t^{\mathrm{T}})^{\mathrm{T}}$, $t = 2,\cdots,T$, implied the filtration $\mathscr{F} = \{\mathscr{F}_2,\cdots,\mathscr{F}_T\}$ in $\mathscr{G}$ with $\mathscr{F}_t = \sigma(\zeta_t)$. Restricting ourselves in this section to the case $T = 2$ allows for the following simplification of this setup.

Assume some probability space $(\Omega,\mathscr{F},P)$ together with a random vector $\xi : \Omega \to \mathbb{R}^r$ to be given, such that $\mathscr{F} = \sigma(\xi)$. Then ξ induces the probability measure $\mathbb{P}_\xi$ on $\mathbb{B}^r$, the Borel σ-algebra in $\mathbb{R}^r$, according to $\mathbb{P}_\xi(B) = P(\xi^{-1}[B])\ \forall B \in \mathbb{B}^r$.

Besides deterministic arrays $A \in \mathbb{R}^{m_1\times n_1}$, $b \in \mathbb{R}^{m_1}$, and $c \in \mathbb{R}^{n_1}$, for the first stage, let the random arrays $T(\xi) \in \mathbb{R}^{m_2\times n_1}$, $W(\xi) \in \mathbb{R}^{m_2\times n_2}$, $h(\xi) \in \mathbb{R}^{m_2}$, and $q(\xi) \in \mathbb{R}^{n_2}$, be defined for the second stage as:

$$\left.\begin{aligned}
T(\xi) &= T + \sum_{j=1}^{r} T^j \xi_j; \quad T, T^j \in \mathbb{R}^{m_2\times n_1} \text{ deterministic},\\
W(\xi) &= W + \sum_{j=1}^{r} W^j \xi_j; \; W,\ W^j \in \mathbb{R}^{m_2\times n_2} \text{ deterministic},\\
h(\xi) &= h + \sum_{j=1}^{r} h^j \xi_j; \quad h,\ h^j \in \mathbb{R}^{m_2} \text{ deterministic},\\
q(\xi) &= q + \sum_{j=1}^{r} q^j \xi_j; \quad q,\ q^j \in \mathbb{R}^{n_2} \text{ deterministic}.
\end{aligned}\right\} \tag{3.6}$$

Then, with $\xi \in \mathscr{L}_r^2$ due to Assumption 3.1. and according to (3.2), the general two-stage SLP with random recourse is formulated as

$$\left.\begin{aligned}
&\min \mathbb{E}_\xi \left\{c^{\mathrm{T}}x + q^{\mathrm{T}}(\xi)y(\xi)\right\}\\
&\qquad Ax = b\\
&T(\xi)x + W(\xi)y(\xi) = h(\xi) \quad \text{a.s.}\\
&\qquad x \geq 0\\
&\qquad y(\xi) \geq 0 \quad \text{a.s.}\\
&\qquad y(\cdot) \in Y \cap M,
\end{aligned}\right\} \tag{3.7}$$

where Y—corresponding to (3.2)—is a linear subspace of $\mathscr{L}^2_{n_2}$ (with respect to $(\Omega, \mathscr{F}, P)$), including the set of $\mathscr{F}$-simple functions; and M is the set of $\mathscr{F}$-measurable functions $\Omega \longrightarrow \mathbb{R}^{n_2}$. To avoid unnecessary formalism, we may just assume, that $Y = \mathscr{L}^2_{n_2}$ which obviously contains the $\mathscr{F}$-simple functions and satisfies $Y \subset M$.

Hence problem (3.7) is equivalent to

$$\left.\begin{array}{lll} \min \mathbb{E}_\xi \left\{c^{\mathrm{T}}x + q^{\mathrm{T}}(\xi)y(\xi)\right\} & & \\ Ax & = b & \\ T(\xi)x + W(\xi)y(\xi) & = h(\xi) & \text{a.s.} \\ x & \geq 0 & \\ y(\xi) & \geq 0 & \text{a.s.} \\ y(\cdot) & \in Y\,. & \end{array}\right\} \tag{3.8}$$

A brief sketch on modeling situations leading to variants of the general two-stage SLP (3.8) is given in Chapter 1 on page 4.

Remark 3.2. *Instead of the constraints $\{Ax = b,\ x \geq 0\}$ in (3.8) we also could consider constraints of the form $\{Ax \propto b,\ l \leq x \leq u\}$ as in (1.1) on page 1, and the constraints $\{W(\xi)y(\xi) = h(\xi) - T(\xi)x,\ y(\xi) \geq 0$ a.s.$\}$ of (3.8) could be replaced as well by $\{W(\xi)y(\xi) \propto h(\xi) - T(\xi)x,\ \tilde{l} \leq y(\xi) \leq \tilde{u}$ a.s.$\}$. However, in order to have a unified presentation, for two-stage programs we stay with the formulation chosen in (3.8).* □

Except for particular cases where it is stated explicitly otherwise, instead of (3.6) we shall restrict ourselves to $W(\cdot) \equiv W$, i.e. to *fixed recourse*. In general, problem (3.8) contains implicitly the *recourse function*

$$\left.\begin{array}{lll} Q(x; T(\xi), h(\xi), W(\xi), q(\xi)) := \inf\limits_y q^{\mathrm{T}}(\xi)y(\xi) & & \\ T(\xi)x + W(\xi)y(\xi) = h(\xi) & \text{a.s.} & \\ \qquad y(\xi) \geq 0 & \text{a.s.} & \\ \qquad y(\cdot) \in Y\,. & & \end{array}\right\} \tag{3.9}$$

To simplify the notation, we shall enter into the recourse function $Q(x; \cdot)$ of (3.9), in addition to the first stage decision variable x, only those parameter arrays being random in the model under consideration. For instance, $Q(x; T(\xi), h(\xi))$ indicates that $T(\cdot)$, $h(\cdot)$ are random arrays defined according to (3.6) whereas $W(\cdot) \equiv W$, $q(\cdot) \equiv q$; and $Q(x; h(\xi))$ stands for $h(\cdot)$ being a random vector due to (3.6) and $T(\cdot) \equiv T$, $W(\cdot) \equiv W$, $q(\cdot) \equiv q$ being deterministic data.

Furthermore, in applications of this model, the selection of a decision $\hat{x}$ feasible for the first stage constraints $Ax = b,\ x \geq 0$, appears to be meaningful only if it allows almost surely to satisfy the second stage constraints $W(\xi)y(\xi) = h(\xi) - T(\xi)\hat{x}$, $y(\xi) \geq 0$ a.s., since otherwise, according to the usual convention, we should get for the recourse function

$$Q(\hat{x};T(\xi),h(\xi),W(\xi),q(\xi)) = \\ = \inf_{y\in Y}\{q^{\mathrm{T}}(\xi)y(\xi) \mid W(\xi)y(\xi) = h(\xi)-T(\xi)\hat{x},\ y(\xi)\geq 0 \text{ a.s.}\} = +\infty$$

with some positive probability. This implies

- either $\mathscr{Q}(\hat{x}) := \mathbb{E}_{\xi}\left[Q(\hat{x};T(\xi),h(\xi),W(\xi),q(\xi))\right] = +\infty$,
- or else the expected recourse $\mathscr{Q}(\hat{x})$ to be undefined if with positive probability $Q(\hat{x};T(\xi),h(\xi),W(\xi),q(\xi)) = -\infty$ results simultaneously.

Clearly in anyone of these situations $\hat{x}$ is not to be chosen since neither an infinite nor an undefined objective value corresponds to our aim to minimize the objective of (3.8). Hence, in general we may be faced with so-called *induced constraints* on x, meaning that we require

$$\hat{x} \in K := \{x \mid x \in \mathbb{R}^{n_1};\ Q(x;T(\xi),h(\xi),W(\xi),q(\xi)) < +\infty \text{ a.s.}\}.$$

For $\Xi = \operatorname{supp}\mathbb{P}_{\xi}$—the support of $\mathbb{P}_{\xi}$, i.e. the smallest closed set in $\mathbb{R}^r$ such that $\mathbb{P}_{\xi}(\Xi) = 1$—being an infinite set, K is described in general by an infinite set of constraints, which is not easy to deal with. If however Ξ is either finite, i.e. $\Xi = \{\xi^1,\cdots,\xi^p\}$, or else a convex polyhedron given by finitely many points as $\Xi = \operatorname{conv}\{\xi^1,\cdots,\xi^p\}$ (see Chapter 1, Def. 1.3. on page 10), then the induced constraints imply $x \in K$ with

$$K := \{x \mid T(\xi^j)x + W(\xi^j)y^j = h(\xi^j),\ y^j \geq 0,\ j=1,\cdots,p\},$$

and, with $\mathscr{B}_1 := \{x \mid Ax = b,\ x \geq 0\} \subset \mathbb{R}^{n_1}$, the first stage decisions have to satisfy $x \in \mathscr{B}_1 \cap K$. A more detailed discussion of induced constraints may be found in Rockafellar–Wets [285] and in Walkup–Wets [340] (see also Kall [154], Ch. III).

3.2.1 The complete fixed recourse problem (CFR)

If for a particular application it does not seem appropriate, that the future outcomes of ξ affect the set of feasible first stage decisions, given as

$$\mathscr{B}_1 = \{x \mid Ax = b,\ x \geq 0\}, \tag{3.10}$$

we might require at least *relatively complete recourse:*

$$\forall x \in \mathscr{B}_1 \Longrightarrow \{y \mid W(\xi)y = h(\xi) - T(\xi)x,\ y \geq 0\} \neq \emptyset \text{ a.s.}. \tag{3.11}$$

Due to the Farkas lemma, Chapter 1, Prop. 1.13. on page 15, condition (3.11) is equivalent to:

$$\forall x \in \mathscr{B}_1 \text{ holds}:\ \left[W^{\mathrm{T}}(\xi)u \leq 0 \Longrightarrow (h(\xi)-T(\xi)x)^{\mathrm{T}}u \leq 0 \text{ a.s.}\right].$$

Hence the requirement of relatively complete recourse is a joint restriction on $\mathscr{B}_1$ and on the range of $h(\xi), T(\xi), W(\xi)$ for $\xi \in \Xi$, simultaneously, which may be difficult to verify, in general.

Therefore, in applications it is often preferred to assume *complete fixed recourse* (CFR), which requires for $W(\xi) \equiv W$ the following condition:

$$\{z \mid z = Wy,\ y \geq 0\} = \mathbb{R}^{m_2}. \tag{3.12}$$

If this condition is satisfied, then for any $\hat{x}$ feasible according to an arbitrary set of first stage constraints in (3.8), and for any realization $\hat{\xi}$ of the random vector ξ, the second stage constraints in (3.9) are feasible. Furthermore, complete fixed recourse is a condition on the matrix W only, and may easily be checked due to

Lemma 3.1. *A matrix $W \in \mathbb{R}^{m_2 \times n_2}$ satisfies the complete recourse condition (3.12) if and only if*

- $\operatorname{rank}(W) = m_2$, *and*
- *for an arbitrary set $\{W_{i_1}, W_{i_2}, \cdots, W_{i_{m_2}}\}$ of linearly independent columns of W, the linear constraints*

$$\left.\begin{aligned} Wy &= 0 \\ y_{i_k} &\geq 1,\ k = 1, \cdots, m_2, \\ y &\geq 0 \end{aligned}\right\} \tag{3.13}$$

are feasible.

Proof: Assume that W is a complete recourse matrix. Then from (3.12) follows that $\operatorname{rank}(W) = m_2$ necessarily holds.

Furthermore, for some selection $\{W_{i_1}, W_{i_2}, \cdots, W_{i_{m_2}}\}$ of linearly independent columns of W, let

$$\hat{z} = -\sum_{k=1}^{m_2} W_{i_k}.$$

By our assumption on W, we have $\{y \mid Wy = \hat{z},\ y \geq 0\} \neq \emptyset$. Hence, with the index set $\{j_1, \cdots, j_{n_2-m_2}\}$ chosen such that

$$\begin{aligned} \{i_1, i_2, \cdots, i_{m_2}\} \cap \{j_1, \cdots, j_{n_2-m_2}\} &= \emptyset \\ \text{and } \{i_1, i_2, \cdots, i_{m_2}\} \cup \{j_1, \cdots, j_{n_2-m_2}\} &= \{1, \cdots, n_2\}, \end{aligned}$$

there exists a feasible solution $\hat{y}$ of

$$\begin{aligned} \sum_{k=1}^{m_2} W_{i_k} \hat{y}_{i_k} + \sum_{l=1}^{n_2-m_2} W_{j_l} \hat{y}_{j_l} &= \hat{z} \\ &= -\sum_{k=1}^{m_2} W_{i_k} \\ \hat{y}_i \geq 0,\ i = 1, \cdots, n_2&. \end{aligned}$$

Hence, with

$$y_\nu = \begin{cases} \hat{y}_\nu + 1\,, & \nu = i_1, i_2, \cdots, i_{m_2}\,, \\ \hat{y}_\nu\,, & \nu = j_1, j_2, \cdots, j_{n_2-m_2}\,, \end{cases}$$

the constraints (3.13) are necessarily satisfied.

Assume now that the conditions of this lemma hold. Choose an arbitrary $\bar{z} \in \mathbb{R}^{m_2}$. Then the linear equation

$$\sum_{k=1}^{m_2} W_{i_k} y_{i_k} = \bar{z}$$

has a unique solution $\{\bar{y}_{i_1}, \cdots, \bar{y}_{i_{m_2}}\}$. If $\bar{y}_{i_k} \geq 0$ for $k = 1, \cdots, m_2$, we have a feasible solution for the recourse equation $Wy = \bar{z}$. Otherwise, set $\gamma := \min\{\bar{y}_{i_1}, \cdots, \bar{y}_{i_{m_2}}\} < 0$. Let $\tilde{y}$ be a feasible solution of (3.13). Then for

$$\hat{y}_\nu = \begin{cases} \bar{y}_\nu - \gamma\tilde{y}_\nu\,, & \nu = i_1, i_2, \cdots, i_{m_2}\,, \\ -\gamma\tilde{y}_\nu\,, & \nu = j_1, j_2, \cdots, j_{n_2-m_2}\,, \end{cases}$$

follows

$$\begin{aligned} W\hat{y} &= \sum_{k=1}^{m_2} W_{i_k}\hat{y}_{i_k} + \sum_{l=1}^{n_2-m_2} W_{j_l}\hat{y}_{j_l} \\ &= \sum_{k=1}^{m_2} W_{i_k}\underbrace{(\bar{y}_{i_k} - \gamma\tilde{y}_{i_k})}_{\geq 0} + \sum_{l=1}^{n_2-m_2} W_{j_l}\underbrace{(-\gamma\tilde{y}_{j_l})}_{\geq 0} \\ &= \bar{z} - \gamma\underbrace{\sum_{r=1}^{n_2} W_r\tilde{y}_r}_{=0} \end{aligned}$$

such that $\hat{y}$ is a feasible solution of $Wy = \bar{z}$, $y \geq 0$. □

Hence, to verify complete fixed recourse, we only have to determine rank(W) and—if rank(W) $= m_2$ is satisfied—to check the feasibility of (3.13) by applying any algorithm for finding a feasible basic solution of this system, as e.g. the method described in Section 1.2.4 on page 19. Throughout our discussion of two-stage SLP's we shall make the

Assumption 3.2. *The recourse matrix W satisfies the complete fixed recourse condition (3.12).*

Even for the complete fixed recourse case if, with $\mathscr{C}^P$ being the polar cone of $\mathscr{C} = \{y \mid Wy = 0, y \geq 0\}$, it happens that

$$\Xi \cap \{\xi \mid -q(\xi) \in \mathscr{C}^P\} \neq \Xi\,,$$

then, due to Prop. 1.6. in Chapter 1 (p. 11) $\{\xi \mid -q(\xi) \in \mathscr{C}^P\} \neq \emptyset$ is closed, such that the definition of the support Ξ implies $\mathbb{P}_\xi(\Xi \cap \{\xi \mid -q(\xi) \in \mathscr{C}^P\}) < 1$.

Hence, with $\Xi_0 = \Xi \setminus \{\xi \mid -q(\xi) \in \mathscr{C}^P\}$, by Prop. 1.7. in Chapter 1 (p. 12) follows $Q(x; T(\xi), h(\xi), q(\xi)) = -\infty$ for $\xi \in \Xi_0$ with probability $\mathbb{P}_\xi(\Xi_0) > 0$, yielding $\mathscr{Q}(x) = -\infty\ \forall x \in \mathscr{B}_1$.

Therefore, for allowing the objective of (3.8) to discriminate among various first stage feasible solutions, we need to assume that $-q(\xi) \in \mathscr{C}^P \; \forall \xi \in \Xi$, i.e. using the Farkas lemma (Chapter 1, Prop. 1.13. on page 15) we add to Assumption 3.2. the further

Assumption 3.3. *The recourse matrix W together with $q(\cdot)$ satisfy*

$$\{u \mid W^{\mathrm{T}} u \leq q(\xi)\} \neq \emptyset \; \forall \xi \in \Xi \,. \tag{3.14}$$

Observe that due to (3.14) the requirement that $-q(\xi) \in \mathscr{C}^P \forall \xi \in \Xi$ is equivalent to dual feasibility of the recourse problem, a.s.

Lemma 3.2. *Given Assumptions 3.2. and 3.3., for any $x \in \mathbb{R}^{n_1}$ there exists an optimal recourse $y(\cdot) \in Y$ such that $Q(x; T(\xi), h(\xi), q(\xi)) = q^{\mathrm{T}}(\xi) y(\xi)$.*

Proof: Due to Assumptions 3.2. and 3.3. the LP

$$\left.\begin{array}{rcl} \min q^{\mathrm{T}}(\xi) y & & \\ \text{s.t. } Wy & = & h(\xi) - T(\xi)x \\ y & \geq & 0 \end{array}\right\} \tag{3.15}$$

is solvable for all $\xi \in \Xi$. Let $B^{(\nu)}$, $\nu = 1, \cdots, K$, denote all bases out of W (i.e. all the regular $m_2 \times m_2$-submatrices of W). Partitioning W into the basic part $B^{(\nu)}$ and the nonbasic part $N^{(\nu)}$ and correspondingly restating $q(\xi) \cong (q_{B^{(\nu)}}(\xi), q_{N^{(\nu)}}(\xi))$ and $y \cong (y_{B^{(\nu)}}, y_{N^{(\nu)}})$, we know from Prop. 1.3. in Chapter 1 (p. 9) that with the convex polyhedral set

$$\mathscr{A}_\nu := \{\xi \mid B^{(\nu)^{-1}}(h(\xi) - T(\xi)x) \geq 0, \; q^{\mathrm{T}}_{B^{(\nu)}}(\xi) B^{(\nu)^{-1}} N^{(\nu)} - q^{\mathrm{T}}_{N^{(\nu)}}(\xi) \leq 0\}$$

$y(\xi) \cong \left(y_{B^{(\nu)}}(\xi) = B^{(\nu)^{-1}}(h(\xi) - T(\xi)x), y_{N^{(\nu)}}(\xi) = 0\right)$ solves (3.15) for any $\xi \in \mathscr{A}_\nu$. Furthermore, from (3.6) follows $y(\cdot) \in \mathscr{L}^2_{n_2}(\mathscr{A}_\nu, \mathbb{B}^r, \mathbb{R}^{n_2})$ for $\nu = 1, \cdots .K$. Since—due to the solvability of (3.15) for all $\xi \in \Xi$—we have that $\bigcup_{\nu=1}^{K} \mathscr{A}_\nu \supset \Xi$, this inclusion also holds for $\bigcup_{\nu=1}^{K} \hat{\mathscr{A}_\nu}$ with the sets $\hat{\mathscr{A}_\nu}$ being defined as $\hat{\mathscr{A}_1} = \mathscr{A}_1$ and $\hat{\mathscr{A}_\nu} = \mathscr{A}_\nu \setminus \bigcup_{\mu=1}^{\nu-1} \mathscr{A}_\mu$ for $\nu = 2, \cdots, K$.

Therefore, $\{\Xi \cap \hat{\mathscr{A}_\nu} \mid \nu = 1, \cdots, K\}$ is a (disjoint) partition of Ξ with $y(\cdot)$ according to

$$y(\xi) \cong \left(y_{B^{(\nu)}}(\xi) = B^{(\nu)^{-1}}(h(\xi) - T(\xi)x), y_{N^{(\nu)}}(\xi) = 0\right) \quad \text{for } \xi \in \hat{\mathscr{A}_\nu}$$

a solution of (3.15), being piecewise linear in ξ and hence belonging to Y, and yielding $Q(x;T(\xi),h(\xi),q(\xi)) = q^{\mathrm{T}}(\xi)y(\xi)$. □

The above convex polyhedral sets $\mathscr{A}_\nu$ depend, by definition, on x, and so do the pairwise disjoint sets $\hat{\mathscr{A}}_\nu$, which we may indicate by denoting them as $\hat{\mathscr{A}}_\nu(x)$. Then for some given $x^{(i)}, i = 1,2$, and any $\xi \in \Xi$ there exist $\nu_i \in \{1,\cdots,K\}$ such that $\xi \in \hat{\mathscr{A}}_{\nu_i}(x^{(i)})$ and hence

$$\left.\begin{aligned} Q(x^{(i)};T(\xi),h(\xi),q(\xi)) &= q^{\mathrm{T}}_{B^{(\nu_i)}}(\xi)B^{(\nu_i)^{-1}}(h(\xi)-T(\xi)x^{(i)}) \\ &= \alpha_{\nu_i}(\xi) + d^{(\nu_i)^{\mathrm{T}}}(\xi)x^{(i)}, \\ \text{where}\quad \alpha_{\nu_i}(\xi) &= q^{\mathrm{T}}_{B^{(\nu_i)}}(\xi)B^{(\nu_i)^{-1}}h(\xi) \in L^1 \\ \text{and}\ -d^{(\nu_i)}(\xi) &= (q^{\mathrm{T}}_{B^{(\nu_i)}}(\xi)B^{(\nu_i)^{-1}}T(\xi))^{\mathrm{T}} \in L^1. \end{aligned}\right\} \tag{3.16}$$

Since, due to the simplex criterion, $u^{(\nu_i)} = B^{(\nu_i)^{-1^{\mathrm{T}}}} q_{B^{(\nu_i)}}(\xi), i = 1,2$, are dual feasible with respect to (3.15), it follows for $i \neq j$

$$\left.\begin{aligned} \alpha_{\nu_i}(\xi) + d^{(\nu_i)^{\mathrm{T}}}(\xi)x^{(j)} &= (h(\xi)-T(\xi)x^{(j)})^{\mathrm{T}}u^{(\nu_i)} \\ &\leq (h(\xi)-T(\xi)x^{(j)})^{\mathrm{T}}u^{(\nu_j)} \\ &= \alpha_{\nu_j}(\xi) + d^{(\nu_j)^{\mathrm{T}}}(\xi)x^{(j)} \\ &= Q(x^{(j)};T(\xi),h(\xi),q(\xi)). \end{aligned}\right\} \tag{3.17}$$

Now we are ready to show that (3.8) under appropriate assumptions is a meaningful optimization problem.

Theorem 3.1. *Let the Assumptions 3.2. and 3.3. be satisfied. Then the recourse function $Q(x;T(\xi),h(\xi),q(\xi))$ is*

a) *finitely valued $\forall x \in \mathscr{B}_1$, $\xi \in \Xi$,*
b) *convex in x $\forall \xi \in \Xi$, and*
c) *Lipschitz continuous in x $\forall \xi \in \Xi$ with a Lipschitz constant $D(\xi) \in \mathscr{L}^1_1$.*

Proof:

a) The LP defining the recourse function $Q(x;T(\xi),h(\xi),q(\xi))$ is given by (3.15) as

$$\min\{q^{\mathrm{T}}(\xi)y \mid Wy = h(\xi) - T(\xi)x,\ y \geq 0\},$$

which due to Assumption 3.2. is primal feasible for arbitrary $x \in \mathbb{R}^{n_1}$ and $\xi \in \mathbb{R}^r$, and according to Assumption 3.3. is also dual feasible $\forall \xi \in \Xi$; therefore it is solvable for all $x \in \mathscr{B}_1$ and for all $\xi \in \Xi$, such that

$$Q(x;T(\xi),h(\xi),q(\xi)) \ \text{ is finitely valued } \forall x \in \mathscr{B}_1 \ \text{ and } \forall \xi \in \Xi.$$

b) Hence for an arbitrary $\hat{\xi} \in \Xi$ and some $x^{(1)}, x^{(2)} \in \mathscr{B}_1$ there exist $y^{(i)}$ for $i = 1,2$ such that

$$Q(x^{(i)};T(\hat{\xi}),h(\hat{\xi}),q(\hat{\xi})) = q^{\mathrm{T}}(\hat{\xi})y^{(i)}, \quad \text{where}$$
$$Wy^{(i)} = h(\hat{\xi}) - T(\hat{\xi})x^{(i)},\ y^{(i)} \geq 0.$$

Then for $\tilde{x} = \lambda x^{(1)} + (1-\lambda)x^{(2)}$ with some $\lambda \in (0,1)$ it follows that

$$\tilde{y} = \lambda y^{(1)} + (1-\lambda)y^{(2)} \ \text{ is feasible for } \ Wy = h(\hat{\xi}) - T(\hat{\xi})\tilde{x},\ y \geq 0.$$

Hence

$$Q(\tilde{x};T(\hat{\xi}),h(\hat{\xi}),q(\hat{\xi})) \leq q^{\mathrm{T}}(\hat{\xi})\tilde{y} = \lambda q^{\mathrm{T}}(\hat{\xi})y^{(1)} + (1-\lambda)q^{\mathrm{T}}(\hat{\xi})y^{(2)},$$

showing the convexity of $Q(x;T(\hat{\xi}),h(\hat{\xi}),q(\hat{\xi}))$ in x.

c) For any two $x^{(1)} \neq x^{(2)}$ and any $\xi \in \Xi$, according to (3.16) there exist $\nu_i \in \{1,\cdots,K\}$, $i = 1,2$, such that

$$Q(x^{(i)};T(\xi),h(\xi),q(\xi)) = \alpha_{\nu_i}(\xi) + d^{(\nu_i)^{\mathrm{T}}}(\xi)x^{(i)},$$

and due to (3.17) holds

$$\begin{aligned}
&[\alpha_{\nu_1}(\xi) + d^{(\nu_1)^{\mathrm{T}}}(\xi)x^{(2)}] - [\alpha_{\nu_1}(\xi) + d^{(\nu_1)^{\mathrm{T}}}(\xi)x^{(1)}] \\
&= d^{(\nu_1)^{\mathrm{T}}}(\xi)(x^{(2)} - x^{(1)}) \\
&\leq Q(x^{(2)};T(\xi),h(\xi),q(\xi)) - Q(x^{(1)};T(\xi),h(\xi),q(\xi)) \\
&\leq [\alpha_{\nu_2}(\xi) + d^{(\nu_2)^{\mathrm{T}}}(\xi)x^{(2)}] - [\alpha_{\nu_2}(\xi) + d^{(\nu_2)^{\mathrm{T}}}(\xi)x^{(1)}] \\
&= d^{(\nu_2)^{\mathrm{T}}}(\xi)(x^{(2)} - x^{(1)}),
\end{aligned}$$

such that

$$\begin{aligned}
&|Q(x^{(2)};T(\xi),h(\xi),q(\xi)) - Q(x^{(1)};T(\xi),h(\xi),q(\xi))| \\
&\leq \max_{i\in\{1,2\}} |d^{(\nu_i)^{\mathrm{T}}}(\xi)(x^{(2)} - x^{(1)})| \leq \max_{i\in\{1,2\}} \|d^{(\nu_i)}(\xi)\|\|(x^{(2)} - x^{(1)})\|.
\end{aligned}$$

Hence, with $D(\xi) = \max_{i\in\{1,\cdots,K\}} \|d^{(\nu_i)}(\xi)\| \in L^1$—due to (3.16)—follows the proposition. □

Due to Chapter 1, Def. 1.10. (p. 54) a vector $g \in \mathbb{R}^n$ is a subgradient of a convex function $\varphi : \mathbb{R}^n \longrightarrow \mathbb{R}$ at a point x if it satisfies

$$g^{\mathrm{T}}(z - x) \leq \varphi(z) - \varphi(x) \ \ \forall z,$$

and the subdifferential $\partial\varphi(x)$ is the set of all subgradients of φ at x. In particular for linear programs we have

Lemma 3.3. *Assume that the LP*

$$\min\{c^{\mathrm{T}}x \mid Ax = b, x \geq 0\}$$

is solvable $\forall b \in \mathbb{R}^m$. *Then its optimal value* $\varphi(b)$ *(obviously convex in* b*) is subdifferentiable at any* b*, and the subdifferential is given as* $\partial\varphi(b) = \operatorname{argmax}\{b^{\mathrm{T}}u \mid A^{\mathrm{T}}u \leq c\}$*, the set of optimal dual solutions at* b*.*

Proof: For a given $\widehat{b}$ let $\hat{u} \in \operatorname{argmax}\{\widehat{b}^{\mathrm{T}}u \mid A^{\mathrm{T}}u \leq c\}$, such that $\varphi(\widehat{b}) = \widehat{b}^{\mathrm{T}}\hat{u}$. Hence $\hat{u}$ is also feasible for the LP $\varphi(\widetilde{b}) = \max\{\widetilde{b}^{\mathrm{T}}u \mid A^{\mathrm{T}}u \leq c\}$ for an arbitrary $\widetilde{b}$ such that $\widetilde{b}^{\mathrm{T}}\hat{u} \leq \varphi(\widetilde{b})$ holds. Hence

$$\hat{u}^{\mathrm{T}}(\widetilde{b} - \widehat{b}) \leq \varphi(\widetilde{b}) - \varphi(\widehat{b})$$

showing that $\operatorname{argmax}\{\widehat{b}^{\mathrm{T}}u \mid A^{\mathrm{T}}u \leq c\} \subset \partial\varphi(\widehat{b})$.

Assume now that $g \in \partial\varphi(\widehat{b})$ for some $\widehat{b}$. Therefore, for any b holds

$$g^{\mathrm{T}}(b - \widehat{b}) \leq \varphi(b) - \varphi(\widehat{b}).$$

With $\hat{x} \in \operatorname{argmin}\{c^{\mathrm{T}}x \mid Ax = \widehat{b}, x \geq 0\}$ and $x^{(i)} = \hat{x} + e_i (\geq 0)$, $i = 1, \cdots, n$, (e_i the i-th unit vector), by our assumption, for all $b^{(i)} = Ax^{(i)}$, the LP's $\varphi(b^{(i)}) = \min\{c^{\mathrm{T}}x \mid Ax = b^{(i)}, x \geq 0\}$ are solvable. Obviously we have $\varphi(b^{(i)}) \leq c^{\mathrm{T}}x^{(i)}$ such that

$$\begin{aligned} g^{\mathrm{T}}Ae_i = g^{\mathrm{T}}A(x^{(i)} - \hat{x}) &= g^{\mathrm{T}}(b^{(i)} - \widehat{b}) \\ &\leq \varphi(b^{(i)}) - \varphi(\widehat{b}) \\ &\leq c^{\mathrm{T}}x^{(i)} - c^{\mathrm{T}}\hat{x} = c^{\mathrm{T}}e_i, \quad i = 1, \cdots, n, \end{aligned}$$

implying $A^{\mathrm{T}}g \leq c$, the dual feasibility of g. Then, due to the weak duality theorem (Chapter 1, Prop. 1.9., page 13), we have $g^{\mathrm{T}}\widehat{b} - \varphi(\widehat{b}) \leq 0$. Assume that with some $\alpha < 0$ holds $g^{\mathrm{T}}\widehat{b} - \varphi(\widehat{b}) \leq \alpha$. For $\widetilde{b} = 0$ obviously follows $\varphi(\widetilde{b}) = 0$ such that the subgradient inequality, valid for all b, yields

$$0 = g^{\mathrm{T}}\widetilde{b} - \varphi(\widetilde{b}) \leq g^{\mathrm{T}}\widehat{b} - \varphi(\widehat{b}) \leq \alpha < 0.$$

This contradiction, implied by the assumption $g^{\mathrm{T}}\widehat{b} - \varphi(\widehat{b}) \leq \alpha < 0$, shows that $g^{\mathrm{T}}\widehat{b} = \varphi(\widehat{b})$ and hence $\partial\varphi(\widehat{b}) \subset \operatorname{argmax}\{\widehat{b}^{\mathrm{T}}u \mid A^{\mathrm{T}}u \leq c\}$. □

Now we get immediately

Theorem 3.2. *Let the Assumptions 3.2. and 3.3. be satisfied. Then the recourse function* $Q(x; T(\xi), h(\xi), q(\xi))$ *is subdifferentiable in* x *for any* $\xi \in \Xi$*. For any* $\hat{x}$ *holds (the subscript at* ∂ *indicating the variable of subdifferentiation)*

$$\begin{aligned} &\partial_x Q(\hat{x}; T(\xi), h(\xi), q(\xi)) = \\ &= \{-T^{\mathrm{T}}(\xi)\hat{u} \mid \hat{u} \in \operatorname{argmax}\{(h(\xi) - T(\xi)\hat{x})^{\mathrm{T}}u \mid W^{\mathrm{T}}u \leq q(\xi)\}\} \ \forall \xi \in \Xi. \end{aligned}$$

Proof: For an arbitrary $\xi \in \Xi$ define $b(x;\xi) := h(\xi) - T(\xi)x$. Introducing

$$\psi(b(x;\xi);\xi) := Q(x; T(\xi), h(\xi), q(\xi))$$

$$= \min\{q^{\mathrm{T}}(\xi)y \mid Wy = b(x;\xi), y \geq 0\},$$

from Lemma 3.3 follows for the subdifferential of $\psi(\cdot;\xi)$ at $b(\hat{x};\xi)$

$$\partial_b \psi(b(\hat{x};\xi);\xi) = \arg\max\{b^{\mathrm{T}}(\hat{x};\xi)u \mid W^{\mathrm{T}}u \leq q(\xi)\}.$$

Then from Prop. 1.25. in Chapter 1 (p. 55) we know that

$$\begin{aligned} \partial_x Q(\hat{x};T(\xi),h(\xi),q(\xi)) &= -T^{\mathrm{T}}(\xi)\partial_b \psi(b(\hat{x};\xi);\xi) \\ &= -T^{\mathrm{T}}(\xi)\arg\max\{b^{\mathrm{T}}(\hat{x};\xi)u \mid W^{\mathrm{T}}u \leq q(\xi)\}. \end{aligned}$$

□

Theorem 3.3. *Since $\xi \in \mathscr{L}_r^2$ (i.e. ξ square-integrable with respect to $\mathbb{P}_\xi$), the expected recourse $\mathscr{Q}(x)$ is*

a) finitely valued $\forall x \in \mathscr{B}_1$, and
b) a convex and Lipschitz continuous function in x.

Hence, (3.8) is a convex optimization problem with a Lipschitz continuous objective function.

Proof:

a) Let $\hat{x} \in \mathbb{R}^{n_1}$ be fixed. Due to Assumptions 3.2. and 3.3., for any $\xi \in \Xi$ there exists an optimal feasible basic solution of the recourse program (3.15), i.e. there is an $(m_2 \times m_2)$-submatrix B of W such that

$$\left.\begin{aligned} B^{-1}(h(\xi) - T(\xi))\hat{x} &\geq 0 \text{ and} \\ Q(\hat{x};T(\xi),h(\xi),q(\xi)) &= q_B(\xi)^{\mathrm{T}}B^{-1}(h(\xi) - T(\xi)\hat{x}) \end{aligned}\right\}, \tag{3.18}$$

where the components of the m_2–subvector $q_B(\xi)$ of $q(\xi)$ correspond to the columns in B selected from W, as mentioned in Chapter 1, Prop. 1.2. (p. 9). Together with the simplex criterion, Prop. 1.3. in Chapter 1 (p. 9), such a particular basis is feasible and optimal on a polyhedral subset $\Xi_B \subset \Xi$, a so-called decision region (also: stability region).
According to (3.6) and (3.18), the recourse function $Q(\hat{x};T(\xi),h(\xi),q(\xi))$ is, in general, a quadratic function in ξ for $\xi \in \Xi_B$, such that, due to the assumption that $\xi \in L_2$, the integral $\int_{\Xi_B} Q(\hat{x};T(\xi),h(\xi),q(\xi))\mathbb{P}_\xi(d\xi)$ exists. By the Assumptions 3.2. and 3.3., the support Ξ is contained in the union of finitely many decision regions, which implies that also

$$\mathscr{Q}(\hat{x}) = \int_\Xi Q(\hat{x};T(\xi),h(\xi),q(\xi))\mathbb{P}_\xi(d\xi) \quad \text{exists.}$$

b) In Theorem 3.1., for any $\xi \in \Xi$, the recourse function $Q(x;T(\xi),h(\xi),q(\xi))$ has been shown to be convex and Lipschitz continuous in x, with a Lipschitz constant $D(\xi) \in \mathscr{L}_1^1$.

Hence the convexity of $\mathcal{Q}(x) = \int_{\Xi} Q(x;T(\xi),h(\xi),q(\xi))\mathbb{P}_{\xi}(d\xi)$ is obvious. And for any two $x^{(1)}$ and $x^{(2)}$ we have

$$\begin{aligned}
&|\mathcal{Q}(x^{(1)}) - \mathcal{Q}(x^{(2)})| \\
&\leq \left| \int_{\Xi} \{Q(x^{(1)};T(\xi),h(\xi),q(\xi)) - Q(x^{(2)};T(\xi),h(\xi),q(\xi))\} \mathbb{P}_{\xi}(d\xi) \right| \\
&\leq \int_{\Xi} \left| Q(x^{(1)};T(\xi),h(\xi),q(\xi)) - Q(x^{(2)};T(\xi),h(\xi),q(\xi)) \right| \mathbb{P}_{\xi}(d\xi) \\
&\leq \int_{\Xi} D(\xi) \|x^{(1)} - x^{(2)}\| \mathbb{P}_{\xi}(d\xi) = D\|x^{(1)} - x^{(2)}\|
\end{aligned}$$

with the Lipschitz constant $D = \int_{\Xi} D(\xi)\mathbb{P}_{\xi}(d\xi)$. □

Corollary 3.1. *Given that the random entries $q(\xi)$ and $(h(\xi),T(\xi))$ are stochastically independent, then with $\xi \in \mathcal{L}_r^1$ (instead of $\xi \in \mathcal{L}_r^2$ as before), the conclusions of Th. 3.3. hold true, as well.*

Proof: Only the existence of $\mathcal{Q}(x) = \int_{\Xi} Q(x;T(\xi),h(\xi),q(\xi))\mathbb{P}_{\xi}(d\xi)$ has to be proved, which follows, with $\xi \in \mathcal{L}_r^1(\Omega,\mathbb{R}^r)$, from the independence of $q(\xi)$ and $(h(\xi),T(\xi))$ according to

$$\begin{aligned}
&\int_{\Xi_B} Q(x;T(\xi),h(\xi),q(\xi))\mathbb{P}_{\xi}(d\xi) = \\
&= \int_{\Xi_B} q_B(\xi)^{\mathrm{T}} B^{-1}(h(\xi) - T(\xi)x)\mathbb{P}_{\xi}(d\xi) \\
&= \left(\int_{\Xi_B} q_B(\xi)\mathbb{P}_{\xi}(d\xi) \right)^{\mathrm{T}} \left(\int_{\Xi_B} B^{-1}(h(\xi) - T(\xi)x)\mathbb{P}_{\xi}(d\xi) \right).
\end{aligned}$$

□

Remark 3.3. *In Theorem 3.2. the subdifferential of the recourse function at any $\hat{x}$ under the Assumptions 3.2. and 3.3. was derived as*

$$\begin{aligned}
&\partial_x Q(\hat{x};T(\xi),h(\xi),q(\xi)) = \\
&= \{-T^{\mathrm{T}}(\xi)\hat{u} \mid \hat{u} \in \operatorname{argmax}\{(h(\xi) - T(\xi)\hat{x})^{\mathrm{T}} u \mid W^{\mathrm{T}} u \leq q(\xi)\}\} \ \forall \xi \in \Xi.
\end{aligned}$$

It can be shown, that then $\mathcal{Q}(\cdot)$ is subdifferentiable at $\hat{x}$ and

$$\partial \mathcal{Q}(\hat{x}) = \int_{\Xi} \partial_x Q(\hat{x};T(\xi),h(\xi),q(\xi))\mathbb{P}_{\xi}(d\xi), \tag{3.19}$$

where this integral is understood as the set $\left\{ \int_{\Xi} G(\xi)\mathbb{P}_{\xi}(d\xi) \right\}$ *for all functions* $G(\cdot)$ *being measurable selections from* $\partial_x Q(\hat{x}; T(\cdot), h(\cdot), q(\cdot))$ *such that the integral* $\int_{\Xi} \|G(\xi)\| \,\mathbb{P}_{\xi}(d\xi)$ *exists.*

Finally, $\mathcal{Q}(\cdot)$ *is differentiable at* $\hat{x}$ *if and only if* $\partial_x Q(\hat{x}; T(\cdot), h(\cdot), q(\cdot))$ *is a singleton a.s. with respect to* $\mathbb{P}_{\xi}$.

To prove statements of this type involves several technicalities, like the existence of measurable selections from subdifferentials or equivalently, from solution sets of optimization problems, integrability statements like Lebesgue's bounded convergence theorem, and so on. Under specific assumptions, these problems were considered for instance in Kall [152], Kall–Oettli [170], Rockafellar [280] (see also Kall [154]), and the general case is dealt with in Ch. 2 of Ruszczyński–Shapiro [295], where a sketch of a proof is presented.

Due to the fact that (sub)gradient methods will—in general—not be a central part of our discussion of solution approaches for recourse problems later on, we omit a proof of the interchangeability of subdifferentiation and integration, as stated in (3.19). □

3.2.1.1 CFR: Direct bounds for the expected recourse $\mathcal{Q}(x)$

Finally, assume that $q(\xi) \equiv q$, i.e. $q(\cdot)$ is deterministic. Then we have

Proposition 3.2. *Given the Assumptions 3.2. and 3.3. (the latter one now reading as* $\{u \mid W^{\mathrm{T}} u \leq q\} \neq \emptyset$*),* $Q(x; T(\cdot), h(\cdot))$ *is a convex function in* ξ *for any* $x \in \mathbb{R}^{n_1}$.

Proof: According to (3.6) for any fixed $x \in \mathbb{R}^{n_1}$ the right–hand–side of the LP

$$Q(x; T(\xi), h(\xi)) := \min\{q^{\mathrm{T}} y \mid Wy = h(\xi) - T(\xi)x, y \geq 0\}$$

is linear in ξ, which implies the asserted convexity. □

In this case we have a lower bound for $\mathcal{Q}(x)$, frequently used in solution methods, which is based on *Jensen's inequality* [148]:

Lemma 3.4. *Let* $\xi \in \mathbb{R}^r$ *be a random vector with probability distribution* $\mathbb{P}_{\xi}$ *such that* $\mathbb{E}_{\xi}[\xi]$ *exists, and assume* $\varphi : \mathbb{R}^r \longrightarrow \mathbb{R}$ *to be a convex function. Then the following inequality holds true:*

$$\varphi(\mathbb{E}_{\xi}[\xi]) \leq \mathbb{E}_{\xi}[\varphi(\xi)]. \tag{3.20}$$

Proof: Due to Chapter 1, Prop. 1.25. (p. 55), at any $\hat{\xi} \in \mathbb{R}^r$ there exists a nonempty, convex, compact subdifferential $\partial\varphi(\hat{\xi})$. Hence for any linear affine function $\ell(\cdot)$ out of the family $\widetilde{\mathcal{L}}_{\hat{\xi}}$ for some $\hat{\xi} \in \mathbb{R}^r$ with

$$\widetilde{\mathscr{L}_{\hat{\xi}}} := \{l(\cdot) \mid l(\xi) := \varphi(\hat{\xi}) + g_{\hat{\xi}}^{\mathrm{T}}(\xi - \hat{\xi}), g_{\hat{\xi}} \in \partial\varphi(\hat{\xi})\}, \; \hat{\xi} \in \mathbb{R}^r,$$

the set of linear support functions to $\varphi(\cdot)$ at $\hat{\xi}$, we have the subgradient inequality

$$\ell(\xi) = \varphi(\hat{\xi}) + g_{\hat{\xi}}^{\mathrm{T}}(\xi - \hat{\xi}) \leq \varphi(\xi) \;\; \forall \xi \in \mathbb{R}^r.$$

By integration with respect to $\mathbb{P}_\xi$ follows

$$\mathbb{E}_\xi\left[\ell(\xi)\right] = \ell(\mathbb{E}_\xi\left[\xi\right]) = \varphi(\hat{\xi}) + g_{\hat{\xi}}^{\mathrm{T}}(\mathbb{E}_\xi\left[\xi\right] - \hat{\xi}) \leq \mathbb{E}_\xi\left[\varphi(\xi)\right]$$

such that $\ell(\mathbb{E}_\xi[\xi])$ yields a lower bound for $\mathbb{E}_\xi[\varphi(\xi)]$.

Since $\mathbb{E}_\xi[\xi] \in \mathbb{R}^r$, due to the subgradient inequality, at any $\hat{\xi} \in \mathbb{R}^r$ holds

$$\ell(\mathbb{E}_\xi\left[\xi\right]) = \varphi(\hat{\xi}) + g_{\hat{\xi}}^{\mathrm{T}}(\mathbb{E}_\xi\left[\xi\right] - \hat{\xi}) \leq \varphi(\mathbb{E}_\xi\left[\xi\right]) \;\; \forall \ell(\cdot) \in \widetilde{\mathscr{L}_{\hat{\xi}}}.$$

Hence, in $\{\widetilde{\mathscr{L}_{\hat{\xi}}}, \; \hat{\xi} \in \mathbb{R}^r\}$, the set of all possible linear support functions to $\varphi(\cdot)$, we get

$$\arg\max_{\hat{\xi}}\{\ell(\mathbb{E}_\xi\left[\xi\right]) \mid \ell(\cdot) \in \widetilde{\mathscr{L}_{\hat{\xi}}}, \; \hat{\xi} \in \mathbb{R}^r\} = \mathbb{E}_\xi\left[\xi\right].$$

Therefore, among all linear support functions to $\varphi(\cdot)$ we get the greatest lower bound for $\mathbb{E}_\xi[\varphi(\xi)]$ by choosing $\hat{\xi} = \mathbb{E}_\xi[\xi]$, i.e. $\ell(\xi) = \varphi(\mathbb{E}_\xi[\xi]) + g_{\mathbb{E}_\xi[\xi]}^{\mathrm{T}}(\xi - \mathbb{E}_\xi[\xi])$, yielding

$$\ell(\mathbb{E}_\xi\left[\xi\right]) = \varphi(\mathbb{E}_\xi\left[\xi\right]) \leq \mathbb{E}_\xi\left[\varphi(\xi)\right].$$

□

Whereas under the assumptions of Lemma 3.4 we know for sure that the integral $\int_{\mathbb{R}^r} \varphi(\xi)\mathbb{P}_\xi(d\xi)$ is bounded below, it cannot be excluded in general that $\mathbb{E}_\xi[\varphi(\xi)] = +\infty$ holds. In contrast, under our assumptions for Prop. 3.2. we know from Cor. 3.1. that $\mathscr{Q}(x) = \mathbb{E}_\xi[Q(x; T(\xi), h(\xi))]$ is finite for all $x \in \mathbb{R}^{n_1}$. From Prop. 3.2. and Lemma 3.4 follows immediately the *Jensen lower bound* for the expected recourse:

Theorem 3.4. *Given the Assumptions 3.2. and 3.3., with* $\bar{\xi} = \mathbb{E}_\xi[\xi]$, *the expected recourse* $\mathscr{Q}(x) = \mathbb{E}_\xi[Q(x; T(\xi), h(\xi))]$ *is bounded below due to*

$$Q(x; T(\bar{\xi}), h(\bar{\xi})) \leq \mathscr{Q}(x). \tag{3.21}$$

Observe that in this case the lower bound for the expected recourse is defined by the one-point distribution $\mathbb{P}_\eta$ with $\mathbb{P}_\eta(\{\eta \mid \eta = \bar{\xi}\}) = 1$, which does not depend on the particular recourse function, since

$$\int Q(x; T(\eta), h(\eta))\mathbb{P}_\eta(d\eta) = Q(x; T(\bar{\xi}), h(\bar{\xi})) \leq \mathscr{Q}(x)$$

holds true for any function $Q(x;T(\cdot),h(\cdot))$ being convex in ξ.

Concerning upper bounds for the expected recourse, the situation is more difficult. The first attempts to derive upper bounds for the expectation of convex functions of random variables are assigned to Edmundson [83] and Madansky [210]. Hence, the basic relation is referred to as *Edmundson–Madansky inequality* (E–M):

Lemma 3.5. *Let* τ *be a random variable with* $\operatorname{supp}\mathbb{P}_\tau \subseteq [\alpha,\beta] \subset \mathbb{R}$ *such that the expectation* $\mu = \mathbb{E}_\tau[\tau] \in [\alpha,\beta]$. *Then, for any convex function* $\psi : [\alpha,\beta] \longrightarrow \mathbb{R}$ *holds*

$$\mathbb{E}_\tau[\psi(\tau)] \le \mathbb{E}_{\hat{\tau}}[\psi(\hat{\tau})], \tag{3.22}$$

where $\hat{\tau}$ *is the discrete random variable with the two-point distribution*

$$\mathbb{P}_{\hat{\tau}}(\{\hat{\tau} \mid \hat{\tau} = \alpha\}) = \frac{\beta-\mu}{\beta-\alpha}, \quad \mathbb{P}_{\hat{\tau}}(\{\hat{\tau} \mid \hat{\tau} = \beta\}) = \frac{\mu-\alpha}{\beta-\alpha}. \tag{3.23}$$

Proof: With $\lambda_\tau = \dfrac{\beta-\tau}{\beta-\alpha}$ we have $\lambda_\tau\alpha + (1-\lambda_\tau)\beta = \tau \; \forall \tau \in [\alpha,\beta]$ and $\lambda_\tau \in [0,1]$. Due to the convexity of ψ follows

$$\psi(\tau) = \psi(\lambda_\tau\alpha + (1-\lambda_\tau)\beta) \le \lambda_\tau\psi(\alpha) + (1-\lambda_\tau)\psi(\beta) \;\; \forall \tau \in [\alpha,\beta]$$

and therefore, integrating both sides of this inequality with respect to $\mathbb{P}_\tau$,

$$\mathbb{E}_\tau[\psi(\tau)] \le \frac{\beta-\mu}{\beta-\alpha} \cdot \psi(\alpha) + \frac{\mu-\alpha}{\beta-\alpha} \cdot \psi(\beta) = \mathbb{E}_{\hat{\tau}}[\psi(\hat{\tau})].$$

□

3.2.1.2 CFR: Moment problems and bounds for $\mathcal{Q}(x)$

It is worthwhile to observe the following relation to the theory of *moment problems* and *semi-infinite programs*.

Under the assumptions of Lemma 3.5 consider, with $\mathcal{P}$ the set of probability measures on $[\alpha,\beta]$, as primal (P) the problem

$$\sup_{\mathbb{P}\in\mathcal{P}} \left\{ \int_\alpha^\beta \psi(\xi)\mathbb{P}(d\xi) \;\Big|\; \int_\alpha^\beta \xi\mathbb{P}(d\xi) = \mu, \; \int_\alpha^\beta \mathbb{P}(d\xi) = 1 \right\}, \tag{3.24}$$

a so-called moment problem, and as its dual problem (D)

$$\inf_{y\in\mathbb{R}^2} \{y_1 + \mu y_2 \mid y_1 + \xi y_2 \ge \psi(\xi) \;\; \forall \xi \in [\alpha,\beta]\}, \tag{3.25}$$

the corresponding semi-infinite program.

Since, as required by the constraints of (D), a linear affine function majorizes a convex function on an interval if and only if it does so on the endpoints, (D) is equivalent to

$$\min_{y \in \mathbb{R}^2} \{ y_1 + \mu y_2 \mid y_1 + \alpha y_2 \geq \psi(\alpha),\ y_1 + \beta y_2 \geq \psi(\beta) \}.$$

Due to the fact that $\alpha \leq \mu \leq \beta$ this LP is solvable, and hence so is its dual (P), which now reads as

$$\max_{p_\alpha, p_\beta} \{ \psi(\alpha) p_\alpha + \psi(\beta) p_\beta \mid p_\alpha + p_\beta = 1,\ \alpha p_\alpha + \beta p_\beta = \mu;\ p_\alpha, p_\beta \geq 0 \}$$

and has, as the unique solution of its constraints, the distribution of $\hat{\tau}$ as given in (3.23). It is worth mentioning that in this case the solution of the moment problem (P), i.e. the E–M distribution yielding the upper bound, is independent of the particular choice of the convex function $\psi : [\alpha, \beta] \longrightarrow \mathbb{R}$.

Suppose now that we have a random vector $\xi \in \mathbb{R}^r$. Then, as mentioned in Kall–Stoyan [171], Lemma 3.5 can immediately be generalized as follows:

Lemma 3.6. *Let* $\operatorname{supp} \mathbb{P}_\xi \subset \Xi = \prod_{i=1}^r [\alpha_i, \beta_i] \subset \mathbb{R}^r$ *and assume the components of* ξ *to be stochastically independent. With* $\mu = \mathbb{E}_\xi[\xi] \in \Xi$ *let* $\mathbb{P}_{\eta_i}, i = 1, \cdots, r$, *be the two-point distributions defined on* $[\alpha_i, \beta_i]$ *as*

$$\mathbb{P}_{\eta_i}(\{\eta_i \mid \eta_i = \alpha_i\}) = \frac{\beta_i - \mu_i}{\beta_i - \alpha_i},\quad \mathbb{P}_{\eta_i}(\{\eta_i \mid \eta_i = \beta_i\}) = \frac{\mu_i - \alpha_i}{\beta_i - \alpha_i}. \tag{3.26}$$

Then for the random vector $\eta \in \mathbb{R}^r$ *with the probability distribution given as*

$$\mathbb{P}_\eta = \mathbb{P}_{\eta_1} \times \mathbb{P}_{\eta_2} \times \cdots \times \mathbb{P}_{\eta_r} \quad on \quad \Xi = \prod_{i=1}^r [\alpha_i, \beta_i] \tag{3.27}$$

it follows for any convex function $\varphi : \Xi \longrightarrow \mathbb{R}$ *that*

$$\mathbb{E}_\xi[\varphi(\xi)] \leq \mathbb{E}_\eta[\varphi(\eta)]. \tag{3.28}$$

Proof: With $\mathbb{P}_{\xi_i}$ the marginal distribution of $\mathbb{P}_\xi$ for $\xi_i \in [\alpha_i, \beta_i]$, the assumed stochastic independence of the components of ξ implies that

$$\mathbb{P}_\xi = \mathbb{P}_{\xi_1} \times \mathbb{P}_{\xi_2} \times \cdots \times \mathbb{P}_{\xi_r}.$$

Hence the asserted inequality (3.28) follows immediately from Lemma 3.5 by induction to r, using the fact that the product measures $\mathbb{P}_\xi$ and $\mathbb{P}_\eta$ allow for iterated integration, as known from Fubini's theorem (see Halmos [131]). □

Also in this case we may assign a moment problem, with $\mathscr{P}$ the set of all product measures on $\Xi = \prod_{i=1}^r \Xi_i = \prod_{i=1}^r [\alpha_i, \beta_i]$, stated as (P)

$$\sup_{\mathbb{P} \in \mathscr{P}} \left\{ \int_\Xi \varphi(\xi) P_{\xi_1}(d\xi_1) \cdots P_{\xi_r}(d\xi_r) \;\middle|\; \begin{array}{l} \int_{\Xi_i} \xi_i P_{\xi_i}(d\xi_i) = \mu_i, \\ \int_{\Xi_i} P_{\xi_i}(d\xi_i) \;\;= 1, \end{array} \forall i \right\}, \tag{3.29}$$

and its dual semi-infinite program (D)

$$\inf_{y \in \mathbb{R}^{2r}} \left\{ \sum_{i=1}^{r} (y_1^i + \mu_i y_2^i) \mid y_1^i + \xi_i y_2^i \geq \tilde{\varphi}_i(\xi_i) \;\; \forall \xi_i \in \Xi_i \forall i \right\} \tag{3.30}$$

where with $\Xi / \Xi_i := \Xi_1 \times \cdots \times \Xi_{i-1} \times \Xi_{i+1} \times \cdots \times \Xi_r$

$$\tilde{\varphi}_i(\xi_i) = \int_{\Xi/\Xi_i} \varphi(\xi_1, \cdots, \xi_r) \mathbb{P}_{\xi_1}(d\xi_1) \cdots \mathbb{P}_{\xi_{i-1}}(d\xi_{i-1}) \mathbb{P}_{\xi_{i+1}}(d\xi_{i+1}) \cdots \mathbb{P}_{\xi_r}(d\xi_r)$$

is obviously a convex function in ξ_i. Therefore again, the constraints of (D) are satisfied if and only if they hold in the endpoints α_i and β_i of all intervals Ξ_i. Hence (D) is equivalent to

$$\inf_{y \in \mathbb{R}^{2r}} \left\{ \sum_{i=1}^{r} (y_1^i + \mu_i y_2^i) \mid y_1^i + \alpha_i y_2^i \geq \tilde{\varphi}_i(\alpha_i);\; y_1^i + \beta_i y_2^i \geq \tilde{\varphi}_i(\beta_i) \;\; \forall i \right\},$$

which due to $\mu_i \in [\alpha_i, \beta_i]$ is solvable again and hence so is its dual, the moment problem

$$\begin{aligned} &\max \left\{ \sum_{i=1}^{r} (\tilde{\varphi}_i(\alpha_i) p_{\alpha_i}^i + \tilde{\varphi}_i(\beta_i) p_{\beta_i}^i) \right\} \\ &\text{s.t.} \quad \alpha_i p_{\alpha_i}^i + \beta_i p_{\beta_i}^i = \mu_i,\; p_{\alpha_i}^i + p_{\beta_i}^i = 1 \;\forall i. \end{aligned}$$

Since the only feasible solution of its constraints coincides with the two-point measures (3.26), the product measure (3.27) solving the moment problem (P) is independent of the particular convex function φ, again.

For later use we just mention the following fact, which due to the above results is evident:

Corollary 3.2. *Let* $\operatorname{supp} \mathbb{P}_\xi \subset \Xi = \prod_{i=1}^r [\alpha_i, \beta_i] \subset \mathbb{R}^r$ *with* $\mu = \mathbb{E}_\xi[\xi]$ *and assume the function* $\varphi : \Xi \longrightarrow \mathbb{R}$ *to be convex separable, i.e.* $\varphi(\xi) = \sum_{i=1}^{r} \varphi_i(\xi_i)$*. Then, with the distributions* $\mathbb{P}_{\eta_i}$ *given in (3.26), it follows that*

$$\mathbb{E}_\xi[\varphi(\xi)] = \sum_{i=1}^{r} \mathbb{E}_\xi[\varphi_i(\xi_i)] \leq \sum_{i=1}^{r} \mathbb{E}_{\eta_i}[\varphi_i(\eta_i)]. \tag{3.31}$$

We shall refer to (3.22), (3.28) and (3.31) as the E–M inequality. For the expected recourse we then get the *E–M upper bound*:

Theorem 3.5. *Assume that the components of* ξ *are stochastically independent and that* $\operatorname{supp} \mathbb{P}_\xi \subset \Xi = \prod_{i=1}^r [\alpha_i, \beta_i]$ *with* $\mu = \mathbb{E}_\xi[\xi] \in \Xi$*. Given the Assumptions 3.2. and 3.3., with the E–M distribution defined by (3.26) and (3.27) the expected recourse* $\mathcal{Q}(x) = \mathbb{E}_\xi[Q(x; T(\xi), h(\xi))]$ *is bounded above according to*

$$\mathscr{Q}(x) \le \mathbb{E}_\eta [Q(x;T(\eta),h(\eta))]. \tag{3.32}$$

According to Lemma 3.6 and Cor. 3.2. we have the E–M inequality for multi-dimensional distributions either for random vectors with independent components or for convex integrands being separable. However this upper bound does not remain valid for arbitrary integrands and dependent components, in general, as shown by the following example:

Example 3.1. *Let ξ be the discrete random vector in $\mathbb{R}^2$ with the* *distribution of ξ:*

realizations:	(0,0)	(1,0)	(0,1)	(1,1)
probabilities:	0.1	0.2	0.1	0.6

yielding the expectation $\bar{\xi} = (0.8, 0.7)$. This implies the *marginal distributions of ξ_1 and ξ_2:*

realizations:	0	1
probabilities $\mathbb{P}_{\xi_1}$:	0.2	0.8
probabilities $\mathbb{P}_{\xi_2}$:	0.3	0.7

being obviously stochastically dependent. Using these marginal distributions to compute the E–M distribution according to Th. 3.5., we get the *E–M distribution of η:*

realizations:	(0,0)	(1,0)	(0,1)	(1,1)
probabilities:	0.06	0.24	0.14	0.56

with the expectation $\bar{\eta} = (0.8, 0.7)$. Then for any convex function $\varphi(\cdot,\cdot)$ such that

$$\varphi(0,0) = \varphi(1,0) = \varphi(0,1) = 0 \;\; and \;\; \varphi(1,1) = 1$$

we get $\mathbb{E}_\xi [\varphi(\xi)] = 0.6$ and $\mathbb{E}_\eta [\varphi(\eta)] = 0.56$. Hence, in this case, with the E–M distribution (3.27) as derived for the independent case, the E–M inequality (3.28) does not hold. □

To generalize the E–M inequality for random vectors with dependent components and $\operatorname{supp} \mathbb{P}_\xi \subset \Xi = \prod_{i=1}^r [\alpha_i, \beta_i]$, and for arbitrary convex integrands, according to Frauendorfer [102] we may proceed as follows:

Assume first that for some $\xi \in \Xi$ we have the random vector ζ with the one-point distribution $\mathbb{P}_\zeta(\{\zeta \mid \zeta = \xi\}) = 1$. Obviously the components of ζ are stochastically independent, and for $\eta_i(\xi_i)$ with the two-point distributions

$$\left.\begin{aligned} \mathbb{P}_{\eta_i(\xi_i)}(\{\eta_i \mid \eta_i = \alpha_i\}) &= \frac{\beta_i - \xi_i}{\beta_i - \alpha_i} \\ \mathbb{P}_{\eta_i(\xi_i)}(\{\eta_i \mid \eta_i = \beta_i\}) &= \frac{\xi_i - \alpha_i}{\beta_i - \alpha_i} \end{aligned}\right\} \tag{3.33}$$

holds

$$\mathbb{E}_{\eta_i(\xi_i)}[\eta_i] = \xi_i = \mathbb{E}_{\zeta_i}[\zeta_i]. \tag{3.34}$$

Hence for the probability measure

$$\mathbb{P}_{\eta(\xi)} = \mathbb{P}_{\eta_1(\xi_1)} \times \mathbb{P}_{\eta_2(\xi_2)} \times \cdots \times \mathbb{P}_{\eta_r(\xi_r)} \quad \text{on} \quad \Xi = \prod_{i=1}^{r} [\alpha_i, \beta_i], \tag{3.35}$$

defined on the vertices v^ν of Ξ, $\nu = 1, \cdots, 2^r$, we have the probabilities

$$\mathbb{P}_{\eta(\xi)}(v^\nu) = \prod_{i \in I_\nu} \frac{\beta_i - \xi_i}{\beta_i - \alpha_i} \cdot \prod_{i \in J_\nu} \frac{\xi_i - \alpha_i}{\beta_i - \alpha_i},$$

where $I_\nu = \{i \mid v_i^\nu = \alpha_i\}$ and $J_\nu = \{1, \cdots, r\} \setminus I_\nu$ (with $\prod_{i \in \emptyset} \{\cdot\} = 1$). Thus we get immediately

Lemma 3.7. *For any convex function $\varphi : \Xi \longrightarrow \mathbb{R}$, Jensen's inequality implies*

$$\left.\begin{aligned} \varphi(\mathbb{E}_\zeta[\zeta]) = \varphi(\xi) &\leq \int_\Xi \varphi(\eta(\xi)) \mathbb{P}_{\eta(\xi)}(d\eta) \\ &= \sum_{\nu=1}^{2^r} \varphi(v^\nu) \mathbb{P}_{\eta(\xi)}(v^\nu). \end{aligned}\right\} \tag{3.36}$$

Hence, with the probability measure $\mathbb{Q}$ defined on the vertices v^ν of Ξ by

$$\left.\begin{aligned} \mathbb{Q}(v^\nu) &= \int_\Xi \mathbb{P}_{\eta(\xi)}(v^\nu) \mathbb{P}_\xi(d\xi) \\ &= \int_\Xi \prod_{i \in I_\nu} \frac{\beta_i - \xi_i}{\beta_i - \alpha_i} \cdot \prod_{i \in J_\nu} \frac{\xi_i - \alpha_i}{\beta_i - \alpha_i} \mathbb{P}_\xi(d\xi), \end{aligned}\right\} \tag{3.37}$$

we get the generalized E–M inequality

$$\mathbb{E}_\xi[\varphi(\xi)] \leq \sum_{\nu=1}^{2^r} \varphi(v^\nu) \mathbb{Q}(v^\nu). \tag{3.38}$$

Remark 3.4. *Observe that for stochastically independent components of ξ, due to (3.37) we get for the generalized E–M distribution*

$$\mathbb{Q}(v^\nu) = \prod_{i \in I_\nu} \frac{\beta_i - \mu_i}{\beta_i - \alpha_i} \cdot \prod_{i \in J_\nu} \frac{\mu_i - \alpha_i}{\beta_i - \alpha_i},$$

such that in this case $\mathbb{Q}$ coincides with the E–M distribution $\mathbb{P}_\eta$ for the independent case as derived in (3.26) and (3.27). □

Hence Theorem 3.5. may be generalized as follows:

Theorem 3.6. *Assume that* $\operatorname{supp}\mathbb{P}_\xi \subset \Xi = \prod_{i=1}^r [\alpha_i, \beta_i]$ *such that also* $\mu = \mathbb{E}_\xi[\xi] \in \Xi$. *Under the Assumptions 3.2. and 3.3. and with the generalized E–M distribution $\mathbb{Q}$ as defined in (3.37), according to (3.38) the expected recourse* $\mathscr{Q}(x) = \mathbb{E}_\xi[Q(x;T(\xi),h(\xi))]$ *is bounded above as*

$$\left.\begin{aligned} \mathscr{Q}(x) &\le \int_\Xi Q(x;T(\eta),h(\eta))\,\mathbb{Q}(d\eta) \\ &= \sum_{\nu=1}^{2^r} Q(x;T(v^\nu),h(v^\nu))\,\mathbb{Q}(v^\nu). \end{aligned}\right\} \tag{3.39}$$

For any $\Lambda \subset \{1,\cdots,r\}$ let $\widehat{m}_\Lambda(\xi) := \prod_{k\in\Lambda} \xi_k$ and denote the joint mixed moments of $\{\xi_k \mid k \in \Lambda\}$ as $\mu_\Lambda := \int_\Xi \widehat{m}_\Lambda(\xi)\mathbb{P}_\xi(d\xi)$ for all $\Lambda \subset \{1,\cdots,r\}$ (with $\widehat{m}_\emptyset(\xi) \equiv 1$ and $\mu_\emptyset = 1$).

Then we have, for any vertex v^ν of Ξ, that $\widehat{m}_\Lambda(v^\nu) = \prod_{k\in\Lambda\cap I_\nu} \alpha_k \cdot \prod_{k\in\Lambda\cap J_\nu} \beta_k$, and from (3.34) and (3.35) follows

$$\int_\Xi \widehat{m}_\Lambda(\eta)\mathbb{P}_{\eta(\xi)}(d\eta) = \widehat{m}_\Lambda(\xi) = \sum_{\nu=1}^{2^r} \widehat{m}_\Lambda(v^\nu)\cdot\mathbb{P}_{\eta(\xi)}(v^\nu), \tag{3.40}$$

such that (3.37) and (3.40) imply

$$\left.\begin{aligned} \sum_{\nu=1}^{2^r} \widehat{m}_\Lambda(v^\nu)\mathbb{Q}(v^\nu) &= \int_\Xi \sum_{\nu=1}^{2^r} \widehat{m}_\Lambda(v^\nu)\mathbb{P}_{\eta(\xi)}(v^\nu)\mathbb{P}_\xi(d\xi) \\ &= \int_\Xi \widehat{m}_\Lambda(\xi)\mathbb{P}_\xi(d\xi) = \mu_\Lambda\,. \end{aligned}\right\} \tag{3.41}$$

Hence the upper bound distribution $\mathbb{Q}$ of Lemma 3.7 preserves all joint moments of the original distribution $\mathbb{P}_\xi$, suggesting to consider, for $\mathscr{P}$ being the set of all probability measures on Ξ, the moment problem (P)

$$\gamma(P) := \sup_{\mathbb{P}\in\mathscr{P}} \left\{ \int_\Xi \varphi(\xi)\mathbb{P}(d\xi) \,\middle|\, \int_\Xi \widehat{m}_\Lambda(\xi)\mathbb{P}(d\xi) = \mu_\Lambda \ \forall \Lambda \subset \{1,\cdots,r\} \right\}. \tag{3.42}$$

For the dual of this problem we assign the variables y_0 to $\Lambda = \emptyset$ ($\mu_\emptyset = 1$) and y_Λ to any nonempty subset $\Lambda \subset \{1,\cdots,r\}$. This yields the semi-infinite program (D)

$$\delta(D) := \inf\left\{ y_0 + \sum_{\Lambda\neq\emptyset} \mu_\Lambda y_\Lambda \,\middle|\, y_0 + \sum_{\Lambda\neq\emptyset} \widehat{m}_\Lambda(\xi) y_\Lambda \geq \varphi(\xi)\ \forall \xi \in \Xi \right\}. \tag{3.43}$$

Requiring the constraints of (D) to hold only at the vertices of Ξ yields the modified problem ($\tilde{\mathrm{D}}$)

$$\delta(\tilde{D}) := \inf\left\{ y_0 + \sum_{\Lambda\neq\emptyset} \mu_\Lambda y_\Lambda \,\middle|\, y_0 + \sum_{\Lambda\neq\emptyset} \widehat{m}_\Lambda(v^\nu) y_\Lambda \geq \varphi(v^\nu),\ \nu = 1,\cdots,2^r \right\}$$

and its dual ($\tilde{\mathrm{P}}$), the moment problem searching for a measure $\mathbb{P}$ in $\mathscr{P}_{\mathrm{ext}\,\Xi}$, the set of probability distributions on the vertices of Ξ, becomes

$$\gamma(\tilde{P}) := \sup_{\mathscr{P}_{\mathrm{ext}\,\Xi}} \left\{ \sum_{\nu=1}^{2^r} \varphi(v^\nu) p_\nu \,\middle|\, \sum_{\nu=1}^{2^r} \widehat{m}_\Lambda(v^\nu) p_\nu = \mu_\Lambda\ \forall \Lambda \subset \{1,\cdots,r\} \right\}.$$

Due to (3.41) the upper bound distribution $\mathbb{Q}$ of Lemma 3.7 is feasible for this moment problem ($\tilde{\mathrm{P}}$). Furthermore, since the matrix of the system of linear constraints of ($\tilde{\mathrm{P}}$), i.e.

$$H := (\widehat{m}_\Lambda(v^\nu);\ \nu = 1,\cdots,2^r,\ \Lambda \subset \{1,\cdots,r\}),$$

is regular, as shown in Kall [157], the generalized E–M distribution $\mathbb{Q}$ is the unique solution of ($\tilde{\mathrm{P}}$) and independent of φ. Finally, according to linear programming duality and since $\mathscr{P}_{\mathrm{ext}\,\Xi} \subset \mathscr{P}$ we have

$$\delta(\tilde{D}) = \gamma(\tilde{P}) \leq \gamma(P).$$

On the other hand for any $\xi \in \Xi$, given the regularity of H, the linear system

$$\sum_{\nu=1}^{2^r} \widehat{m}_\Lambda(v^\nu) q_\nu(\xi) = \widehat{m}_\Lambda(\xi),\ \Lambda \subset \{1,\cdots,r\} \tag{3.44}$$

has the unique solution $\{q_\nu(\xi) = \mathbb{P}_{\eta(\xi)}(v^\nu);\ \nu = 1,\cdots,2^r\}$ due to (3.40), being continuous in ξ. Then for any $\mathbb{P}$ feasible in (P) follows

$$\begin{aligned}
\forall \Lambda \subset \{1,\cdots,r\}:\ \mu_\Lambda &= \int_\Xi \widehat{m}_\Lambda(\xi)\mathbb{P}(d\xi) \\
&= \int_\Xi \sum_{\nu=1}^{2^r} \widehat{m}_\Lambda(v^\nu) q_\nu(\xi)\mathbb{P}(d\xi) \\
&= \sum_{\nu=1}^{2^r} \widehat{m}_\Lambda(v^\nu)\hat{q}_\nu \ \text{ with } \hat{q}_\nu = \int_\Xi q_\nu(\xi)\mathbb{P}(d\xi).
\end{aligned}$$

Hence $\{\hat{q}_\nu;\ \nu = 1,\cdots,2^r\}$ is a probability distribution on the vertices of Ξ which is feasible for the moment problem (P). Since (3.44) also includes $\sum_{\nu=1}^{2^r} v^\nu q_\nu(\xi) = \xi$, by the convexity of φ follows for the objective of (P)

$$\sum_{\nu=1}^{2^r} \varphi(v^\nu)\hat{q}_\nu = \int_\Xi \sum_{\nu=1}^{2^r} \varphi(v^\nu)q_\nu(\xi)\mathbb{P}(d\xi) \geq \int_\Xi \varphi(\xi)\mathbb{P}(d\xi).$$

Therefore we have

$$\gamma(\tilde{P}) \geq \gamma(P) \Longrightarrow \gamma(\tilde{P}) = \gamma(P),$$

such that the generalized E–M distribution $\mathbb{Q}$ solves the moment problem (P), and as shown in Kall [157], it is the unique solution of (P).

Remark 3.5. *In the above cases we could reduce particular moment problems* (P), *as e.g. (3.42), stated on $\mathscr{P}$, the set of all probability measures on some support Ξ, to moment problems* ($\tilde{\mathrm{P}}$) *on $\mathscr{P}_d$, some sets of probability measures with finite discrete supports $\Xi_d \subset \Xi$, such that a solution of* ($\tilde{\mathrm{P}}$) *was simultaneously a solution of* (P).

This observation is not surprising in view of a very general result, mentioned in Kemperman [181] and assigned to Richter [276] and Rogosinski [289], stated as follows:

"Let $f_1,\cdots,f_N$ be integrable functions on the probability space $(\Omega,\mathscr{G},P)$. Then there exists a probability measure $\tilde{P}$ with finite support in Ω such that

$$\int_\Omega f_i(\omega)P(d\omega) = \int_\Omega f_i(\omega)\tilde{P}(d\omega),\ i=1,\cdots,N.$$

Even $\operatorname{card}(\operatorname{supp}\tilde{P}) \leq N+1$ *may be achieved."*

Hence we can take advantage of the theory of semi-infinite programming. With

S, an arbitrary (usually infinite) index set, and
$a : S \longrightarrow \mathbb{R}^n$, $b : S \longrightarrow \mathbb{R}$, $c \in \mathbb{R}^n$ arbitrary,

the problem

$$v(P) := \inf\{c^{\mathrm{T}}y \mid a^{\mathrm{T}}(s)y \geq b(s)\ \ \forall s \in S\}$$

is called a (primal) semi-infinite program. Its dual program requires, for some $s_i \in S$, $i=1,\cdots,q \geq 1$, to determine a positive finite discrete measure μ with $\mu(s_i) = x_i$ as a solution of the generalized moment problem

$$v(D) := \sup\left\{\sum_{i=1}^{q} b(s_i)x_i \mid \sum_{i=1}^{q} a(s_i)x_i = c,\, x_i \geq 0,\, s_i \in S,\, q \geq 1\right\}.$$

Whereas weak duality, i.e. $v(D) \leq v(P)$, is evident, a detailed discussion of statements on (strong) duality as well as on existence of solutions for these two problems under various regularity assumptions may be found in textbooks like Glasshoff–Gustafson [126] and Goberna–López [128] (or in reviews as e.g. in Kall [158]).

Moment problems have been considered in detail in probability theory (see e.g. Krein–Nudel'man [196]) and in other areas of applied mathematics (like e.g.

Karlin–Studden [176]), and a profound geometric approach was presented in Kemperman [181].

In connection with stochastic programs with recourse moment problems were investigated to find upper bounds for the expected recourse, also under assumptions on the set Ξ *containing* $\operatorname{supp}\mathbb{P}_\xi$ *and moment conditions being different from those mentioned above.*

For instance, for a convex function φ, Ξ *being a (bounded) convex polyhedron, and the feasible set of probability measures* $\mathscr{P}$ *given by the moment conditions* $\int_\Xi \xi \mathbb{P}(d\xi) = \bar{\xi}\ (= \mathbb{E}_\xi[\xi])$, *the moment problem* $\sup_{\mathbb{P}\in\mathscr{P}} \int_\Xi \varphi(\xi)\mathbb{P}(d\xi)$ *turns out to be the linear program to determine an optimal discrete measure on the vertices of* Ξ *where, in contrast to the above E–M measures, the solution depends on* φ *in general (see e.g. Dupačová [74, 75]).*

Furthermore, for a lower semi-continuous proper convex function φ *and* Ξ *being an arbitrary closed convex set, and again with*

$$\mathscr{P} = \left\{ \mathbb{P} \,\middle|\, \int_\Xi \xi \mathbb{P}(d\xi) = \bar{\xi} \right\},$$

the moment problem $\sup_{\mathbb{P}\in\mathscr{P}} \int_\Xi \varphi(\xi)\mathbb{P}(d\xi)$, *considered by Birge–Wets [28], amounts to determine a finite discrete probability measure* $\mathbb{P}$ *on* $\operatorname{ext}\Xi$ *and a finite discrete nonnegative measure* ν *on* $\operatorname{ext}\operatorname{rc}\Xi$ *(with* $\operatorname{rc}\Xi$ *the recession cone of* Ξ*, see Rockafellar [281]), which for infinite sets* $\operatorname{ext}\Xi$ *and* $\operatorname{ext}\operatorname{rc}\Xi$ *appears to be a difficult task, whereas it seems to become somewhat easier if* Ξ *is assumed to be a convex polyhedral set as discussed e.g. in Edirisinghe–Ziemba [81], Gassmann–Ziemba [122], Huang–Ziemba–Ben-Tal [144]). Also in these cases, the solutions of the moment problems, i.e. the optimal measures, depend on* φ*, in general. For the special situation where* φ *is convex and* Ξ *is a regular simplex, i.e.*

$$\Xi = \operatorname{conv}\{v^0, v^1, \cdots, v^r\} \subset \mathbb{R}^r,\ \operatorname{rank}(v^1 - v^0, v^2 - v^0, \cdots, v^r - v^0) = r,$$

mentioned in Birge–Wets [27] and later investigated and used extensively by Frauendorfer [103], the moment problem under the above first order moment conditions has the unique solution of a regular system of linear equations, independent of φ *again.*

Finally, for $\Xi = \mathbb{R}^r$ *with* $\int_\Xi \xi \mathbb{P}_\xi(d\xi) = \mu$ *and* $\int_\Xi \|\xi\|^2 \mathbb{P}_\xi(d\xi) = \rho$, *(with* $\|\cdot\|$ *the Euclidean norm) moment problems with the nonlinear moment conditions*

$$\int_\Xi \xi\, \mathbb{P}(d\xi) = \mu \quad \textit{and} \quad \int_\Xi \|\xi\|^2\, \mathbb{P}(d\xi) = \rho$$

have been discussed, first for simplicial recourse functions φ *by Dulá [73], and then for more general nonlinear recourse functions in Kall [159]. In these cases, the solutions of the moment problems depend on* φ*, in general. Under appropriate assumptions on the recourse functions these moment problems turn out to be non-*

smooth optimization problems, solvable with bundle-trust methods as described in Schramm–Zowe [299], for instance.

We have sketched possibilities to derive upper bounds for the expected recourse using results from the theory on semi-infinite programming and moment problems. Similarly, the theory on partial orderings of spaces of probability measures, as described in Stoyan [313] and Müller–Stoyan [237], could be used. Attempts in this direction may be found e.g. in Frauendorfer [103] and in Kall–Stoyan [171]. □

3.2.1.3 CFR: Approximation by successive discretization

Assuming that, for the given random vector ξ, we have $\operatorname{supp} \mathbb{P}_\xi \subset \Xi = \prod_{i=1}^r [\alpha_i, \beta_i]$, due to Jensen and Edmundson–Madansky there follow for any convex function φ and $\bar{\xi} = \mathbb{E}_\xi[\xi]$ the bounds

$$\varphi(\bar{\xi}) \le \mathbb{E}_\xi[\varphi(\xi)] \le \mathbb{E}_\eta[\varphi(\eta)] = \int_\Xi \varphi(\eta)\mathbb{Q}(d\eta), \tag{3.45}$$

where η has the discrete distribution $\mathbb{Q}$ defined on the vertices of Ξ, as described in Lemma 3.7. Hence these bounds result from finitely many arithmetic operations provided the joint moments $\mu_\Lambda := \int_\Xi \widehat{m}_\Lambda(\xi)\mathbb{P}_\xi(d\xi) = \mathbb{E}_\xi[\widehat{m}_\Lambda(\xi)]$ are known for all $\Lambda \subset \{1,\cdots,r\}$.

The following observation is the basis of a method of discrete approximations (of the distribution) to solve complete recourse problems.

Assume that,with half-open or closed intervals Ξ_k as the *cells*, a partition $\mathscr{X}$ of the interval Ξ is given satisfying

$$\mathscr{X} = \{\Xi_k;\ k = 1,\cdots,K\},\ \text{such that}\ \Xi_k \cap \Xi_\ell = \emptyset,\ k \ne \ell,\ \text{and}\ \bigcup_{k=1}^K \Xi_k = \Xi. \tag{3.46}$$

Then there follows

Lemma 3.8. *Under the above assumptions holds, with $\pi_k = \mathbb{P}_\xi(\Xi_k)$, for the lower bounds of $\mathbb{E}_\xi[\varphi(\xi)]$*

$$\left.\begin{aligned} \varphi(\bar{\xi}) &\le \sum_{k=1}^K \pi_k \varphi(\mathbb{E}_\xi[\xi \mid \xi \in \Xi_k]) \\ &\le \sum_{k=1}^K \pi_k \mathbb{E}_\xi[\varphi(\xi) \mid \xi \in \Xi_k] \\ &= \mathbb{E}_\xi[\varphi(\xi)] \end{aligned}\right\} \tag{3.47}$$

whereas for the upper bounds we get the inequalities

$$\left.\begin{aligned} \mathbb{E}_\xi\left[\varphi(\xi)\right] &= \sum_{k=1}^{K} \pi_k \mathbb{E}_\xi\left[\varphi(\xi) \mid \xi \in \Xi_k\right] \\ &\le \sum_{k=1}^{K} \pi_k \int_{\Xi_k} \varphi(\eta)\mathbb{Q}_k(d\eta) \\ &\le \int_{\Xi} \varphi(\eta)\mathbb{Q}(d\eta), \end{aligned}\right\} \tag{3.48}$$

where $\mathbb{Q}_k$ is the E–M distribution on Ξ_k yielding $\mu_\Lambda^k := \mathbb{E}_\xi\left[\widehat{m}_\Lambda(\xi) \mid \xi \in \Xi_k\right]$ for all $\Lambda \subset \{1,\cdots,r\}$ and $k = 1,\cdots,K$, and $\mathbb{Q}$ is the E–M distribution on Ξ as described in Lemma 3.7.

Proof: For any $\mathbb{P}_\xi$-integrable function $\psi : \Xi \longrightarrow \mathbb{R}^p$, $p \in \mathbb{N}$, we have the equality

$$\sum_{k=1}^{K} \pi_k \mathbb{E}_\xi\left[\psi(\xi) \mid \xi \in \Xi_k\right] = \mathbb{E}_\xi\left[\psi(\xi)\right]. \tag{3.49}$$

Hence, with ψ the identity, we have $\sum_{k=1}^{K} \pi_k \mathbb{E}_\xi\left[\xi \mid \xi \in \Xi_k\right] = \bar{\xi}$. Then, the convexity of φ implies the first inequality of (3.47), whereas the second one follows from the fact that Jensen's inequality holds true for conditional expectations, as well (see Pfanzagl [253]).

The first equation in (3.48) follows from (3.49) with $\psi = \varphi$. The following inequality holds true due to the fact, that the E–M inequality is valid for conditional expectations, as well. For the probability measure $\mathbb{Q}_k$ holds for all $\Lambda \subset \{1,\cdots,r\}$

$$\int_{\Xi_k} \widehat{m}_\Lambda(\xi)\mathbb{Q}_k(d\xi) = \mu_\Lambda^k = \mathbb{E}_\xi\left[\widehat{m}_\Lambda(\xi) \mid \xi \in \Xi_k\right],\ k = 1,\cdots,K,$$

such that with $\psi = \widehat{m}_\Lambda$ due to (3.49)

$$\sum_{k=1}^{K} \int_{\Xi_k} \pi_k \widehat{m}_\Lambda(\xi)\mathbb{Q}_k(d\xi) = \sum_{k=1}^{K} \pi_k \mathbb{E}_\xi\left[\widehat{m}_\Lambda(\xi) \mid \xi \in \Xi_k\right] = \mathbb{E}_\xi\left[\widehat{m}_\Lambda(\xi)\right] = \mu_\Lambda .$$

Hence, the probability measure $\sum_{k=1}^{K} \pi_k \mathbb{Q}_k$ is feasible for the moment problem (3.42) which is solved by $\mathbb{Q}$, thus implying the last inequality of (3.48). □

Hence, with any arbitrary convex function $\varphi : \Xi \longrightarrow \mathbb{R}$ on the interval $\Xi \subset \mathbb{R}^r$, for any probability distribution $\mathbb{P}_\xi$ on Ξ and for each choice of a partition $\mathscr{X} = \{\Xi_k;\ k = 1,\cdots,K\}$ of Ξ, we have bounds on $\mathbb{E}_\xi\left[\varphi(\xi)\right]$ by

– a discrete random vector η with distribution $\mathbb{P}_{\eta_{\mathscr{X}}}$ yielding

$$\int_{\Xi} \varphi(\eta)\mathbb{P}_{\eta_{\mathscr{X}}}(d\eta) \le \mathbb{E}_\xi\left[\varphi(\xi)\right],$$

the Jensen lower bound due to (3.47), and

– a discrete random vector η with distribution $\mathbb{Q}_{\eta\mathscr{X}}$ yielding

$$\mathbb{E}_\xi[\varphi(\xi)] \le \int_\Xi \varphi(\eta)\mathbb{Q}_{\eta\mathscr{X}}(d\eta),$$

the (generalized) E–M upper bound according to (3.48) (with the measure $\mathbb{Q}_{\eta\mathscr{X}} = \sum_{k=1}^{K} \pi_k \mathbb{Q}_k$ in the above notation).

Let a further partition $\mathscr{Y} = \{\Upsilon_l;\ l = 1,\cdots,L\}$ of Ξ be a refinement of $\mathscr{X}$, i.e. each cell of $\mathscr{X}$ is the union of one or several cells of $\mathscr{Y}$, then as an immediate consequence of Lemma 3.8 follows

Corollary 3.3. *Under the above assumptions, the partition $\mathscr{Y}$ of Ξ being a refinement of the partition $\mathscr{X}$ implies*

$$\int_\Xi \varphi(\eta)\mathbb{P}_{\eta\mathscr{X}}(d\eta) \le \int_\Xi \varphi(\eta)\mathbb{P}_{\eta\mathscr{Y}}(d\eta) \le \mathbb{E}_\xi[\varphi(\xi)]$$

and

$$\mathbb{E}_\xi[\varphi(\xi)] \le \int_\Xi \varphi(\eta)\mathbb{Q}_{\eta\mathscr{Y}}(d\eta) \le \int_\Xi \varphi(\eta)\mathbb{Q}_{\eta\mathscr{X}}(d\eta)$$

and hence an increasing lower and a decreasing upper bound.

Proof: Since $\mathscr{Y}$ is a refinement of $\mathscr{X}$, for $\mathscr{Y} \neq \mathscr{X}$ there is at least one cell Ξ_k of $\mathscr{X}$ being partitioned into some cells $\Upsilon_{l_k 1},\cdots,\Upsilon_{l_k s_k}$ of $\mathscr{Y}$, such that $s_k > 1$ and $\bigcup_{\nu=1}^{s_k} \Upsilon_{l_k\nu} = \Xi_k$. Observing that with $p_{l_k\nu} = \mathbb{P}_\xi(\Upsilon_{l_k\nu})$ holds

$$\mathbb{E}_\xi[\xi \mid \xi \in \Xi_k] = \frac{1}{\pi_k}\sum_{\nu=1}^{s_k} p_{l_k\nu}\mathbb{E}_\xi[\xi \mid \xi \in \Upsilon_{l_k\nu}],$$

due to $\sum_{\nu=1}^{s_k} p_{l_k\nu} = \pi_k$ the convexity of φ implies

$$\varphi(\mathbb{E}_\xi[\xi \mid \xi \in \Xi_k]) \le \frac{1}{\pi_k}\sum_{\nu=1}^{s_k} p_{l_k\nu}\varphi(\mathbb{E}_\xi[\xi \mid \xi \in \Upsilon_{l_k\nu}]).$$

Therefore, this increases in (3.47) the k-th term

$$\pi_k\varphi(\mathbb{E}_\xi[\xi \mid \xi \in \Xi_k]) \text{ to } \sum_{\nu=1}^{s_k} p_{l_k\nu}\varphi(\mathbb{E}_\xi[\xi \mid \xi \in \Upsilon_{l_k\nu}]).$$

In a similar way, the monotone decreasing of the upper bound may be shown, following the arguments in the proof of Lemma 3.8. □

Hence, refining the partitions of Ξ successively improves the approximation of $\mathbb{E}_\xi[\varphi(\xi)]$, by the Jensen bound from below and by the E–M bound from above. Defining in some partition $\mathscr{X} = \{\Xi_k;\ k = 1, \cdots, K\}$ of Ξ the diameter of any cell $\Xi_k \in \mathscr{X}$ as

$$\operatorname{diam} \Xi_k := \sup\{\|\xi - \eta\| \mid \xi, \eta \in \Xi_k\}$$

and then introducing the grid width of this partition $\mathscr{X}$ as

$$\operatorname{grid} \mathscr{X} := \max_{k=1,\cdots,K} \operatorname{diam} \Xi_k ,$$

we may prove convergence of the above bounds to $\mathbb{E}_\xi[\varphi(\xi)]$ under appropriate assumptions (see Kall [153]).

Lemma 3.9. *Let* $\operatorname{supp} \mathbb{P}_\xi \subseteq \Xi = \prod_{i=1}^r [\alpha_i, \beta_i]$ *and* $\varphi : \Xi \longrightarrow \mathbb{R}$ *be continuous. Assume a sequence* $\{\mathscr{X}^\nu\}$ *of successively refined partitions of* Ξ *to be given such that* $\lim_{\nu\to\infty} \operatorname{grid} \mathscr{X}^\nu = 0$. *Then, for* $\{\mathbb{P}_{\eta_{\mathscr{X}^\nu}}\}$ *and* $\{\mathbb{Q}_{\eta_{\mathscr{X}^\nu}}\}$ *the corresponding sequences of Jensen distributions and E–M distributions, respectively, follows*

$$\lim_{\nu\to\infty} \int_\Xi \varphi(\xi)\mathbb{P}_{\eta_{\mathscr{X}^\nu}}(d\xi) = \lim_{\nu\to\infty} \int_\Xi \varphi(\xi)\mathbb{Q}_{\eta_{\mathscr{X}^\nu}}(d\xi) = \int_\Xi \varphi(\xi)\mathbb{P}_\xi(d\xi).$$

Proof: Due to our assumptions φ is uniformly continuous on Ξ implying

$$\forall \varepsilon > 0\ \exists \delta_\varepsilon > 0 \text{ such that } |\varphi(\xi) - \varphi(\eta)| < \varepsilon\ \forall \xi, \eta \in \Xi :\ \|\xi - \eta\| < \delta_\varepsilon .$$

According to the assumptions on $\{\mathscr{X}^\nu\}$ there exists some $\nu(\delta_\varepsilon)$ such that $\operatorname{grid} \mathscr{X}^\nu < \delta_\varepsilon\ \forall \nu > \nu(\delta_\varepsilon)$. Hence, for $\nu > \nu(\delta_\varepsilon)$ and any cell $\Xi_k^\nu \in \mathscr{X}^\nu$ holds $|\varphi(\xi) - \varphi(\eta)| < \varepsilon\ \forall \xi, \eta \in \Xi_k^\nu$. The Jensen distribution $\mathbb{P}_{\eta_{\mathscr{X}^\nu}}$ assigns the probability $\pi_k^\nu = \mathbb{P}_\xi(\Xi_k^\nu) = \int_{\Xi_k^\nu} \mathbb{P}_\xi(d\xi)$ to the realization $\bar{\xi}_k^\nu = \mathbb{E}_\xi[\xi \mid \xi \in \Xi_k^\nu]$. Hence we get

$$\begin{aligned}
&\Big| \int_\Xi \varphi(\xi)\mathbb{P}_{\eta_{\mathscr{X}^\nu}}(d\xi) - \int_\Xi \varphi(\xi)\mathbb{P}_\xi(d\xi) \Big| \\
&= \Big| \sum_{k=1}^{K^\nu} \int_{\Xi_k^\nu} (\varphi(\bar{\xi}_k^\nu) - \varphi(\xi))\mathbb{P}_\xi(d\xi) \Big| \\
&\le \sum_{k=1}^{K^\nu} \int_{\Xi_k^\nu} |\varphi(\bar{\xi}_k^\nu) - \varphi(\xi)|\mathbb{P}_\xi(d\xi) \le \sum_{k=1}^{K^\nu} \varepsilon \cdot \pi_k^\nu = \varepsilon
\end{aligned}$$

such that $\int_\Xi \varphi(\xi)\mathbb{P}_{\eta_{\mathscr{X}^\nu}}(d\xi) \longrightarrow \int_\Xi \varphi(\xi)\mathbb{P}_\xi(d\xi)$.

The convergence of the E–M bound may be shown similarly. □

This result gives rise to introduce the following convergence concepts:

Definition 3.1. *A sequence of probability measures $\mathbb{P}_\xi^\nu$ on $\mathbb{B}^r$ (the Borel σ-algebra on $\mathbb{R}^r$) is said to converge weakly to the measure $\mathbb{P}_\xi$ if for the corresponding distribution functions F_ν and F, respectively, holds*

$$\lim_{\nu \longrightarrow \infty} F_\nu(\xi) = F(\xi) \quad \textit{for every continuity point } \xi \textit{ of } F.$$

Definition 3.2. *Let $\{\psi;\ \psi_\nu,\ \nu \in \mathbb{N}\}$ be a set of functions on $\mathbb{R}^r$. The sequence $\{\psi_\nu,\ \nu \in \mathbb{N}\}$ is said to epi-converge to ψ if for any $\xi \in \mathbb{R}^r$*

- *there exists a sequence $\{\eta_\nu \longrightarrow \xi\}$ such that $\limsup_{\nu \longrightarrow \infty} \psi_\nu(\eta_\nu) \le \psi(\xi)$,*
- *for all sequences $\{\eta_\nu \longrightarrow \xi\}$ holds $\psi(\xi) \le \liminf_{\nu \longrightarrow \infty} \psi_\nu(\eta_\nu)$.*

Lemma 3.9 ensures that the sequences of measures $\{\mathbb{P}_{\eta_{\mathscr{X}^\nu}}\}$ and $\{\mathbb{Q}_{\eta_{\mathscr{X}^\nu}}\}$ converge weakly to $\mathbb{P}_\xi$, as shown in Billingsley [20, 21]. Under the Assumptions 3.2. and 3.3., for the recourse function $Q(x;T(\xi),h(\xi))$ (with $\xi \in \Xi$, the above interval) and for any sequence of probability measures $\mathbb{P}_\xi^\nu$ on Ξ converging weakly to $\mathbb{P}_\xi$, it follows that the approximating expected recourse functions $\mathscr{Q}^\nu(x) = \int_\Xi Q(x;T(\xi),h(\xi))\mathbb{P}_\xi^\nu(d\xi)$ epi-converge to the true expected recourse $\mathscr{Q}(x) = \int_\Xi Q(x;T(\xi),h(\xi))\mathbb{P}_\xi(d\xi)$, as has been shown e.g. in Wets [343]; related investigations are found in Robinson–Wets [279] and Kall [156]. The epi-convergence of the $\mathscr{Q}^\nu$ has the following desirable consequence:

Theorem 3.7. *Assume that $\{\mathscr{Q}^\nu\}$ epi-converges to $\mathscr{Q}$. Then, with some convex polyhedral set $X \subset \mathbb{R}^n$, for the two-stage SLP with recourse we have*

$$\limsup_{\nu \longrightarrow \infty}[\inf_X\{c^{\mathrm{T}}x + \mathscr{Q}^\nu(x)\}] \le \inf_X\{c^{\mathrm{T}}x + \mathscr{Q}(x)\}.$$

If

$$\hat{x}^\nu \in \arg\min_X\{c^{\mathrm{T}}x + \mathscr{Q}^\nu(x)\}\ \forall \nu \in \mathbb{N},$$

then for any accumulation point $\hat{x}$ of $\{\hat{x}^\nu\}$ it follows that

$$c^{\mathrm{T}}\hat{x} + \mathscr{Q}(\hat{x}) = \min_X\{c^{\mathrm{T}}x + \mathscr{Q}(x)\};$$

and for any subsequence $\{\hat{x}^{\nu_\kappa}\} \subset \{\hat{x}^\nu\}$ with $\lim_{\kappa \longrightarrow \infty} \hat{x}^{\nu_\kappa} = \hat{x}$ we have

$$c^{\mathrm{T}}\hat{x} + \mathscr{Q}(\hat{x}) = \lim_{\kappa \longrightarrow \infty}\{c^{\mathrm{T}}\hat{x}^{\nu_\kappa} + \mathscr{Q}(\hat{x}^{\nu_\kappa})\}.$$

A proof of this statement may be found for instance in Wets [343] (see also Kall [155]).

Due to this result discrete approximation algorithms (DAPPROX) for the solution of two-stage SLP's with recourse may be designed, based on successive partitions $\{\mathscr{X}^{\nu}\}$ of Ξ, yielding lower bounds

$$\mathscr{Q}^{\nu}_{LB}(x) = \int_{\Xi} Q(x;T(\xi),h(\xi))\mathbb{P}_{\eta_{\mathscr{X}^{\nu}}}(d\xi) \tag{3.50}$$

and upper bounds

$$\mathscr{Q}^{\nu}_{UB}(x) = \int_{\Xi} Q(x;T(\xi),h(\xi))\mathbb{Q}_{\eta_{\mathscr{X}^{\nu}}}(d\xi) \tag{3.51}$$

for $\mathscr{Q}(\cdot) = \mathbb{E}_{\xi}\left[Q(\cdot;T(\xi),h(\xi))\right]$ due to Jensen and Edmundson–Madansky, respectively. In other words, a solution of

$$\hat{\gamma} := \min_{X}\{c^{\mathrm{T}}x + \mathscr{Q}(x)\} \tag{3.52}$$

may be approximated by an approach like

DAPPROX: Approximating CFR solutions

With Ξ an interval and $\operatorname{supp}\mathbb{P}_{\xi} \subset \Xi$, let $\mathscr{X}^1 := \{\Xi\}$ be the first (trivial) partition of Ξ and $\mathbb{P}_{\eta_{\mathscr{X}^1}}$, $\mathbb{Q}_{\eta_{\mathscr{X}^1}}$ the corresponding Jensen– and E–M – distributions determining, due to (3.50) and (3.51), the approximating expected recourse functions $\mathscr{Q}^1_{LB}(x)$ and $\mathscr{Q}^1_{UB}(x)$.

With $\nu = 1$ iterate the cycle of the following steps I.–III. until achieving the required accuracy $\varepsilon > 0$ of an approximate solution.

I. Analyze the approximating problems

$$\text{a) } \hat{\gamma}_{LB} := \min_{X}\{c^{\mathrm{T}}x + \mathscr{Q}^{\nu}_{LB}(x)\} \quad \text{and} \quad \text{b) } \hat{\gamma}_{UB} := \min_{X}\{c^{\mathrm{T}}x + \mathscr{Q}^{\nu}_{UB}(x)\}.$$

$\{$*Observe that, since* $\mathbb{P}_{\eta_{\mathscr{X}^{\nu}}}$ *and* $\mathbb{Q}_{\eta_{\mathscr{X}^{\nu}}}$ *are finite discrete distributions, problems* a) *and* b) *are LP's with decomposition structures.*$\}$

II. If the prescribed accuracy is achieved, stop the procedure; otherwise, go on to step III.

$\{$*As an example, with a solution* $\hat{x}^{\nu}$ *of problem* I.a) *and its optimal value* $\hat{\gamma}^{\nu}_{LB}$, *the error estimate* $\hat{\gamma} - \hat{\gamma}^{\nu}_{LB} \le c^{\mathrm{T}}\hat{x} + \mathscr{Q}^{\nu}_{UB}(\hat{x}^{\nu}) - \hat{\gamma}^{\nu}_{LB} =: \delta^{\nu}$ *might be used to check whether* $\delta^{\nu} \le \varepsilon$, *if this corresponds to the required accuracy.*$\}$

III. To improve the approximation, choose a partition $\mathscr{X}^{\nu+1}$ as an appropriate refinement of $\mathscr{X}^{\nu}$. With the corresponding Jensen– and E–M – distributions $\mathbb{P}_{\eta_{\mathscr{X}^{\nu+1}}}$ and $\mathbb{Q}_{\eta_{\mathscr{X}^{\nu+1}}}$, defining $\mathscr{Q}^{\nu+1}_{LB}(x)$ and $\mathscr{Q}^{\nu+1}_{UB}(x)$ due to (3.50) and (3.51), let $\nu := \nu + 1$ and return to step I. above.

{As mentioned above, due to the finite discrete Jensen– and E–M – distributions the problems I.a) *and* I.b) *generated in each cycle are LP's with decomposition structure as shown in (1.10) on page 4. For the error estimate mentioned at* II. *the LP* I.a) *has to be solved, suggesting to apply an appropriate decomposition algorithm. In general and due to many experiments, QDECOM (see p. 47) can be considered as a proven reliable solver for this purpose. Nevertheless, keep in mind Remark 1.2. (p. 48).}* □

Obviously, this conceptual description of DAPPROX gives rise to quite a variety of algorithms, depending on various strategies of refining the partitions. For instance, the selection of the particular cells to be refined is relevant for the effectiveness of the method. Or for a cell $\Xi_k^\nu \in \mathscr{X}^\nu$ to be refined, in order to maintain the assumed interval structure of the successive partitions, through this cell we need a cut being perpendicular to one of the coordinate axes; but which coordinate axis is to be preferred, and where is the cut to be located? These and further strategies, playing a significant role for the efficiency of DAPPROX–solvers implementations, will be discussed later in Section 4.7.2.

Exercises

3.1. Consider the two–stage SLP

$$\begin{array}{llr}
\max\{2x_1 + x_2 + \sum_{k=1}^{3} p_k q^{\mathrm{T}} y^{(k)}\} & & \\
x_1 \ \ +x_2 \qquad\qquad\qquad\qquad\qquad \le 10 & & \\
x_1 + 2x_2 \ - y_1^{(k)} \ + y_2^{(k)} \ + 2y_3^{(k)} = h_1^{(k)} & & \\
x_1 \ - x_2 + 2y_1^{(k)} + 3y_2^{(k)} \ \ + y_3^{(k)} = h_2^{(k)} & & k = 1,2,3, \\
\qquad\qquad\qquad\qquad\quad x_j, y_\nu^{(k)} \ge 0 & & \forall\, j,k,\nu
\end{array}$$

with $q = (-2,-3,-2)^{\mathrm{T}}$, $h^{(1)} = (5,4)^{\mathrm{T}}$, $h^{(2)} = (3,5)^{\mathrm{T}}$, $h^{(3)} = (2,2)^{\mathrm{T}}$.

(a) Has the problem the (relatively) complete recourse property?
(b) If not, determine the induced constraints for $(x_1,x_2)^{\mathrm{T}}$ (see page 198).
(c) Compute the first stage solution and its first stage objective of the problem.

You may verify your answers to items (a) and (c) using SLP-IOR.

3.2. Change the recourse matrix W of exercise **3.1** to the new recourse matric

$$\widetilde{W} = \begin{pmatrix} -1 & 1 & -1 \\ 2 & 3 & -5 \end{pmatrix};$$

(a) check the complete recourse property;
(b) compute the first stage solution (including its objective) and compare it to the result of exercise **3.1**.

You may use SLP-IOR to confirm your answers.

3.3. Let the function $\psi : [-1,1] \times [0,2] \longrightarrow \mathbb{R}$ be defined as $\psi(\xi,\eta) := \varphi(\xi) + \theta(\eta)$ with $\varphi(\xi) := \begin{cases} -\xi & \text{for } -1 \le \xi \le -\frac{1}{4} \\ 0.5+\xi & \text{for } -\frac{1}{4} \le \xi \le 1 \end{cases}$ and $\theta(\eta) := 2\eta$. Assume ξ and η to be independent random variables with the densities $g_\xi(\zeta) \equiv \dfrac{1}{2}$ for $\zeta \in [-1,1]$ and $h_\eta(\zeta) = \dfrac{e^{-\zeta}}{1-e^{-2}}$. Due to the independence of ξ and η holds

$$\begin{aligned}\mathbb{E}[\psi(\xi,\eta)] &= \int_0^2 \int_{-1}^1 \psi(\xi,\eta)d\xi d\eta \\ &= \int_{-1}^1 \varphi(\xi)d\xi + \int_0^2 \theta(\eta)d\eta = 0.781250 + 1.373929 = 2.155179.\end{aligned}$$

(a) Compute $(\bar{\xi},\bar{\eta}) = \mathbb{E}[(\xi,\eta)]$ and Jensen's bound $\psi(\bar{\xi},\bar{\eta}) = \varphi(\bar{\xi}) + \theta(\bar{\eta})$.
(b) Compute the E–M upper bound of $\mathbb{E}[\psi(\xi,\eta)]$.
(c) Subdivide the support $\Xi = [-1,1] \times [0,2]$ into two rectangles,

(c1) either by dividing the ξ-interval $I^{(\xi)} = [-1,1]$ at $\bar{\xi}$ into $I_1^{(\xi)}$ and $I_2^{(\xi)}$
(c2) or by dividing the η-interval $I^{(\eta)} = [0,2]$ at $\bar{\eta}$ into $I_1^{(\eta)}$ and $I_2^{(\eta)}$.

Compute alternatively the two new Jensen bounds of $\mathbb{E}[\psi(\xi,\eta)]$, as either
$lb_{|\bar{\xi}} := \mathbb{P}_\xi(I_1^{(\xi)}) \cdot \varphi(\mathbb{E}_\xi[\xi \mid I_1^{(\xi)}]) + \mathbb{P}_\xi(I_2^{(\xi)}) \cdot \varphi(\mathbb{E}_\xi[\xi \mid I_2^{(\xi)}]) + \theta(\mathbb{E}_\eta[\eta])$,
or else
$lb_{|\bar{\eta}} := \varphi(\mathbb{E}_\xi[\xi]) + \mathbb{P}_\eta(I_1^{(\eta)}) \cdot \theta(\mathbb{E}_\eta[\eta \mid I_1^{(\eta)}]) + \mathbb{P}_\eta(I_2^{(\eta)}) \cdot \theta(\mathbb{E}_\eta[\eta \mid I_2^{(\eta)}])$.

(d) How do the new Jensen bounds $lb_{|\bar{\xi}}$ and $lb_{|\bar{\eta}}$ compare to the first bound $\psi(\bar{\xi},\bar{\eta})$, and how much does the above error estimate decrease at best?

3.4. Concerning the moment problem (3.24) and its dual, the semi-infinite program (3.25), it was claimed, that

(a) a linear affine function majorizes a convex function $\psi(\cdot)$ on an interval, if and only if it does so at the endpoints of the interval;
(b) due to the relation $\alpha \le \mu \le \beta$ (with the natural assumption that $\alpha < \beta$) the LP corresponding to (3.25) (due to (a)) is solvable and hence its dual, the LP equivalent to the moment problem (3.24), is uniquely solvable.

Show that these claims hold true.

3.5. Let F be a convex function on a convex polyhedron $\mathscr{B} = \text{conv}\{z^{(1)},\cdots,z^{(k)}\}$, the support of some distribution $\mathbb{P}_\zeta$ with expectation $EX[\zeta] = \mu \in \mathscr{B}$. There is a lower bound for $\bar{F} = \int_{\mathscr{B}} F(z)\mathbb{P}_\zeta(dz)$ given by $F(\mu)$ due to Jensen, and as mentioned on page 217 referring to Dupačová, an upper bound can be determined by solving an LP, which maximizes $EX[F]$ on the class $\tilde{P}$ of dicrete distributions on the vertices of $\mathscr{B}$, satisfying the moment conditions $\sum_i p_i \cdot z^{(i)} = \mu$, $\sum_i p_i = 1$, $p_i \ge 0 \; \forall i$.

As an example, define $\mathscr{B} := \{\xi \mid \xi_1 + 2\xi_2 \leq 10\,,\, 2\xi_1 + \xi_2 \leq 8\,,\, \xi_i \geq 0\}$. Assume some distribution $\mathbb{P}_\xi$ on $\mathscr{B}$ with first moments $\mu = \mathbb{E}[\xi] = (2;2)^{\mathrm{T}}$. Finally define on $\mathscr{B}$ a function $F(\xi) = \xi^{\mathrm{T}} M \xi + c^{\mathrm{T}} \xi$ with $M = \begin{pmatrix} 3 & 2 \\ 2 & 7 \end{pmatrix}$ and $c = (-18;\,-46)^{\mathrm{T}}$.

(a) Is $F(\xi)$ a convex function on $\mathscr{B}$ (and why)? If so:
(b) Compute the Jensen lower bound of $EX[F(\xi)]$ re $\mathbb{P}_\xi$.
(c) Find an upper bound of $EX[F(\xi)]$ re $\mathbb{P}_\xi$ as an LP-solution as described above.

3.6. Consider the recourse problem

$$\begin{aligned} &\min_x \{c^{\mathrm{T}} x + \mathbb{E}[Q(x;\,\xi,\eta)] \mid Ax \leq b,\, x \geq 0\} \;\text{ where} \\ &Q(x;\xi,\eta) := \min_y \{q^{\mathrm{T}} y \mid Wy = h(\xi,\eta) - Tx,\, y \geq 0\} \;\text{ with} \end{aligned}$$

the data: $c = (3,5)^{\mathrm{T}}$; $b = (18,18)^{\mathrm{T}}$; $q = (2,3,2,1)^{\mathrm{T}}$; $h = (12+\xi, 22+\eta)^{\mathrm{T}}$; the arrays $A = \begin{pmatrix} 1 & 3 \\ 3 & 2 \end{pmatrix}$; $W = \begin{pmatrix} 1 & 1 & 1 & -3 \\ 2 & 1 & -4 & 2 \end{pmatrix}$; $T = \begin{pmatrix} 2 & 2 \\ 5 & 3 \end{pmatrix}$; and the random variables ξ and η, with ξ distributed with density $\varphi(\zeta) = \lambda \cdot e^{-\lambda\zeta}/(1 - e^{9\lambda})$ (exponential, conditional to the interval [0,9], or else truncated at the confidence interval of $\hat{p} = 0.95$), and η distributed as $\mathscr{U}[-10,10]$ (uniform).

(a) Is this problem of complete fixed recourse?
(b) Compute on the support $\Xi = [0,9] \times [-10,10]$ of (ξ,η) the lower (Jensen) and upper (E–M) bound for the optimal value and the resulting error estimate.
(c) Compute (e.g with SLP-IOR) the corresponding bounds for partitioning the support Ξ into two (dividing the η-interval) and four subintervals (dividing the ξ- and the η-interval once, each).

3.2.2 The simple recourse case

For the special complete recourse case with $q(\xi) \equiv (q^{+\mathrm{T}}, q^{-\mathrm{T}})^{\mathrm{T}}$ and $W = (I, -I)$, we get the *generalized simple recourse* (GSR) function

$$\left.\begin{aligned} Q^G(x,\xi) := \min\; & q^{+\mathrm{T}} y^+ + q^{-\mathrm{T}} y^- \\ & Iy^+ - \quad Iy^- = h(\xi) - T(\xi)x \\ & y^+, \quad y^- \geq 0. \end{aligned}\right\} \tag{3.53}$$

Given that ξ is a random vector in $\mathbb{R}^R$ such that $\mathbb{E}_\xi[\xi]$ exists, we have the *expected generalized simple recourse* (EGSR)

$$\mathscr{Q}^G(x) := \mathbb{E}_\xi\left[Q^G(x,\xi)\right], \tag{3.54}$$

yielding the two-stage SLP with *generalized simple recourse* (GSR)

$$\left.\begin{array}{l}\min\{c^{\mathrm{T}}x+\mathscr{Q}^{G}(x)\}\\ Ax=b\\ \quad x\geq 0.\end{array}\right\}\qquad(3.55)$$

Before dealing with GSR problems, it is meaningful to discuss first the original version of simple recourse problems, as first analyzed in detail by Wets [342].

3.2.2.1 The standard simple recourse problem (SSR)

In contrast to (3.53) it is now assumed in addition that $T(\xi)\equiv T$. Then it is obviously meaningful to let $h(\xi)\equiv\xi\in\mathbb{R}^{m_2}$ such that instead of (3.53) the *standard simple recourse* (SSR) function is given as

$$\left.\begin{array}{rl}Q(x;\xi):=\min & {q^{+}}^{\mathrm{T}}y^{+}+{q^{-}}^{\mathrm{T}}y^{-}\\ & Iy^{+}-Iy^{-}=\xi-Tx\\ & y^{+},\quad y^{-}\geq 0.\end{array}\right\}\qquad(3.56)$$

This implies the expected simple recourse $\mathscr{Q}(x)=\mathbb{E}_{\xi}\left[Q(x;\xi)\right]$.

Obviously, problem (3.56) is always feasible; and it is solvable iff its dual program

$$\left.\begin{array}{l}\max(\xi-Tx)^{\mathrm{T}}u\\ u\leq\ q^{+}\\ u\geq -q^{-}\end{array}\right\}\qquad(3.57)$$

is feasible, which in turn is true iff $q^{+}+q^{-}\geq 0$. Considering (3.57), we get immediately the optimal recourse value as

$$Q(x,\xi)=\sum_{i=1}^{m_2}[(\xi-Tx)_i]^{+}q_i^{+}+\sum_{i=1}^{m_2}[(\xi-Tx)_i]^{-}q_i^{-}\qquad(3.58)$$

where, for $\rho\in\mathbb{R}$,

$$[\rho]^{+}=\begin{cases}\rho \text{ if } \rho>0\\ 0 \text{ else}\end{cases}\quad\text{and}\quad[\rho]^{-}=\begin{cases}-\rho \text{ if } \rho<0\\ 0 \quad\text{else.}\end{cases}$$

This optimal recourse value $Q(x,\xi)$ is achieved in (3.56) by choosing

$$y_i^{+}=[(\xi-Tx)_i]^{+}\ \text{ and }\ y_i^{-}=[(\xi-Tx)_i]^{-},\ i=1,\cdots,m_2.\qquad(3.59)$$

Introducing $\chi:=Tx$, we get from (3.59) the optimal value of (3.56) as

$$\left.\begin{array}{rl}\tilde{Q}(\chi,\xi)&:=\displaystyle\sum_{i=1}^{m_2}\left\{q_i^{+}[\xi_i-\chi_i]^{+}+q_i^{-}[\xi_i-\chi_i]^{-}\right\}\\ &=:\displaystyle\sum_{i=1}^{m_2}\tilde{Q}_i(\chi_i,\xi_i)\end{array}\right\}\qquad(3.60)$$

with

$$\begin{aligned}\tilde{Q}_i(\chi_i,\xi_i) &= q_i^+[\xi_i-\chi_i]^+ + q_i^-[\xi_i-\chi_i]^- \\ &= \min\{q_i^+y_i^+ + q_i^-y_i^- \mid y_i^+ - y_i^- = \xi_i-\chi_i;\ y_i^+,y_i^- \ge 0\}.\end{aligned} \tag{3.61}$$

Hence the recourse function $Q(x,\xi)$ of (3.56) may be rewritten as a function $\tilde{Q}(\chi,\xi)$ being separable in (χ_i,ξ_i), implying the expected recourse $\mathscr{Q}(x)$ to be equivalent to a function $\widetilde{\mathscr{Q}}(\chi)$, separable in χ_i (see Wets [342]), according to

$$\left.\begin{aligned}\widetilde{\mathscr{Q}}(\chi) &= \sum_{i=1}^{m_2}\widetilde{\mathscr{Q}}_i(\chi_i), \quad \text{where} \\ \widetilde{\mathscr{Q}}_i(\chi_i) &:= \mathbb{E}_\xi\,[\tilde{Q}_i(\chi_i,\xi_i)] = \mathbb{E}_{\xi_i}\,[\tilde{Q}_i(\chi_i,\xi_i)]\,,\ i=1,\cdots,m_2\,,\end{aligned}\right\} \tag{3.62}$$

such that (3.55) may now be rewritten as

$$\left.\begin{aligned}&\min\{c^{\mathrm{T}}x + \sum_{i=1}^{m_2}\widetilde{\mathscr{Q}}_i(\chi_i)\} \\ &Ax \qquad = b \\ &Tx - \chi = 0 \\ &\quad x \qquad \ge 0\end{aligned}\right\} \tag{3.63}$$

with

In this case, as indicated by the operator $\mathbb{E}_{\xi_i}$, to compute the expected simple recourse we may restrict ourselves to the marginal distributions of the single components ξ_i instead of the joint distribution of $\xi = (\xi_1,\cdots,\xi_{m_2})^{\mathrm{T}}$. From (3.61) obviously follows that $\tilde{Q}_i(\cdot,\xi_i)$ is a convex function in χ_i (and hence in x) for any fixed value of ξ_i. Hence, the expected recourse $\widetilde{\mathscr{Q}}_i(\cdot)$ is convex in χ_i as well.

If $\mathbb{P}_\xi$ happens to be a finite discrete distribution with the marginal distribution of any component given by $p_{ij} = \mathbb{P}_\xi(\{\xi \mid \xi_i = \hat{\xi}_{ij}\}),\ j=1,\cdots,k_i$, then (3.63) is equivalent to the linear program

$$\left.\begin{aligned}&\min\{c^{\mathrm{T}}x + \sum_{i=1}^{m_2}\sum_{j=1}^{k_i} p_{ij}(q_i^+y_{ij}^+ + q_i^-y_{ij}^-)\} \\ &Ax \qquad\qquad = b \\ &Tx - \chi \qquad = 0 \\ &\qquad y_{ij}^+ - y_{ij}^- = \hat{\xi}_{ij} - \chi_i\ \ \forall i,j \\ &x, \quad y_{ij}^+,\ y_{ij}^- \ge 0\end{aligned}\right\} \tag{3.64}$$

which due to its special data structure can easily be solved.

If, on the other hand, $\mathbb{P}_\xi$ or at least some of its marginal distributions $\mathbb{P}_{\xi_i}$ are of the continuous type, the corresponding expected recourse $\widetilde{\mathscr{Q}}_i(\cdot)$ and hence the program (3.63) may be expected to be nonlinear. Nevertheless, the simple recourse functions $\tilde{Q}_i(\chi_i,\xi_i)$ and their expectations $\widetilde{\mathscr{Q}}_i(\chi_i)$ have some special properties, advantageous in solution procedures and not shared by complete recourse functions in general. To point out these particular properties we introduce *simple recourse type* functions (referred to as SRT functions) and discuss some of their properties advantageous for their approximation.

Definition 3.3. *For a real variable z, a random variable ξ with distribution $\mathbb{P}_{\xi}$, and real constants α, β, γ with $\alpha + \beta \geq 0$, the function $\varphi(\cdot,\cdot)$ given by*

$$\varphi(z,\xi) := \alpha \cdot [\xi - z]^{+} + \beta \cdot [\xi - z]^{-} - \gamma$$

is called a simple recourse type function (see Fig. 3.2).

Then, $\mathbb{E}_{\xi}[\xi]$ provided to exist,

$$\Phi(z) := \mathbb{E}_{\xi}[\varphi(z,\xi)] = \int_{-\infty}^{\infty} (\alpha \cdot [\xi - z]^{+} + \beta \cdot [\xi - z]^{-}) \mathbb{P}_{\xi}(d\xi) - \gamma$$

is the expected SRT function (ESRT function).

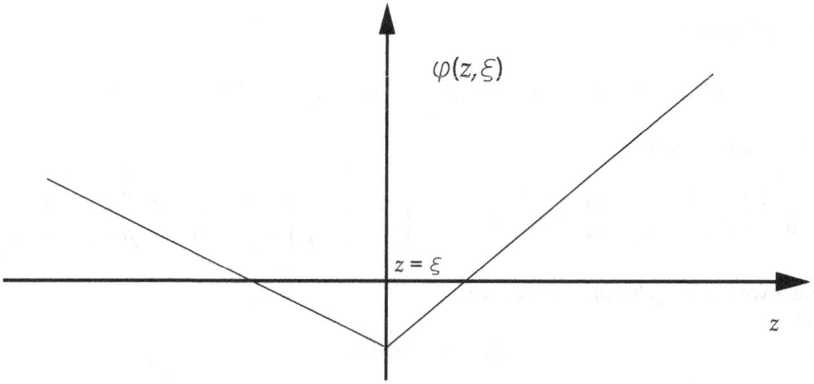

Fig. 3.2 SRT function.

Obviously, the functions $\tilde{Q}_i(\chi_i,\xi_i)$ and $\widetilde{\mathscr{Q}}_i(\chi_i)$ considered above are SRT and ESRT functions, respectively; however, SRT functions may also appear in models different from (3.61)–(3.63), as we shall see later.

From Definition 3.3. follows immediately

Lemma 3.10. *Let $\varphi(\cdot,\cdot)$ be a SRT function and $\Phi(\cdot)$ the corresponding expected SRT function. Then*

- *$\varphi(z,\cdot)$ is convex in ξ for any fixed $z \in \mathbb{R}$;*
- *$\varphi(\cdot,\xi)$ is convex in z for any fixed $\xi \in \mathbb{R}$;*
- *$\Phi(\cdot)$ is convex in z.*

Since (3.61)–(3.63) describes a particular complete fixed recourse problem, we know already from Section 3.2.1 that, ξ provided to be integrable and $q^{+} + q^{-} \geq 0$, the functions $\tilde{Q}_i(\chi_i,\xi_i)$ and $\widetilde{\mathscr{Q}}_i(\chi_i)$ are SRT and ESRT functions, respectively.

Assuming $\mu := \mathbb{E}_\xi[\xi]$ to exist, Jensen's inequality for SRT functions obviously holds:

$$\varphi(z,\mu) = \varphi(z,\mathbb{E}_\xi[\xi]) \le \mathbb{E}_\xi[\varphi(z,\xi)] = \Phi(z).$$

Furthermore, for ξ being integrable (with F_ξ the distribution function of ξ), the asymptotic behaviour of the ESRT function may immediately be derived:

Lemma 3.11. *For*

$$\begin{aligned}\Phi(z) &:= \mathbb{E}_\xi[\varphi(z,\xi)]\\ &= \int_{-\infty}^{\infty}(\alpha\cdot[\xi-z]^+ + \beta\cdot[\xi-z]^-)\,dF_\xi(\xi) - \gamma\\ &= \left\{\alpha\cdot\int_z^{\infty}[\xi-z]\,dF_\xi(\xi) + \beta\cdot\int_{-\infty}^{z}[z-\xi]\,dF_\xi(\xi)\right\} - \gamma\end{aligned}$$

holds:

$$\Phi(z) - \varphi(z,\mu) = \Phi(z) - [\alpha\cdot(\mu-z)-\gamma] \longrightarrow 0 \text{ as } z\to-\infty$$

and analogously

$$\Phi(z) - \varphi(z,\mu) = \Phi(z) - [\beta\cdot(z-\mu)-\gamma] \longrightarrow 0 \text{ as } z\to+\infty.$$

In particular follows:

$$\text{If } \begin{cases}\mathbb{P}_\xi(\xi<a)=0\\ \mathbb{P}_\xi(\xi>b)=0\end{cases} \text{ then } \Phi(z) = \begin{cases}\alpha\cdot(\mu-z)-\gamma = \varphi(z,\mu) \text{ for } z\le a\\ \beta\cdot(z-\mu)-\gamma = \varphi(z,\mu) \text{ for } z\ge b.\end{cases}$$

Hence we have, as mentioned above,

$$\varphi(z,\mu) \le \Phi(z)\ \forall z$$

and, furthermore (see Fig. 3.3),

$$\begin{aligned}a &:= \inf\operatorname{supp}\mathbb{P}_\xi > -\infty \Longrightarrow \Phi(z) = \varphi(z,\mu)\ \forall z\le a\\ b &:= \sup\operatorname{supp}\mathbb{P}_\xi < +\infty \Longrightarrow \Phi(z) = \varphi(z,\mu)\ \forall z\ge b.\end{aligned}$$

Consider now an interval $I = \{\xi \mid a < \xi \le b\} \not\supseteq \operatorname{supp}\mathbb{P}_\xi$—implying at least one of the bounds a,b to be finite—with $\mathbb{P}_\xi(I) > 0$. Then Jensen's inequality holds as well for the corresponding conditional expectations.

Lemma 3.12. *With* $\mu_{|I} = \mathbb{E}_\xi[\xi \mid \xi\in I]$ *and* $\Phi_{|I}(z) = \mathbb{E}_\xi[\varphi(z,\xi) \mid \xi\in I]$*, for all* $z\in\mathbb{R}$ *holds*

$$\varphi(z,\mu_{|I}) \le \Phi_{|I}(z) = \frac{1}{\mathbb{P}_\xi(I)}\int_a^b \varphi(z,\xi)\,dF_\xi(\xi).$$

As shown in Kall-Stoyan [171], in analogy to Lemma 3.11 follows also

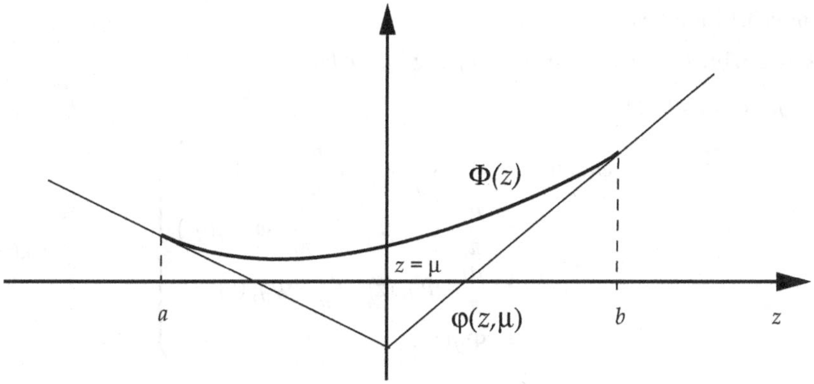

Fig. 3.3 SRT and expected SRT function ($\operatorname{supp}\mathbb{P}_\xi$ bounded).

Lemma 3.13. *For any finite a and/or b, for* $I = (a,b]$ *holds*

$$\Phi_{|I}(z) = \begin{cases} \varphi(z,\mu_{|I}) \text{ for } z \le a \\ \varphi(z,\mu_{|I}) \text{ for } z \ge b. \end{cases}$$

If in particular $J := \operatorname{supp}\mathbb{P}_\xi = [a,b]$ is a finite interval, then Lemma 3.11 yields

$$\Phi(z) = \Phi_{|J}(z) = \varphi(z,\mu_{|J}) = \varphi(z,\mu) \quad \text{for } z \le a \text{ or } z \ge b, \tag{3.65}$$

and for $z \in (a,b)$ Jensen's inequality implies $\varphi(z,\mu) \le \Phi(z)$. To get an upper bound for $z \in (a,b)$ and hence an estimate for $\Phi(z)$, the E–M inequality may be used:

$$\Phi_{|J}(z) = \Phi(z) \le \frac{b-\mu}{b-a}\varphi(z,a) + \frac{\mu-a}{b-a}\varphi(z,b) = \frac{b-\mu_{|J}}{b-a}\varphi(z,a) + \frac{\mu_{|J}-a}{b-a}\varphi(z,b).$$

Analogously, for an interval $I = \{\xi \mid a < \xi \le b\} \not\supseteq \operatorname{supp}\mathbb{P}_\xi$ and $z \in \operatorname{int} I$ follows

$$\varphi(z,\mu_{|I}) \le \Phi_{|I}(z) \le \frac{b-\mu_{|I}}{b-a}\varphi(z,a) + \frac{\mu_{|I}-a}{b-a}\varphi(z,b). \tag{3.66}$$

If $\varphi(z,\cdot)$ happens to be linear on I, the lower and upper bounds of these inequalities coincide such that $\Phi_{|I}(z) = \varphi(z,\mu_{|I})\ \forall z$. If, on the other hand, $\varphi(z,\cdot)$ is nonlinear (convex) in I, the approximation of $\Phi_{|I}(\hat{z})$ for any $\hat{z} \in (a,b)$ due to (3.66) can be improved as follows: Partition $I = (a,b]$ at $a_1 := \hat{z}$ into the two intervals $I_1 := (a_0,a_1]$ and $I_2 := (a_1,a_2]$, where $a_0 := a$ and $a_2 := b$. Observing that, with $\pi_I = \mathbb{P}_\xi(I)$ and $p_\nu := \mathbb{P}_\xi(I_\nu)$, $\nu = 1,2$, we have $\frac{p_1}{\pi_I}\cdot\mu_{|I_1} + \frac{p_2}{\pi_I}\cdot\mu_{|I_2} = \mu_{|I}$ as well as for arbitrary $\mathbb{P}_\xi$-integrable functions $\psi(\cdot)$ the relation

$$\mathbb{E}_\xi[\psi(\xi) \mid \xi \in I] = \frac{p_1}{\pi_I}\cdot\mathbb{E}_\xi[\psi(\xi) \mid \xi \in I_1] + \frac{p_2}{\pi_I}\cdot\mathbb{E}_\xi[\psi(\xi) \mid \xi \in I_2], \tag{3.67}$$

Lemma 3.12 implies

Lemma 3.14. *Due to the convexity of $\varphi(z,\cdot)$, we have*

a) *for arbitrary $z \in (a_0, a_2)$*

$$\left.\begin{aligned} \varphi(z,\mu_{|I}) &= \varphi(z, \frac{p_1}{\pi_I}\cdot\mu_{|I_1} + \frac{p_2}{\pi_I}\cdot\mu_{|I_2}) \\ &\le \frac{p_1}{\pi_I}\cdot\varphi(z,\mu_{|I_1}) + \frac{p_2}{\pi_I}\cdot\varphi(z,\mu_{|I_2}) \\ &\le \frac{p_1}{\pi_I}\cdot\Phi_{|I_1}(z) + \frac{p_2}{\pi_I}\cdot\Phi_{|I_2}(z) \\ &= \Phi_{|I}(z); \end{aligned}\right\} \tag{3.68}$$

b) *for $a_\kappa \in \{a_0, a_1, a_2\}$*

$$\left.\begin{aligned} \Phi_{|I_\nu}(a_\kappa) &= \varphi(a_\kappa,\mu_{|I_\nu}) \qquad \text{for } \nu = 1,2 \\ \Phi_{|I}(a_\kappa) &= \sum_{\nu=1}^{2}\frac{p_\nu}{\pi_I}\Phi_{|I_\nu}(a_\kappa) = \sum_{\nu=1}^{2}\frac{p_\nu}{\pi_I}\varphi(a_\kappa,\mu_{|I_\nu}). \end{aligned}\right\} \tag{3.69}$$

Proof: The above relations are consequences of previously mentioned facts:

a) The two equations reflect (3.67), the first inequality follows from the convexity of $\varphi(z,\cdot)$, and the second inequality applies Lemma 3.12.
b) The first two equations apply Lemma 3.13, the last equation uses (3.67) again.

□

Based on Lemmas 3.12–3.14, similar to the general complete recourse case, approximation schemes with successively refined discrete distributions may be designed.

3.2.2.2 SSR: Approximation by successive discretization

Eq. (3.68) yields with $\varphi^\star(z,\mu_{|I_1},\mu_{|I_2}) = \frac{p_1}{\pi_I}\varphi(z,\mu_{|I_1}) + \frac{p_2}{\pi_I}\varphi(z,\mu_{|I_2}) \le \Phi_{|I}(z)$ an increased lower bound of $\Phi_{|I}(z)$ as

$$\varphi^\star(z,\mu_{|I_1},\mu_{|I_2}) = \begin{cases} \varphi(z,\mu_{|I}) \text{ for } z \in (-\infty,\mu_{|I_1}) \cup (\mu_{|I_2},\infty) \\ \left\{\left(\frac{p_1}{\pi_I}\beta - \frac{p_2}{\pi_I}\alpha\right)z - \frac{p_1}{\pi_I}\beta\mu_{|I_1} + \frac{p_2}{\pi_I}\alpha\mu_{|I_2} - \gamma\right\} \\ \text{for } z \in [\mu_{|I_1},\mu_{|I_2}], \end{cases} \tag{3.70}$$

and instead of the general upper bound (3.66) of $\Phi_{|I}(z)$, with $\hat{z} = a_1$ (3.69) yields the exact value $\Phi_{|I}(\hat{z}) = \frac{p_1}{\pi_I}\varphi(\hat{z},\mu_{|I_1}) + \frac{p_2}{\pi_I}\varphi(\hat{z},\mu_{|I_2}) = \varphi^\star(\hat{z},\mu_{|I_1},\mu_{|I_2})$ (see Fig. 3.4).

If, on the other hand, the partition $I = (a_0,a_1] \cup (a_1,a_2] = I_1 \cup I_2$ is given, from Lemma 3.13 and 3.14 together with (3.67) follows

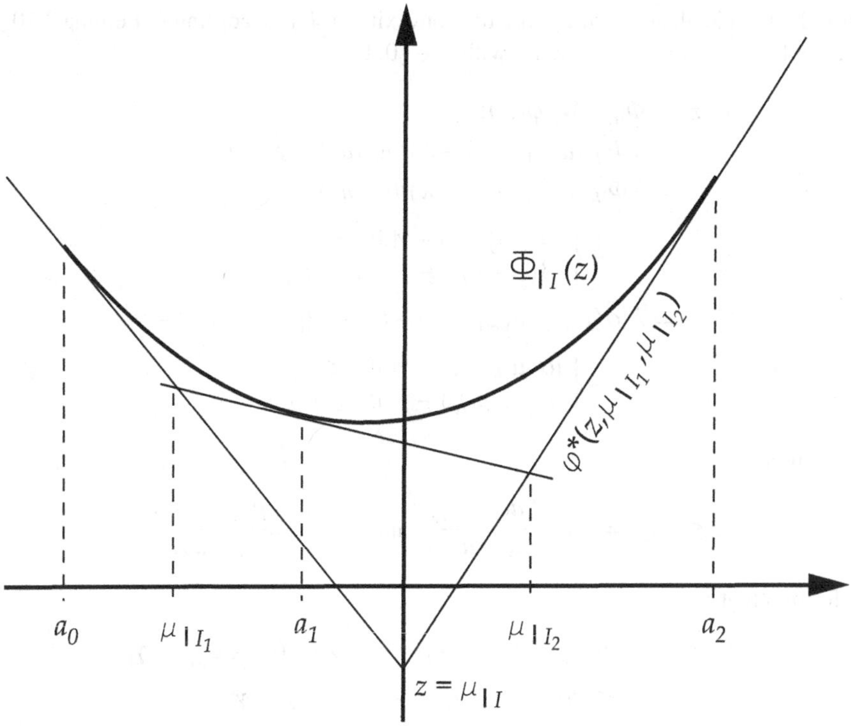

Fig. 3.4 Expected SRT function: Increasing lower bounds.

$$\Phi_{|I}(z) = \frac{p_1}{\pi_I}\varphi(z,\mu_{|I_1}) + \frac{p_2}{\pi_I}\varphi(z,\mu_{|I_2}) \text{ for } z \le a_0 \text{ or } z \ge a_2 \text{ or } z = a_1\,; \qquad (3.71)$$

hence $\Phi_{|I}(z) > \frac{p_1}{\pi_I}\varphi(z,\mu_{|I_1}) + \frac{p_2}{\pi_I}\varphi(z,\mu_{|I_2})$ may occur only if $z \in \operatorname{int} I_1 \cup \operatorname{int} I_2$, which implies that $\Phi_{|I_\nu}(z) > \varphi(z,\mu_{|I_\nu})$ for $z \in \operatorname{int} I_\nu$ with $\nu = 1$ or $\nu = 2$. Then we may derive the following rather rough error estimate:

Lemma 3.15. *For $z \in \operatorname{int} I_\nu$, $\nu = 1,2$, we have the parameter-free error estimate $\Delta_\nu(z)$ satisfying*

$$0 \le \Delta_\nu(z) := \Phi_{|I_\nu} - \varphi(z,\mu_{|I_\nu}) \le \frac{1}{2}(\alpha+\beta)\,\frac{a_\nu - a_{\nu-1}}{2}\,.$$

Proof: Using the relations $\varphi(z,\mu_{|I_\nu}) = \alpha[\mu_{|I_\nu} - z]^+ + \beta[\mu_{|I_\nu} - z]^- - \gamma$ from Definition 3.3. as well as the relations

$$\Phi_{|I_\nu}(a_{\nu-1}) = \varphi(a_{\nu-1},\mu_{|I_\nu}) \text{ and } \Phi_{|I_\nu}(a_\nu) = \varphi(a_\nu,\mu_{|I_\nu})$$

from Lemma 3.14, and furthermore the convexity of $\Phi_{|I_\nu}$ according to Lemma 3.10, we get for $z = \lambda a_{\nu-1} + (1-\lambda) a_\nu$ with $\lambda \in (0,1)$

$$\begin{aligned}
\Delta_\nu(z) &= \Phi_{|I_\nu}(z) - \varphi(z, \mu_{|I_\nu}) \\
&\le \lambda \Phi_{|I_\nu}(a_{\nu-1}) + (1-\lambda)\Phi_{|I_\nu}(a_\nu) - \varphi(z, \mu_{|I_\nu}) \\
&= \lambda \Phi_{|I_\nu}(a_{\nu-1}) + (1-\lambda)\Phi_{|I_\nu}(a_\nu) \\
&\quad - \begin{cases} [\alpha(\mu_{|I_\nu} - z) - \gamma] & \text{if } z < \mu_{|I_\nu} \\ [\beta(z - \mu_{|I_\nu}) - \gamma] & \text{if } z \ge \mu_{|I_\nu} \end{cases} \\
&= \lambda[\alpha(\mu_{|I_\nu} - a_{\nu-1}) - \gamma] + (1-\lambda)[\beta(a_\nu - \mu_{|I_\nu}) - \gamma] \\
&\quad - \begin{cases} [\alpha(\mu_{|I_\nu} - z) - \gamma] & \text{if } z < \mu_{|I_\nu} \\ [\beta(z - \mu_{|I_\nu}) - \gamma] & \text{if } z \ge \mu_{|I_\nu}. \end{cases}
\end{aligned}$$

Assuming

$$z \le \mu_{|I_\nu} \Longleftrightarrow \lambda \ge \frac{a_\nu - \mu_{|I_\nu}}{a_\nu - a_{\nu-1}} \quad \text{and} \quad 1 - \lambda \le \frac{\mu_{|I_\nu} - a_{\nu-1}}{a_\nu - a_{\nu-1}}$$

it follows that

$$\begin{aligned}
\Delta_\nu(z) &\le \lambda[\alpha(\mu_{|I_\nu} - a_{\nu-1}) - \gamma] + (1-\lambda)[\beta(a_\nu - \mu_{|I_\nu}) - \gamma] \\
&\qquad - [\alpha(\mu_{|I_\nu} - \lambda a_{\nu-1} - (1-\lambda)a_\nu) - \gamma] \\
&= (1-\lambda)(\alpha+\beta)(a_\nu - \mu_{|I_\nu}) \\
&\le \frac{\mu_{|I_\nu} - a_{\nu-1}}{a_\nu - a_{\nu-1}} (\alpha+\beta)(a_\nu - \mu_{|I_\nu}),
\end{aligned}$$

the maximum of the last term being assumed for $\mu_{|I_\nu} = \dfrac{a_{\nu-1} + a_\nu}{2}$ such that

$$\Delta_\nu(z) \le \frac{1}{2}(\alpha+\beta)\frac{a_\nu - a_{\nu-1}}{2}.$$

For $z \ge \mu_{|I_\nu}$ the result follows analogously. □

Taking the probabilities p_ν associated with the partition intervals I_ν into account yields an improved (global) error estimate:

Lemma 3.16. *Given the interval partition $\{I_\nu;\ \nu = 1,2\}$ of I and $z \in I_\kappa$, then the (global) error estimate $\Delta(z)$ satisfies*

$$0 \le \Delta(z) = \Phi_{|I}(z) - \sum_{\nu=1}^{2} \frac{p_\nu}{\pi_I} \varphi(z, \mu_{|I_\nu}) \le \frac{1}{2}\frac{p_\kappa}{\pi_I}(\alpha+\beta)\frac{a_\kappa - a_{\kappa-1}}{2}$$

for $z \in \operatorname{int} I_\kappa$, whereas for $z \notin (\operatorname{int} I_1 \cup \operatorname{int} I_2)$ we have $\Delta(z) = 0$.

Proof: For $z \in I_\kappa$ Lemma 3.13 yields $\Phi_{|I_\nu}(z) - \varphi(z, \mu_{|I_\nu}) = 0$ for $\nu \neq \kappa$; hence from Lemmas 3.14 and 3.15 follows for $z \in \text{int} I_\kappa$

$$\begin{aligned}\Delta(z) &= \Phi_{|I}(z) - \sum_{\nu=1}^{2} \frac{p_\nu}{\pi_I} \varphi(z, \mu_{|I_\nu}) \\ &= \sum_{\nu=1}^{2} \frac{p_\nu}{\pi_I} (\Phi_{|I_\nu}(z) - \varphi(z, \mu_{|I_\nu})) \\ &= \frac{p_\kappa}{\pi_I} (\Phi_{|I_\kappa}(z) - \varphi(z, \mu_{|I_\kappa})) \\ &\leq \frac{1}{2} \frac{p_\kappa}{\pi_I} (\alpha + \beta) \frac{a_\kappa - a_{\kappa-1}}{2},\end{aligned}$$

and for $z \notin (\text{int} I_1 \cup \text{int} I_2)$ from (3.71) follows that $\Delta(z) = 0$. □

Due to (3.60) and (3.63) the simple recourse function $\tilde{Q}(\chi, \xi) = \sum_{i=1}^{m_2} \tilde{Q}^{(i)}(\chi_i, \xi_i)$ as well as the expected simple recourse function $\tilde{\mathscr{Q}}(\chi) = \sum_{i=1}^{m_2} \tilde{\mathscr{Q}}^{(i)}(\chi_i)$ are separable, and their additive components $\tilde{Q}^{(i)}(\chi_i, \xi_i)$ and $\tilde{\mathscr{Q}}^{(i)}(\chi_i)$ are SRT and ESRT functions, respectively. Therefore, the properties derived for these functions allow for modifications of the discrete approximation algorithms of the type DAPPROX, as described on page 223 for the more general complete recourse case. This leads for the standard simple recourse case to special algorithms (named SRAPPROX), being more efficient than the general DAPPROX approach since, for an interval $I^{(i)} \supset \text{supp}\, \mathbb{P}_{\xi_i}$, at any partitioning point $\hat{\xi}_i := \hat{\chi}_i \in \text{int} I^{(i)}$, instead of the E–M upper bound of $\tilde{\mathscr{Q}}^{(i)}_{|I}(\hat{\chi})$ its exact value is—due to (3.71) and Lemma 3.14—easily computed.

SRAPPROX: Approximating SSR solutions

Assume that $\text{supp}\, \mathbb{P}_\xi \subset \Xi := \prod_{i=1}^{m_2} I^{(i)}$ for some intervals $I^{(i)} = (a^{(i)}, b^{(i)}]$, $i = 1, \cdots, m_2$. For each component ξ_i of ξ choose as a first partition $\mathscr{X}^{(i)} = \{I^{(i)}\}$ corresponding for $\Xi \subset \mathbb{R}^{m_2}$ to the first partition $\mathscr{X} = \{\mathscr{X}^{(1)} \times \mathscr{X}^{(2)} \times \cdots \times \mathscr{X}^{(m_2)}\}$. For the trivial discrete ditribution $\pi_{i1} = \mathbb{P}_{\xi_i}(\{\xi_i \in I_1^{(i)}\}) = 1\ \forall i$, with $I_1^{(i)} = (a_0^{(i)}, a_1^{(i)}] = I^{(i)}$ and with $\tilde{Q}^{(i)}(\chi_i, \xi_i) = q_i^+ [\xi_i - \chi_i]^+ + q_i^- [\xi_i - \chi_i]^-$ due to (3.61), it follows for $\mu_{|I_1^{(i)}} := \mathbb{E}_{\xi_i}[\xi_i \mid \xi_i \in I_1^{(i)}]$ that $\tilde{Q}^{(i)}(\chi_i, \mu_{|I_1^{(i)}}) \leq \tilde{\mathscr{Q}}^{(i)}_{|I_1^{(i)}}(\chi_i) = \mathbb{E}_{\xi_i}[\tilde{Q}_i(\chi_i, \xi_i) \mid \xi_i \in I_1^{(i)}]$. With $K_i = 1\ \forall i$ iterate the following cycle:

I. Find a solution $(\hat{x}, \hat{\chi})$ of

$$\min\{c^{\mathrm{T}}x + \sum_{i=1}^{m_2}\sum_{\nu=1}^{K_i} \pi_{i\nu}\tilde{Q}^{(i)}(\chi_i, \mu_{|I_\nu^{(i)}}) \mid Ax = b, Tx - \chi = 0, x \geq 0\}.$$

If $\hat{\chi}_i \notin \cup_{\nu=1}^{K_i} \mathrm{int} I_\nu^{(i)}$ for all $i \in \{1, \cdots, m_2\}$, then $(\hat{x}, \hat{\chi})$ solves problem (3.63) due to Lemma 3.13 and (3.67) and hence stop; otherwise continue.

II. With $I_\nu^{(i)} = (a_{\nu-1}^{(i)}, a_\nu^{(i)}]$, $\nu = 1, \cdots, K_i$, let $\Lambda := \{i \mid \hat{\chi}_i \in \mathrm{int} I_{\nu_i}^{(i)}$ for one $\nu_i\}$. For $i \in \Lambda$ split up $I_{\nu_i}^{(i)}$ as $I_{\nu_i}^{(i)} = (a_{\nu_i - 1}^{(i)}, \hat{\chi}_i] \cup (\hat{\chi}_i, a_{\nu_i}^{(i)}] =: I_{\nu_i 1}^{(i)} \cup I_{\nu_i 2}^{(i)}$ and determine the conditional expectations $\mu_{|I_{\nu_i j}^{(i)}} := \mathbb{E}_{\xi_i}[\xi_i \mid \xi_i \in I_{\nu_i j}^{(i)}]$, $j = 1, 2$.

Due to Lemma 3.14 this implies for $\widetilde{\mathscr{Q}}^{(i)}_{|I_{\nu_i}^{(i)}}(\hat{\chi}_i)$ a lower bound $\ell_{\nu_i}^{(i)}$ and the exact value, respectively, to be given as

$$\ell_{\nu_i}^{(i)} = \tilde{Q}^{(i)}(\hat{\chi}_i, \mu_{|I_{\nu_i}^{(i)}}) \leq \widetilde{\mathscr{Q}}^{(i)}_{|I_{\nu_i}^{(i)}}(\hat{\chi}_i)$$

$$\widetilde{\mathscr{Q}}^{(i)}_{|I_{\nu_i}^{(i)}}(\hat{\chi}_i) = \sum_{j=1}^{2} \frac{p_{ij}}{\pi_{\nu_i}} \tilde{Q}^{(i)}(\hat{\chi}_i, \mu^{(i)}_{|I_{\nu_i j}})$$

with $p_{ij} = \mathbb{P}_{\xi_i}(\{\xi_i \in I_{\nu_i j}^{(i)}\})$, $j = 1, 2$.

If for $\delta^{(i)} := \pi_{\nu_i} \cdot (\widetilde{\mathscr{Q}}^{(i)}_{|I_{\nu_i}^{(i)}}(\hat{\chi}_i) - \ell_{\nu_i}^{(i)})$ follows that $\delta^{(i)} < \varepsilon \; \forall i \in \Lambda$ with ε a prescribed tolerance, then stop with the required accuracy achieved; otherwise continue with $\tilde{\Lambda} := \{i \in \Lambda \mid \delta^{(i)} \geq \varepsilon\}$.

III. For (some) $i \in \tilde{\Lambda}$ extend $\mathscr{X}^{(i)}$ to the new partition $\mathscr{Y}^{(i)}$ by splitting up the interval $I_{\nu_i}^{(i)}$ into the two subintervals $I_{\nu_i j}^{(i)}$ with $\pi_{\nu_i j} := p_{ij}$, $j = 1, 2$, and adjust $K_i := K_i + 1$. With the new data $I_{\nu_i j}^{(i)}, \pi_{\nu_i j}, \mu_{|I_{\nu_i j}^{(i)}}$ (for $j = 1, 2$) and K_i, update the extended partitions to $\mathscr{X}^{(i)} := \mathscr{Y}^{(i)}$ and return to step I.

□

This conceptual algorithm does, in contrast to DAPPROX, leave no choice of where to split an interval $I_{\nu_i}^{(i)}$, $i \in \tilde{\Lambda}$, as long as the true value $\widetilde{\mathscr{Q}}^{(i)}_{|I_{\nu_i}^{(i)}}(\hat{\chi}_i)$, and thus also the exact value of $\widetilde{\mathscr{Q}}^{(i)}(\hat{\chi}_i)$, are of interest. On the other hand there are various strategies for the selection of components $i \in \tilde{\Lambda}$, for which the respective subintervals $I_{\nu_i}^{(i)}$ are splitted up. A detailed description of an executable version of SRAPPROX, including the presentation of the implemented algorithm, can be found in Section 4.7.2 of the next chapter.

3.2.2.3 The multiple simple recourse problem

The simple recourse function (3.56) was extended by Klein Haneveld [188] to the multiple simple recourse function. Here, instead of (3.61), for any single recourse constraint the following value is to be determined:

$$
\left.\begin{array}{rcll}
\psi(z,\xi) := \min \left\{ \sum_{k=1}^{K} q_k^+ y_k^+ + \sum_{k=1}^{K} q_k^- y_k^- \right\} & & & \\
\sum_{k=1}^{K} y_k^+ - \sum_{k=1}^{K} y_k^- & = & \xi - z & \\
\left.\begin{array}{rcl} y_k^+ & \leq & u_k - u_{k-1} \\ y_k^- & \leq & l_k - l_{k-1} \end{array}\right\}, & & & k = 1,\cdots,K-1, \\
y_k^+, y_k^- & \geq & 0, & k = 1.\cdots,K,
\end{array}\right\} \quad (3.72)
$$

where

$$
\begin{array}{l}
u_0 = 0 < u_1 < \cdots < u_{K-1} \\
l_0 = 0 < l_1 < \cdots < l_{K-1},
\end{array}
$$

and

$$
q_k^+ \geq q_{k-1}^+,\; q_k^- \geq q_{k-1}^-,\; k = 2,\cdots,K,
$$

with $q_1^+ \geq -q_1^-$ and $q_K^+ + q_K^- > 0$ (to ensure convexity and prevent from linearity of this modified recourse function).

According to these assumptions, for any value of $\tau := \xi - z$ it is obvious to specify a feasible solution of (3.72), namely for any $\kappa \in \{1,\cdots,K\}$ (with $u_K = \infty$ and $l_K = \infty$)

$$
\tau \in [u_{\kappa-1}, u_\kappa) \Longrightarrow \begin{cases}
y_k^+ = u_k - u_{k-1}, & 1 \leq k \leq \kappa - 1 \\
y_\kappa^+ = \tau - u_{\kappa-1} & \\
y_k^+ = 0 & \forall k > \kappa \\
y_k^- = 0 & k = 1,\cdots,K;
\end{cases}
$$

$$
\tau \in (-l_\kappa, -l_{\kappa-1}] \Longrightarrow \begin{cases}
y_k^+ = 0 & k = 1,\cdots,K \\
y_k^- = l_k - l_{k-1}, & 1 \leq k \leq \kappa - 1 \\
y_\kappa^- = \tau - l_{\kappa-1} & \\
y_k^- = 0 & \forall k > \kappa.
\end{cases}
$$

Furthermore, this feasible solution is easily seen to be optimal along the following arguments:

- Due to the increasing marginal costs (for surplus as well as for shortage), assuming $\tau \in [u_{\kappa-1}, u_\kappa)$ and $y_k^- = 0\ \forall k$, it is certainly meaningful to exhaust the available capacities for the variables $y_1,\cdots,y_{\kappa-1}$ first. The same argument holds true if $\tau \in (-l_\kappa, -l_{\kappa-1}]$ and $y_k^+ = 0\ \forall k$.
- Assuming a feasible solution of (3.72) with some $y_{k_1}^+$ as well as some $y_{k_2}^-$ simultaneously being greater than some $\delta > 0$, allows to reduce these variables to $\widehat{y}_{k_1}^+ = y_{k_1}^+ - \delta$ and $\widehat{y}_{k_2}^- = y_{k_2}^- - \delta$, yielding a new feasible solution with the objec-

tive changed by $(-\delta)\cdot(q_{k_1}^+ + q_{k_2}^-)$ with $(q_{k_1}^+ + q_{k_2}^-) \geq 0$ due to the assumptions. Therefore, the modified feasible solution is at least as good as the original one as far as minimization of the objective is concerned.

Hence, for $\tau = \xi - z \in [u_{\kappa-1}, u_\kappa)$ with $\kappa \in \{1,\cdots,K\}$ we get

$$\begin{aligned}
\psi(z,\xi) &= \left\{ \sum_{k=1}^{K} q_k^+ y_k^+ + \sum_{k=1}^{K} q_k^- y_k^- \right\} \\
&= \sum_{k=1}^{\kappa-1} q_k^+ (u_k - u_{k-1}) + q_\kappa^+ (\tau - u_{\kappa-1}) \\
&= \sum_{k=1}^{\kappa-1} q_k^+ u_k - \sum_{k=0}^{\kappa-2} q_{k+1}^+ u_k + q_\kappa^+ (\tau - u_{\kappa-1}) \\
&= \sum_{k=1}^{\kappa-2} (q_k^+ - q_{k+1}^+) u_k + q_{\kappa-1}^+ u_{\kappa-1} - q_1^+ u_0 + q_\kappa^+ (\tau - u_{\kappa-1}) \\
&= \sum_{k=0}^{\kappa-1} (q_k^+ - q_{k+1}^+) u_k + q_\kappa^+ \tau \quad \text{with} \quad u_0 = 0,\ q_0^+ = 0 .
\end{aligned}$$

Defining

$$\alpha_0 := q_1^+\,,\ \alpha_k := q_{k+1}^+ - q_k^+\,, k = 1,\cdots,K-1\,,$$

it follows immediately that

$$q_k^+ = \sum_{\nu=1}^{k} \alpha_{\nu-1} \quad \text{for} \quad k = 1,\cdots,K$$

such that

$$\psi(z,\xi) = -\sum_{k=0}^{\kappa-1} \alpha_k u_k + \sum_{k=0}^{\kappa-1} \alpha_k \cdot \tau = \sum_{k=0}^{\kappa-1} \alpha_k (\tau - u_k) = \sum_{k=0}^{K-1} \alpha_k [\tau - u_k]^+ .$$

Analogously, for $\tau = \xi - z \in (-l_\kappa, -l_{\kappa-1}]$ with $\kappa \in \{1,\cdots,K\}$ we get

$$\psi(z,\xi) = \sum_{k=0}^{K-1} \beta_k [\tau + l_k]^-$$

with $\beta_0 := q_1^-\,,\ \beta_k := q_{k+1}^- - q_k^-\,, k = 1,\cdots,K-1$, such that in general

$$\psi(z,\xi) = \sum_{k=0}^{K-1} \alpha_k [\tau - u_k]^+ + \sum_{k=0}^{K-1} \beta_k [\tau + l_k]^- .$$

Due to the assumptions on (3.72), we have $\alpha_0 + \beta_0 \geq 0$ as well as

$$\alpha_k \geq 0,\; \beta_k \geq 0,\; \forall k \in \{1,\cdots,K-1\} \text{ and } \sum_{k=0}^{K-1}(\alpha_k+\beta_k) = q_K^+ + q_K^- > 0.$$

Hence, whereas the SRT function

$$\varphi(z,\xi) := \alpha \cdot [\xi - z]^+ + \beta \cdot [\xi - z]^- - \gamma$$

according to Definition 3.3. represents the optimal objective value with a simple recourse constraint and implies for some application constant marginal costs for shortage and surplus, respectively, we now have the objective's optimal value for a so-called multiple simple recourse constraint, allowing to model increasing marginal costs for shortage and surplus, respectively, which may be more appropriate for particular real life problems.

To study properties of this model in more detail it is meaningful to introduce *multiple simple recourse type* functions (referred to as MSRT functions) as follows.

Definition 3.4. *For real constants* $\{\alpha_k, \beta_k, u_k, l_k; k = 0,\cdots,K-1\}$ *and* γ*, such that* $\alpha_0 + \beta_0 \geq 0$ *and*

$$\begin{aligned} &\alpha_k \geq 0,\; \beta_k \geq 0 \;\textit{for}\; k = 1,\cdots,K-1 \;\textit{with}\; \sum_{k=0}^{K-1}(\alpha_k+\beta_k) > 0\,, \\ &u_0 = 0 < u_1 < \cdots < u_{K-1}\,, \\ &l_0 = 0 < l_1 < \cdots < l_{K-1}\,, \end{aligned}$$

the function $\psi(\cdot,\cdot)$ *given by*

$$\psi(z,\xi) := \sum_{k=0}^{K-1}\{\alpha_k \cdot [\xi - z - u_k]^+ + \beta_k \cdot [\xi - z + l_k]^-\} - \gamma$$

is called a multiple simple recourse type function (see Fig. 3.5).

$$\begin{aligned} \Psi(z) &= \mathbb{E}_\xi\left[\psi(z,\xi)\right] \\ &= \int_{-\infty}^{\infty}\sum_{k=0}^{K-1}\{\alpha_k \cdot [\xi - z - u_k]^+ + \beta_k \cdot [\xi - z + l_k]^-\}\,dF_\xi(\xi) - \gamma \end{aligned}$$

is the expected MSRT function.

Remark 3.6. *In this definition the number of "shortage pieces" and of "surplus pieces" is assumed to coincide (with K). Obviously this is no restriction. If, for instance, we had for the number L of "surplus pieces" that L < K, with the trivial modification*

$$l_k = l_{k-1} + 1,\; \beta_k = 0 \;\; \textit{for} \;\; k = L,\cdots,K-1$$

we would have that

$$\begin{aligned}\psi(z,\xi) &:= \sum_{k=0}^{K-1} \alpha_k \cdot [\xi - z - u_k]^+ + \sum_{k=0}^{L-1} \beta_k \cdot [\xi - z + l_k]^- \} - \gamma \\ &= \sum_{k=0}^{K-1} \{\alpha_k \cdot [\xi - z - u_k]^+ + \beta_k \cdot [\xi - z + l_k]^- \} - \gamma.\end{aligned}$$

□

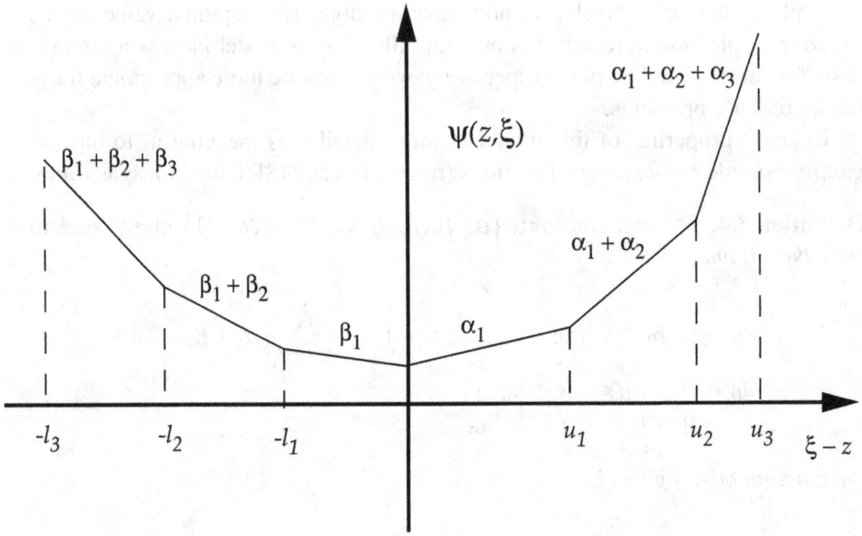

Fig. 3.5 MSRT function.

For the expected MSRT function we have

$$\begin{aligned}&\Psi(z) + \gamma = \\ &= \sum_{k=0}^{K-1} \left\{ \alpha_k \int_{-\infty}^{\infty} [\xi - z - u_k]^+ dF_\xi(\xi) + \beta_k \int_{-\infty}^{\infty} [\xi - z + l_k]^- dF_\xi(\xi) \right\} \\ &= \sum_{k=0}^{K-1} \left\{ \alpha_k \int_{z+u_k}^{\infty} (\xi - z - u_k)\, dF_\xi(\xi) + \beta_k \int_{-\infty}^{z-l_k} (z - l_k - \xi)\, dF_\xi(\xi) \right\} \\ &= \sum_{k=0}^{K-1} \alpha_k \int_{z}^{\infty} (\eta - z)\, dF_\xi(\eta + u_k) + \sum_{k=0}^{K-1} \beta_k \int_{-\infty}^{z} (z - \zeta)\, dF_\xi(\zeta - l_k) \\ &= \sum_{k=0}^{K-1} \alpha_k \int_{-\infty}^{\infty} (\xi - z)^+ dF_\xi(\xi + u_k) + \sum_{k=0}^{K-1} \beta_k \int_{-\infty}^{\infty} (\xi - z)^- dF_\xi(\xi - l_k),\end{aligned}$$

using the substitutions $\eta = \xi - u_k$ and $\zeta = \xi + l_k$ (and $\xi = \eta$ and $\xi = \zeta$ in the last expression).

The last one of the above relations for $\Psi(z)+\gamma$, i.e.

$$\begin{aligned}\Psi(z)+\gamma = &\sum_{k=0}^{K-1}\alpha_k\int_{-\infty}^{\infty}(\xi-z)^+\,dF_\xi(\xi+u_k)\\ &+\sum_{k=0}^{K-1}\beta_k\int_{-\infty}^{\infty}(\xi-z)^-\,dF_\xi(\xi-l_k),\end{aligned} \tag{3.73}$$

indicates a formal similarity with an expected SRT function using a positive mixture of the distribution functions $F_\xi(\xi+u_k)$ and $F_\xi(\xi-l_k)$, $k=0,\cdots,K-1$,

$$H(\xi)=\sum_{k=0}^{K-1}\alpha_k F_\xi(\xi+u_k)+\sum_{k=0}^{K-1}\beta_k F_\xi(\xi-l_k).$$

Due to Definition 3.4., $H(\cdot)$ is monotonically increasing, right-continuous, and satisfies

$$H(\xi)\geq 0\ \forall\xi,\ \lim_{\xi\to-\infty}H(\xi)=0,\ \text{ and }\ \lim_{\xi\to\infty}H(\xi)=\sum_{k=0}^{K-1}(\alpha_k+\beta_k)>0,$$

such that standardizing $H(\cdot)$, i.e. dividing by $W:=\sum_{k=0}^{K-1}(\alpha_k+\beta_k)$, yields a new distribution function as the mixture

$$G(\xi):=\frac{H(\xi)}{W}=\frac{\sum_{k=0}^{K-1}\alpha_k F_\xi(\xi+u_k)+\sum_{k=0}^{K-1}\beta_k F_\xi(\xi-l_k)}{W}. \tag{3.74}$$

Assuming now that $\Psi(\cdot)$ may be represented as an expected SRT function using the distribution function $G(\cdot)$ we get, with constants A, B and C to be determined later, using the trivial relations $\rho^+=\rho+\rho^-$ and $\rho^-=-\rho+\rho^+$, and writing $\int$ instead of $\int_{-\infty}^{\infty}$ for simplicity,

$$\begin{aligned}\Psi(z)+C &= A\int(\xi-z)^+\,dG(\xi)+B\int(\xi-z)^-\,dG(\xi)\\ &=\frac{A}{W}\left\{\sum_{k=0}^{K-1}\alpha_k\int(\xi-z)^+\,dF_\xi(\xi+u_k)+\sum_{k=0}^{K-1}\beta_k\int(\xi-z)^+\,dF_\xi(\xi-l_k)\right\}\\ &\quad+\frac{B}{W}\left\{\sum_{k=0}^{K-1}\alpha_k\int(\xi-z)^-\,dF_\xi(\xi+u_k)+\sum_{k=0}^{K-1}\beta_k\int(\xi-z)^-\,dF_\xi(\xi-l_k)\right\}\\ &=\frac{A}{W}\left\{\sum_{k=0}^{K-1}\alpha_k\int(\xi-z)^+\,dF_\xi(\xi+u_k)+\sum_{k=0}^{K-1}\beta_k\int(\xi-z)\,dF_\xi(\xi-l_k)\right.\\ &\qquad\left.+\sum_{k=0}^{K-1}\beta_k\int(\xi-z)^-\,dF_\xi(\xi-l_k)\right\}\end{aligned}$$

$$+\frac{B}{W}\left\{\sum_{k=0}^{K-1}\alpha_k\int(z-\xi)\,dF_\xi(\xi+u_k)+\sum_{k=0}^{K-1}\beta_k\int(\xi-z)^-\,dF_\xi(\xi-l_k)\right.$$
$$\left.+\sum_{k=0}^{K-1}\alpha_k\int(\xi-z)^+\,dF_\xi(\xi+u_k)\right\}$$
$$=\frac{A}{W}\left\{\sum_{k=0}^{K-1}\alpha_k\int(\xi-z)^+\,dF_\xi(\xi+u_k)+\sum_{k=0}^{K-1}\beta_k(\mu+l_k-z)\right.$$
$$\left.+\sum_{k=0}^{K-1}\beta_k\int(\xi-z)^-\,dF_\xi(\xi-l_k)\right\}$$
$$+\frac{B}{W}\left\{\sum_{k=0}^{K-1}\alpha_k(z-\mu+u_k)+\sum_{k=0}^{K-1}\beta_k\int(\xi-z)^-\,dF_\xi(\xi-l_k)\right.$$
$$\left.+\sum_{k=0}^{K-1}\alpha_k\int(\xi-z)^+\,dF_\xi(\xi+u_k)\right\}.$$

Hence we have

$$\Psi(z)+C=$$
$$=\frac{A+B}{W}\sum_{k=0}^{K-1}\left\{\alpha_k\int(\xi-z)^+\,dF_\xi(\xi+u_k)+\beta_k\int(\xi-z)^-\,dF_\xi(\xi-l_k)\right\}$$
$$+\frac{A}{W}\sum_{k=0}^{K-1}\beta_k(\mu+l_k-z)+\frac{B}{W}\sum_{k=0}^{K-1}\alpha_k(z-\mu+u_k).$$

To get coincidence with equation (3.73) we ought to have, with $W_\alpha=\sum_{k=0}^{K-1}\alpha_k$ and $W_\beta=\sum_{k=0}^{K-1}\beta_k$,

$$\frac{A+B}{W}=1 \quad\text{and}$$
$$\frac{A}{W}\left(W_\beta(\mu-z)+\sum_{k=0}^{K-1}\beta_k l_k\right)+\frac{B}{W}\left(W_\alpha(z-\mu)+\sum_{k=0}^{K-1}\alpha_k u_k\right)=C.$$

To assure that the left-hand side of the last equation is constant (in z), we have the condition

$$A\cdot W_\beta-B\cdot W_\alpha=0,$$

which together with $A+B=W=W_\alpha+W_\beta$ implies that

$$A=W_\alpha \quad\text{and}\quad B=W_\beta,$$

such that

$$C = \frac{W_\alpha \sum_{k=0}^{K-1} \beta_k l_k + W_\beta \sum_{k=0}^{K-1} \alpha_k u_k}{W}.$$

Hence, for the multiple simple recourse problem (with one recourse constraint)

$$\left. \begin{array}{ll} \min\{c^{\mathrm{T}}x + \Psi(z)\} & \\ Ax & = b \\ t^{\mathrm{T}}x - z & = 0 \\ x & \geq 0 \end{array} \right\} \tag{3.75}$$

we have derived in an elementary way the following result, deduced first in Van der Vlerk [334], based on a statement proved in Klein Haneveld–Stougie–Van der Vlerk [189]:

Theorem 3.8. *The multiple simple recourse problem (3.75) with the expected MSRT function*

$$\Psi(z) = \tag{3.76}$$
$$= \sum_{k=0}^{K-1} \alpha_k \int (\xi - z)^+ \, dF_\xi(\xi + u_k) + \sum_{k=0}^{K-1} \beta_k \int (\xi - z)^- \, dF_\xi(\xi - l_k)$$

is equivalent to the simple recourse problem with the expected SRT function

$$\Psi(z) = \tag{3.77}$$
$$\left(\sum_{k=k}^{K-1} \alpha_k\right) \int (\xi - z)^+ \, dG(\xi) + \left(\sum_{k=k}^{K-1} \beta_k\right) \int (\xi - z)^- \, dG(\xi) - C$$

using the distribution function

$$G(\xi) = \frac{\sum_{k=0}^{K-1} \alpha_k F_\xi(\xi + u_k) + \sum_{k=0}^{K-1} \beta_k F_\xi(\xi - l_k)}{\sum_{k=0}^{K-1} (\alpha_k + \beta_k)} \tag{3.78}$$

and the constant

$$C = \frac{\left(\sum_{k=0}^{K-1} \alpha_k\right) \sum_{k=0}^{K-1} \beta_k l_k + \left(\sum_{k=0}^{K-1} \beta_k\right) \sum_{k=0}^{K-1} \alpha_k u_k}{\sum_{k=0}^{K-1} (\alpha_k + \beta_k)}. \tag{3.79}$$

As shown in Van der Vlerk [334], if F_ξ represents a finite discrete distribution

$$\{(\xi_\nu, p_\nu);\ \nu = 1,\cdots,N\} \text{ with } p_\nu > 0\ \forall\nu,\ \sum_{\nu=1}^{N} p_\nu = 1, \tag{3.80}$$

then G corresponds to a finite discrete distribution with at most $N \cdot (2K-1)$ pairwise different realizations (with positive probabilities). This distribution, disregarding possible coincidences of some of its realizations, according to (3.78) and (3.80) is given by the following set of realizations and their corresponding probabilities

$$\left.\begin{array}{ll} \xi_\nu, & \pi_{\nu 0} = \dfrac{(\alpha_0+\beta_0)p_\nu}{\gamma};\ \nu = 1,\cdots,N;\ (\kappa = 0); \\ \xi_\nu - u_\kappa, & \pi_{\nu\kappa}^- = \dfrac{\alpha_\kappa p_\nu}{\gamma};\ \nu = 1,\cdots,N;\ \kappa = 1,\cdots,K-1; \\ \xi_\nu + l_\kappa, & \pi_{\nu\kappa}^+ = \dfrac{\beta_\kappa p_\nu}{\gamma};\ \nu = 1,\cdots,N;\ \kappa = 1,\cdots,K-1; \\ & \text{with } \gamma = \displaystyle\sum_{k=0}^{K-1}(\alpha_k+\beta_k). \end{array}\right\} \tag{3.81}$$

3.2.2.4 The generalized simple recourse problem (GSR)

GSR functions according to (3.53) on page 226 are defined as

$$\left.\begin{array}{rl} Q^G(x,\xi) := \min & q^{+\mathrm{T}}y^+ + q^{-\mathrm{T}}y^- \\ & Iy^+ - \quad Iy^- = h(\xi) - T(\xi)x \\ & y^+, \qquad y^- \geq 0. \end{array}\right\}$$

In contrast to (3.60) and (3.62) on page 227, neither GSR functions nor the corresponding EGSR functions $\mathscr{Q}^G(x) := \mathbb{E}_\xi\,[Q^G(x,\xi)]$ can be converted in a similar manner into separable functions in (χ_i, ξ_i) and in (χ_i), respectively.

Requiring Assumption 3.3., and hence in this case presuming that $q^+ + q^- \geq 0$, implies problem (3.53) to have the optimal value

$$\left.\begin{array}{l} Q^G(x,\xi) = \displaystyle\sum_{i=1}^{m_2} Q_i^G(x,\xi^{(i)}) \quad \text{with} \\ Q_i^G(x,\xi^{(i)}) = q_i^+[(\eta_i(x,\xi^{(i)})]^+ + q_i^-[(\eta_i(x,\xi^{(i)})]^-,\ \ i = 1,\cdots,m_2, \end{array}\right\} \tag{3.82}$$

where $\eta(x,\xi) = h(\xi) - T(\xi)x$, and $\xi^{(i)}$ is the subvector of ξ with those components (of ξ) affecting $(h_i(\xi) - T_i(\xi)x)$, the i-th row of $(h(\xi) - T(\xi)x)$.

Observing that

$$\eta_i(x,\xi) = [\eta_i(x,\xi)]^+ - [\eta_i(x,\xi)]^- \Longrightarrow [\eta_i(x,\xi)]^+ = \eta_i(x,\xi) + [\eta_i(x,\xi)]^-$$

and denoting by $\mathbb{E}_{\xi^{(i)}}$ integration with respect to the marginal distribution $\mathbb{P}_{\xi^{(i)}}$ of $\xi^{(i)}$, it follows with $\overline{q} = q^+ + q^-$ and $(\overline{h}, \overline{T}) = \mathbb{E}_\xi[(h(\xi), T(\xi)]$, that

$$\mathcal{Q}^G(x) = \mathbb{E}_\xi[Q^G(x,\xi)] = \sum_{i=1}^{m_2} \mathbb{E}_{\xi^{(i)}}[Q_i^G(x,\xi^{(i)})] = \sum_{i=1}^{m_2} \mathcal{Q}_i^G(x), \tag{3.83}$$

where

$$\left.\begin{aligned}
\mathcal{Q}_i^G(x) &= \mathbb{E}_{\xi^{(i)}}[Q_i^G(x,\xi^{(i)})] \\
&= q_i^+ \mathbb{E}_{\xi^{(i)}}\left[[(\eta_i(x,\xi^{(i)})]^+\right] + q_i^- \mathbb{E}_{\xi^{(i)}}\left[[(\eta_i(x,\xi^{(i)})]^-\right] \\
&= q_i^+ \mathbb{E}_{\xi^{(i)}}[(\eta_i(x,\xi^{(i)})] \\
&\quad + q_i^+ \mathbb{E}_{\xi^{(i)}}\left[[(\eta_i(x,\xi^{(i)})]^-\right] + q_i^- \mathbb{E}_{\xi^{(i)}}\left[[(\eta_i(x,\xi^{(i)})]^-\right] \\
&= q_i^+(\overline{h}_i - \overline{T}_i x) + \overline{q}_i \mathbb{E}_{\xi^{(i)}}\left[[(\eta_i(x,\xi^{(i)})]^-\right].
\end{aligned}\right\}$$

As shown in Corollary 3.1. (p. 206) the expected recourse $\mathcal{Q}_i^G(x)$ is a convex function $\forall \overline{q} \geq 0$, and hence also $\mathbb{E}_{\xi^{(i)}}\left[[(\eta_i(x,\xi^{(i)})]^-\right]$ is convex in x.

By defining $S^{(i)}(x) := \{\xi^{(i)} \mid \eta_i(x,\xi^{(i)}) < 0\}$ for arbitrary $x \in \mathbb{R}^n$, it follows

$$\begin{aligned}
\mathbb{E}_{\xi^{(i)}}\left[[(\eta_i(x,\xi^{(i)})]^-\right] &= \int_{S^{(i)}(x)} -\eta_i(x,\xi^{(i)})\,\mathbb{P}_{\xi^{(i)}}(d\xi^{(i)}) \\
&= \int_{S^{(i)}(x)} (T_i(\xi)x - h_i(\xi))\,\mathbb{P}_{\xi^{(i)}}(d\xi^{(i)})
\end{aligned}$$

and, with fixed $\tilde{x}$, arbitrary x, and with $\overline{S}^{(i)}(x)$ the complement of $S^{(i)}(x)$, for

$$\left.\begin{aligned}
L_i(x;\tilde{x}) &:= \int_{S^{(i)}(\tilde{x})} -\eta_i(x,\xi^{(i)})\,\mathbb{P}_{\xi^{(i)}}(d\xi^{(i)}) \\
&= \int_{S^{(i)}(\tilde{x})\cap S^{(i)}(x)} -\eta_i(x,\xi^{(i)})\,\mathbb{P}_{\xi^{(i)}}(d\xi^{(i)}) \\
&\quad + \underbrace{\int_{S^{(i)}(\tilde{x})\cap \overline{S}^{(i)}(x)} -\eta_i(x,\xi^{(i)})\,\mathbb{P}_{\xi^{(i)}}(d\xi^{(i)})}_{\leq 0} \\
&\leq \int_{S^{(i)}(x)} -\eta_i(x,\xi^{(i)})\,\mathbb{P}_{\xi^{(i)}}(d\xi^{(i)}) = \mathbb{E}_{\xi^{(i)}}\left[[(\eta_i(x,\xi^{(i)})]^-\right].
\end{aligned}\right\} \tag{3.84}$$

Hence, the function

$$\left.\begin{aligned} L_i(x;\tilde{x}) &= \int_{S^{(i)}(\tilde{x})} -\eta_i(x,\xi^{(i)})\,\mathbb{P}_{\xi^{(i)}}(d\xi^{(i)}) \\ &= \int_{S^{(i)}(\tilde{x})} (T_i(\xi^{(i)})x - h_i(\xi^{(i)}))\,\mathbb{P}_{\xi^{(i)}}(d\xi^{(i)}) \\ &= \left\{\int_{S^{(i)}(\tilde{x})} T_i(\xi^{(i)})\,\mathbb{P}_{\xi^{(i)}}(d\xi^{(i)})\right\}x - \int_{S^{(i)}(\tilde{x})} h_i(\xi^{(i)})\,\mathbb{P}_{\xi^{(i)}}(d\xi^{(i)}) \end{aligned}\right\} \tag{3.85}$$

is a lower bound for $\mathbb{E}_{\xi^{(i)}}\left[[(\eta_i(x,\xi^{(i)})]^-\right]$, due to (3.85) linear affine in x, and sharp for $x=\tilde{x}$, since $L_i(\tilde{x};\tilde{x}) = \mathbb{E}_{\xi^{(i)}}\left[[(\eta_i(\tilde{x},\xi^{(i)})]^-\right]$ by (3.84). Due to $\overline{q} \geq 0$ follows that

$$\left.\begin{aligned} \mathscr{L}_i(x;\tilde{x}) &:= q_i^+(\overline{h}_i - \overline{T}_i x) + \overline{q}_i L_i(x;\tilde{x}) \\ &\leq q_i^+(\overline{h}_i - \overline{T}_i x) + \overline{q}_i \mathbb{E}_{\xi^{(i)}}\left[[(\eta_i(x,\xi^{(i)})]^-\right] = \mathscr{Q}_i^G(x). \end{aligned}\right\} \tag{3.86}$$

Furthermore $\mathscr{L}_i(\tilde{x};\tilde{x}) = \mathscr{Q}_i^G(\tilde{x})$, since $L_i(\tilde{x};\tilde{x}) = \mathbb{E}_{\xi^{(i)}}\left[[(\eta_i(\tilde{x},\xi^{(i)})]^-\right]$, such that

$$\left.\begin{aligned} &\mathscr{Q}_i^G(x) - \mathscr{Q}_i^G(\tilde{x}) \geq \\ &\quad \geq \mathscr{L}_i(x;\tilde{x}) - \mathscr{L}_i(\tilde{x};\tilde{x}) \\ &\quad = q_i^+\overline{T}_i(\tilde{x}-x) + \overline{q}_i\{L_i(x;\tilde{x}) - L_i(\tilde{x};\tilde{x})\} \\ &\quad = \left\{-q_i^+\overline{T}_i + \overline{q}_i \int_{S^{(i)}(\tilde{x})} T_i(\xi^{(i)})\,\mathbb{P}_{\xi^{(i)}}(d\xi^{(i)})\right\}(x-\tilde{x}), \end{aligned}\right\} \tag{3.87}$$

thus yielding a linear support function of $\mathscr{Q}_i^G(\cdot)$ at $\tilde{x}$ as

$$\mathscr{L}_i(x;\tilde{x}) = \mathscr{Q}_i^G(\tilde{x}) + g_i(\tilde{x})(x-\tilde{x}) = \mathscr{L}_i(\tilde{x};\tilde{x}) + g_i(\tilde{x})(x-\tilde{x}) \tag{3.88}$$

with $g_i(\tilde{x})$ a subgradient (row vector) of $\mathscr{Q}_i^G(\cdot)$ at $\tilde{x}$ given as

$$g_i(\tilde{x}) := -q_i^+\overline{T}_i + \overline{q}_i \int_{S^{(i)}(\tilde{x})} T_i(\xi^{(i)})\,\mathbb{P}_{\xi^{(i)}}(d\xi^{(i)}) \in \partial\mathscr{Q}_i^G(\tilde{x}).$$

Assume the first stage feasible set

$$\mathscr{B}_1 := \{x \mid Ax = b,\ x \geq 0\}$$

to be nonempty and compact. As mentioned above $\mathscr{Q}_i^G(\cdot)$, and thus also the related EGSR function $\mathscr{Q}^G(\cdot) = \sum_{i=1}^{m_2} \mathscr{Q}_i^G(\cdot)$, are convex functions and hence, according to Prop. 1.24. (p. 54), continuous. Therefore, $\hat{\Theta} := \min_{x\in\mathscr{B}_1} \mathscr{Q}^G(x)$ exists.

Then problem (3.55) (see p. 227) can be written as

$$\min\{c^{\mathrm{T}}x + \mathscr{Q}^G(x) \mid x \in \mathscr{B}_1\}$$

or equivalently as

$$\min\{c^{\mathrm{T}}x + \Theta \mid x \in \mathscr{B}_1,\ \mathscr{Q}^G(x) - \Theta \leq 0\}. \tag{3.89}$$

Obviously, $\Theta \geq \hat{\Theta}$, and a fortiori $\Theta \geq \hat{\Theta} - \gamma$ with some $\gamma > 0$, holds for all (x, Θ) being feasible in (3.89). On the other hand, to add the constraint $\Theta \leq \hat{\Theta} + \gamma$ has no impact on the solution of problem (3.89). Thus

$$\mathscr{B}^+ := \{(x^{\mathrm{T}}, \Theta)^{\mathrm{T}} \mid x \in \mathscr{B}_1\,,\ \hat{\Theta} - \gamma \leq \hat{\Theta} \leq \Theta \leq \hat{\Theta} + \gamma\} \subset \mathbb{R}^{n+1},$$

instead of $\mathscr{B}_1 \subset \mathbb{R}^n$, is nonempty and compact again. Hence with $z := (x^{\mathrm{T}}, \Theta)^{\mathrm{T}}$, the contraint function $F(z) := \mathscr{Q}^G(x) - \Theta$ (convex in z as well), and with the objective $d^{\mathrm{T}}z := (c^{\mathrm{T}}, 1)z = c^{\mathrm{T}}x + \Theta$, the program (3.89) has the same set of solutions as

$$\min\{d^{\mathrm{T}}z \mid z \in \mathscr{B}^+\,,\ F(z) \leq 0\}. \tag{3.90}$$

Finally, since $\mathscr{Q}^G(\cdot) = \sum_{i=1}^{m_2} \mathscr{Q}_i^G(\cdot)$ due to (3.83), from (3.88) follows obviously that $g(\tilde{x}) = \sum_{i=1}^{m_2} g_i(\tilde{x}) \in \partial\mathscr{Q}^G(\tilde{x})$ at any arbitrary $\tilde{x}$, such that the function $F(\cdot)$ has at any $\tilde{z} = (\tilde{x}^{\mathrm{T}}, \tilde{\Theta})^{\mathrm{T}}$ a subgradient (row vector), given as

$$\left.\begin{aligned} f(\tilde{z}) &:= (g(\tilde{x}), -1) \\ &= \left(\sum_{i=1}^{m_2} g_i(\tilde{x}), -1\right) \\ &= \left(\sum_{i=1}^{m_2} \left\{-q_i^+ T_i + \overline{q}_i \int_{S^{(i)}(\tilde{x})} T_i(\xi^{(i)})\, \mathbb{P}_{\xi^{(i)}}(d\xi^{(i)})\right\}, -1\right) \in \partial F(\tilde{z}). \end{aligned}\right\} \tag{3.91}$$

With these requisites the following procedure can be formulated:

GSR-CUT: Approximating GSR solutions by successive cuts

Find a solution $\hat{x}$ of the LP $\min\{c^{\mathrm{T}}x \mid x \in \mathscr{B}_1\}$.

With $\hat{u}^{(1)} := \hat{x}$ and $J := 1$ go to Step I..

I. Find a solution $(\hat{x}^{\mathrm{T}}, \hat{\Theta})^{\mathrm{T}}$ of the LP

$$\left.\begin{aligned} &\min\{c^{\mathrm{T}}x + \Theta\} \\ &x \in \mathscr{B}_1 \\ &\sum_{i=1}^{m_2} \mathscr{L}_i(x; \hat{u}^{(j)}) \leq \Theta,\ \ j = 1, \cdots, J, \end{aligned}\right\} \tag{3.92}$$

and denote this solution as $\hat{z}^{(J)} := (\hat{x}^{\mathrm{T}}, \hat{\Theta})^{(J)\mathrm{T}}$.

II. If

$$\Delta := F(\hat{z}^{(J)}) = \sum_{i=1}^{m_2} \mathscr{Q}_i^G(\hat{x}) - \hat{\Theta} = \sum_{i=1}^{m_2} \mathscr{L}_i(\hat{x}; \hat{x}) - \hat{\Theta} \leq 0,$$

stop (in practice: if $\Delta \leq \varepsilon$ with a prescribed tolerance ε, stop);

else, with $J := J+1$ and $\hat{u}^{(J)} := \hat{x}$, return to Step I..

□

Remark 3.7. *The following observations on the above procedure GSR-CUT may be useful:*

1) *Due to (3.87), the $\mathscr{L}_i(x;\hat{u}^{(j)})$ are linear support functions of $\mathscr{Q}_i^G(x)$ at $\hat{u}^{(j)}$, and their gradients $\nabla_x \mathscr{L}_i(x;\hat{u}^{(j)})$ coincide due to (3.88) with the subgradients of $\mathscr{Q}_i^G(\hat{u}^{(j)})$ given as $g_i(\hat{u}^{(j)}) \in \partial \mathscr{Q}_i^G(\hat{u}^{(j)})$.*
2) *It follows immediately that, due to (3.91),*

$$\sum_{i=1}^{m_2} \mathscr{L}_i(x;\hat{u}^{(j)}) - \Theta =$$
$$= \{\mathscr{Q}^G(\hat{u}^{(j)}) - \hat{\Theta}^{(j)}\} + g(\hat{u}^{(j)})(x-\hat{u}^{(j)}) + (-1)(\Theta - \hat{\Theta}^{(j)})$$
$$= F(\hat{z}^{(j)}) + f(\hat{z}^{(j)})(z-\hat{z}^{(j)})$$

is a linear support function of $F(z) = \mathscr{Q}^G(x) - \Theta$ at $\hat{z}^{(j)} = (\hat{x}^{(j)\mathrm{T}}, \hat{\Theta}^{(j)})^{\mathrm{T}}$, and since by (3.92) obviously holds $\hat{\Theta}^{(J)} = \max_{1\leq j\leq J} \mathscr{L}(\hat{x}^{(J)};\hat{u}^{(j)})$ for all J, from the compactness of $\mathscr{B}_1$, the continuity of $\mathscr{Q}^G(\cdot)$ as well as the uniform boundedness of the subgradients $g(\cdot) \in \partial \mathscr{Q}^G(\cdot)$ (see the proof of Prop. 1.29., p. 61), follows the existence of an appropriate compact (polyhedral) set $\mathscr{B}^+ \subset \mathbb{R}^{n+1}$ such that $\hat{z}^{(j)} \in \mathscr{B}^+$ for all solutions of (3.92) generated within the above iteration. In other words, in the above iteration we deal simultaneously with problem (3.89) as well as with problem (3.90).
3) *The standard convergence statements—convergence of $\varphi_J = c^{\mathrm{T}}\hat{x}^{(J)} + \hat{\Theta}^{(J)}$, the optima of (3.92), to the optimal value of (3.89), and any accumulation point of iterates $\{\hat{z}^{(J)}\}$, generated by (3.92), being a solution of (3.89)—follow immediately from Prop. 1.29. (p. 61), observing that procedure GSR-CUT is just the application of Kelley's cutting plane method (on page 61) to problem (3.90).*
4) *In (3.92) the evaluation of $\mathscr{L}_i(x;\hat{u}^{(j)}) = q_i^+\overline{h}_i - (q_i^+\overline{T}_i)x + \overline{q}_i L_i(x;\hat{u}^{(j)})$ requires according to (3.6) (p. 196) for $\overline{h}_i$ and $\overline{T}_i$ the expectations $\mathbb{E}_{\xi^{(i)}}[\xi^{(i)}]$ and due to (3.85) in particular the computation of the integrals*

$$\left\{\int_{S^{(i)}(\hat{u}^{(j)})} T_i(\xi^{(i)})\,\mathbb{P}_{\xi^{(i)}}(d\xi^{(i)})\right\} \quad \textit{and} \quad \left\{-\int_{S^{(i)}(\hat{u}^{(j)})} h_i(\xi^{(i)})\,\mathbb{P}_{\xi^{(i)}}(d\xi^{(i)})\right\}.$$

Since in general for multivariate distributions of continuous type (described by densities) there is no algebraic formula for these integrals, they need to be approximated by some simulation approach, e.g. an appropriate variant of the Monte Carlo method.
For a finite discrete distribution $\mathbb{P}_{\xi^{(i)}}(\xi^{(i)} = \xi^{(i)\nu}) = p_\nu^{(i)}$, $\nu = 1,\cdots,N^{(i)}$, the sets $S^{(i)}(\hat{u}^{(j)}) := \{\xi^{(i)} \mid \eta_i(\hat{u}^{(j)},\xi^{(i)}) < 0\}$ are replaced by the index sets

$K^{(i)}(\hat{u}^{(j)}) := \{\nu \mid \eta_i(\hat{u}^{(j)}, \xi^{(i)\nu}) < 0\}$, *thus yielding* $\mathbb{E}_{\xi^{(i)}}[\xi^{(i)}] = \sum_{\nu=1}^{N^{(i)}} p_\nu^{(i)} \xi^{(i)\nu}$ *and*

$$\mathscr{L}_i(x;\hat{u}^{(j)}) = q_i^+ \bar{h}_i - (q_i^+ \bar{T}_i)x + \bar{q}_i \sum_{\nu \in K^{(i)}(\hat{u}^{(j)})} p_\nu^{(i)} \{h_i(\xi^{(i)\nu}) - T_i(\xi^{(i)\nu})x\}.$$

□

Exercises

3.7. Consider the following two simple recourse problems:

$$\left.\begin{array}{lll}
\min\{c^{\mathrm{T}}x + \mathbb{E}[q^{\mathrm{T}}y(\zeta)]\} & & \text{with } c = (3,1,2,4)^{\mathrm{T}}, \quad q = (2,1,1,3,2,1,2,1)^{\mathrm{T}} \\
Ax & \le b & \text{with } A = \begin{pmatrix} 2\ 1\ 3\ 5 \\ 3\ 4\ 3\ 2 \end{pmatrix}, \quad b = \begin{pmatrix} 32 \\ 35 \end{pmatrix} \\
Tx + Wy(\zeta) = h(\zeta) & \text{a.s.} & \text{with } T = \begin{pmatrix} 2\ 0\ 3\ 2 \\ 3\ 5\ 0\ 2 \\ 0\ 2\ 4\ 0 \\ 2\ 1\ 0\ 3 \end{pmatrix}, \quad h(\zeta) = \begin{pmatrix} 25+\xi_1 \\ 15+\xi_2 \\ 17+\xi_3 \\ 23+\xi_4 \end{pmatrix} \\
x, y(\zeta) \ge 0 & \text{a.s.} &
\end{array}\right\} \quad (3.93)$$

where ζ with independent components has either a uniform or a normal distribution as
$\mathscr{U}\{[-5,5]\times[-7,7]\times[-3,3]\times[-8,8]\}$ or
$\mathscr{N}\{(0;2),(0;1.5),(0;2.2),(0;1.7)\}$ with truncation probabilities of 0.999, each.

Solve both problems using SLP-IOR, applying SRAPPROX as well as DAPPROX.

(a) For each of the two problems compare, as indicators for the efficiency of the two solvers, the number of iterations as well as the number of splits (or subintervals, respectively) used by SLP-IOR to get the solutions with the preset accuracy.

(b) How do you explain the difference with respect to the above indicators, in particular the remarkable discrepancy of the numbers of splits/subintervals?

3.8. For the SRT function $\varphi(z,\xi) := \alpha[\xi - z]^+ + \beta[\xi - z]^-$, $\alpha + \beta \ge 0$, and the corresponding ESRT function $\Phi(z) = \alpha(\bar{\xi} - z) - (\alpha + \beta)\int_{-\infty}^{z}(\xi - z)\mathbb{P}_\xi(d\xi)$ assume the distribution $\mathbb{P}_\xi$ to be bounded to the interval $[a,b]$.

(a) Show that then the integral $\Psi(z) := \int_{-\infty}^{z}(z - \xi)\mathbb{P}_\xi(d\xi)$ may be computed with $\Theta(z_1,z_2,z_3) := z_3 + \int_{\xi \le a+z_2}(z_2 + a - \xi)\mathbb{P}_\xi(d\xi)$ as

$$\left.\begin{array}{l}
\Psi(z) := \min \Theta(z_1,z_2,z_3) \text{ subject to:} \\
-z_1 + z_2 + z_3 = z - \bar{\xi},\ z_1 \ge \bar{\xi} - a,\ z_2 \le b - a,\ (z_1,z_2,z_3) \ge 0
\end{array}\right\}. \quad (3.94)$$

(b) How does $\Psi(z)$ and hence the ESRT function $\Phi(z)$ look like, if $\mathbb{P}_\xi$ is given as $\mathscr{U}\{[a,b]\}$?

3.9. Assume for (3.93) of Exercise 3.7 the uniform distributions mentioned for the right-hand sides $h_i(\zeta_i)$ and formulate problem (3.93) according to the result of the previous exercise as a quadratic program. If you have access to any convex programming software package, than solve the quadratic program and compare the solution with that one you have got with SLP-IOR applying SRAPPROX (and/or DAPPROX).

3.2.3 CVaR and recourse problems

Assume the result of some process to be a loss, modelled as a random variable $\vartheta \in \mathscr{L}_1^1$ with a distribution function $F_\vartheta(z)$. As mentioned in Section 2.1, an example from finance could be a portfolio optimization problem with $t^{\mathrm{T}}(\xi)x$ as random return of a portfolio $x \in \mathbb{R}^n$ (usually represented as the mixture of different assets) compared to $h(\xi)$, the random return of some benchmark portfolio. In this case $\vartheta := (t^{\mathrm{T}}(\xi)x - h(\xi))$ is considered as loss if $\vartheta^- > 0$. With the α–VaR (value at risk) v_α, defined in Section 2.3 (p. 137) as $v_\alpha := v_\alpha(\vartheta) := \min\{z \mid F_\vartheta(z) \geq \alpha\}$, $\alpha \in (0,1)$, the α–CVaR (conditional value at risk) $v_\alpha^c := v_\alpha^c(\vartheta)$was introduced in Section 2.4.3 (p. 152) as

$$v_\alpha^c(\vartheta) := v_\alpha + \frac{1}{1-\alpha}\mathbb{E}_\vartheta[(\vartheta - v_\alpha)^+] = \min_z \left\{ z + \frac{1}{1-\alpha}\mathbb{E}_\vartheta[(\vartheta - z)^+] \right\}. \tag{3.95}$$

It is well known, that—in spite of the naming—for the α–CVaR holds the inequality $v_\alpha^c(\vartheta) \geq \mathbb{E}_\vartheta[\vartheta \mid \vartheta \geq v_\alpha]$, where equality can only be ensured for continuous distribution functions $F_\vartheta(\cdot)$. Nevertheless, $v_\alpha^c(\vartheta)$ is widely used in finance applications as risk measure. Whereas the VaR $v_\alpha(\vartheta)$ is by definition the (smallest) threshold for a realization $\hat{\vartheta}$ not being exceeded with a probability of at least α, for continuous distributions the α–CVaR $v_\alpha^c(\vartheta)$ is then the conditional expectation of ϑ given that $\vartheta \geq \hat{\vartheta}$. Moreover, due to Prop. 2.48. the α–CVaR satisfies the axioms for coherent risk measures presented in Artzner, Delbaen et al. [7], which is in general not true for the α–VaR. A more detailed discussion of the concept of CVaR can be found in Rockafellar–Uryasev [283].

Due to (3.95), computing the α–CVaR $v_\alpha^c(\vartheta)$ can be considered as solving a single-stage stochastic program. However, $v_\alpha^c(\vartheta)$ can also be considered as the optimal value of a particular two-stage stochastic program with simple recourse.

Proposition 3.3. *The α–CVaR as defined in (3.95) is the optimal value of the SSR problem*

$$v_\alpha^c = \min_z (z + \mathbb{E}_\vartheta[Q(z;\vartheta)]), \tag{3.96}$$

where

$$Q(z;\vartheta) = \min_{\eta} \left\{ \frac{1}{1-\alpha}\eta \,\middle|\, z+\eta \geq \vartheta,\ \eta \geq 0 \right\}.$$

Proof: Obviously holds

$$(\vartheta - z)^{+} = \min_{\eta}\{\eta \mid \eta \geq \vartheta - z,\ \eta \geq 0\} = (1-\alpha)Q(z;\vartheta),$$

thus yielding the proposition and allowing for the interpretation, that after the first-stage decision on z a realization of ϑ has to be observed before taking the second-stage decision on η. □

Assuming now that, instead of $\vartheta : \Omega \longrightarrow \mathbb{R}$, a random vector $\xi : \Omega \longrightarrow \Xi \subset \mathbb{R}^r$ is given with $\Xi = \operatorname{supp}\xi$, and $f(x,\xi) : X \times \Xi \longrightarrow \mathbb{R}$ is defined as decision-dependent (loss) function, where

- $X \subset \mathbb{R}^n$ is a closed convex set of feasible decisions,
- $f(\cdot,\xi)$ is continuous in x $\forall \xi \in \Xi$,
- $f(x,\cdot)$ is ξ–measurable $\forall x \in X$, and
- $\mathbb{E}_{\xi}[|f(x,\xi)|] < \infty\ \forall x \in X$.

With the distribution function $\Phi(x,z) := \mathbb{P}(\{\xi \mid f(x,\xi) \leq z\})$ the α–VaR of $f(x,\xi)$ is $v_{\alpha}(x) = \min\{z \mid \Phi(x,z) \geq \alpha\}$, yielding in analogy to (3.95) the α–CVaR of $f(x,\xi)$ as

$$v_{\alpha}^{c}(x) := \min_{z} \left\{ z + \frac{1}{1-\alpha}\mathbb{E}_{\xi}[(f(x,\xi)-z)^{+}] \right\}.$$

If in addition to the above assumptions $f(\cdot,\xi)$ is convex in x $\forall \xi \in \Xi$, then it follows that $v_{\alpha}^{c}(x)$ is convex in x as well. In this case Prop. 3.3. is modified to

Proposition 3.4. *For $f(\cdot,\xi)$ convex $\forall \xi \in \Xi$ the α–CVaR denoted as $v_{\alpha}^{c}(x)$ is a convex function in x, computable as the optimal value of the convex CFR program*

$$\left.\begin{aligned} v_{\alpha}^{c}(x) &:= \min_{z\in\mathbb{R}}\{z + \mathbb{E}_{\xi}[Q(x,z;\xi)]\} \\ \textit{with}\ \ Q(x,z;\xi) &:= \min_{\eta\in\mathbb{R}} \left\{ \frac{1}{1-\alpha}\eta \,\middle|\, z+\eta \geq f(x,\xi),\ \eta \geq 0 \right\}, \end{aligned}\right\} \tag{3.97}$$

or equivalently, the optimum of

$$\left.\begin{aligned} v_{\alpha}^{c}(x) := \min_{z\in\mathbb{R},\eta(x,z;\cdot)\in\mathcal{L}_1^1} & \left\{ z + \mathbb{E}_{\xi}\left[\frac{1}{1-\alpha}\eta(x,z;\xi)\right] \right\} \\ z + \eta(x,z;\xi) &\geq f(x,\xi)\ \textit{a.s.} \\ \eta(x,z;\xi) &\geq 0 \qquad \textit{a.s.} \end{aligned}\right\} \tag{3.98}$$

Proof: Introducing x as a parameter and replacing ϑ by $f(x;\xi)$, (3.97) follows immediately from (3.96). The integrability of $f(x;\xi)$ with respect ot ξ implies the well known fact, that for each (x,z) there exists an $\eta(x,z;\cdot) \in \mathcal{L}_1^1$ such that $\eta(x,z;\xi) =$

$(1-\alpha)Q(x,z;\xi)$ a.s. Finally, the convexity of $Q(x,z;\xi)$ in (x,z) $\forall \xi \in \Xi$ follows trivially from (3.97), implying the convexity of $\mathbb{E}_\xi[Q(x,z;\xi)]$ in x for any fixed z and thus the convexity of $v_\alpha^c(x)$ in x. □

To be more specific, assume that $X := \{x \mid Ax = b,\ x \geq 0\} \neq \emptyset$ is compact, and that the loss function is defined as $f(x,\xi) := \lambda(h(\xi) - t^{\mathrm{T}}(\xi)x)$ with some coefficient $\lambda > 0$. As an interpretation, think of a linear production function, transforming a vector x of input factors with a random vector $t(\xi)$ of productivities into a random output $t^{\mathrm{T}}(\xi)x$; on the other hand let $h(\xi)$ be a random demand to be covered by that output, such that, given that $h(\xi) - t^{\mathrm{T}}(\xi)x > 0$, the above loss function $f(x,\xi)$ is just proportional to this excess demand.

Different types of models may be set up in this situation, as for instance:

1) In addition to the linear constraints of an LP a further constraint, restricting the α–CVaR $v_\alpha^c(x) := v_\alpha^c(f(x,\xi)) = \min_z \left\{ z + \frac{1}{1-\alpha}\mathbb{E}_\xi[(f(x,\xi)-z)^+] \right\}$ of the above loss function, may be inserted yielding the model

$$\begin{aligned} & \min c^{\mathrm{T}}x \\ \text{s.t.} \quad & Ax = b \\ & v_\alpha^c(x) \leq \gamma \\ & x \geq 0, \end{aligned}$$

which according to (2.152), (2.153) on pages157/157 coincides with the convex NLP

$$\begin{aligned} & \min c^{\mathrm{T}}x \\ \text{s.t.} \quad & Ax = b \\ & z + \frac{1}{1-\alpha}\mathbb{E}_\xi[(\lambda(h(\xi) - t^{\mathrm{T}}(\xi)x) - z)^+] \leq \gamma \\ & x \geq 0, \end{aligned}$$

a single stage problem.

2) Extending instead the linear term of an LP's objective by adding the α–CVaR $v_\alpha^c(x) := v_\alpha^c(f(x,\xi))$ of the loss $f(x;\xi)$ yields the NLP

$$\begin{aligned} & \min\{c^{\mathrm{T}}x + v_\alpha^c(x)\} \\ \text{s.t.} \quad & Ax = b \\ & x \geq 0, \end{aligned}$$

which due to Prop. 3.4. can be restated as

$$
\left.
\begin{array}{l}
\min\{c^{\mathrm{T}}x + z + \mathbb{E}_{\xi}[Q(x,z;\xi)]\} \\
\quad Ax = b \\
\quad x \in \mathbb{R}^n_+,\ z \in \mathbb{R} \\
\text{where} \\
Q(x,z;\xi) := \min\limits_{\eta \in \mathbb{R}} \left\{ \dfrac{1}{1-\alpha}\eta \,\middle|\, z+\eta \geq f(x,\xi),\ \eta \geq 0 \right\} \\
\qquad = \min\limits_{\eta \in \mathbb{R}} \left\{ \dfrac{1}{1-\alpha}\eta \,\middle|\, \lambda t^{\mathrm{T}}(\xi)x + z + \eta \geq \lambda h(\xi),\ \eta \geq 0 \right\},
\end{array}
\right\} \quad (3.99)
$$

a particular two-stage generalized simple recourse SLP with the first stage variables $x \in \mathbb{R}^n$ and $z \in \mathbb{R}$ and the recourse variable η (which, as mentioned above, can be chosen for each (x,z) as a function $\eta(x,z;\cdot) \in \mathscr{L}_1^1$); in (2.150) on page 156 this model was derived for the special case of a finite discrete probability distribution.
Solution methods of the GSR-CUT type were considered for this problem by Klein Haneveld and van der Vlerk [191] and by Künzi–Bay and Mayer [198], assuming ξ to have a finite discrete distribution; since in this case (3.99) is a special LP with decomposition structure, in accordance with Remark 1.2. (p. 48) the main concern of the authors was to find appropriate cut generation strategies for the corresponding decomposition algorithm to be as efficient as possible.

Due to the above discussion on general GSR-Cut procedures, for continuous distributions the cutting plane methods described on page 247 can be designed to solve (3.99) as well.

3) For the two-stage model (3.99) it is assumed that the *first-stage* decision on x implies the deterministic *first-stage* outcome $c^{\mathrm{T}}x$, and that the loss $f(x,\xi)$, given the first-stage decision x, is the random *second-stage* outcome determined by the realization of ξ (unknown when deciding on x). To take into account the risk (due to the random loss $f(x,\xi)$), this model aims at determining a minimizer $\hat{x}$ for the overall objective given as the sum of the first-stage outcome $c^{\mathrm{T}}x$ with the α–CVaR of the second-stage outcome $f(x,\xi)$.
Another two-stage model is based on the following view: With a convex polyhedral set $X \subset \mathbb{R}^n_+$ of feasible first-stage decisions x, causing $c^{\mathrm{T}}x$ as the deterministic part of of the first-stage outcome, and with a very general recourse function

$$
Q(x;\xi) := \min_y \{q^{\mathrm{T}}(\xi)y \mid T(\xi)x + W(\xi)y = h(\xi),\ y \geq 0\}, \qquad (3.100)
$$

as the random part of the first-stage outcome, the *overall first-stage objective* is defined as

$$
f(x;\xi) := c^{\mathrm{T}}x + Q(x;\xi).
$$

Now the decision maker wants to find any $\hat{x} \in X$ which minimizes some mixture of the mean of this outcome and some risk measure of it, e.g. the α–CVaR of f. Thus, observing that due to Prop. 2.48. (page 177) the α–CVaR is translation invariant, with some $\lambda > 0$ the problem to solve would be

$$\begin{aligned}\min_{x\in X}\{\mathbb{E}_\xi[f(x;\xi)]+\lambda v_\alpha^c(f(x;\xi))\} &= \\ =\min_{x\in X}\{(1+\lambda)c^{\mathrm{T}}x+\mathbb{E}_\xi[Q(x;\xi)]+\lambda v_\alpha^c(Q(x;\xi))\}.\end{aligned}$$

This can be rewritten as the two-stage SLP

$$\left.\begin{array}{ll}\min\limits_{x,\eta,y,\theta}\left[(1+\lambda)c^{\mathrm{T}}x+\mathbb{E}_\xi[q^{\mathrm{T}}(\xi)y(\xi)]+\lambda\left(\eta+\dfrac{1}{1-\alpha}\mathbb{E}_\xi[\theta(\xi)]\right)\right] & \\ \quad x\in X,\ \eta\in\mathbb{R} & \\ \quad y(\cdot)\in\mathscr{L}_{n_2}^2,\ \theta(\cdot)\in\mathscr{L}_1^1 & \\ \quad W(\xi)y(\xi)=h(\xi)-T(\xi)x & \text{a.s.} \\ \quad \theta(\xi)\geq q^{\mathrm{T}}(\xi)y(\xi)-\eta & \text{a.s.} \\ \quad y(\xi)\geq 0 & \text{a.s.} \\ \quad \theta(\xi)\geq 0 & \text{a.s.}\end{array}\right\}\qquad(3.101)$$

with first-stage variables (x,η) and second-stage decisions $(y(\xi),\theta(\xi))$, or more precisely $y(\xi)$, since due to the objective of (3.101) automatically $\theta(\xi)=(q^{\mathrm{T}}(\xi)y(\xi)-\eta)^+$ a.s. will result. At present, it seems unlikely to find an efficient solver for the above problem in this generality for continuous distributions $\mathbb{P}_\xi$. Obviously one might think of approximating solutions via constructing sequences of discrete distributions $\mathbb{P}_\xi^\nu$, weakly converging to $\mathbb{P}_\xi$ and thus taking advantage of known results on the use of epi-convergence in optimization, as presented for instance in Pennanen [251, 252], Robinson and Wets [279], Wets [343], and Kall [156]. However, since in this generality the recourse function $Q(x;\xi)$ is not convex in ξ, neither Jensen's inquality nor the Edmundson–Madansky inequality apply. Hence, there seems to be no efficient tool to check the approximation error (as e.g. in DAPPROX) and thus to verify a prescribed accuracy during such an iterative procedure. Obviously this would change substantially, if for the recourse (3.100) holds $q(\xi)\equiv q\in\mathbb{R}^{n_2}$ and $W(\xi)\equiv W$, a constant $(m_2\times n_2)$–matrix, thus allowing for an approximation via successive discretization.

In the general case the situation becomes much better manageable for finite discrete distributions of ξ given by $\mathbb{P}_\xi(\xi=\xi_i)=p_i,\ i=1,\cdots,N$. Then (3.101) reads as

$$\left.\begin{array}{ll}\min\limits_{x,\eta,y,\theta}\left[(1+\lambda)c^{\mathrm{T}}x+\sum\limits_{i=1}^{N}p_i\cdot {q_i}^{\mathrm{T}}y_i+\lambda\left(\eta+\dfrac{1}{1-\alpha}\sum\limits_{i=1}^{N}p_i\cdot\theta_i\right)\right] & \\ \quad x\in X,\ \eta\in\mathbb{R} & \\ \quad W_iy_i=h_i-T_ix & i=1,\cdots,N \\ \quad \theta_i\geq {q_i}^{\mathrm{T}}y_i-\eta & i=1,\cdots,N \\ \quad y_i\geq 0 & i=1,\cdots,N \\ \quad \theta_i\geq 0 & i=1,\cdots,N.\end{array}\right\}\qquad(3.102)$$

This model, a linear program with decomposition structure, was recently analyzed by Noyan [246], providing two variants of appropriate cuts within decomposition procedures for solving this problem efficiently.

3.2.4 Some characteristic values for two-stage SLP's

Among various paradigms of modeling two-stage stochastic linear programs we have discussed so far the general (two-stage) stochastic program with recourse with the optimal value RS given due to (3.8), (3.9) as

$$\left.\begin{aligned} RS := \min_x & \left\{c^{\mathrm{T}}x + \mathbb{E}_\xi\left[Q(x;T(\xi),h(\xi),W(\xi),q(\xi))\right]\right\} \\ \text{s.t. } & Ax = b \\ & x \geq 0, \end{aligned}\right\} \tag{3.103}$$

where

$$\begin{aligned} Q(x;T(\xi),h(\xi),W(\xi),q(\xi)) := & \inf q^{\mathrm{T}}(\xi)y(\xi) \\ \text{s.t. } & W(\xi)y(\xi) = h(\xi) - T(\xi)x \quad \text{a.s.} \\ & y(\xi) \geq 0 \qquad\qquad\qquad\quad \text{a.s.} \\ & y(\cdot) \in Y \end{aligned}$$

with $Y = \mathscr{L}^2_{n_2}$. As in (3.6), we assume that the random parameters in these problems are defined as linear affine mappings on $\Xi = \mathbb{R}^r$ by

$$\begin{aligned} T(\xi) &= T + \sum_{j=1}^r T^j \xi_j; \quad T,\ T^j \in \mathbb{R}^{m_2 \times n_1} \text{ deterministic,} \\ W(\xi) &= W + \sum_{j=1}^r W^j \xi_j; \; W,\ W^j \in \mathbb{R}^{m_2 \times n_2} \text{ deterministic,} \\ h(\xi) &= h + \sum_{j=1}^r h^j \xi_j; \quad h,\ h^j \in \mathbb{R}^{m_2} \text{ deterministic,} \\ q(\xi) &= q + \sum_{j=1}^r q^j \xi_j; \quad q,\ q^j \in \mathbb{R}^{n_2} \text{ deterministic.} \end{aligned}$$

Remark 3.8. *Whereas by the modeling paradigm of problem (3.103), the second stage decision on $y(\xi)$ is to be taken after observing the realization of ξ, and knowing the first stage decision on x, which was taken before having knowledge of the realization of ξ—one possible interpretation being (see Fig. 3.1, p. 191) that, in time, the decision on the first stage variables x is taken before the observation of a realization of ξ, and the second stage variables $y(\xi)$ are determined afterwards—other paradigms could be either to replace the random vector ξ in advance by its expectation $\bar{\xi}$, thus yielding the expected value problem (3.104), or else to delay the first stage decision until a realization of ξ is known, such that now the second stage* *decision $y(\cdot)$ as well as the first stage decision $x(\cdot)$ depend on ξ, which leads to the wait-and-see model (3.105).* □

As just mentioned, replacing the random vector ξ by its expectation $\bar{\xi} = \mathbb{E}_\xi[\xi]$, yields instead of *RS* the optimal value *EV* of the *expected value problem*,

$$\left.\begin{array}{rl} EV := \min\limits_{x,y}\{c^{\mathrm{T}}x + q^{\mathrm{T}}(\bar{\xi})y\} & \\ \text{s.t.} \quad Ax \qquad\qquad\quad = b & \\ T(\bar{\xi})x + W(\bar{\xi})y = h(\bar{\xi}) & \\ x, \qquad\quad y \geq 0. & \end{array}\right\} \tag{3.104}$$

Except for the first moment $\bar{\xi}$, this model does not take at all into account the distribution of ξ. Hence the solution will always be the same, no matter of the distribution being discrete or continuous, skew or symmetric, flat or concentrated, as long as the expectation remains the same. In other words, the randomness of ξ does not play an essential role in this model.

In contrast to the recourse model (3.103), in the *wait-and-see* model both, the decisions on the first stage variables x and the second stage variables y, are taken simultaneously only when the outcome of ξ is known, with the optimal values of the family of LP's

$$\left.\begin{array}{rl} \forall \xi \in \Xi : \gamma(\xi) := & \min\limits_{x,y}\{c^{\mathrm{T}}x + q^{\mathrm{T}}(\xi)y\} \\ \text{s.t.} & Ax \qquad\qquad\quad = b \\ & T(\xi)x + W(\xi)y = h(\xi) \\ & x, \qquad\quad y \geq 0 \end{array}\right\} \tag{3.105}$$

the so-called *wait-and-see* value WS is the expected value

$$WS := \mathbb{E}_\xi[\gamma(\xi)]. \tag{3.106}$$

Finally, with the first stage solution fixed as any optimal first stage solution $\hat{x}$ of the EV problem (3.104), we may ask for the objective's value of (3.103), the *expected result of the EV solution*

$$\begin{aligned} EEV &:= \\ &= c^{\mathrm{T}}\hat{x} + \mathbb{E}_\xi\left[\min_y\{q^{\mathrm{T}}(\xi)y \mid W(\xi)y = h(\xi) - T(\xi)\hat{x}, y \geq 0\}\right]. \end{aligned} \tag{3.107}$$

Observe that, in contrast to the values *RS*, EV, and WS, the value EEV may not be uniquely determined by (3.107): If the expected value problem (3.104) happens to have two different solutions $\hat{x} \neq \tilde{x}$, this may lead to $EEV(\hat{x}) \neq EEV(\tilde{x})$.

For the above values assigned in various ways to the two-stage stochastic programming situations mentioned, several relations are known which, essentially, can be traced back to Madansky [211].

Proposition 3.5. *For an arbitrary recourse problem (3.103) and the associated problems (3.106) and (3.107) the following inequalities hold:*

$$WS \leq RS \leq EEV\,. \tag{3.108}$$

Furthermore, with the recourse function $Q(x;T(\xi),h(\xi))$, allowing only for the matrix $T(\cdot)$ and the right–hand–side $h(\cdot)$ to contain random data, it follows that

$$EV \leq RS \leq EEV\,. \tag{3.109}$$

Proof: Let $x^\star$ be an optimal first stage solution of (3.103). Then obviously the inequality

$$\gamma(\xi) \leq c^{\mathrm{T}}x^\star + Q(x^\star;T(\xi),h(\xi),W(\xi),q(\xi)) \;\; \forall \xi \in \Xi$$

holds, and therefore

$$WS = \mathbb{E}_\xi\left[\gamma(\xi)\right] \leq \{c^{\mathrm{T}}x^\star + \mathbb{E}_\xi\left[Q(x^\star;T(\xi),h(\xi),W(\xi),q(\xi))\right]\} = RS\,.$$

The second inequality in (3.108) is obvious.

To show the second part, for any fixed $\tilde{x}$ the recourse function

$$Q(\tilde{x};T(\xi),h(\xi)) = \min\{q^{\mathrm{T}}y \mid Wy = h(\xi) - T(\xi)\tilde{x},\; y \geq 0\}$$

is convex in ξ. In particular, for the optimal first stage solution $x^\star$ of (3.103) follows with Jensen's inequality and the definition (3.104) of EV, that

$$\begin{aligned} RS &= c^{\mathrm{T}}x^\star + \mathbb{E}_\xi\left[Q(x^\star;T(\xi),h(\xi))\right] \\ &\geq c^{\mathrm{T}}x^\star + Q(x^\star;T(\bar{\xi}),h(\bar{\xi})) \\ &\geq EV \end{aligned}$$

which implies (3.109). □

Proposition 3.6. *Given the recourse function $Q(x;h(\xi))$ (i.e. only the right–hand–side $h(\cdot)$ is random) it follows that*

$$EV \leq WS.$$

Proof: For the wait-and-see situation we have

$$\gamma(\xi) = \min_{x,y}\{c^{\mathrm{T}}x + q^{\mathrm{T}}y \mid Ax = b,\; Tx + Wy = h(\xi);\; x,y \geq 0\}\,,$$

which is obviously convex in ξ. Then by Jensen's inequality follows

$$\gamma(\bar{\xi}) = EV \leq \mathbb{E}_\xi\left[\gamma(\xi)\right] = WS\,.$$

□

For more general recourse functions the inequality of Prop. 3.6. cannot be expected to hold true; for a counterexample see Birge–Louveaux [26].

Furthermore, in Avriel–Williams [9] the *expected value of perfect information* $EVPI$ was introduced as

$$EVPI := RS - WS \tag{3.110}$$

and may be understood in applications as the maximal amount a decision maker would be willing to pay for the exact information on future outcomes of the random vector ξ. Obviously due to Prop. 3.5. we have $EVPI \geq 0$. However, to compute this value exactly would require by (3.110) to solve the original recourse problem (3.103) as well as the wait-and-see problem (3.106), both of which may turn out to be hard tasks. Hence the question of easier computable and still sufficiently tight bounds on the $EVPI$ was widely discussed. As may be expected, the results on bounding the expected recourse function mentioned earlier are used for this purpose as well as approaches especially designed for bounding the $EVPI$ as presented e.g. in Huang–Vertinsky–Ziemba [143] and some of the references therein.

Finally, the *value of the stochastic solution* was introduced in Birge [22] as the quantity

$$VSS := EEV - RS, \tag{3.111}$$

which in applications may be given the interpretation of the expected loss for neglecting stochasticity in determining the first stage decision, as mentioned with the EV solution of (3.104). Obviously it measures the extra cost for using, instead of the "true" first stage solution for the recourse problem (3.103), the first stage solution of the expected value problem (3.104). Also in this case Prop. 3.5. implies $VSS \geq 0$.

If in the problem at hand there is no randomness around, in other words if with some fixed $\hat{\xi} \in \mathbb{R}^r$ we have $\mathbb{P}_\xi(\xi = \hat{\xi}) = 1$, then obviously follows $EVPI = VSS = 0$. In turn, if one of these characteristic values is strictly positive, it is often considered as a "measure of the degree of stochasticity" of the recourse problem. However, one must be careful with this interpretation; it should be observed that examples can be given for which either $EVPI = 0$ and $VSS > 0$ or, on the other side, $EVPI > 0$ and $VSS = 0$ (see Birge–Louveaux [26]). Hence, the impact of stochasticity to the $EVPI$ and the VSS may be rather different. Although these values are not comparable in general, there are at least some joint bounds:

Proposition 3.7. *With the recourse function $Q(x;T(\xi),h(\xi))$, allowing only for the matrix $T(\cdot)$ and the right–hand–side $h(\cdot)$ to contain random data, the value of the stochastic solution has the upper bound*

$$VSS \leq EEV - EV. \tag{3.112}$$

With the recourse function $Q(x;h(\xi))$, i.e. with only the right–hand–side $h(\cdot)$ being random, the expected value of perfect information is bounded above as

$$EVPI \leq EEV - EV. \tag{3.113}$$

Proof: Due to (3.109) in Prop. 3.5., we have $RS \geq EV$ and therefore

$$VSS = EEV - RS \leq EEV - EV.$$

From Prop. 3.6. we know that with the recourse function $Q(x;h(\xi))$ holds $EV \leq WS$. Hence, together with Prop. 3.5. we get

$$EVPI = RS - WS \leq EEV - EV\,.$$

□

The above bounds are due to Avriel–Williams [9] for the $EVPI$ and Birge [22] for the VSS.

In the literature, you may occasionally find statements claiming that the bounds given in (3.112) and (3.113) hold true without the restrictions made in Prop. 3.7.. There are obvious reasons to doubt those claims. Concerning VSS the above argument for (3.109) using Jensen's inequality fails as soon as we loose the convexity of the recourse function in ξ for any fixed $\tilde{x}$. For the $EVPI$ we present again the following example (as mentioned in Kall [154]):

Example 3.2. *With* $X = \mathbb{R}_+$ *let* $c = 2$, $W = (1,-1)$, $q = (1,0)^{\mathrm{T}}$ *and*

$$\mathbb{P}_\xi\,\{(T^{(1)},h^{(1)}) = (1,2)\} = \mathbb{P}_\xi\,\{(T^{(2)},h^{(2)}) = (3,12)\} = \frac{1}{2}\,.$$

Then we have $\bar{T} = 2$, $\bar{h} = 7$ *and*

$$EV = \min\{2x + y_1 \mid 2x + y_1 - y_2 = 7;\ x \geq 0, y \geq 0\} = 7 \ \text{with}\ \hat{x} = \frac{7}{2}\,.$$

With

$$Q(\hat{x};T^{(1)},h^{(1)}) = \min_y\{y_1 \mid y_1 - y_2 = 2 - \hat{x}, y \geq 0\} = 0$$

and

$$Q(\hat{x};T^{(2)},h^{(2)}) = \min_y\{y_1 \mid y_1 - y_2 = 12 - 3\hat{x}, y \geq 0\} = \frac{3}{2}$$

follows

$$EEV = 2 \cdot \frac{7}{2} + \frac{1}{2} \cdot \frac{3}{2} = 7.75$$

and hence $EEV - EV = 0.75$. *On the other hand we get* RS *as optimal value from*

$$\begin{array}{llll} \min\{2\cdot x + 0.5\cdot y_1^{(1)} + 0.5\cdot y_1^{(2)}\} & & & \\ 1\cdot x + 1\cdot y_1^{(1)} - 1\cdot y_2^{(1)} & & = & 2 \\ 3\cdot x + & 1\cdot y_1^{(2)} - 1\cdot y_2^{(2)} & = & 12 \\ & x,\ y^{(1)},\ y^{(2)} & \geq & 0, \end{array}$$

yielding $RS = 7$ *with* $x^\star = 2$, $y_1^{(2)} = 6$. *To get the* WS *we compute*

$$\gamma_1 := \min\{2\cdot x + y_1 \mid 1\cdot x + 1\cdot y_1 - 1\cdot y_2 = 2;\ x, y \geq 0\} = 2$$

and

$$\gamma_2 := \min\{2\cdot x + y_1 \mid 3\cdot x + 1\cdot y_1 - 1\cdot y_2 = 12;\ x, ,y \geq 0\} = 8$$

yielding $WS = 0.5\cdot 2 + 0.5\cdot 8 = 5$ *such that*

$$EVPI = RS - WS = 2 > EEV - EV = 0.75\,.$$

□

3.3 The multi-stage SLP

According to (3.1) on page 192 the general MSLP may be stated as

$$\left.\begin{array}{l} \min\{c_1^{\mathrm{T}} x_1 + \mathbb{E} \sum\limits_{t=2}^{T} c_t^{\mathrm{T}}(\zeta_t) x_t(\zeta_t)\} \\ A_{11} x_1 \qquad\qquad\qquad\qquad\quad = b_1 \\ A_{t1}(\zeta_t) x_1 \;+ \sum\limits_{\tau=2}^{t} A_{t\tau}(\zeta_t) x_\tau(\zeta_\tau) \;=\; b_t(\zeta_t) \text{ a.s., } t = 2,\cdots,T, \\ \qquad\qquad x_1 \geq 0,\; x_t(\zeta_t) \geq 0 \qquad \text{a.s., } t = 2,\cdots,T, \end{array}\right\} \tag{3.114}$$

where on a given probability space $(\Omega, \mathscr{G}, P)$ random vectors $\xi_t : \Omega \longrightarrow \mathbb{R}^{r_t}$ are defined, with $\xi = (\xi_2^{\mathrm{T}}, \cdots, \xi_T^{\mathrm{T}})^{\mathrm{T}}$ inducing the probability distribution $\mathbb{P}_\xi$ on $\mathbb{R}^{r_2+\cdots+r_T}$, and $\zeta_t = (\xi_2^{\mathrm{T}}, \cdots, \xi_t^{\mathrm{T}})^{\mathrm{T}}$ the state variable at stage t.

Remark 3.9. *Not to overload the notation, for the remainder of this section, instead of* $\xi = (\xi_2^{\mathrm{T}}, \cdots, \xi_T^{\mathrm{T}})^{\mathrm{T}}$ *and* $\zeta_t = (\xi_2^{\mathrm{T}}, \cdots, \xi_t^{\mathrm{T}})^{\mathrm{T}}$, *we shall write* $\xi = (\xi_2, \cdots, \xi_T)$ *and* $\zeta_t = (\xi_2, \cdots, \xi_t)$, *understanding that* $\xi = (\xi_2, \cdots, \xi_T) \in \mathbb{R}^{r_2+\cdots+r_T}$ *and* $\zeta_t = (\xi_2, \cdots, \xi_t) \in \mathbb{R}^{r_2+\cdots+r_t}$, *as before.* □

Furthermore, the (random) decisions $x_t(\cdot)$ are required to be $\mathscr{F}_t$-measurable, with $\mathscr{F}_t = \sigma(\zeta_t) \subset \mathscr{G}$. Since $\{\mathscr{F}_1, \cdots, \mathscr{F}_T\}$ is a filtration, this implies the nonanticipativity of the feasible policies $\{x_1(\cdot), \cdots, x_T(\cdot)\}$. Finally, Assumption 3.1., page 192, prescribes the square-integrability of $\xi_t(\cdot)$ w.r.t. P for $t = 1, \cdots, T$, and $A_{t\tau}(\cdot), b_t(\cdot), c_t(\cdot)$ are assumed to be linear affine in ζ_t. In addition, we have required the square-integrability of the decisions $x_t(\cdot)$.

Obviously, for ξ having a non-discrete distribution, to solve problem (3.114) means to determine decision functions $x_t(\cdot)$ (instead of decision variables) satisfying infinitely many constraints, which appears to be a very hard task to achieve, in general. The problem becomes more tractable for the case of ξ having a finite discrete distribution, a situation found or assumed in most applications of this model.

3.3.1 MSLP with finite discrete distributions

Let $\xi : \Omega \longrightarrow \mathbb{R}^R$, $R = \sum_{t=2}^{T} r_t$, be a random vector with a finite discrete distribution, having the realizations $\widehat{\xi}^1, \widehat{\xi}^2, \cdots, \widehat{\xi}^S$ with the probabilities $q_1, q_2, \cdots, q_S$, respectively.

Anyone of these realizations is also denoted as a *scenario* $\widehat{\xi}^s = (\widehat{\xi}_2^s, \cdots, \widehat{\xi}_T^s)$ with the probability $\mathbb{P}_\xi\{\xi = \widehat{\xi}^s\} = q_s$, $s \in \mathscr{S} := \{1, \cdots, S\}$. Then the time discrete stochastic process $\{\zeta_t;\ t = 2, \cdots, T\}$ with discretely distributed state variables ζ_t may be assigned to a *scenario tree* as follows:

- The (deterministic) state of the system at stage 1 is assigned to node 1, the unique root of the tree.
- Among all scenarios $\widehat{\xi}^s$, $s = 1, \cdots, S$, there are a finite number k_2 having pairwise different realizations $\widehat{\zeta}_2^s$ of the stage 2 state variables, denoted as $\widehat{\zeta}_2^{\rho(n)} = \widehat{\xi}_2^{\rho(n)}$, $n = 2, \cdots, 1 + k_2$, and assigned to the nodes numbered as $n = 2, \cdots, 1 + k_2 =: K_2$. Here $\rho(n)$ refers to the first of the scenarios $\widehat{\xi}^s$, $s = 1, \cdots, S$, passing through the particular state $\widehat{\zeta}_2^s$. Node 1 is connected by an arc to each of the k_2 nodes in stage 2 due to the fact, that the corresponding states in stage 2 are realized by at least one scenario.
- Having assigned, according to all scenarios, up and until stage $t < T$ the nodes and arcs to all states and implied transitions between consecutive states (i.e. given a scenario $\widehat{\xi}^s = (\widehat{\xi}_2^s, \cdots, \widehat{\xi}_{t-1}^s, \widehat{\xi}_t^s, \cdots, \widehat{\xi}_T^s)$, implies a transition from state $\widehat{\zeta}_{t-1}^s = (\widehat{\xi}_2^s, \cdots, \widehat{\xi}_{t-1}^s)$ to $\widehat{\zeta}_t^s = (\widehat{\xi}_2^s, \cdots, \widehat{\xi}_t^s)$ at least once), we consider for each scenario $\widehat{\xi}^s$ the state $\widehat{\zeta}_{t+1}^s = (\widehat{\xi}_2^s, \cdots, \widehat{\xi}_{t+1}^s)$. Again, in stage $t + 1$ there is a finite number k_{t+1} of different states denoted as $\widehat{\zeta}_{t+1}^{\rho(n)}$, $n = K_t + 1, \cdots, K_t + k_{t+1} =: K_{t+1}$, and assigned to the nodes $K_t + 1, \cdots, K_t + k_{t+1} =: K_{t+1}$ (with $\rho(n)$ referring again to the first scenario passing through this particular state). Finally, we insert the arcs from stage t to stage $t + 1$ according to the implied transitions.

With this scenario tree, representing graphically the possible developments of the stochastic process $\{\xi_2, \cdots, \xi_T\}$ over time, we may combine probabilistic information to get a complete description of the process (see Fig. 3.6).

To this end, we may identify the leaf nodes of the tree (the stage T nodes) $K_{T-1} + 1, \cdots, K_T$ with the scenarios $\widehat{\xi}^s$, $s = 1, \cdots, S$, and assign to these nodes the probabilities q_s of the respective scenario. Hence we have first the probabilities to reach the leaf nodes $n = K_{T-1} + 1, \cdots, K_T$ as $p_n = q_{n-K_{T-1}}$.

For all other nodes, i.e. for $n \leq K_{T-1}$, we then compute the probabilities p_n to pass through these nodes: Given node n, by the above construction of the scenario tree we know the stage t_n of this node as well as its corresponding state $\widehat{\zeta}_{t_n}^{\rho(n)}$; then with $\mathscr{S}(n) = \{s \mid \widehat{\zeta}_{t_n}^s = \widehat{\zeta}_{t_n}^{\rho(n)}\}$ we have $\{\widehat{\xi}^s \mid s \in \mathscr{S}(n)\}$, the set of scenarios passing through this state, called the *scenario bundle* of node n, and we get p_n, the total probability of this scenario bundle, as $p_n = \sum_{s \in \mathscr{S}(n)} q_s$.

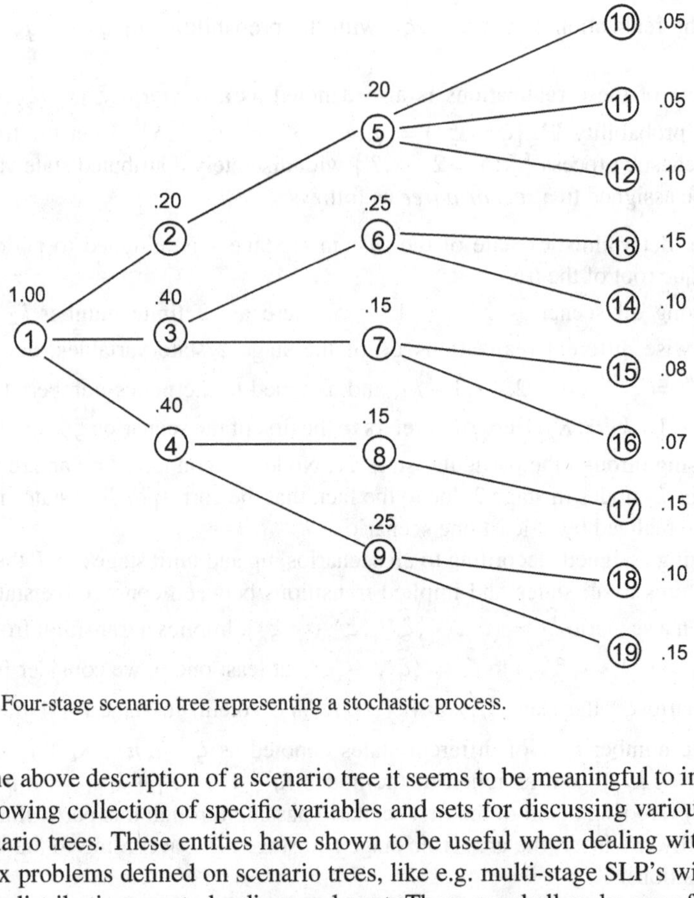

Fig. 3.6 Four-stage scenario tree representing a stochastic process.

After the above description of a scenario tree it seems to be meaningful to introduce the following collection of specific variables and sets for discussing various issues on scenario trees. These entities have shown to be useful when dealing with rather complex problems defined on scenario trees, like e.g. multi-stage SLP's with finite discrete distributions, as to be discussed next. There we shall make use of the following

<u>Notation for scenario trees:</u>

$(\mathscr{N}, \mathscr{A})$: rooted tree with nodes $\mathscr{N} \subset \mathbb{N}$ ($n = 1$ the unique root), and $\mathscr{A}$ the set of arcs.
The nodes $n \in \mathscr{N}$ are assigned to stages $t = 1, \cdots, T$, with $n = 1$ in stage $t = 1$, and with $k_t > 0$ nodes for $t = 2, \cdots, T$, and $|\mathscr{N}| = 1 + \sum_{t=2}^{T} k_t$.
The arcs in $\mathscr{A}$ connect selected nodes of stage t and stage $t+1$, $t = 1, \cdots, T-1$, such that each node in

some stage $t < T$ has at least one immediate successor, and each node in some stage $t > 1$ has exactly one immediate predecessor.
Any path $n_1, \cdots, n_T$, with $n_1 = 1$, $t_{n_t} = t$ for $t \geq 2$, and $(n_t, n_{t+1}) \in \mathscr{A}$ for $t = 1, \cdots, T-1$, corresponds one-to-one to the scenario $\widehat{\xi}^s$, $s \in \mathscr{S} = \{1, \cdots, S\}$, identified with the leaf node n_T.

$q_s, s \in \mathscr{S}$: $q_s = \mathbb{P}_\xi\{\xi = \widehat{\xi}^s\}$, the probability of scenario $\widehat{\xi}^s$, and hence the probability to reach the leaf node identified with this scenario;

t_n : the stage of node $n \in \mathscr{N}$;

$\rho(n)$: the smallest $s \in \mathscr{S}$ such that scenario $\widehat{\xi}^{\rho(n)}$ passes through the state $\widehat{\zeta}^s_{t_n}$ assigned to node n;

$\widehat{\zeta}^n$: $\widehat{\zeta}^n := \widehat{\zeta}^{\rho(n)}_{t_n}$, the state in stage t_n uniquely assigned to n;

$\mathscr{D}(t) \subset \mathscr{N}$: the set of nodes in stage t with $|\mathscr{D}(t)| = k_t$;

h_n : parent node (immediate predecessor) of $n \in \mathscr{N}$, $n \geq 2$;

$\mathscr{H}(n) \subset \mathscr{N}$: set of nodes in the unique path from $n \in \mathscr{N}$ through the successive predecessors back to the root, ordered by stages, the *history* of n (including n);

$\mathscr{S}(n)$: $\mathscr{S}(n) = \{s \mid \widehat{\zeta}^s_{t_n} = \widehat{\zeta}^{\rho(n)}_{t_n}\}$, the index set identifying the scenario bundle of node n;

p_n : $p_n = \sum\limits_{s \in \mathscr{S}(n)} q_s$, the probability to pass node n;

$\mathscr{C}(n) \subset \mathscr{N}$: the set of children (immediate successors) of node n;

$\mathscr{G}_s(n) \subseteq \mathscr{N}$: the future of node n along scenario $\widehat{\xi}^s : s \in \mathscr{S}(n)$, including node n, i.e. the nodes $n_{t_n} = n, \cdots, n_T$ provided the path $\{n_1, \cdots, n_{t_n}, \cdots, n_T\}$ corresponds to scenario $\widehat{\xi}^s$ (hence $\mathscr{G}_s(n) = \emptyset$ if $s \notin \mathscr{S}(n)$);

$\mathscr{G}(n) \subseteq \mathscr{N}$: the future of $n \in \mathscr{N}$, $\mathscr{G}(n) = \bigcup\limits_{s \in \mathscr{S}(n)} \mathscr{G}_s(n)$;

$q_{n \to m}$: $q_{n \to m} = \dfrac{p_m}{p_n}$ $\forall m \in \mathscr{G}(n)$, the conditional probability to reach node m given node n (provided that $p_n > 0$).

To keep the following problem formulations simple, we introduce

Assumption 3.4. *For any MSLP with a finite discrete distribution of the scenarios* ξ *holds*

$$q_s = \mathbb{P}_\xi\{\xi = \widehat{\xi}^s\} > 0 \;\; \forall s \in \mathscr{S}. \tag{3.115}$$

By construction the following facts are obvious:

- Through each node passes at least one scenario, i.e. $\mathscr{S}(n) \neq \emptyset \; \forall n \in \mathscr{N}$;
- given any stage t, each scenario passes through exactly one node in stage t, i.e. $\bigcup_{n\in\mathscr{D}(t)} \mathscr{S}(n) = \mathscr{S}$ and $\mathscr{S}(n) \cap \mathscr{S}(m) = \emptyset \; \forall n,m \in \mathscr{D}(t) : n \neq m$.

Hence, it follows in general that

$$\sum_{n\in\mathscr{D}(t)} p_n = 1,\; t = 1,\cdots,T, \tag{3.116}$$

and due to Assumption 3.4. holds

$$p_n = \sum_{s\in\mathscr{S}(n)} q_s > 0 \;\; \forall n \in \mathscr{N}. \tag{3.117}$$

For the general MSLP (3.114), the decisions $x_t(\zeta_t)$ in stage t are required to be $\mathscr{F}_t$-measurable with $\mathscr{F}_t = \sigma(\zeta_t) \subset \mathscr{G}$. For ξ having a finite discrete distribution, $\sigma(\zeta_t)$ is generated by the k_t atoms $\zeta_t^{-1}[\widehat{\zeta}_{t_n}^{\rho(n)}]$, $n = K_{t-1}+1,\cdots,K_t$. Then $x_t(\cdot)$ has to be constant on each of these atoms or equivalently, to each node n we have to determine the decision vector $x_n := x_{t_n}(\widehat{\zeta}^n)$. Observing that the expected values $\mathbb{E}\,[c_t^{\mathrm{T}}(\zeta_t)x_t(\zeta_t)]$ may now be written as $\sum_{n=K_{t-1}+1}^{K_t} p_n c_{t_n}^{\mathrm{T}}(\widehat{\zeta}^n)x_n$, problem (3.114) for a discrete distribution reads as

$$\left.\begin{aligned} \min \sum_{m\in\mathscr{N}} p_m c_{t_m}^{\mathrm{T}}(\widehat{\zeta}^m)x_m & \\ \sum_{m\in\mathscr{H}(n)} A_{t_n t_m}(\widehat{\zeta}^n)x_m &= b_{t_n}(\widehat{\zeta}^n) \;\forall n \in \mathscr{N} \\ x_m &\geq 0 \qquad\;\; \forall m \in \mathscr{N} \end{aligned}\right\} \tag{3.118}$$

with $p_1 = 1$ and $c_{t_1}^{\mathrm{T}}(\widehat{\zeta}^1) = c_1$, $A_{t_1t_1}(\widehat{\zeta}^1) = A_{11}$, $b_{t_1}(\widehat{\zeta}^1) = b_1$ being constant.
With an obvious simplification of the notation problem (3.118) may be rewritten equivalently as

$$\left.\begin{aligned} \min \sum_{m\in\mathscr{N}} p_m c_{t_m}^{\mathrm{T}}(m)x_m & \\ \sum_{m\in\mathscr{H}(n)} A_{t_n t_m}(n)x_m &= b_n \;\forall n \in \mathscr{N} \\ x_m &\geq 0 \;\;\forall m \in \mathscr{N}. \end{aligned}\right\} \tag{3.119}$$

As the dual LP of (3.119) we have

$$\left.\begin{aligned} \max \sum_{n\in\mathscr{N}} b_n^{\mathrm{T}} u_n & \\ \sum_{n\in\mathscr{G}(m)} A_{t_n t_m}^{\mathrm{T}}(n)u_n &\leq p_m c_{t_m}(m) \;\forall m \in \mathscr{N}. \end{aligned}\right\} \tag{3.120}$$

Remark 3.10. *If in particular, $\forall n \in \mathcal{N} \setminus \{1\}$ and for each node $m \in \mathcal{H}(n) : t_m < t_n - 1$, we have that $A_{t_n t_m}(n) = A_{t_n t_m}(\widehat{\zeta}^n) = 0$, then with $W_1 := A_{11}$ and*

$$T_n := A_{t_n t_n - 1}(n) \text{ and } W_n := A_{t_n t_n}(n) \quad \forall n \in \mathcal{N} \setminus \{1\}$$

problem (3.119) reads as

$$\left.\begin{array}{rl} \min \sum\limits_{m \in \mathcal{N}} p_m c_{t_m}^{\mathrm{T}}(m) x_m & \\ W_1 x_1 & = b_1 \\ T_n x_{h_n} + W_n x_n & = b_n \ \forall n \in \mathcal{N} \setminus \{1\} \\ x_n & \geq 0 \ \forall n \in \mathcal{N}. \end{array}\right\} \tag{3.121}$$

Hence we have the same problem structure as assumed when discussing the nested decomposition in section 1.2.7 of Chapter 1, in particular the structure of problem (1.29) on page 33.

The general MSLP problem (3.114) can always be transformed to an equivalent problem where $A_{t\tau} = 0$ holds for $\tau < t - 1$, thus assuming the following staircase form

$$\left.\begin{array}{lll} \min\{c_1^{\mathrm{T}} x_1 + \mathbb{E} \sum\limits_{t=2}^{T} c_t^{\mathrm{T}}(\zeta_t) z_t(\zeta_t)\} & & \\ W_1 z_1 & = b_1 & \\ T_t(\zeta_t) z_{t-1}(\zeta_{t-1}) \ + W_t(\zeta_t) z_t(\zeta_t) & = b_t(\zeta_t) & \text{a.s., } t = 2, \cdots, T, \\ x_1 \geq 0, \ x_t(\zeta_t) & \geq 0 & \text{a.s., } t = 2, \cdots, T, \end{array}\right\} \tag{3.122}$$

formally corresponding to (3.121), where now z_t is an $n_1 + \ldots + n_t$–dimensional variable and T_t and W_t have $m_t + n_1 + \ldots + n_t$ rows. For specifying the transformation which maps (3.114) into (3.122) we will employ double indices. The transformation is as follows. Let

$$z_t^{\mathrm{T}}(\zeta_t) = (z_{t,1}(\zeta_t), \cdots, z_{t,t-1}(\zeta_t), z_{t,t}(\zeta_t))$$

with $z_{t,\tau}$ being an n_τ–dimensional variable, $\tau = 1, \ldots, t$, and with z_{tt} corresponding to x_t in (3.114). The matrices are defined as follows. Let $W_1 = A_{1,1}$. For $1 < t < T$ we define

$$T_t(\zeta_t) = \begin{pmatrix} A_{t,1}(\zeta_t) & \ldots & A_{t,t-1}(\zeta_t) \\ I & & \\ & \ddots & \\ & & I \end{pmatrix}$$

and

$$W_t(\zeta_t) = \begin{pmatrix} 0 & \dots & 0 & A_{t,t}(\zeta_t) \\ -I & & & 0 \\ & \ddots & & \vdots \\ & & -I & 0 \end{pmatrix}$$

and for $t = T$ let

$$T_T(\zeta_T) = (A_{T,1}(\zeta_T), \dots A_{T,T-1}(\zeta_T)) \;\; \text{and} \;\; W_T(\zeta_T) = A_{TT}(\zeta_T).$$

Loosely speaking, the auxiliary variables $(z_{t,1}, \dots, z_{t,t-1})$ serve for "forwarding" the solution to later stages. As an example let us consider an MSLP with $T = 4$ and let us drop in the notation the dependency on ζ_t. The original structure is

$$\left.\begin{array}{lllll} A_{11}x_1 & & & & = b_1 \\ A_{21}x_1 & +A_{22}x_2 & & & = b_2 \\ A_{31}x_1 & +A_{32}x_2 & +A_{33}x_3 & & = b_3 \\ A_{41}x_1 & +A_{42}x_2 & +A_{43}x_3 & +A_{44}x_4 & = b_4 \end{array}\right\}$$

which transforms into

$A_{11}z_{11}$							$=$	b_1
$A_{21}z_{11}$		$+A_{22}z_{22}$					$=$	b_2
z_{11}	$-z_{21}$						$=$	0
	$A_{31}z_{21}$	$+A_{32}z_{22}$			$+A_{33}z_{33}$		$=$	b_3
	z_{21}		$-z_{31}$				$=$	0
		z_{22}		$-z_{32}$			$=$	0
			$A_{41}z_{31}$	$+A_{42}z_{32}$	$+A_{43}z_{33}$	$+A_{44}z_{44}$	$=$	b_4

In the literature, multi-stage SLP's are often presented just in the so-called staircase formulation (3.121). Although problems of this form, at the first glance, look simpler than problems in the lower block triangular formulation like (3.119), this does not imply a computational advantage in general. Indeed, if the staircase formulation results from the above transformation of (3.114) into (3.122), then the numbers of variables and of constraints are increased. □

3.3.2 MSLP with non-discrete distributions

In Section 3.2.1 we have discussed two-stage SLP's with complete fixed recourse and with bounded distributions, i.e. with $\text{supp}\,\mathbb{P}_\xi \subseteq \Xi = \prod_{i=1}^r [\alpha_i, \beta_i]$. In particular, we considered the recourse function $Q(x; T(\xi), h(\xi))$, which according to our notation (see page 197) implies for the second stage problem (3.9) that only $T(\cdot)$ and $h(\cdot)$ (or some elements of these arrays) are random. In this case, we could apply Jensen's inequality to get in Theorem 3.4. a lower bound for the expected recourse $\mathscr{Q}(x) = \int_{\Xi} Q(x;T(\xi),h(\xi))\mathbb{P}_{\xi}(d\xi)$ as $Q(x;T(\bar{\xi}),h(\bar{\xi})) \le \mathscr{Q}(x)$, where $\bar{\xi} := \mathbb{E}_{\xi}[\xi]$. In other words, introducing the Jensen distribution $\mathbb{P}_{\eta}$ as the one-point distribution with $\mathbb{P}_{\eta}\{\eta = \mathbb{E}_{\xi}[\xi]\} = 1$, the Jensen inequality can formally be written as

$$\int_{\Xi} Q(x;T(\eta),h(\eta))\mathbb{P}_{\eta}(d\eta) \le \mathscr{Q}(x).$$

On the other hand, we have derived particular discrete probability distributions $\mathbb{Q}_{\eta}$ on the vertices v^{ν} of Ξ, the E–M distribution for stochastically independent components of ξ in Lemma 3.6 and the generalized E–M distribution for stochastically dependent components of ξ in Lemma 3.7, respectively, which were shown to solve two special types of moment problems. According to Theorems 3.5. and 3.6., using these distributions the E–M inequality provides an upper bound for the expected recourse as

$$\begin{aligned}\mathscr{Q}(x) &\le \int_{\Xi} Q(x;T(\eta),h(\eta))\,\mathbb{Q}_{\eta}(d\eta)\\ &= \sum_{\nu=1}^{2^r} Q(x;T(v^{\nu}),h(v^{\nu}))\,\mathbb{Q}_{\eta}(v^{\nu}).\end{aligned}$$

For any disjoint interval partition $\mathscr{X} = \{\Xi_k;\ k = 1,\cdots,K\}$ of Ξ, we apply Jensen's inequality for the conditional expectations, meaning to introduce on the set of conditional expectations $\{\bar{\xi}_k := \mathbb{E}_{\xi}[\xi \mid \xi \in \Xi_k] \mid k = 1,\cdots,K\}$ the corresponding discrete distribution $\mathbb{P}_{\eta_{\mathscr{X}}}$, defined by $\mathbb{P}_{\eta_{\mathscr{X}}}\{\bar{\xi}_k\} = \mathbb{P}_{\xi}\{\Xi_k\}$, and to compute $\int_{\Xi} Q(x;T(\eta),h(\eta))\mathbb{P}_{\eta_{\mathscr{X}}}(d\eta)$ to get a lower bound for $\mathscr{Q}(x)$. Similarly, we apply the E–M inequality using the distribution $\mathbb{Q}_{\eta_{\mathscr{X}}} = \sum_{k=1}^{K} \mathbb{P}_{\xi}\{\Xi_k\} \cdot \mathbb{Q}_{\eta_{\Xi_k}}$, where $\mathbb{Q}_{\eta_{\Xi_k}}$ is either the E–M distribution or else the generalized E–M distribution solving the corresponding conditional moment problems, conditioned with respect to the cell $\Xi_k \in \mathscr{X}$. This way, according to Lemma 3.8 we get an increased lower bound as well as a decreased upper bound.

For any sequence of appropriately refined interval partitions $\{\mathscr{X}^{\nu}\}$ the corresponding sequences of discrete distributions $\{\mathbb{P}_{\eta_{\mathscr{X}^{\nu}}}\}$ and $\{\mathbb{Q}_{\eta_{\mathscr{X}^{\nu}}}\}$ of Jensen distributions and E–M distributions, respectively, are shown in Lemma 3.9 to converge weakly to the original distribution $\mathbb{P}_{\xi}$. For the corresponding sequences $\{\widetilde{\mathscr{Q}}^{\nu}\}$ and $\{\widehat{\mathscr{Q}}^{\nu}\}$ of Jensen lower bounds and E–M upper bounds, respectively, of the expected recourse function $\mathscr{Q}$, this implies epi-convergence of both sequences to $\mathscr{Q}$. This convergence behaviour, however, provides due to Theorem 3.7. promising conditions to design approximation schemes for the solution of two-stage SLP's with complete fixed recourse.

The question arises whether we may expect a similar approach to be applicable for the solution of multi-stage SLP's with more than two stages. To get a first impression let us take a look at a rather simple three-stage example.

Example 3.3. *Consider the complete fixed recourse problem*

$$\begin{array}{rl} V^\star := \min\{2x + \mathbb{E}[y_1(\xi_2) + 2y_2(\xi_2)] + \mathbb{E}[z_1(\xi_2,\xi_3) + z_2(\xi_2,\xi_3)]\} & \\ \text{s.t. } x + y_1(\xi_2) - y_2(\xi_2) \qquad\qquad\qquad\qquad\qquad = & \xi_2 \\ x + y_1(\xi_2) - y_2(\xi_2) + z_1(\xi_2,\xi_3) - z_2(\xi_2,\xi_3) = & \xi_3 \\ x, y_1, y_2, z_1, z_2 \geq & 0 \end{array}$$

with $\zeta := (\xi_2, \xi_3)^{\mathrm{T}}$ *having the (joint) probability distribution* $\mathbb{P}_\zeta$ *on* $\operatorname{supp}\mathbb{P}_\zeta := \Xi = [0,1] \times [0,1]$, *given by the density*

$$f(\xi_2,\xi_3) = \begin{cases} 1+\varepsilon \; \textit{for} & 0 \leq \xi_2, \xi_3 \leq 0.5 \\ 1+\varepsilon \; \textit{for} & 0.5 \leq \xi_2, \xi_3 \leq 1 \\ 1-\varepsilon \; \textit{for} & 0 \leq \xi_2 < 0.5 < \xi_3 \leq 1 \\ 1-\varepsilon \; \textit{for} & 0 \leq \xi_3 < 0.5 < \xi_2 \leq 1 \\ 0 & \textit{else}, \end{cases}$$

where ε *is some constant such that* $\varepsilon \in (-1, +1)$.

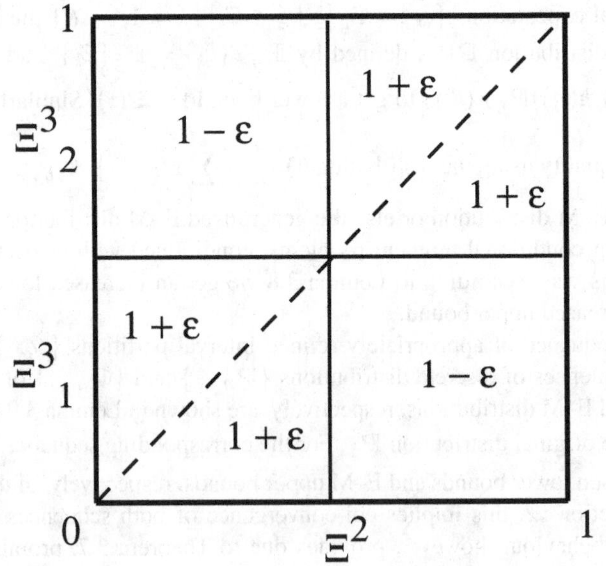

Fig. 3.7 $\operatorname{supp}\mathbb{P}_\zeta = \Xi^2 \times \Xi^3 = \Xi^2 \times (\Xi^3_1 \cup \Xi^3_2)$ with density $f(\xi_2,\xi_3)$.

For the marginal distribution of ξ_2 *we obviously get the marginal density as*

$$f_2(\xi_2) = \int_0^1 f(\xi_2,\xi_3)d\xi_3 = \begin{cases} 1 \; for \; \xi_2 \in [0,1] \\ 0 \; else, \end{cases}$$

such that the corresponding distribution $\mathbb{P}_{\xi_2}$ *is* $\mathscr{U}[0,1]$, *the uniform distribution on the interval* $[0,1]$. *According to the definition of* $f(\xi_2,\xi_3)$, *for* ξ_3 *follows the same marginal distribution.*

Considering, for instance, the interval $\widetilde{\Xi} := \{[0,0.5]\times[0,0.5]\} \subset \mathbb{R}^2$, *we get*

$$\mathbb{P}_\zeta(\widetilde{\Xi}) = \int_{\widetilde{\Xi}} f(\zeta)d\zeta = \frac{1}{4}(1+\varepsilon),$$

whereas for the marginal distributions in $\mathscr{U}[0,1]$ *follows*

$$\mathbb{P}_{\xi_2}([0,0.5])\cdot\mathbb{P}_{\xi_3}([0,0.5]) = \frac{1}{4}.$$

Hence, for $\varepsilon \neq 0$ *the random variables* ξ_2 *and* ξ_3 *are dependent.*

Due to the objective of our recourse problem, for any given first stage solution $x \geq 0$ *the second stage solution* $y_i(\xi_2), i=1,2$, *minimizing the second stage objective* $y_1(\xi_2)+2y_2(\xi_2)$, *has to satisfy the rules*

a) $\xi_2 < x \Longrightarrow y_1(\xi_2) = 0,\; y_2(\xi_2) = x-\xi_2$
b) $\xi_2 \geq x \Longrightarrow y_1(\xi_2) = \xi_2 - x,\; y_2(\xi_2) = 0.$

Minimizing the third stage objective $z_1(\xi_2,\xi_3)+z_2(\xi_2,\xi_3)$ *then yields, for both of the cases* a) *and* b) *above,*

$$\begin{aligned} x+y_1(\xi_2)-y_2(\xi_2) \leq \xi_3 &\Longrightarrow z_1(\xi_2,\xi_3) = \xi_3-\xi_2,\; z_2(\xi_2,\xi_3) = 0 \\ x+y_1(\xi_2)-y_2(\xi_2) > \xi_3 &\Longrightarrow z_1(\xi_2,\xi_3) = 0,\; z_2(\xi_2,\xi_3) = \xi_2-\xi_3. \end{aligned}$$

Observe that a first stage decision $x<0$ *is not feasible. On the other hand,* $x>1$ *cannot be optimal, since this would increase unnecessarily the overall objective, more precisely the first stage cost* $2x$ *plus the expected second stage cost* $\mathbb{E}[y_1(\xi_2)+2y_2(\xi_2)]$ *due to* a) *by at least* $2(x-1)+2\mathbb{E}[(x-\xi_2)] > 4(x-1)$. *Hence we compute the objective value, for* $0 \leq x \leq 1$, *as*

$$\begin{aligned} V(x) &= 2x + \int_{\xi_2=0}^{x} 2(x-\xi_2)d\xi_2 + \int_{\xi_2=x}^{1}(\xi_2-x)d\xi_2 + \\ &\quad + \int_{\Xi} |\xi_3-\xi_2| f(\xi_2,\xi_3)d\xi_2 d\xi_3 \\ &= 2x + \frac{3}{2}x^2 - x + \frac{1}{2} + \int_{\Xi} |\xi_3-\xi_2| f(\xi_2,\xi_3)d\xi_2 d\xi_3. \end{aligned}$$

For the last integral we get

$$\int_{\Xi} |\xi_3 - \xi_2| f(\xi_2, \xi_3) d\xi_2 d\xi_3 = \underbrace{\int_{\xi_2=0}^{1} \int_{\xi_3=\xi_2}^{1} (\xi_3 - \xi_2) f(\xi_2, \xi_3) d\xi_2 d\xi_3}_{A}$$
$$+ \underbrace{\int_{\xi_3=0}^{1} \int_{\xi_2=\xi_3}^{1} (\xi_2 - \xi_3) f(\xi_2, \xi_3) d\xi_2 d\xi_3}_{B},$$

where $A = B$ for symmetry reasons (see Fig. 3.7). For A, the integral taken over the triangle above the line $\xi_3 = \xi_2$ in Fig. 3.7, we get by integration of $(\xi_3 - \xi_2) f(\xi_2, \xi_3)$

$$A = \frac{1}{2}(1+\varepsilon)\frac{1}{24} + \frac{1}{2}(1-\varepsilon)\frac{1}{4} + \frac{1}{2}(1+\varepsilon)\frac{1}{24} = \frac{2-\varepsilon}{12}$$

such that $A + B = \dfrac{2-\varepsilon}{6}$ *and hence*

$$V(x) = \frac{3}{2}x^2 + x + \frac{1}{2} + \frac{2-\varepsilon}{6} = \frac{3}{2}x^2 + x + \frac{5-\varepsilon}{6}.$$

Obviously, $\min_{x \geq 0} V(x)$ *is achieved at* $\hat{x} = 0$ *such that the optimal value of our problem turns out to be*

$$V^{\star} = \min_{x \geq 0} V(x) = \frac{5-\varepsilon}{6}.$$

Let us now discretize the distributions of ξ_2 in stage two and $\zeta = (\xi_2, \xi_3)^{\mathrm{T}}$ in stage three by choosing the partitions $\mathscr{X}^2$ of Ξ^2 and $\mathscr{X}^3$ of $\Xi^2 \times \Xi^3$, respectively, as follows:

<u>Stage 2:</u> $\mathscr{X}^2 = \{\Xi^2\}$ *yielding for ξ_2 the realization*

$$\bar{\xi}_2 = \mathbb{E}_{\xi_2}[\xi_2] = \frac{1}{2} \quad \text{with } p_2 = \mathbb{P}(\{\xi_2 \in [0,1]\}) = 1;$$

<u>Stage 3:</u> $\mathscr{X}^3 = \left\{ \Xi^2 \times \left[0, \frac{1}{2}\right), \Xi^2 \times \left[\frac{1}{2}, 1\right] \right\}$ *yielding for ξ_3 the realizations*

$$\bar{\xi}_{31} = \mathbb{E}\left[\xi_3 \mid \xi_2 \in [0,1], \xi_3 \in \left[0, \frac{1}{2}\right)\right] = \mathbb{E}\left[\xi_3 \mid \xi_3 \in \left[0, \frac{1}{2}\right)\right] = \frac{1}{4}$$
$$\bar{\xi}_{32} = \mathbb{E}\left[\xi_3 \mid \xi_2 \in [0,1], \xi_3 \in \left[\frac{1}{2}, 1\right]\right] = \mathbb{E}\left[\xi_3 \mid \xi_3 \in \left[\frac{1}{2}, 1\right]\right] = \frac{3}{4}$$

with

$$p_{31} = \mathbb{P}\left(\left\{\xi_3 \in \left[0, \frac{1}{2}\right)\right\}\right) = \frac{1}{2} \quad \text{and}$$
$$p_{32} = \mathbb{P}\left(\left\{\xi_3 \in \left[\frac{1}{2}, 1\right]\right\}\right) = \frac{1}{2}.$$

Then the discretized problem reads as

$$\begin{aligned}
\overline{V} := \min \Big\{ 2x + y_1 + 2y_2 + \frac{1}{2}(z_1^1 + z_2^1) + \frac{1}{2}(z_1^2 + z_2^2) \Big\} \\
s.t.\ x + y_1 - y_2 \qquad\qquad\qquad &= \frac{1}{2} \\
x + y_1 - y_2 + z_1^1 - z_2^1 \qquad\qquad &= \frac{1}{4} \\
x + y_1 - y_2 \qquad\quad + z_1^2 - z_2^2 &= \frac{3}{4} \\
x, y_1, y_2, z_1^1, z_2^1, z_1^2, z_2^2 &\geq 0 .
\end{aligned}$$

Also in this case the optimum is achieved for $\tilde{x} = 0$ *with* $\overline{V} = \frac{3}{4}$. *Comparing this value with the optimum* $V^{\star} = \frac{5 - \varepsilon}{6}$ *of the original problem, we see that*

$$\overline{V} \begin{cases} \leq V^{\star} \text{ if } \varepsilon \leq \frac{1}{2} \\ > V^{\star} \text{ if } \varepsilon > \frac{1}{2} . \end{cases}$$

In conclusion, even for a rather simple situation like three stages, randomness in the right–hand–sides only, and complete fixed recourse, we cannot expect in general to get a lower bound of the optimum by discretization of the distributions in an analogous manner as in the two-stage case. □

This example as well as the following considerations are essentially based on discussions related to an idea, originally due to S. Sen, concerning refinements of discretizations in order to improve discrete approximations for MSLP problems. The outcome of these endeavours was reported in Fúsek–Kall–Mayer–Sen–Siegrist [108].

Obviously, with appropriate successive refinements of partitions $\mathscr{X}_\nu^t$ of the sets $[\Xi^2 \times \cdots \times \Xi^t] \supseteq \operatorname{supp} \mathbb{P}_{\zeta_t}$, $t = 2, \cdots, t$; $\nu = 1, 2, \cdots$, we may expect weak convergence of the associated discrete distributions $\{\mathbb{P}_{\eta_t \mathscr{X}_\nu^t}\}$ and hence epi-convergence of the related objective functions of the general MSLP (3.114), as shown by Pennanen [251, 252]. Thus Th. 3.7. (page 222) suggests that a solution could be approximated by this kind of successive discretization of the distributions. However it seems difficult to control this procedure since, in difference to the two-stage case, for the general MSLP we do not have error bounds on the optimal value. According to Ex. 3.3., even for the much simpler problem

$$\left.\begin{array}{lrl} \min\{c_1^{\mathrm{T}}x_1 + \mathbb{E}\sum_{t=2}^{T} c_t^{\mathrm{T}}x_t(\zeta_t)\} & & \\ A_{11}x_1 & = & b_1 \\ A_{t1}x_1 \ + \sum_{\tau=2}^{t} A_{t\tau}x_\tau(\zeta_\tau) & = & b_t(\zeta_t) \text{ a.s.}, t = 2,\cdots,T, \\ x_1 \geq 0,\ x_t(\zeta_t) & \geq & 0 \qquad \text{a.s.}, t = 2,\cdots,T, \end{array}\right\} \tag{3.123}$$

with complete fixed recourse and only the right–hand–sides being random, we cannot expect to get at least lower bounds, in general.

Nevertheless, we shall discuss first, for the purpose of defining a fully aggregated problem instead of the MSLP (3.114), how an arbitrary finite subfiltration $\widehat{\mathscr{F}}$ and the corresponding scenario tree can be generated. Again, we assume the supports of the stagewise distributions to be bounded. Hence there exist intervals $\Xi^t \subset \mathbb{R}^{r_t}$ such that $\operatorname{supp} \mathbb{P}_{\xi_t} \subseteq \Xi^t$, $t = 2,\cdots,T$. Then we proceed as follows:

Subfiltration and the corresponding scenario tree

- With $\Omega^{(1)} := \Omega$ and $\widehat{\mathscr{F}_1} := \{\Omega, \emptyset\}$ define $\mathscr{N}_1 := \{1\}$.
- For the stages $\nu = 1,\cdots,T-1$ repeat:

  Let $\mathscr{N}_{\nu+1} := \emptyset$.
  Then for each node n in stage ν (i.e. $t_n = \nu$) and some $r_n \geq 1$:
  Define a finite set C_n of children of n such that $|C_n| = r_n$ and, for any m with $m \neq n$, $t_m = t_n = \nu$, that $C_m \cap C_n = \emptyset$ as well as $C_n \cap \mathscr{N}_\mu = \emptyset\ \forall \mu \leq \nu$ holds.
  Furthermore, let $\mathscr{N}_{\nu+1} := \mathscr{N}_{\nu+1} \cup C_n$ and associate individually with the set $C_n := \{k_1^{(n)},\cdots,k_{r_n}^{(n)}\}$ a partition of $\Xi^{\nu+1}$ into subintervals as

$$\Xi^{\nu+1} = \bigcup_{l=1}^{r_n} \Xi^{\nu+1}_{k_l^{(n)}}\,. \tag{3.124}$$

- To generate the subfiltration, for $t = 2,\cdots,T$ repeat:
  For each $n \in \mathscr{N}_t$ and $h_n \in \mathscr{N}_{t-1}$, its unique parent node, and Ξ^t_n the subinterval corresponding to node n in the partition of Ξ^t associated with C_{h_n}, let $\Omega^{(n)} := \Omega^{(h_n)} \cap \xi_t^{-1}[\Xi^t_n]$.
  Define the subfiltration $\widehat{\mathscr{F}}$ by $\widehat{\mathscr{F}_t} := \sigma\{\Omega^{(n)} \mid n \in \mathscr{N}_t\}$, $t = 2,\cdots,T$, with $\sigma\{\Omega^{(n)} \mid n \in \mathscr{N}_t\}$ the σ-algebra generated by the sets $\Omega^{(n)}$, $n \in \mathscr{N}_t$.
- The defining elements of the discretely distributed stochastic process corresponding to the above finite subfiltration, i.e. the realizations $\widehat{\zeta}^n$ at node n and their probabilities p_n, may be assigned to the nodes as follows:
  For any $n \in \mathscr{N} \setminus \{1\}$ let $\mathscr{H}(n) = \{\ell_1 = 1,\cdots,\ell_{t_n-1},\ell_{t_n} = n\}$ be the history of node n. By the above construction, each node $\ell_\nu \in \mathscr{H}(n)$ corresponds uniquely to a particular subinterval $\Xi^\nu_{l_\nu}$ of Ξ^ν. Then for the discrete process we choose the state $\widehat{\zeta}^n$ at node n and the corresponding probability p_n as

$$\left.\begin{aligned} \widehat{\zeta}^n &= \mathbb{E}\left[\zeta_{t_n} \mid \zeta_{t_n} \in \prod_{\nu=2}^{t_n} \Xi_{\ell_\nu}^\nu\right] \\ p_n &= \mathbb{P}_{\zeta_{t_n}}\left(\left\{\zeta_{t_n} \in \prod_{\nu=2}^{t_n} \Xi_{\ell_\nu}^\nu\right\}\right). \end{aligned}\right\} \tag{3.125}$$

Using this discrete process we may then replace the general MSLP (3.114), defined with respect to the filtration $\mathscr{F}$, by the fully aggregated problem with respect to the subfiltration $\widehat{\mathscr{F}}$, as represented by the LP (3.118).

Whereas, according to Ex. 3.3., for problem (3.123) we cannot expect to achieve lower bounds for the optimal value by discretization of the underlying stochastic process in general, the situation will be better if Assumption 3.1. is modified as follows:

Assumption 3.5. *Let*

- *only the right–hand–sides b_t be random (and linear affine in ζ_t);*
- *the distributions of ξ_t be bounded within some intervals $\Xi^t \subset \mathbb{R}^{r_t}$, i.e.* $\operatorname{supp}\mathbb{P}_{\xi_t} \subseteq \Xi^t$*;*
- *the random vectors $\xi_2, \cdots, \xi_T$ be stochastically independent;*
- *the A_{tt} be complete fixed recourse matrices $\forall t$.*

With $\mathscr{H}(n) = \{\ell_1 = 1, \cdots, \ell_{t_n-1}, \ell_{t_n} = n\}$ the history of node n as before, the assumed stochastic independence of $\xi_2, \cdots, \xi_T$ implies the distribution (3.125) to be modified to

$$\left.\begin{aligned} \widehat{\zeta}^n &= \mathbb{E}\left[\zeta_{t_n} \mid \zeta_{t_n} \in \prod_{\nu=2}^{t_n} \Xi_{\ell_\nu}^\nu\right] \\ &= \mathbb{E}\begin{bmatrix} \xi_2 \mid \xi_\nu \in \Xi_{\ell_\nu}^\nu,\ \nu = 2, \cdots, t_n \\ \vdots \quad \vdots \\ \xi_{t_n} \mid \xi_\nu \in \Xi_{\ell_\nu}^\nu,\ \nu = 2, \cdots, t_n \end{bmatrix} \\ &= \begin{pmatrix} \mathbb{E}[\xi_2 \mid \xi_2 \in \Xi_{\ell_2}^2] \\ \vdots \\ \mathbb{E}[\xi_{t_n} \mid \xi_{t_n} \in \Xi_{\ell_{t_n}}^{t_n}] \end{pmatrix} \\ p_n &= \mathbb{P}_{\zeta_{t_n}}\left(\left\{\zeta_{t_n} \in \prod_{\nu=2}^{t_n} \Xi_{\ell_\nu}^\nu\right\}\right) = \prod_{\nu=2}^{t_n} \mathbb{P}_{\xi_\nu}(\Xi_{\ell_\nu}^\nu). \end{aligned}\right\} \tag{3.126}$$

Hence we replace problem (3.123) by the fully aggregated problem

$$\left.\begin{aligned} \min \sum_{m \in \mathscr{N}} p_m c_{t_m}^{\mathrm{T}} x_m & \\ \sum_{m \in \mathscr{H}(n)} A_{t_n t_m} x_m = b_{t_n}(\widehat{\zeta}^n) & \quad \forall n \in \mathscr{N} \\ x_m \geq 0 & \quad \forall m \in \mathscr{N} \end{aligned}\right\} \tag{3.127}$$

using the distribution (3.126). Then we get

Lemma 3.17. *Let problem (3.123) satisfy Assumption 3.5.. Then for any subfiltration $\widehat{\mathscr{F}}$ constructed as above, the optimal value of the aggregated problem (3.127) is a lower bound of the optimum in (3.123).*

Proof: It is well known that problem (3.123) can be formulated as a recursive sequence of optimization problems (see Olsen [247] and Rockafellar–Wets [286]). For this purpose we use the following notation:

$z_t := \{x_1, \cdots, x_t\}$ for the sequence of decision vectors up to stage t;
$\zeta_t := (\xi_2, \cdots, \xi_t)$ for the state variable at stage t, as before;
$\widehat{\zeta}_t$ for any realization of ζ_t;
$\Xi_n^t \subseteq \Xi^t$ for node n in stage t due to (3.124), and $\bar{\xi}_t^n := \mathbb{E}[\xi_t \mid \xi_t \in \Xi_n^t]$.

Now the above mentioned recursion may be formulated as follows:

Let $\Phi_{T+1}(z_T; \widehat{\zeta}_T) \equiv 0 \; \forall z_T, \widehat{\zeta}_T$. Determine iteratively for $t = T, T-1, \cdots, 2$, and for all nodes n in stage $t_n = t$, using the assumed stagewise independence by applying Fubini's theorem (see Halmos [131]),

$$\left.\begin{aligned}
r_t(z_{t-1}; \widehat{\zeta}_t) \quad &:= \min_{x_t} \{c_t^{\mathrm{T}} x_t + \Phi_{t+1}(z_t; \widehat{\zeta}_t)\} \\
&\qquad \text{s.t. } A_{tt} x_t = b_t(\widehat{\zeta}_t) - \sum_{\tau=1}^{t-1} A_{t\tau} x_\tau, \; x_t \geq 0 \\
\Phi_t(z_{t-1}; \widehat{\zeta}_{t-1}) &:= \mathbb{E}[r_t(z_{t-1}; \zeta_t) \mid \zeta_{t-1} = \widehat{\zeta}_{t-1}] \\
&= \mathbb{E}_{\xi_t}[r_t(z_{t-1}; \widehat{\zeta}_{t-1}, \xi_t)] \\
&= \sum_{\nu \in C_{h_n}} \mathbb{P}_{\xi_t}(\Xi_\nu^t) \mathbb{E}_{\xi_t}[r(z_{t-1}; \widehat{\zeta}_{t-1}, \xi_t) \mid \xi_t \in \Xi_\nu^t],
\end{aligned}\right\} \qquad (3.128)$$

which finally yields

$$\begin{aligned}
r_1 &= \min_{x_1} \{c_1^{\mathrm{T}} x_1 + \Phi_2(x_1; \widehat{\zeta}_1)\} \\
&\qquad \text{s.t. } A_{11} x_1 = b_1(\widehat{\zeta}_1) \equiv b_1, \; x_1 \geq 0,
\end{aligned}$$

the optimal value of (3.123), with $\widehat{\zeta}_1$ being the realization of $\xi_1 \equiv \text{const}$ due to the fact that in the first stage there is only one (deterministic) state. The notation "$\mathbb{E}_{\xi_t}$" just indicates that the integral is taken with respect to $\mathbb{P}_{\xi_t}$ only.

If $\Phi_{t+1}(z_t, \widehat{\zeta}_t)$ is jointly convex in $(z_t, \widehat{\zeta}_t)$, as is trivially true for Φ_{T+1}, then it follows immediately, that

$$\begin{aligned}
r_t(z_{t-1}; \widehat{\zeta}_t) &= \min_{x_t} \{c_t^{\mathrm{T}} x_t + \Phi_{t+1}(z_t; \widehat{\zeta}_t)\} \\
&\qquad \text{s.t. } A_{tt} x_t = b_t(\widehat{\zeta}_t) - \sum_{\tau=1}^{t-1} A_{t\tau} x_\tau, \; x_t \geq 0
\end{aligned}$$

is jointly convex in $(z_{t-1};\widehat{\zeta}_t)$ (recall that $b_t(\widehat{\zeta}_t)$ is linear affine in $\widehat{\zeta}_t$). Thus, from (3.128) follows that

$$\Phi_t(z_{t-1};\widehat{\zeta}_{t-1}) = \mathbb{E}_{\xi_t}[r_t(z_{t-1};\widehat{\zeta}_{t-1},\xi_t)]$$

is jointly convex in $(z_{t-1};\widehat{\zeta}_{t-1})$ as well. Hence, by Jensen's inequality holds

$$r_t(z_{t-1};\widehat{\zeta}_{t-1},\mathbb{E}[\xi_t]) \leq \mathbb{E}_{\xi_t}[r_t(z_{t-1};\widehat{\zeta}_{t-1},\xi_t)] = \Phi_t(z_{t-1};\widehat{\zeta}_{t-1}). \tag{3.129}$$

In analogy to (3.128), for the discretized problem (3.127) with $\Psi_{T+1} \equiv 0$ we define for $t = T, T-1, \cdots, 2$, and for all nodes n in stage $t_n = t$, the recursion

$$\left.\begin{aligned} q_t(z_{t-1};\widehat{\zeta}_{t-1},\bar{\xi}_t^n) &:= \min_{x_t}\{c_t^{\mathrm{T}}x_t + \Psi_{t+1}(z_t;\widehat{\zeta}_{t-1},\bar{\xi}_t^n)\} \\ &\text{s.t. } A_{tt}x_t = b_t(\widehat{\zeta}_{t-1},\bar{\xi}_t^n) - \sum_{\tau=1}^{t-1} A_{t\tau}x_\tau,\ x_t \geq 0 \\ \Psi_t(z_{t-1};\widehat{\zeta}_{t-1}) &:= \sum_{\nu\in\mathcal{C}_{h_n}} \mathbb{P}_{\xi_t}(\Xi_\nu^t) q_t(z_{t-1};\widehat{\zeta}_{t-1},\bar{\xi}_t^\nu), \end{aligned}\right\} \tag{3.130}$$

we'll get

$$\begin{aligned} q_1 &:= \min_{x_1}\{c_1^{\mathrm{T}}x_1 + \Psi_2(x_1;\widehat{\zeta}_1)\} \\ &\text{s.t. } A_{11}x_1 = b_1(\widehat{\zeta}_1) \equiv b_1,\ x_1 \geq 0 \end{aligned}$$

as the optimal value of (3.127).

Provided that $\Psi_{t+1}(z_t;\widehat{\zeta}_t) \leq \Phi_{t+1}(z_t;\widehat{\zeta}_t)$, as it is obviously the case for $t = T$, we conclude from (3.128) and (3.130), using Jensen's inequality (3.129) (for conditional expectations), that

$$\begin{aligned} q_t(z_{t-1};\widehat{\zeta}_{t-1},\bar{\xi}_t^n) &\leq r_t(z_{t-1};\widehat{\zeta}_{t-1},\bar{\xi}_t^n) \\ &\leq \mathbb{E}_{\xi_t}[r_t(z_{t-1};\widehat{\zeta}_{t-1},\xi_t) \mid \xi_t \in \Xi_n^t] \end{aligned}$$

and hence

$$\begin{aligned} \Psi_t(z_{t-1};\widehat{\zeta}_{t-1}) &:= \sum_{\nu\in\mathcal{C}_{h_n}} \mathbb{P}_{\xi_t}(\Xi_\nu^t) q_t(z_{t-1};\widehat{\zeta}_{t-1},\bar{\xi}_t^\nu) \\ &\leq \sum_{\nu\in\mathcal{C}_{h_n}} \mathbb{P}_{\xi_t}(\Xi_\nu^t) \mathbb{E}_{\xi_t}[r_t(z_{t-1};\widehat{\zeta}_{t-1},\xi_t) \mid \xi_t \in \Xi_\nu^t] \\ &= \Phi_t(z_{t-1};\widehat{\zeta}_{t-1}), \end{aligned}$$

such that finally

$$q_1 := \min_{x_1\in\mathcal{B}}\{c_1^{\mathrm{T}}x_1 + \Psi_2(x_1;\widehat{\zeta}_1)\} \leq \min_{x_1\in\mathcal{B}}\{c_1^{\mathrm{T}}x_1 + \Phi_2(x_1;\widehat{\zeta}_1)\} =: r_1$$

with $\mathscr{B} := \{x_1 \mid A_{11}x_1 = b_1(\widehat{\zeta}_1) \equiv b_1,\ x_1 \geq 0\}$. □

As seen above, with Assumption 3.5., and observing Assumption 3.4. when generating a finite subfiltration $\widehat{\mathscr{F}}$ and the corresponding scenario tree for problem (3.123), as described on page 272, we get the fully aggregated problem (see (3.127))

$$\left.\begin{array}{r} \min \displaystyle\sum_{m\in\mathscr{N}} p_m c_{t_m}^{\mathrm{T}} x_m \\ \displaystyle\sum_{m\in\mathscr{H}(n)} A_{t_n t_m} x_m = b_n \ \forall n \in \mathscr{N} \\ x_m \geq 0 \ \ \forall m \in \mathscr{N} \end{array}\right\} \tag{3.131}$$

with $b_n = b_{t_n}(\widehat{\zeta}^n)$ and $p_n > 0\ \forall n \in \mathscr{N}$.

As the dual LP of (3.131) we have

$$\left.\begin{array}{r} \max \displaystyle\sum_{n\in\mathscr{N}} b_n^{\mathrm{T}} u_n \\ \displaystyle\sum_{n\in\mathscr{G}(m)} A_{t_n t_m}^{\mathrm{T}} u_n \leq p_m c_{t_m} \ \forall m \in \mathscr{N}. \end{array}\right\} \tag{3.132}$$

With the substitution $u_n = p_n \pi_n$ (3.132) is equivalent to

$$\left.\begin{array}{r} \max \displaystyle\sum_{n\in\mathscr{N}} p_n b_n^{\mathrm{T}} \pi_n \\ \displaystyle\sum_{n\in\mathscr{G}(m)} q_{m\to n} A_{t_n t_m}^{\mathrm{T}} \pi_n \leq c_{t_m} \ \forall m \in \mathscr{N} \end{array}\right\} \tag{3.133}$$

with $q_{m\to n}$ the conditional probability to reach node n given node m.

For $\{\hat{x}_m, \hat{\pi}_n\}$ to be a primal-dual pair of optimal solutions, according to Chapter 1, Prop. 1.12., the complementarity conditions

$$(c_{t_m} - \sum_{n\in\mathscr{G}(m)} q_{m\to n} A_{t_n t_m}^{\mathrm{T}} \hat{\pi}_n)^{\mathrm{T}} \hat{x}_m = 0 \ \forall m \in \mathscr{N} \tag{3.134}$$

have to hold (with $q_{m\to m} = 1$).

Discretization under special assumptions

Under Assumption 3.5. on problem (3.123) and Assumption 3.4. on the discretized distributions (implying positive probabilities for all scenarios generated) we shall discuss now, how a successive refinement of the partitions and hence a correspondingly growing scenario tree can be designed, such that the approximation of (3.123) by the generated problem (3.131) is improved.

To begin with, let $\widehat{\mathscr{F}}$ be the coarse subfiltration with each $\widehat{\mathscr{F}}_t$ being generated by the elementary events $\{\xi_\tau^{-1}[\Xi^\tau], \emptyset \mid \tau = 1, \cdots, t\}$ i.e. by $\{\Omega, \emptyset\}$. Then for node $n = t$ holds $t_n = n = t$ and $\Xi_n^{t_n} = \Xi^t$, such that by (3.126) follows $\widehat{\zeta}^n = \widehat{\zeta}^t = \mathbb{E}[\zeta_t]$, yielding the aggregated problem

$$\left.\begin{array}{r} \min \sum_{t=1}^{T} c_t x_t \\ \sum_{\tau=1}^{t} A_{t\tau} x_\tau = b_t(\widehat{\zeta}^t) \ \forall t \\ x_t \geq 0 \ \forall t\,. \end{array}\right\} \tag{3.135}$$

The corresponding basic scenario tree is shown in Fig. 3.8.

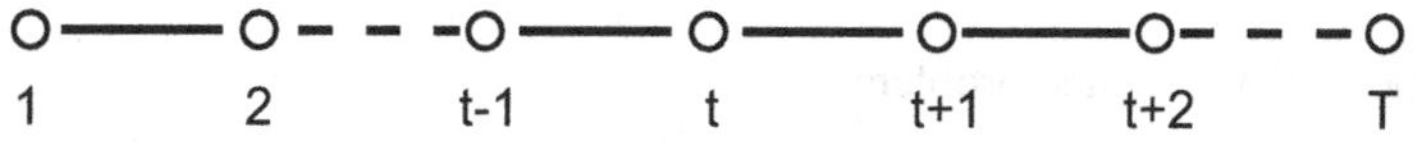

Fig. 3.8 Basic scenario tree.

In the coarse subfiltration, $\widehat{\mathscr{F}_t}$ was generated by $\{\Omega, \emptyset\}$ $\forall t \in \{1, \cdots, T\}$. Let this subfiltration be refined into $\widetilde{\mathscr{F}}$ by partitioning Ξ^t for a particular $t > 1$ into two subintervals Ξ_1^t, Ξ_2^t (whereas in all other stages the trivial partitions $\{\Xi^s,\ s \neq t\}$remain unchanged). Then it follows that

$$\widetilde{\mathscr{F}_s} \text{ is generated by } \begin{cases} \{\Omega, \emptyset\} & \text{for } s < t \\ \{\Omega, \xi_{st}^{-1}[\Xi_1^t], \xi_{st}^{-1}[\Xi_2^t], \emptyset\} & \text{for } s = t \\ \{\Omega, \xi_{st}^{-1}[\Xi_1^t], \xi_{st}^{-1}[\Xi_2^t], \xi_{ss}^{-1}[\Xi^s], \emptyset\} & \text{for } s > t\,. \end{cases}$$

The modification of the scenario tree, corresponding to splitting node $n = t$, is shown in Fig. 3.9.

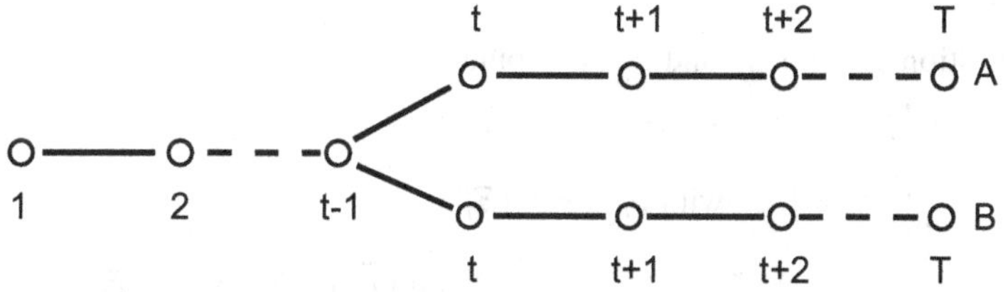

Fig. 3.9 Basic scenario tree: First split.

Obviously we have now two branches from stage t onwards, corresponding to the subintervals Ξ_1^t and Ξ_2^t of the partition of Ξ^t. Denoting the nodes of the two scenarios as $(t, A), t = 1, \cdots, T$, and $(t, B), t = 1, \cdots, T$, the respective components

$\widehat{\zeta}_{\tau}^{(s,A)}$ of $\widehat{\zeta}^{(s,A)}$, $s=2,\cdots,T$, are, due to (3.126), determined as

$$\widehat{\zeta}_{\tau}^{(s,A)} = \begin{cases} \mathbb{E}[\xi_\tau] & \text{for } \tau \neq t \\ \mathbb{E}[\xi_t \mid \xi_t \in \Xi_1^t] & \text{for } \tau = t, \end{cases}$$

and analogously for $\widehat{\zeta}^{(s,B)}$, $s=2,\cdots,T$, follows

$$\widehat{\zeta}_{\tau}^{(s,B)} = \begin{cases} \mathbb{E}[\xi_\tau] & \text{for } \tau \neq t \\ \mathbb{E}[\xi_t \mid \xi_t \in \Xi_2^t] & \text{for } \tau = t. \end{cases}$$

The corresponding node probabilities are

$$p_{(s,A)} = \begin{cases} 1 & \text{if } s < t \\ \mathbb{P}_{\xi_t}(\Xi_1^t) & \text{if } s \geq t \end{cases} \text{ and } p_{(s,B)} = \begin{cases} 1 & \text{if } s < t \\ \mathbb{P}_{\xi_t}(\Xi_2^t) & \text{if } s \geq t. \end{cases}$$

Hence the new aggregated problem is

$$\left.\begin{aligned} &\min \left\{ \sum_{\tau=1}^{t-1} c_\tau^{\mathrm{T}} x_{(\tau,A)} + \sum_{\tau=t}^{T} c_\tau^{\mathrm{T}} \left[p_{(\tau,A)} x_{(\tau,A)} + p_{(\tau,B)} x_{(\tau,B)} \right] \right\} \\ &\sum_{\tau=1}^{s} A_{s\tau} x_{(\tau,A)} = b_s(\widehat{\zeta}^{(s,A)}) \;\; \forall s \\ &\sum_{\tau=1}^{s} A_{s\tau} x_{(\tau,B)} = b_s(\widehat{\zeta}^{(s,B)}) \;\; \forall s \geq t \\ &\qquad x_{(s,B)} = x_{(s,A)} \;\; \forall s < t \\ &x_{(s,A)}, x_{(s,B)} \geq 0 \;\; \forall s. \end{aligned}\right\} \tag{3.136}$$

Assume now that $\mathscr{T} = (\mathscr{N}, \mathscr{A})$ is the scenario tree associated with problem (3.131). To split in this tree some node $i > 1$ into the nodes i_1 and i_2, or equivalently to subdivide the corresponding $\Xi_i^{t_i} \subseteq \Xi^{t_i}$ into two subintervals $\Xi_{i_1}^{t_i}$ and $\Xi_{i_2}^{t_i}$ (observing Assumption 3.4.), we have to run the following *node splitting procedure*:

Cut and paste

S1 Partition $\Xi_i^{t_i}$ into $\Xi_{i_1}^{t_i}$ and $\Xi_{i_2}^{t_i}$; compute
$\widetilde{p}_{i_\nu} = \mathbb{P}_{\xi_{t_i}}(\Xi_{i_\nu}^{t_i})$, $\nu = 1,2$,
$r_\nu = \dfrac{\widetilde{p}_{i_\nu}}{\widehat{p}_i}$, $\nu = 1,2$, with $\widehat{p}_i = \mathbb{P}_{\xi_{t_i}}(\Xi_i^{t_i})$,
$b_{i_\nu} = \mathbb{E}_{\xi_{t_i}}[\widetilde{b}_i(\xi_{t_i}) \mid \xi_{t_i} \in \Xi_{i_\nu}^{t_i}]$, $\nu = 1,2$, with $\widetilde{b}_i(\xi_{t_i}) := b_{t_i}(\widehat{\zeta}^{h_i}, \xi_{t_i})$,
such that $r_1 + r_2 = 1$ and $r_1 b_{i_1} + r_2 b_{i_2} = b_i$.

S2 Let $\mathscr{T}_1 = (\mathscr{N}_1, \mathscr{A}_1)$ with $\mathscr{N}_1 \subset \mathscr{N}$, $\mathscr{A}_1 \subset \mathscr{A}$ be the maximal subtree of $\mathscr{T} = (\mathscr{N}, \mathscr{A})$ rooted at node $i \in \mathscr{N}$.
Let $\mathscr{T}_2 = (\mathscr{N}_2, \mathscr{A}_2)$ be a copy of $\mathscr{T}_1$, with its root denoted as $j \notin \mathscr{N}$ and all other node labels modified such that $\mathscr{N}_2 \cap \mathscr{N} = \emptyset$, $\mathscr{A}_2 \cap \mathscr{A} = \emptyset$.

Assign to the nodes of $\mathscr{T}_2$ the same quantities as associated with the corresponding nodes of $\mathscr{T}_1$.

S3 With $\mathscr{H}(i)$ the history of node i in $\mathscr{T}$, and $\widetilde{\mathscr{H}}(n)$ the history within $\mathscr{T}_\nu$ for $n \in \mathscr{N}_\nu$, $\nu = 1,2$ respectively, update the values of the subtrees $\mathscr{T}_1$ and $\mathscr{T}_2$ as follows:

$\mathscr{T}_1$: Set $b_i^{(1)} := b_{i_1}$, and for $n \in \mathscr{G}(i) \setminus \{i\}$, the future of i in $\mathscr{T}_1$, let $b_n^{(1)} := b_{t_n}(\widehat{\zeta}^n)$, with $\widehat{\zeta}^n$ computed according to (3.126), with the history of n being composed as $\{\mathscr{H}(h_i), i, \widetilde{\mathscr{H}}(n)\}$;
multiply the node probabilities by r_1.

$\mathscr{T}_2$: Set $b_j^{(2)} = b_{i_2}$, and for $m \in \mathscr{G}(j) \setminus \{j\}$, the future of j in $\mathscr{T}_2$, let $b_m^{(2)} := b_{t_m}(\widehat{\zeta}^m)$, with $\widehat{\zeta}^m$ computed according to (3.126), with the history of m being composed as $\{\mathscr{H}(h_i), j, \widetilde{\mathscr{H}}(m)\}$, implying that $b_m^{(2)}$ equals the right–hand–side for the corresponding node in $\mathscr{N}_1$;
multiply the node probabilities by r_2.
(Observe that $\mathscr{H}(h_i) = \mathscr{H}(h_j)$ will be enforced in step **S4**.)

S4 Introduce a new edge from the parent node h_i of i to the node j, the root of $\mathscr{T}_2$, thus pasting $\mathscr{T}_2$ to $\mathscr{T}$ and yielding the new tree.

$$
\begin{aligned}
\mathscr{T}^+ &= (\mathscr{N}^+, \mathscr{A}^+), \text{ with} \\
\mathscr{N}^+ &= \mathscr{N} \cup \mathscr{N}_2 \qquad \text{and} \\
\mathscr{A}^+ &= \mathscr{A} \cup \mathscr{A}_2 \cup \{(h_i, j)\}.
\end{aligned}
$$

In Fig. 3.10 one cycle of this procedure is illustrated.

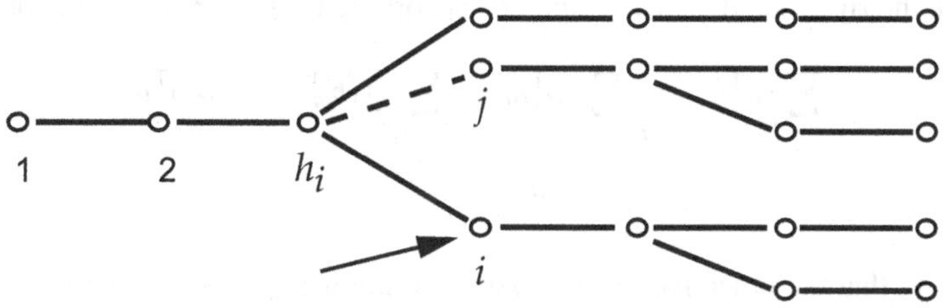

Fig. 3.10 Cut and paste.

It is easy to see that with the above procedure of cut and paste the optimal values of the related primal LP's are non-decreasing.

Proposition 3.8. *With V being the optimal value of the fully aggregated problem (3.131) corresponding to the scenario tree $\mathscr{T}$, and V^+ being the optimal value for the corresponding LP on $\mathscr{T}^+$ as generated by cut and paste, it follows that $V^+ \geq V$.*

Proof: Let $\{u_n,\ n \in \mathscr{N}\}$ be a solution of the dual program (3.132) associated with $\mathscr{T}$. To each node $n \in \mathscr{N}_2$ assign the vector u_n as determined for the corresponding node $n \in \mathscr{N}_1$.

Now define for $n \in \mathscr{N}^+$, with r_ν from step **S1**,

$$\widetilde{u}_n := \begin{cases} r_1 u_n \text{ if } n \in \mathscr{N}_1 \\ r_2 u_n \text{ if } n \in \mathscr{N}_2 \\ u_n \text{ else.} \end{cases}$$

In order to show that $\{\widetilde{u}_n,\ n \in \mathscr{N}^+\}$ is a feasible solution to the dual program (3.132) associated with $\mathscr{T}^+$, we have to distinguish the following cases:

1) $\underline{m \in \mathscr{N}_1}$

$$\sum_{n \in \mathscr{G}(m)} A^{\mathrm{T}}_{t_n t_m} \widetilde{u}_n = r_1 \left(\sum_{n \in \mathscr{G}(m)} A^{\mathrm{T}}_{t_n t_m} u_n \right) \leq r_1 p_m c_{t_m} = \widetilde{p}_m c_{t_m}$$

with $\widetilde{p}_m$ as defined in step **S3** for $m \in \mathscr{N}_1$.

2) $\underline{m \in \mathscr{N}_2}$

The analogous argument holds, with r_2 instead of r_1.

3) $\underline{m \in \mathscr{N}^+ \setminus (\mathscr{N}_1 \cup \mathscr{N}_2) =: \Delta\mathscr{N}}$

$$\begin{aligned} \sum_{n \in \mathscr{G}(m)} A^{\mathrm{T}}_{t_n t_m} \widetilde{u}_n &= \\ &= \sum_{n \in \mathscr{G}(m) \cap \Delta\mathscr{N}} A^{\mathrm{T}}_{t_n t_m} u_n + \sum_{n \in \mathscr{G}(m) \cap \mathscr{N}_1} (r_1 + r_2) A^{\mathrm{T}}_{t_n t_m} u_n \\ &\leq p_m c_{t_m} . \end{aligned}$$

Hence, $\{\widetilde{u}_n,\ n \in \mathscr{N}^+\}$ is feasible for the dual program (3.132) corresponding to $\mathscr{T}^+$ and, with the right–hand–sides $\widetilde{b}_n$ updated according to step **S3**, yields the objective value

$$\begin{aligned} \sum_{n \in \mathscr{N}^+} \widetilde{b}^{\mathrm{T}}_n \widetilde{u}_n &= \sum_{n \in \Delta\mathscr{N}} b^{\mathrm{T}}_n u_n + \sum_{m \in \mathscr{N}_1} (r_1 b^{(1)}_m + r_2 b^{(2)}_m)^{\mathrm{T}} u_n \\ &= \sum_{n \in \mathscr{N}} b^{\mathrm{T}}_n u_n . \end{aligned}$$

This shows that the objective of the feasible solution $\{\widetilde{u}_n,\ n \in \mathscr{N}^+\}$ for $\mathscr{T}^+$ coincides with the optimal value for $\mathscr{T}$, such that $V^+ \geq V$ obviously has to hold. □

Corollary 3.4. *Let $\widehat{V}$ be the optimal value of problem (3.123). If Assumption 3.5. is satisfied, then each method, splitting succesively any nodes (except the root) in the scenario tree according to the cut and paste procedure, converges to a value $V^\star \leq \widehat{V}$.*

Proof: Under the given assumptions, the optimal objective values of the aggregated problems are

– monotonically nondecreasing according to Prop. 3.8., and

– they are lower bounds of the optimal value of (3.123) due to Lemma 3.17. □

Although this cut and paste procedure seems to have a promising behaviour, we are still left with two open questions:

1) Is there any criterion (even a heuristic one, maybe) for deciding on the next node to be split?
2) Given this criterion, may it happen that for the limit $V^\star$ in Corollary 3.4. holds $V^\star < \widehat{V}$?

As to the first question, for a fixed node $n > 1$ let $\{\hat{x}_m \mid m \in \mathscr{H}(n) \setminus \{n\}\}$ and $\{\hat{\pi}_m \mid m \in \mathscr{G}(n)\}$ be parts of solutions of (3.131) and (3.133), respectively, and consider the LP

$$\left.\begin{aligned} \varphi_n(b_n) := \min(c_{t_n} - &\sum_{m\in\mathscr{G}(n)} q_{n\to m} A_{t_m t_n}^{\mathrm{T}} \hat{\pi}_m)^{\mathrm{T}} x_n \\ A_{t_n t_n} x_n &= b_n - \sum_{m\in\mathscr{H}(n)\setminus\{n\}} A_{t_n t_m} \hat{x}_m \\ x_n &\geq 0 . \end{aligned}\right\} \tag{3.137}$$

Since $\{\hat{x}_k;\ k \in \mathscr{N}\}$ solves (3.131), in particular $\hat{x}_n$ is feasible in (3.137). Furthermore, the $\{\hat{\pi}_\ell;\ \ell \in \mathscr{N}\}$ being optimal in (3.133) and $\hat{x}_n \geq 0$ due to (3.137), we conclude, observing (3.134), that

$$0 \leq (c_{t_n} - \sum_{m\in\mathscr{G}(n)} q_{n\to m} A_{t_m t_n}^{\mathrm{T}} \hat{\pi}_m)^{\mathrm{T}} \hat{x}_n = 0,$$

showing that $\hat{x}_n$ with the optimal value $\varphi_n(b_n) = 0$ solves the LP (3.137). Using (3.126) we have that $\widehat{\zeta}^n = (\widehat{\zeta}^{h_n}, \mathbb{E}[\xi_{t_n} \mid \xi_{t_n} \in \Xi_n^{t_n}])$. Replacing $b_n = b_{t_n}(\widehat{\zeta}^n)$ by the random $\widetilde{b}_n(\xi_{t_n}) := b_{t_n}(\widehat{\zeta}^{h_n}, \xi_{t_n})$, it is obvious that the optimal value

$$\left.\begin{aligned} \varphi_n(\widetilde{b}_n(\xi_{t_n})) := \min(c_{t_n} - &\sum_{m\in\mathscr{G}(n)} q_{n\to m} A_{t_m t_n}^{\mathrm{T}} \hat{\pi}_m)^{\mathrm{T}} x_n \\ A_{t_n t_n} x_n &= \widetilde{b}_n(\xi_{t_n}) - \sum_{m\in\mathscr{H}(n)\setminus\{n\}} A_{t_n t_m} \hat{x}_m \\ x_n &\geq 0 . \end{aligned}\right\} \tag{3.138}$$

is a convex function in ξ_{t_n}, such that due to Jensen

$$\begin{aligned} \mathbb{E}[\varphi_n(\widetilde{b}_n(\xi_{t_n})) \mid \xi_{t_n} \in \Xi_n^{t_n}] &\geq \varphi_n(\widetilde{b}_n(\mathbb{E}[\xi_{t_n} \mid \xi_{t_n} \in \Xi_n^{t_n}]) \\ &= \varphi_n(b_{t_n}(\widehat{\zeta}^{h_n}, \mathbb{E}[\xi_{t_n} \mid \xi_{t_n} \in \Xi_n^{t_n}]) \\ &= \varphi_n(b_{t_n}(\widehat{\zeta}^n)) \\ &= \varphi_n(b_n) = 0, \end{aligned}$$

and we have the lower bound $l_n = 0$ for $\mathbb{E}[\varphi_n(\widetilde{b}_n(\xi_{t_n})) \mid \xi_{t_n} \in \Xi_n^{t_n}]$. On the other hand, according to Lemma 3.7 (on page 213), we can determine the E–M upper bound u_n for $\mathbb{E}[\varphi_n(\widetilde{b}_n(\xi_{t_n})) \mid \xi_{t_n} \in \Xi_n^{t_n}]$. If, with some prescribed tolerance $\Delta > 0$, the *splitting criterion*

$$u_n - l_n > \Delta \tag{3.139}$$

is satisfied, we may decide to split node n as described in the cut and paste procedure, in order to increase the lower bound and thereby to improve the approximative solution. Observe however, that this criterion ($u_n - l_n > \Delta$) to increase the lower bound and thereby to improve the solution in a particular node, is based on a heuristic argument. But it is one positive answer to the first question, at least. Moreover, test runs with this criterion did work out surprisingly well.

To come to the second question, consider the following example:

Example 3.4. *Assume the following problem to be given:*

$$\begin{array}{ll} \min\{x_1 + x_2 + \mathbb{E}[y_1 + y_2 + z_1 + z_2]\} & \\ x_1 - x_2 & = 0 \\ x_1 + 2x_2 + 3y_1 - 3y_2 & = \xi_2 \\ x_1 + 3x_2 + \quad y_1 - y_2 + 4z_1 - 4z_2 & = \xi_3 \\ & x_i, y_i, z_i \geq 0, \end{array}$$

where $\xi_2 \sim \mathscr{U}[0,6]$ and $\xi_3 \sim \mathscr{U}[1,1.5]$, with $\mathscr{U}$ being the uniform distribution. The fully aggregated problem with $\mathbb{E}[\xi_2] = 3$ and $\mathbb{E}[\xi_3] = 1.25$ as right–hand–sides is easily seen to have the optimal solution

$$(\hat{x}_1, \hat{x}_2, \hat{y}_1, \hat{y}_2, \hat{z}_1, \hat{z}_2) = (0, 0, 1, 0, \frac{1}{16}, 0)$$

with the optimal value

$$V = \frac{17}{16}$$

and the dual solution

$$\hat{\pi}^{\mathrm{T}} = \left(\frac{1}{4}, \frac{1}{4}, \frac{1}{4}\right).$$

Considering problem (3.138) for $n = 2$, we find that $\varphi_2(\widetilde{b_2}(\xi_{t_2})) \equiv 0$ for $\xi_2 \in [0,6]$, i.e. φ_2 is linear on Ξ^2 implying that $u_2 - l_2 = 0$. Analogously $\varphi_3(\widetilde{b_3}(\xi_{t_3})) \equiv 0$ for $\xi_3 \in [1,1.5]$ such that also φ_3 is linear on Ξ^3 and therefore $u_3 - l_3 = 0$. Hence the above splitting criterion (3.139) cannot be satisfied, and the procedure would stop with the above solution, with $V^\star = V$.

However, subdividing $\Xi^2 = [0,6]$ into the intervals $[0,3)$ and $[3,6]$ and solving the corresponding LP, would yield the optimal value

$$V^+ = \frac{18}{16} > V,$$

and the same result would be achieved with splitting, instead of Ξ^2, the interval $\Xi^3 = [1,1.5]$ into $[1,1.25)$ and $[1.25,1.5]$. □

Hence, in this example the procedure, using the above splitting criterion (3.139), had to be finished with $u_n - l_n = 0$ for all nodes $n > 1$, although there was a substantial difference $\widehat{V} - V^{\star} > 0$. This fact could (and can in general) only be discovered by analyzing (sub)sets of nodes simultaneously in detail. In other words: For the approach using the splitting criterion (3.139) so far there is not known any simple stopping rule stating the (near-)optimality of the present iterative solution for problem (3.123).

Chapter 4
Algorithms

4.1 Introduction

The discussion of algorithms in this chapter is organized according to the framework of different SLP model classes, as presented in the previous chapters. A computer implementation of an algorithm will be called a *solver.*

For the algorithms presented in detail in this chapter, sufficient and reproducible empirical evidence is available concerning the numerical efficiency of a corresponding solver. On the one hand, this means that results of computational experiments with several test problems or test problem batteries are available in the literature. On the other hand, reproducibility presupposes the public availability of the solver. With most of the algorithms, discussed in detail in this chapter, we have our own computational experience; several solvers have been implemented and tested by ourselves. These solvers, along with further solvers provided by their authors, are publicly available as connected to our modeling system SLP–IOR, see Section 4.9.2.

4.2 Single–stage models with separate probability functions

In this section we discuss algorithms for SLP models with separate probability functions, presented in Chapter 2, Section 2.2.3. If only the right–hand–side is stochastic then the models can be transformed into deterministic LP models, as discussed in Chapter 2, Section 2.2.3. In this section we have also pointed out some pitfalls which have to be taken into account in this approach. The equivalent LP models do not have any SLP–specific structure, thus the recommended approach is to employ general–purpose LP solvers.

In the general case we will consider probability functions of the form $G(x) = \mathbb{P}_\xi(\ x \mid \eta^{\mathrm{T}}x - \xi \geq 0)$ where η is an n–dimensional random vector and ξ is a random variable. We concentrate on constraints

$$\mathbb{P}_\xi(\ x \mid \eta^{\mathrm{T}}x - \xi \geq 0) \geq \alpha.$$

If the joint distribution of (η, ξ) is multivariate normal then the above constraint can be written in the following equivalent form, see (2.60) on page 102

$$\Phi^{-1}(\alpha)\|D^{\mathrm{T}}x - d\| - \mu^{\mathrm{T}}x \leq -\mu_{n+1}. \tag{4.1}$$

Assuming that $\alpha > \frac{1}{2}$ holds, we can write (4.1) as

$$\|D^{\mathrm{T}}x - d\| \leq \frac{1}{\Phi^{-1}(\alpha)}\,\mu^{\mathrm{T}}x - \frac{\mu_{n+1}}{\Phi^{-1}(\alpha)}, \tag{4.2}$$

which is called a *second–order cone constraint.* Models involving this type of constraints are called second–order cone programs (SOCP). Such models have been first studied by Nesterov and Nemirovsky [242], who also proposed interior–point methods for their solution.

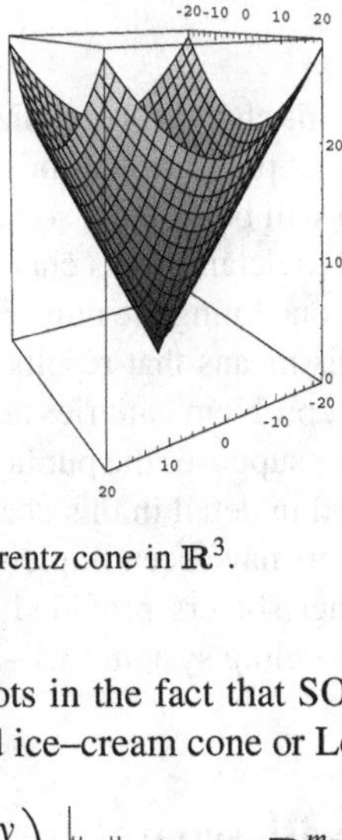

Fig. 4.1 The ice–cream cone, or Lorentz cone in $\mathbb{R}^3$.

The terminology has its roots in the fact that SOCP is intimately related to the second–order cone (also called ice–cream cone or Lorentz cone)

$$\mathscr{C}_m := \left\{ \begin{pmatrix} y \\ t \end{pmatrix} \middle| \, \|y\| \leq t, y \in \mathbb{R}^{m-1}, t \in \mathbb{R} \right\},$$

where $\|\cdot\|$ stands for the Euclidean norm. See Figure 4.1 for a second–order cone in $\mathbb{R}^3$. A general SOCP–constraint has the form

$$\|Ax + b\| \leq d^{\mathrm{T}}x + f$$

with A denoting an $m \times n$ matrix, $x \in \mathbb{R}^n$, the other arrays having compatible dimensions. This constraint can equivalently be written as the following cone–constraint

$$\begin{pmatrix} A \\ d^{\mathrm{T}} \end{pmatrix} x + \begin{pmatrix} b \\ f \end{pmatrix} \in \mathscr{C}_{m+1}.$$

For an overview on SOCP see Lobo et al. [206] and the references therein. The state–of–the–art solution methods are primal–dual interior point methods, for a nice unified presentation see Peng, Roos, and Terlaky [250].

Assuming additionally to the normal distribution that the components of (η, ξ) are stochastically independent, Seppälä and Orpana [304] propose a successive linearization algorithm, which is also based on the second–order cone structure as discussed above.

Weintraub and Vera [341] propose a different approach by applying the supporting hyperplane method of Veinott (see Section 1.3.2 in Chapter 1) for the general normally distributed case.

In the case when (η, ξ) has a multivariate Cauchy distribution and assuming that $\alpha \geq \frac{1}{2}$ holds, the probabilistic constraint can be written in the following equivalent form, see (2.73) on page 110

$$\|D^{\mathrm{T}}x - d\|_1 \leq \frac{1}{\Psi^{-1}(\alpha)} m^{\mathrm{T}}x - \frac{m_{n+1}}{\Psi^{-1}(\alpha)} \tag{4.3}$$

which can be interpreted as a first–order cone constraint. This problem can be formulated equivalently as an LP problem, see (2.74) on page 110, which can then be solved by general–purpose LP solvers. Alternatively, special–purpose interior point algorithms might be more efficient.

4.2.1 A guide to available software

The straightforward approach is to solve the deterministic equivalent problems by employing a general–purpose solver. This is the only approach presently available for the case of the Cauchy distribution.

For the case of the non–degenerate multivariate normal distribution, a much better approach is to employ solvers for SOCP. There are several solvers available in the public domain:

- SOCP (C and Matlab), developed by Miguel S. Lobo, Lieven Vandenberghe, and Stephen Boyd, [206]
  http://stanford.edu/~boyd/old_software/SOCP.html.
- SeDuMi (MatLab toolbox) Jos F. Sturm [316],
  http://sedumi.ie.lehigh.edu/.
- SDPT3 version 3.02 (Matlab) Kim C. Toh, Reha Tütüncü, and Michael J. Todd [327],
  http://www.math.nus.edu.sg/~mattohkc/sdpt3.html.

For implementing your own solver see, for instance, Andersen et al. [6], Kuo and Mittelmann [199], Lobo et al. [206], or Peng et al. [250].

Commercial solvers: MOSEK and LOQO, for further information see the Decision Tree for Optimization Software at node
http://plato.asu.edu/sub/nlores.html#semidef.

For selecting an appropriate solver, see the benchmarks of Hans Mittelmann, [236]; http://plato.asu.edu/bench.html.

If, additionally, the components of (η, ξ) are stochastically independent, the solver CHAPS, developed by Seppälä and Orpana, T. [304], could prove to be an efficient alternative.

4.3 Single–stage models with joint probability functions

This section is devoted to algorithms for solving models which involve joint probability functions, under the assumption that only the right–hand–side is stochastic.

The general case, where the technology matrix is also stochastic, has been discussed in Chapter 2, Section 2.2.6. In this case the probability function G is not quasi–concave in general, implying that the SLP problems are non–convex optimization problems. This is in general so, even if ξ has a multivariate normal distribution. However, under some assumptions concerning the structure of the correlation matrices, convex optimization problems arise, as discussed in Section 2.2.6. According to our knowledge, there are no specialized deterministic algorithms available for this type of problems. Consequently, presently there are two available approaches for such problems. Either they can be treated as nonlinear optimization problems and one can try to apply general–purpose algorithms of nonlinear optimization, or in the non–convex case techniques of non–convex programming. The other possibility is to employ stochastic algorithms for which the interested reader is referred to Luedtke and Ahmed [207] and to Pagnoncelli et al. [248] (cf. Section 4.7.3 for two–stage recourse problems).

Under the assumption that only the right–hand–side is stochastic, the joint probability function is defined as

$$G(x) = \mathbb{P}_\xi(Tx \geq \xi) = \mathbb{P}_\xi(t_i^{\mathrm{T}} x \geq \xi_i, i = 1, \ldots, s), \tag{4.4}$$

where T is an $(s \times n)$ matrix, ξ is an s–dimensional random vector, $x \in \mathbb{R}^n$, and the components of t_i are the elements of the ith row of T. For separate probability functions ($s = 1$), the corresponding SLP–problems are equivalent to LP–problems, see Section 2.2.3 in Chapter 2. Consequently, we assume that $s > 1$ holds.

Concerning the probability distribution of ξ, we will discuss algorithms for two cases.

On the one hand, we will assume that ξ has a continuous distribution, with a logconcave probability distribution function F. The presentation will mainly be focused on the case when ξ has a non–degenerate multivariate normal distribution; possible extensions to other logconcave distributions will be indicated via remarks. Algorithmic issues for this case are the subject of the subsequent sections 4.3.1–4.3.5.

On the other hand, in Section 4.3.6 we assume that ξ has a finite discrete distribution, and discuss different algorithmic approaches for this case.

The model formulations considered in this section are the following (see Chapter 2, (2.29) and (2.30) on page 88 and (2.86) on page 115):

$$\left.\begin{array}{rl} \min & c^{\mathrm{T}}x \\ \text{s.t.} & F(Tx) \geq \alpha \\ & x \in \mathscr{B}, \end{array}\right\} \tag{4.5}$$

where $0 \leq \alpha \leq 1$ is a probability level, and

$$\left.\begin{array}{rl} \max & F(Tx) \\ \text{s.t.} & x \in \mathscr{B}. \end{array}\right\} \tag{4.6}$$

In both cases F denotes the joint probability distribution function of the random right–hand–side ξ. It is sometimes advantageous to recast (4.5) as

$$\left.\begin{array}{lrrl} \min c^{\mathrm{T}}x & & & \\ \text{s.t.} & & F(y) & \geq \alpha \\ & Tx & -y & \geq 0 \\ & x & & \in \mathscr{B}. \end{array}\right\} \tag{4.7}$$

To see the equivalence of (4.5) and (4.7), let $(\bar{x},\bar{y})$ be a feasible solution of (4.7) and let $\widehat{y} := T\bar{x}$. Due to the monotonicity properties of probability distribution functions, $(\bar{x},\widehat{y})$ is also a feasible solution of (4.7), with the same objective value. From this the equivalence follows readily.

Taking the algorithmic point of view, let us consider, for instance, cutting plane methods. When applying these methods, the matrix of cuts is usually dense. Assuming that cuts are stored in the rows, in formulation (4.5) this matrix would involve n columns whereas in formulation (4.7) the number of columns is s, where s is the number of inequalities involved in the joint probability function. The point is that usually $s << m$ holds, therefore formulation (4.7) is more suitable from the point of view of implementation, than (4.5).

4.3.1 Numerical considerations

In this section the general assumption will be that ξ has a non–degenerate multivariate normal distribution.

Notice, that we can assume that the distribution of ξ is standardized.

$$G(x) = \mathbb{P}_{\xi}\left(\frac{1}{\sigma_i}(t_i^{\mathrm{T}}x - \mu_i) \geq \frac{1}{\sigma_i}(\xi_i - \mu_i),\, i = 1,\ldots,s\right)$$

is equivalent to the formulation (4.4), where μ_i and σ_i are the expected value and the standard deviation of ξ_i, respectively.

Notice, that both problems (4.5) and (4.6) are nonlinear programming problems. Although (4.6) is linearly constrained and (4.5) involves a linear objective and a single nonlinear constraint, problems of the above type are hard to solve numerically in general. Next we summarize the main sources of difficulties, see also Mayer [231].

Let us note, that the nonlinear function F, involved in the model formulations, is in general not given via an algebraic formula. For computing function values $F(x)$ and gradient values $\nabla F(x)$, numerical integration is needed. For higher dimensions of the random right–hand–side ξ, for instance for $s = 20$, the only way for computing $F(x)$ and $\nabla F(x)$ is utilizing Monte Carlo integration methods. This implies on the one hand, that computing these quantities is relatively time–consuming compared to the evaluation of algebraic formulas. On the other hand, the approximation error is relatively large. Consequently, for higher values of s (for instance, $s = 20$), there is no chance to obtain a solution of (4.5) or (4.6) with a high accuracy. Therefore, according to our opinion, the main requirement which solution methods should fulfill, is robustness.

Considering our optimization problems from the purely nonlinear programming point of view, these problems are convex programming problems for a large class of probability distributions, including the non–degenerate multivariate normal distribution, see Chapter 2 Section 2.2.5. We observe, however, some quite unfavorable features. Figure 4.2 shows the graph of the bivariate standard normal distribution function.

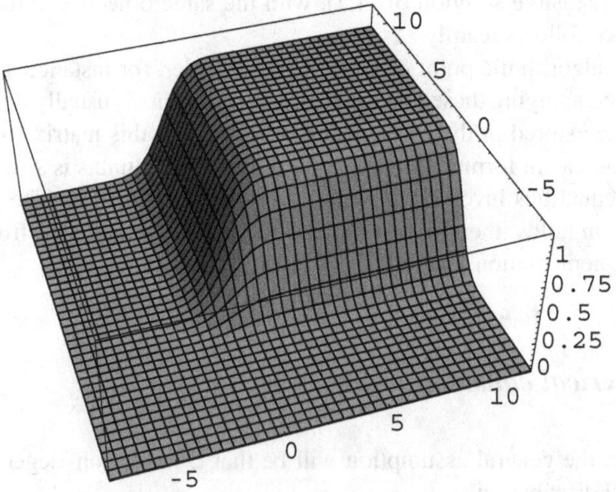

Fig. 4.2 The probability distribution function of the bivariate standard normal distribution.

Notice the large flat regions in the figure. Assume that we have an iteration point $\hat{x}$ somewhere in the flat region. For finding the next iteration point, the vast majority of NLP algorithms utilizes local information, based on $\nabla F(\hat{x})$, and perhaps also requires curvature information which originate in higher–order derivatives. The difficulty is that derivatives are very small in that region and vary largely depending on the location of $\hat{x}$. The latter property indicates that this deficiency is difficult to overcome, if not impossible, by scaling the problem. To illustrate, how derivatives behave, let us consider the standard n–variate normal distribution with independent random variables, in which case it is easy to compute derivatives. The order of magnitude of $\frac{\partial F(\hat{x})}{\partial x_1}$ is displayed in Table 4.1, where $\hat{x}_i = \lambda$ for all i. The different λ–values correspond to the rows of the table. Let us take, for instance, the entry -3945, corresponding to $\lambda = -30$ and $s = 20$. The interpretation is that the magnitude of the partial derivative in the 20–dimensional case, in the point with all coordinates being equal to -30, is 10^{-3945}. This phenomenon can be interpreted as some kind of

	$s=2$	$s=10$	$s=20$	$s=30$
-30	-394	-1972	-3945	-5918
-10	-46	-231	-462	-693
0	-1	-3	-7	-10
10	-23	-23	-23	-23
30	-196	-196	-196	-196

Table 4.1 Order of magnitude of derivatives of the multivariate normal distribution function.

hidden non–convexity of the convex optimization problem. The region, where the derivatives have reasonable magnitude and thus iteration points can be well dealt with by algorithms, is non–convex as can be seen in Figure 4.2.

As noted in Mayer [231], an additional difficulty is that the steepness of the function between the lower– and upper almost–horizontal parts becomes rather high with increasing dimension s, as displayed in Figure 4.3. This implies that the region, where the derivatives have manageable values, becomes narrower for higher–dimensional random vectors ξ.

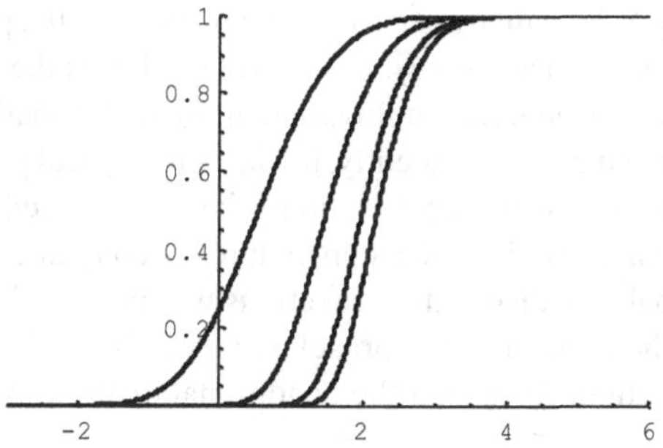

Fig. 4.3 The probability distribution function of the standard normal distribution along the line $y_i = \lambda\ \forall i$, for dimensions $s = 2,\ 10,\ 30,$ and 50.

All algorithms for the solution of (4.5) and (4.6), discussed in this book, require the computation of the gradient ∇F. Assuming a continuous distribution and that the density function f is a positive and continuous function, we have the following well–known expression (see, for instance, Cramér [47]):

$$F(x_1,x_2,\dots,x_s) = \int_{-\infty}^{x_1}\int_{-\infty}^{x_2}\dots\int_{-\infty}^{x_s} f(t_2,\dots t_s \mid t_1) f_1(t_1) dt_1 dt_2,\dots dt_s,$$

where $f(t_2,\dots t_s \mid t_1)$ is the conditional density function of $(\xi_2,\dots,\xi_s)$, given $\xi_1 = t_1$, and f_1 is the marginal density function of ξ_1. By differentiating both sides with respect to x_1 we obtain

$$\begin{aligned} \frac{\partial F(x)}{\partial x_1} &= f_1(x_1) \int_{-\infty}^{x_2} \dots \int_{-\infty}^{x_s} f(t_2,\dots t_s \mid x_1) dt_2,\dots dt_s \\ &= F(x_2,\dots,x_s \mid x_1) f_1(x_1), \end{aligned} \tag{4.8}$$

where $F(x_2,\dots,x_s \mid x_1)$ stands for the conditional distribution function of $(\xi_2,\dots,\xi_s)$, given $\xi_1 = x_1$.

Analogous expressions hold for the partial derivatives with respect to the variables $x_2,\dots,x_s$.

In the case of a non–degenerate multivariate standard normal distribution the above formula takes an especially simple form. On the one hand, $f_1(x_1) = \varphi(x_1)$ with φ denoting the density function of the univariate standard normal distribution. On the other hand, it is well–known, see Mardia et al. [216], that the conditional distribution of $(\xi_2,\dots,\xi_s)$, given $\xi_1 = x_1$, is also non–degenerate multivariate normal. Denoting the correlation matrix of ξ by $R = (\rho_{ij})$, the parameters of this normal distribution are

$$\begin{aligned} \hat{\mu}_i &= \rho_{i1} x_1, \ i = 2,\dots,s \\ \hat{\Sigma}_{ij} &= \rho_{ij} - \rho_{i1}\rho_{j1}, \ i,j = 2,\dots,s, \end{aligned}$$

where $\hat{\mu} \in \mathbb{R}^{s-1}$ is the expected value vector and the $((s-1)\times(s-1))$ covariance matrix is denoted by $\hat{\Sigma}$. The matrix $\hat{\Sigma}$ is in fact nonsingular, the proof of this is left as an exercise for the reader. Thus, $F(x_2,\dots,x_s \mid x_1)$ in (4.8) is the probability distribution function of a non–degenerate multivariate normal distribution, specified by the above parameters $\hat{\mu}$ and $\hat{\Sigma}$. Consequently, having a numerical procedure for evaluating multivariate normal distribution functions, the same procedure can also be used to compute the partial derivatives. This procedure for computing the gradient vector of multivariate probability distribution functions was proposed by Prékopa [257].

Another case, where the above approach is especially well–suited is the case of the Dirichlet distribution, for which the conditional distributions are also Dirichlet distributions.

As pointed out above, our problems are nonlinear optimization problems. Provided that for computing $F(x)$ and $\nabla F(x)$ we have numerical procedures at our disposal, implemented in the appropriate programming language, a possible approach to solve these problems could be employing general–purpose NLP solvers. In fact,

Dupačová, Gaivoronski, Kos, and Szántai [77] report on a successful application of the general–purpose solver *Minos*, for the solution of a problem of our type. The point is, however, that the starting point for the iterations has been quite close to the optimal solution (by far not somewhere in the flat regions). According to our experience concerning the same numerical problem, the solver *Minos* gets stuck in the starting point, as soon as the starting point is not that close to the optimum. Nevertheless, in practical problems there are frequently good starting points available, thus, for such problems, the approach via a general–purpose NLP solver might work.

In general, however, special–purpose algorithms and their implementation in solvers is needed. The usual way for developing such algorithms is adapting general nonlinear programming algorithms to the special structure and properties of problems involving joint probability functions.

In the next section we present approaches based on cutting–plane algorithms and will summarize the other approaches in Section 4.3.3.

4.3.2 Cutting plane methods

Cutting plane methods are discussed in detail in Section 1.3.2. In this section we restrict ourselves to pointing out those features, which are taken into account in the development of methods, adapted to the special structure and properties of (4.5). The problem will be considered in the equivalent form (4.7).

We begin by considering the classical outer approximation methods of Kelley [180], Kleibohm [187], and Veinott [338]. These methods involve outer polyhedral approximations $\mathscr{B}_k$ of the feasible domain $\mathscr{B}$ of (4.5) and generate a sequence $\hat{x}^{(k)} \in \mathscr{B}_k$ as solutions of the LP $\min\{c^{\mathrm{T}}x \mid x \in \mathscr{B}_k\}$. If $\hat{x}^{(k)} \in \mathscr{B}$ then the algorithms stop, otherwise $\mathscr{B}_{k+1}$ is constructed by appending a cutting plane to the set of constraints in $\mathscr{B}_k$.

In the algorithm proposed by Kelley, the cutting plane, based on $\nabla F(\hat{y}^{(k)})$ (with $\hat{y}^{(k)} = T\hat{x}^{(k)}$) is computed by linearizing F at the infeasible point $\hat{y}^{(k)}$. Infeasible points correspond to the lower plateau in Figure 4.2 where the components of ∇F are very small (see Table 4.1) and become practically zero not too far away from the feasible domain.

In the algorithm of Kleibohm [187] and Veinott [338] a Slater–point x^S, lying on the upper plateau in Figure 4.2, is utilized as a navigation point. The intersection $z^{(k)}$ of the line segment $[y^{(k)}, x^S]$ and the feasible domain's boundary is computed and the cut is constructed as a supporting hyperplane of the feasible domain at $z^{(k)}$. This fits well the properties of F in our problem: at the feasible point $z^{(k)}$, ∇F behaves well from the numerical point of view. On the basis of this supporting hyperplane method, Szántai [320] developed an algorithm for solving (4.7), with the additional feature of a moving Slater–point, as described in Section 1.3.2.

In the central cutting plane method of Elzinga and Moore [86], the sequence of approximating polyhedra $\mathscr{B}_k$ and iteration points $x^{(k)}$ are computed differently: instead of solving the LP $\min\{c^{\mathrm{T}}x \mid x \in \mathscr{B}_k\}$, the center of the largest inscribed sphere of $\mathscr{B}_k$ is taken as the next iteration point $x^{(k)}$. The cut is constructed as follows. If, with $\hat{y}^{(k)} = T\hat{x}^{(k)}$, $\hat{y}^{(k)} < \alpha$ holds, then a Kelley cut is applied, otherwise a central (objective) cut $c^{\mathrm{T}}x \geq c^{\mathrm{T}}\hat{x}^{(k)}$ is utilized. For the same reason, as discussed above concerning the cutting plane method of Kelley, this algorithm is unsuitable for solving (4.7).

The remedy is obvious: in the case $\hat{y}^{(k)} < \alpha$, instead of the Kelley cut, a supporting hyperplane should be applied. This presupposes again the availability of a Slater point. A further idea concerns the moving of the Slater points. In the case when a central cut is applied, and additionally $\hat{y}^{(k)} > \alpha$ holds then $(\hat{x}^{(k)}, \hat{y}^{(k)})$ can be employed as the next Slater point in the algorithm. This leads to a central cutting plane method for solving (4.7), as proposed by Mayer [230]. For the details of the algorithm see Section 1.3.2.

For both algorithms we need the existence of a Slater point, thus we require:

Assumption 4.6. *Problem (4.7) is Slater regular, that means that there exists a feasible solution (x^S, y^S) of (4.7), for which $F(y^S) > \alpha$ holds.*

Notice that for starting up the algorithms an initial starting point is needed. For computing this, we employ problem (4.6) which involves maximizing the probability. The problem can equivalently written as

$$\left.\begin{array}{lrrl} \max \tau & & & \\ \text{s.t.} & \log F(y) & -\tau & \geq 0 \\ Tx & -y & & \geq 0 \\ x & & & \in \mathscr{B} \end{array}\right\} \quad (4.9)$$

where the function in the nonlinear constraint is concave, due to our assumptions. Notice that for getting a convex optimization problem, we took $\log F(x)$ in the constraint. This is necessary in general, see the discussion concerning (2.25) in Chapter 2, Section 2.1. Problem (4.9) is obviously Slater regular and for any $x \in \mathscr{B}$ it is easy to construct a Slater point by appropriately choosing τ. Thus, theoretically, both cutting–plane methods can be applied for the solution of this problem. Considering our problem, we still have to overcome the difficulty that, depending on the choice of x, $y = Tx$ may be in the domain where ∇F is practically zero. This can be overcome by imposing individual lower bounds on the components of y. In the case of a normal distribution these can be, for instance, $y_i \geq \mu_i - 3 \cdot \sigma_i$, with μ_i and σ_i being the expected value and standard deviation of the ith one–dimensional marginal distribution of ξ, respectively. If the goal is just to find a Slater point for (4.7) then the iterations can be stopped, when the current iterate is already a Slater point for that problem.

Notice that the vehicle of imposing lower bounds on the components of y can also be utilized when applying the algorithms for solving (4.7). Some care is needed in this case, however. Applying too narrow lower bounds may result in a largely increased number of iterations and possibly also in almost parallel supporting hyperplanes. The reason is that in this case the iteration points may lay in a narrow region along the boundary of the feasible domain.

An important ingredient of both algorithms is the line–search procedure, for computing the intersection of the line segment $[y^{(k)}, x^S]$ and the boundary of the feasible domain. Introducing the notation $\Psi(\lambda) := F(y^{(k)} + \lambda(x^S - y^{(k)}))$, the problem is to find a λ^*, for which $\alpha \geq \Psi(\lambda) \geq \alpha - \varepsilon$ holds, for some prescribed tolerance $\varepsilon > 0$. The line–search is an important part of many nonlinear programming techniques and the overall performance of the algorithm may critically depend on the proper choice of the line–search procedure. There are several algorithms available for this purpose, see, for instance, Bazaraa and Shetty [11]. In our case, computing F and ∇F is relatively time–consuming and can only be performed with a rather limited accuracy, in general. As we will see later, for F there are some easily computable lower– and upper bounds available. The idea is to utilize these bounds in the line–search for reducing the number of steps where the value of F has to be computed.

For illustrating, let us consider bisection search which would run as follows: Initially we have $\Psi(0) < \alpha - \varepsilon$ and $\Psi(1) > \alpha$. We consider $[0,1]$ as our starting interval. Compute Ψ at the midpoint of the current interval, that is, compute $\Psi(\frac{1}{2})$. If $\Psi(\frac{1}{2}) < \alpha - \varepsilon$ choose $[\frac{1}{2}, 1]$ as the next interval, otherwise take $[0, \frac{1}{2}]$. Repeat the procedure till the length of the interval becomes small enough.

Let us assume now that we have bounds $F_L(x) \leq F(x) \leq F_U(x)$ available, with Ψ_L and Ψ_U denoting the corresponding bounds on Ψ. If in the above procedure $\Psi_U(\frac{1}{2}) < \alpha - \varepsilon$ holds, then we can safely choose $[\frac{1}{2}, 1]$ as our next interval. If this is not the case, then we check the inequality $\Psi_L(\frac{1}{2}) \geq \alpha$. If this holds then we can choose $[0, \frac{1}{2}]$ as the successor interval in the search. If none of these two inequalities hold then we are forced to compute $\Psi(\frac{1}{2})$ and to decide on that basis. For the details concerning implementation and further computational issues see Kall and Mayer [166], Mayer [230], and Szántai [320].

4.3.3 Other algorithms

Several authors have proposed further algorithms, based on some general framework of nonlinear programming.

The first algorithm for joint probabilistic constraints is due to Prékopa and Deák, see Prékopa et al. [269]. This method was based on a feasible direction method of Zoutendijk.

In the case, when F is a logconcave function, a natural idea is to work with logarithmic barrier functions by taking $\Psi(x, \kappa) := c^{\mathrm{T}}x + \kappa \log F(x) - \alpha$ as the objective function in the barrier subproblem. For fixed κ, $\Psi(x, \kappa)$ is a concave function of x on the set $\{x \mid F(x) \geq \alpha\}$. This fact is by no means obvious, for a proof see Prékopa [266]. An algorithm based on this idea has been developed by Rapcsák [275]. For variants and applications see Prékopa [266] and for penalty and barrier methods in general see, for example, Bazaraa and Shetty [11].

Komáromi [193] proposed a dual method, based on an appropriately constructed dual problem, for a detailed exposition see also Prékopa [266].

Mayer [228] constructed a reduced gradient type algorithm, with a suitably chosen direction finding subproblem. For details see, for instance, Kall and Wallace [172], Prékopa [266], and Mayer [230].

Gröwe [130] has developed algorithms for the case when the components of ξ are stochastically independent and the marginal distributions are logconcave. The algorithms are sample based and use techniques of non–parametric statistics for building LP–approximations to the problem.

Deák [57] proposes a regression–based algorithm for the case when the probability distribution has a logconcave density function; the basic idea is to approximate the probability distribution function $F(x)$ via quadratic regression and to work with a sequence of the corresponding approximating problems.

Gaivoronski [109] proposes quasigradient methods and reports on their implementation. Prékopa [267] presents an approach for obtaining approximate solutions by incorporating the bounds on the probability distribution function into the model formulation.

For overviews on existing methods see Prékopa [263], [268] and Mayer [229]; for detailed exposition of the methods see Kall and Wallace [172], Mayer [230] and Prékopa [266].

4.3.4 Bounds for the probability distribution function

The bounds in this section are distribution–free, meaning that they are valid irrespective of the probability distribution of the random vector $\xi : \Omega \to \mathbb{R}^r$, being a random vector on a probability space $(\Omega, \mathcal{F}, P)$. Our goal is to find lower– and upper bounds on the probability distribution function F of ξ.

We consider a fixed $x \in \mathbb{R}^r$ and will derive bounds on $F(x)$. We will proceed as follows. In a step–by–step fashion we derive several alternative formulas and methods for computing $F(x)$. It will turn out that, in general, none of them can be used in practice for computing $F(x)$ numerically. Nevertheless, finally we arrive at a formulation which offers a natural framework for constructing numerically computable bounds on $F(x)$.

We introduce the notation

$$\begin{aligned} A_i(x) &:= \{\omega \mid \xi_i(\omega) \le x_i\}, \quad i = 1, \ldots, r \\ B_i(x) &:= A_i^c(x) = \{\omega \mid \xi_i(\omega) > x_i\}, \quad i = 1, \ldots, r, \end{aligned} \tag{4.10}$$

where superscript c denotes the complement of a set. Notice that these sets depend on x. Having a fixed x, for the ease of presentation we will suppress this dependency in the notation, concerning also notions derived on the basis of the above sets.

Using the newly introduced notation, for the probability distribution function we get

$$\begin{aligned} F(x_1, \ldots, x_s) &= \mathbb{P}(A_1 \cap \ldots \cap A_r) = \mathbb{P}((B_1 \cup \ldots \cup B_r)^c) = \\ &= 1 - \mathbb{P}(B_1 \cup \ldots \cup B_r). \end{aligned} \tag{4.11}$$

Let furthermore $\nu : \Omega \to \{0,1,\dots,r\}$ be a random variable which counts the number of events which occur out of $B_1,\dots,B_r$. Formally we have the definition $\nu(\omega) := |I(\omega)|$, with $I(\omega) := \{1 \leq i \leq r \mid \omega \in B_i\}$. Employing this random variable, we obtain for the distribution function the following expression

$$F(x) = 1 - \mathbb{P}(\{\omega \mid \nu(\omega) \geq 1\}). \tag{4.12}$$

The question remains open, how to compute the probability on the right–hand–side of the above expression. We introduce the notation

$$S_0 := 1, \quad S_k := \sum_{1 \leq i_1 < \dots < i_k \leq r} \mathbb{P}(B_{i_1} \cap \dots \cap B_{i_k}) \text{ for } 1 \leq k \leq r \tag{4.13}$$

and will call S_k the kth *binomial moment* of ν, $k = 0,1,\dots,r$. Notice that for computing all binomial moments, we have to evaluate all probabilities in (4.13), the number of which grows exponentially with r. Anyhow, presupposing that all binomial moments are known, the probability in question can be computed according to the inclusion–exclusion formula, see Feller [91], as follows:

$$\mathbb{P}(\nu \geq 1) = S_1 - S_2 + \dots + (-1)^{r-1} S_r. \tag{4.14}$$

Using mathematical induction it is easy to prove that, for every even integer $0 \leq m < r$, we have the inequalities

$$\sum_{j=1}^{m} (-1)^{j-1} S_j \leq \mathbb{P}(\nu \geq 1) \leq \sum_{j=1}^{m+1} (-1)^{j-1} S_j.$$

For $m = 0$ we get the well–known inequality concerning probabilities

$$0 \leq \mathbb{P}(\nu \geq 1) = \mathbb{P}(B_1 \cup \dots \cup B_r) \leq \sum_{j=1}^{r} \mathbb{P}(B_j)$$

and for $m = 2$ we obtain the inequalities

$$S_1 - S_2 \leq \mathbb{P}(\nu \geq 1) \leq S_1 - S_2 + S_3.$$

We wish to derive sharp bounds of this type. Let us associate with ν the random variables

$$\binom{\nu}{k} = \frac{\nu(\omega)!}{k!(\nu(\omega)-k)!}, \quad k = 0,1,\dots,r.$$

The following fact explains the term *binomial moments* concerning S_k:

Proposition 4.9. *For $k = 0,\dots,r$ holds*

$$\mathbb{E}\left[\binom{\nu}{k}\right] = S_k.$$

Proof: Let $\chi_i : \Omega \to \mathbb{R}$ be indicator variables defined as

$$\chi_i(\omega) = \begin{cases} 1 \text{ if } \omega \in B_i \\ 0 \text{ otherwise.} \end{cases}$$

Then we obviously have that $\nu(\omega) = \chi_1(\omega) + \ldots + \chi_r(\omega)$ holds, for all $\omega \in \Omega$. Consequently,

$$\binom{\nu(\omega)}{k} = \binom{\chi_1 + \ldots + \chi_r}{k} = \sum_{1 \le i_1 < \ldots < i_k \le r} \chi_{i_1}\chi_{i_2}\cdots\chi_{i_k}$$

holds. Taking expectation

$$\begin{aligned} \mathbb{E}\left[\binom{\nu}{k}\right] &= \sum_{1 \le i_1 < \ldots < i_k \le r} \mathbb{E}[\chi_{i_1}\chi_{i_2}\cdots\chi_{i_k}] \\ &= \sum_{1 \le i_1 < \ldots < i_k \le r} \mathbb{P}[B_{i_1} \cap \ldots \cap B_{i_k}] = S_k \end{aligned}$$

yields the result. □

Utilizing that ν has a finite discrete distribution, the above result can also be written as

$$S_k = \mathbb{E}\left[\binom{\nu}{k}\right] = \sum_{i=k}^{r} \binom{i}{k} \mathbb{P}(\nu(\omega) = i), \ k = 0, 1, \ldots r. \tag{4.15}$$

Assuming that the binomial moments S_k are known, (4.15) can be viewed as a system of linear equations for the unknown probabilities $\mathbb{P}(\nu(\omega) = i)$, $i = 0, 1, \ldots, r$. Let us consider this system with added nonnegativity requirements concerning the unknowns:

$$\begin{array}{rcl} \nu_0 + \nu_1 + \nu_2 + \ \cdots & +\nu_r &= S_0 \\ \nu_1 + 2\nu_2 + \ \cdots & +r\nu_r &= S_1 \\ \nu_2 + \ \cdots + & \binom{r}{2}\nu_r &= S_2 \\ \ddots & \vdots & \\ & \nu_r &= S_r \\ & \nu_i \ge 0, & i = 0, \ldots, r. \end{array} \tag{4.16}$$

The coefficient matrix of the equation part of the system has an upper–triangular structure with non–zeros along the main diagonal. Consequently, this matrix is nonsingular implying that the equation part of (4.16) has the unique solution $\nu_i^* = \mathbb{P}(\nu(\omega) = i) \ge 0$, $i = 0, \ldots, r$ (cf. (4.15)).

Thus we get

$$\mathbb{P}(\nu \ge 1) = \sum_{i=1}^{r} \mathbb{P}(\nu = i) = \sum_{i=1}^{r} \nu_i^*.$$

Theoretically, this approach offers a possibility for computing $\mathbb{P}(\nu \geq 1)$ as follows. Compute all binomial moments S_k, $k = 0, 1, \ldots, r$. Subsequently set up and solve (4.16) and compute $\mathbb{P}(\nu \geq 1)$ according to the formula above. Finally compute $F(x)$ according to (4.12). From the numerical point of view the difficulty is the very first step in this procedure: computing the binomial moments involves the computation of probabilities according to (4.13), the number of which grows exponentially with r. On the other hand, having computed the binomial moments, there is no need to take the roundabout way via solving (4.16), because $\mathbb{P}(\nu \geq 1)$ can be directly computed using the exclusion–inclusion formula (4.14).

The formulation via (4.14) offers, however, an elegant way for constructing bounds by employing relaxation as follows. The idea is keeping only the equations corresponding to the first few binomial moments. With the first and second binomial moments, the LP formulation for the lower bound is the following:

$$\begin{array}{lll} V_{min} = & & \\ \min & v_1 + v_2 + \ldots + v_r & \\ \text{s.t.} & v_1 + 2v_2 + \ldots + rv_r & = S_1 \\ & v_2 + \ldots + \binom{r}{2} v_r & = S_2 \\ & v_i & \geq 0, \; i = 1, \ldots, r. \end{array} \tag{4.17}$$

Notice that the system of linear equations in (4.16) has a unique solution $v^* = (v_0^*, \ldots, v_r^*)^{\mathrm{T}}$. Therefore, when formulating it as an LP with the same objective as in (4.17), the resulting LP has the optimal solution v^*. Observing that (4.17) is a relaxation of that LP, we immediately get that

$$V_{min} \leq \sum_{i=1}^{r} v_i^* = \mathbb{P}(\nu \geq 1)$$

holds, showing that the optimal objective value of (4.17) in fact provides a lower bound. An upper bound V_{max} can be obtained analogously, by simply changing in (4.17) the direction of optimization to maximization.

Observe that both (4.17) and its counterpart for the upper bound are LP–problems just having two equality constraints and for both problems the feasible domain being non–empty and bounded, both problems have optimal solutions. By taking into account the special structure, closed form solutions can be derived for both LP problems as explained in detail in Kall and Wallace [172]. We get

$$\begin{aligned} V_{min} &= \tfrac{2}{k^*+1} S_1 - \tfrac{2}{k^*(k^*+1)} S_2, \text{ with } k^* := \left\lfloor \tfrac{2S_2}{S_1} \right\rfloor + 1, \\ V_{max} &= S_1 - \tfrac{2}{r} S_2, \end{aligned} \tag{4.18}$$

where for any real number λ, $\lfloor \lambda \rfloor$ denotes the *floor* of λ, meaning the greatest integer which is less than or equal to λ. The bounds (4.18) are called Boole–Bonferroni bounds.

The above way for deriving these bounds is due to Kwerel [201] and Prékopa [262]. Bounds in explicit form, involving higher order binomial moments, have been obtained by Kwerel [201] and Boros and Prékopa [32]. Algorithmically computable bounds are presented in Prékopa [264]. For the details see Prékopa [266] and for Boole–Bonferroni–type bounds in general see also Galambos and Simonelli [111].

Taking into account (4.12) and (4.18), we get for the probability distribution $F(x)$ the following bounds

$$1 - S_1(x) + \frac{2}{r} S_2(x) \le F(x) \le 1 - \frac{2}{k^*(x)+1} S_1(x) + \frac{2}{k^*(x)(k^*(x)+1)} S_2(x),$$

where we now explicitly indicate the dependency of the binomial moments on x (see the remarks concerning (4.10)) and $k(x)$ is given as specified in (4.18).

The final step in presenting an algorithm for computing the bounds consists of specifying how the binomial moments $S_1(x)$ and $S_2(x)$ can be computed. For $S_1(x)$ we have (cf. (4.13))

$$\begin{aligned} S_1(x) &= \sum_{i=1}^{r} \mathbb{P}(B_i(x)) = r - \sum_{i=1}^{r} \mathbb{P}(A_i(x)) \\ &= r - \sum_{i=1}^{r} F_i(x_i), \end{aligned}$$

where $F_i(x_i)$ is the distribution function of the ith one dimensional marginal distribution of ξ. Considering now $S_2(x)$, for fixed i and j we have

$$\mathbb{P}(B_i \cap B_j) = 1 - \mathbb{P}(A_i) - \mathbb{P}(A_j) + \mathbb{P}(A_i \cap A_j),$$

where $\mathbb{P}(A_i) = F_i(x_i)$ holds for all i and furthermore

$$\mathbb{P}(A_i \cap A_j) = \mathbb{P}(\xi_i \le x_i, \xi_j \le x_j) = F_{ij}(x_i, x_j)$$

holds for all i and j. Here $F_{ij}(x_i, x_j)$ is the probability distribution function corresponding to the two–dimensional marginal distribution of (ξ_i, ξ_j). Thus we get for the binomial moment $S_2(x)$ the expression

$$\begin{aligned} S_2(x) &= \sum_{1 \le i < j \le r} \mathbb{P}(A_i \cap A_j) \\ &= \binom{r}{2} - (r-1) \sum_{i=1}^{r} F_i(x_i) + \sum_{1 \le i < j \le r} F_{ij}(x_i, x_j). \end{aligned}$$

If ξ has a non–degenerate multivariate normal distribution then all marginal distributions are non–degenerate normal distributions, see, for instance, Mardia et al. [216]. Similar results hold for the Dirichlet and gamma distributions, see Prékopa [266], cf. also Theorem 2.9. on page 121. For computing the value of univariate and bivariate normal distribution functions see the next section.

An alternative way for deriving bounds on $F(x)$ is based on graphs. Our starting point is the formulation (4.11). According to this relation, for deriving bounds on $F(x)$ it is sufficient to construct bounds on $\mathbb{P}(B_1 \cup \ldots \cup B_r)$. We will discuss an upper bound for this probability, due to Hunter [145], which results in a lower bound on $F(x)$. The following relations obviously hold:

$$\begin{aligned}\mathbb{P}(B_1 \cup \ldots \cup B_r) &= \mathbb{P}(B_1) + \sum_{j=2}^{r} \mathbb{P}(B_1^c \cap \ldots \cap B_{j-1}^c \cap B_j) \\ &\leq \mathbb{P}(B_1) + \sum_{j=2}^{r} \mathbb{P}(B_{[j]}^c \cap B_j) \\ &= \sum_{j=1}^{r} \mathbb{P}(B_j) - \sum_{j=2}^{r} \mathbb{P}(B_{[j]} \cap B_j),\end{aligned} \tag{4.19}$$

where $[j]$ is any index in $\{1,\ldots,j-1\}$. Thus, depending on the choice of the $[j]$'s, (4.19) provides altogether $(r-1)!$ upper bounds. We would be interested in the best of these bounds. A convenient way for dealing with the upper bounds in (4.19) is via the following construction:

Let $G = (V,E)$ be an undirected complete weighted graph with r nodes (vertices). We associate the event B_i with vertex i and the intersection $B_i \cap B_j$ with edge $(i,j) \in E$, for all $1 \leq i,j \leq r$, $i \neq j$. The weights are associated to the edges via $(i,j) \mapsto \mathbb{P}(B_i \cap B_j)$ for all $(i,j) \in E$. The idea is to represent the second term on the right–hand–side of (4.19) as the weight of a *spanning tree* in G. A spanning tree is a subgraph T of G, which is a tree and has the same set of vertices as G. Consequently, T has $r-1$ edges, it is connected, and it contains no cycles (see, for instance Ahuja et al. [5]). The weight of the spanning tree, denoted by $w(T)$, is defined as the sum of weights over all edges of T.

We observe that, for any fixed choice of $[j]$ for all j, the second sum in the right–hand–side of (4.19) is equal to the weight of the following spanning tree of G. Choose all edges $([j],j)$ and consider the subtree T of G which has this set of edges and the corresponding set of nodes. Notice that, for $j=2$, $[2]=1$ is the only available choice. Consequently, all nodes of G appear also as nodes of T. Furthermore, due to its construction, T is obviously a tree with its weight equal to the sum under consideration in (4.19).

Thus we have associated to each one of the $(r-1)!$ bounds in (4.19) a spanning tree in G. However, the number of different spanning trees of G is r^{r-2}, see Knuth [192] Volume 1, which is in general much higher than the number of possible bounds considered so far. Our next observation is the following. While the left–hand–side and the first term in the right–hand–side of the inequality (4.19) are both independent on the assignment of indices to the events, the second term on the right–hand–side depends not only on the choice of the $[j]$'s but also on the numbering of the events. Thus we can get further bounds by renumbering these events.

For accomplishing this let us now consider an *arbitrary* spanning tree T of G. We associate with this tree a reordering of the indices of the events by the following r–

step process. In step 1 choose any node i and assign the index 1 to B_i. In general, in step ν ($1 < \nu \leq r$) proceed as follows. Select one of the already renumbered nodes, which has a not yet renumbered neighbor B_k. Let κ be the already assigned new index of this node. Assign the index ν to B_k and set $[\nu] := \kappa$. Due to the fact that T is a tree, it is easy to see that this procedure can be carried out in r steps and that the weight of the tree is equal to the corresponding sum in (4.19), according to the new indexing of events.

Consequently, the best upper bound can be obtained by solving the following optimization problem:

$$\max_{T \in \mathscr{T}} \sum_{(i,j) \in E_T} \mathbb{P}(B_i(x) \cap B_j(x)), \tag{4.20}$$

where $\mathscr{T}$ is the set of all spanning trees of G and E_T is the set of edges of T. Let us denote by $T^*(x)$ an optimal solution of (4.20). We obtain the following lower bound for $F(x)$ (see (4.11) and (4.19)):

$$1 - S_1(x) + w(T^*(x)) \leq F(x), \quad \forall x \in \mathbb{R}^r. \tag{4.21}$$

Problem (4.20) is a classical problem in combinatorial optimization, where it is usually formulated as a minimization problem and is called the *minimum spanning tree problem.* There are several thoroughly studied and well-tested algorithms available, see Ahuja et al. [5]. It is easy to see that the direction of optimization does not matter; the same algorithms can be used for both variants, with obvious modifications. In our case we have a dense graph (G is a complete graph), therefore Prim's method is well–suited for the solution of the problem, see Ahuja et al. [5]. The algorithm builds the minimum spanning tree in a greedy manner in $r-1$ iterations, by adding a new edge to the tree at each of the iterations. Wee will keep two lists: at iteration ν, V_ν will be the list of vertices and E_ν will be the list of edges of the current subtree. The general framework of this algorithm for solving (4.20) is the following:

Step 1. *Initialization*
Look for a longest edge $(i,j) = \text{argmax}_{(k,l)} \mathbb{P}(B_k \cap B_l)$. Set $\nu = 1$, $E_1 = \{(i,j)\}$, and $V_1 = \{i\}$.

Step 2. *Choose the next edge*
If $\nu = r-1$ then **Stop**, the current graph with set of edges E_ν is a maximum weight spanning tree of G. Otherwise look for the longest edge with one of its vertices in V_ν and the other one in $V \setminus V_\nu$:

$$(p,q) := \text{argmax}_{(k,l)} \{ \mathbb{P}(B_k \cap B_l) \mid k \in V_\nu, l \in V \setminus V_\nu \}.$$

Step 3. *Add a new edge to the tree*
Let $E_{\nu+1} = E_\nu \cup \{(p,q)\}$; $V_{\nu+1} = V_\nu \cup \{q\}$; set $\nu := \nu + 1$ and **continue with *Step 2*.**

Let us point out, that the above scheme is just the framework of the method; the efficiency largely depends on the details of the implementation, especially on the organization of the heap, see Ahuja et al. [5].

It is well–known that the Hunter bound is always at least as good, as the Boole–Bonferroni bound, see Prékopa [266]. The bounds can be further improved by employing hypergraphs and hypertrees, see Bukszár and Prékopa [37], Szántai [321], Szántai and Bukszár [322], and the references therein.

4.3.5 Computing probability distribution functions

The main goal of this section is to discuss algorithms for computing the value of the multivariate normal probability distribution function. Besides this, we will also outline ideas for computing the probability distribution function of some other multivariate distributions.

For computing the probability distribution function of the univariate normal distribution, there are ready–made functions available in almost all computing environments. For computing the bivariate normal distribution function $F_{ij}(x_i, x_j)$ several well–tested procedures are available. One of the simplest tricks is based on the following reformulation:

$$\begin{aligned} F_{ij}(x_i, x_j) &= \int_{-\infty}^{x_i} \int_{-\infty}^{x_j} f_{ij}(x, y) dx dy \\ &= \int_{-\infty}^{x_i} \int_{-\infty}^{x_j} f_{ij}(y \mid x) f_i(x) dx dy \\ &= \int_{-\infty}^{x_i} f_i(x) \left(\int_{-\infty}^{x_j} f_{ij}(y \mid x) dy \right) dx, \end{aligned} \tag{4.22}$$

where $f_{ij}(y \mid x)$ is the conditional density function of ξ_j, given $\xi_i = x$. For the normal distribution, the conditional distributions are also normal distributions, see Mardia et al. [216]. We obtain that $f_{ij}(y \mid x)$ is the density function of a univariate normal distribution $\mathscr{N}(\mu_j + \left(\rho_{ij} \frac{\sigma_j}{\sigma_i}\right)(x - \mu_i), \sigma_j^2(1 - \rho_{ij}^2))$, where μ_i denotes the expected value and σ_i stands for the standard deviation of ξ_i, and ρ_{ij} denotes the correlation coefficient between ξ_i and ξ_j, for all i and j. Consequently the inner integrand in (4.22) is just a normal univariate probability distribution function. Thus, $F_{ij}(x_i, x_j)$ can be evaluated by employing a univariate numerical quadrature for integration. For a state–of–the art review for computing bivariate normal probabilities see Genz [124]. This paper also presents methods for computing trivariate normal probabilities.

Let us turn our attention to the multivariate case. Recall that the standard non-degenerate multivariate normal distribution function has the following form (cf. (2.52) on page 99):

$$F(x) = \frac{1}{(2\pi)^{\frac{r}{2}}|R|^{\frac{1}{2}}} \int\limits_{\mathbb{R}^r} e^{-\frac{1}{2}y^{\mathrm{T}}R^{-1}y}\,dy, \tag{4.23}$$

where R is the nonsingular correlation matrix of ξ. From the numerical point of view the problem is to evaluate the above multivariate integral. In principle, this can be done by standard nested quadrature methods of numerical analysis. For higher dimensions, however, the specialities of the problem are to be taken into account. Algorithms of this kind have been developed by several authors, mainly for the cases of multivariate normal– and t–distributions, see the review papers of Genz and Bretz [125] and Gassmann, Deák and Szántai [116], and the references therein.

In this book we will restrict ourselves to the Monte–Carlo approach and will discuss two basic techniques for computing $F(x)$. The two algorithms can also be combined; for the resulting hybrid method see Gassmann et al. [116]. This paper also provides a review on methods for computing multivariate normal probabilities.

A Monte–Carlo approach with antithetic variates

For the non–degenerate multivariate normal distribution, this method has been developed by Deák [55]. Recently, Genz and Bretz [125] extended the method to multivariate t–distributions. We will discuss the multivariate normal case.

The starting point is to transform the integral in (4.23) to a polar form. Let $R = LL^{\mathrm{T}}$ be the Cholesky–factorization of R with L being a lower–triangular matrix (see Section 2.2.3). Applying the transformation $y = Lz$ first results in

$$F(x) = \frac{1}{(2\pi)^{\frac{r}{2}}} \int\limits_{\{z:\, Lz \le x\}} e^{-\frac{1}{2}z^{\mathrm{T}}z}\,dz.$$

For changing to polar coordinates apply the transformation $z = ru$ with $\|u\| = 1$ which results in

$$F(x) = \frac{1}{(2\pi)^{\frac{r}{2}}} \int\limits_{\{u:\, \|u\|=1\}} \int_{\rho_1(u)}^{\rho_2(u)} \rho^{r-1} e^{-\frac{1}{2}\rho}\,d\rho\,du, \tag{4.24}$$

where

$$\rho_1(u) = \min\{\rho \mid \rho \ge 0, \rho Lu \le x\}$$
$$\rho_2(u) = \max\{\rho \mid \rho \ge 0, \rho Lu \le x\}$$

Notice that, apart of a normalizing constant, the integrand in (4.24) is the probability density function of the χ–distribution with r degrees of freedom, see Johnson et al. [150]. In fact, the method can also be derived in a purely probabilistic fashion, as it has been done in the original paper of Deák [55], see also Deák [56] and Theorem 4.1.1 in Tong [328].

Normalizing (4.24) leads to the equivalent form

$$F(x) = \frac{1}{\gamma} \int_{\{u:\, \|u\|=1\}} \int_{\rho_1(u)}^{\rho_2(u)} g(\rho)\, d\rho\, du, \tag{4.25}$$

where

$$g(\rho) = \kappa \rho^{r-1} e^{-\frac{1}{2}\rho}$$

is the probability distribution function of the χ–distribution with r degrees of freedom and the normalizing constants γ and κ are

$$\kappa = 2^{\frac{r}{2}-1}\Gamma\left(\frac{r}{2}\right), \quad \gamma = \frac{2\pi^{\frac{1}{2}r}}{\Gamma(\frac{r}{2})},$$

where γ is the surface area of the r–dimensional unit sphere, see, for instance, Price [272]. (4.25) can also be written as follows

$$F(x) = \frac{1}{\gamma} \int_{\{u:\, \|u\|=1\}} h(u)\, du \quad \text{with} \quad h(u) := \int_{\rho_1(u)}^{\rho_2(u)} g(\rho)\, d\rho.$$

The idea is to evaluate the first (surface) integral by Monte–Carlo methods, whereas for the second (univariate) integral numerical integration is used. Choosing a sample–size N, the framework of the method is the following:

Step 1. *Generating points on the unit sphere*
Generate N sample points $\hat{u}_1, \ldots, \hat{u}_N$ uniformly distributed on the unit sphere in $\mathbb{R}^r$.

Step 2. *Compute h*
For each of the sample points $\hat{u}_k$, $k = 1, \ldots, N$, in turn do:
- compute $\rho_1(\hat{u}_k)$ and $\rho_2(\hat{u}_k)$;
- evaluate $h_k := h(\hat{u}_k)$ by numerical integration.

Step 3. *Compute the Monte–Carlo estimator*

$$\hat{F} = \frac{1}{N} \sum_{k=1}^{N} h_k.$$

For computing uniformly distributed points on the r–dimensional unit sphere, the standard method is the following:

Step 1. Generate r (i.i.d.) random numbers $d_1, \ldots, d_r$ according to the standard univariate normal distribution;

Step 2. compute $d = \sqrt{d_1^2 + \ldots + d_r^2}$;

Step 3. deliver $u^{\mathrm{T}} = (\frac{d_1}{d}, \ldots, \frac{d_r}{d})$.

For this method and further methods for generating uniformly distributed points on the r–dimensional unit sphere see Devroye [69]. The method discussed so far would correspond to the "crude" Monte–Carlo method, with the estimator having a variance proportional to $\frac{1}{N}$. As in Monte–Carlo methods in general, it is of vital importance to include some variance reduction techniques, see, for instance, Ripley [278] or Ross [291]. Deák proposes the following variant of the method of antithetic variates, with $m \le r$ being a parameter of the algorithm:

Step 1. *Generate points on the unit sphere*
Generate $N \cdot r$ sample points $\hat{u}_{k,j}$, $k = 1,\ldots,N$, $j = 1,\ldots,r$ uniformly distributed on the unit sphere in $\mathbb{R}^r$.

Step 2. *Compute h*
For each $k = 1,\ldots,N$, in turn do:

- Convert $\hat{u}_{k1},\ldots,\hat{u}_{kr}$ into an orthonormal system $v_1,\ldots,v_r$, by employing, for instance, the standard Gram–Schmidt procedure. For a possible (but very unlikely) linear dependency among the generated vectors, apply a heuristics based on dropping and recomputing some of the vectors.
- Compute $M := 2^m \binom{r}{m}$ vectors on the unit sphere according to

$$w(s, j_1,\ldots, j_m) := \frac{1}{\sqrt{m}} \sum_{l=1}^{m} s_l v_{j_l}$$

with $1 \le j_1 < \ldots < j_m \le r$ and with
$s \in S$, where $S := \{ s \in \mathbb{R}^m \mid s_i = 1 \text{ or } s_i = -1, \ \forall i \}$.
- For each of these vectors compute $\rho_1(w(s, j_1,\ldots, j_m))$ and $\rho_2(w(s, j_1,\ldots, j_m))$ and
- evaluate $h_k(s, j_1,\ldots, j_m) := h(w(s, j_1,\ldots, j_m))$ by numerical integration.
- Compute

$$h_k := \frac{1}{M} \sum_{s \in S} \sum_{1 \le j_1 < \ldots < j_m \le r} h_k(s, j_1,\ldots, j_m).$$

Step 3. *Compute the Monte–Carlo estimator*

$$\hat{F} = \frac{1}{N} \sum_{k=1}^{N} h_k.$$

Concerning the parameter m of the procedure, best results were reported for the choices $m = 1$ and $m = 3$, see Gassmann, Deák, and Szántai [116]. For the implementation of the algorithm and recent improvements see Deák [58].

A Monte–Carlo approach based on probability bounds

This approach has been developed by Szántai [317], [318]. We discuss the technique for computing the probability

$$P_V := \mathbb{P}(\{\omega \mid \nu(\omega) \ge 1\}) = \mathbb{P}(B_1 \cup \ldots \cup B_r),$$

where the random variable ν counts the number of events which occur out of $B_1,\ldots,B_r$, see page 297. Recall that according to Proposition 4.9. we have for all $k \geq 1$

$$\mathbb{E}\left[\binom{\nu}{k}\right] = S_k. \tag{4.26}$$

The method will be based on the Boole–Bonferroni bounds

$$P_L := \frac{2}{k^*+1}S_1 - \frac{2}{k^*(k^*+1)}S_2 \leq \hat{P} \leq S_1 - \frac{2}{r}S_2 =: P_U, \tag{4.27}$$

see (4.18) where the definition of k^* can also be found.

Having computed an estimate for $\hat{P}$, the estimate for the probability distribution function $F(x)$ can be obtained according to (4.11).

The algorithm is based on the inclusion–exclusion formula (4.14) and on the Boole–Bonferroni bounds. The idea is to compute three unbiased estimators for $\hat{P}$. Using these, a linear combination of them is computed with minimal variance, which will be the final unbiased estimator.

The first estimator is the crude Monte–Carlo estimator, concerning the random variable

$$\vartheta_0 = \begin{Bmatrix} 0 \text{ if } \nu = 0 \\ 1 \text{ if } \nu > 0 \end{Bmatrix}$$

for which we obviously have $\mathbb{E}[\vartheta_0] = P_V$.

For the second estimator we consider the difference between $\hat{P}$ and the Boole–Bonferroni lower bound

$$\Delta_{\hat{L}} := P_V - P_L.$$

Substituting the inclusion–exclusion formula (4.14) for the probability, and the expression for P_L according to (4.27), we get

$$\Delta_{\hat{L}} = \sum_{k=1}^{r}(-1)^{k^*-1}S_k - \frac{2}{k^*+1}S_1 + \frac{2}{k^*(k^*+1)}S_2.$$

The method is based on the following observation: if we substitute the binomial moments by the random variables from (4.26), then cancellation occurs according to

$$\begin{aligned}\Delta_L &:= \sum_{k=1}^{r}(-1)^{k-1}\binom{\nu}{k} - \frac{2}{k+1}\binom{\nu}{1} + \frac{2}{k(k+1)}\binom{\nu}{2} \\ &= \frac{1}{k^*(k^*+1)}(\nu - k^*)(\nu - k^* - 1),\end{aligned}$$

where we assumed $\nu \geq 2$ and have utilized the obvious relation

$$\sum_{k=0}^{\nu}(-1)^{k-1}\binom{\nu}{k} = (1-1)^{\nu} \equiv 0.$$

Thus we will consider the random variable

$$\theta_1 := \left\{ \begin{array}{cl} 0 & \text{if } \nu \le 1 \\ \dfrac{1}{k^*(k^*+1)}(\nu - k^*)(\nu - k^* - 1) & \text{if } \nu \ge 2 \end{array} \right\}.$$

According to the above considerations, $\vartheta_1 := \theta_1 + P_L$ is an unbiased estimator for P_V, that is, $\mathbb{E}[\vartheta_1] = P_V$ holds.

For the third estimator we consider the difference between $\hat{P}$ and the Boole–Bonferroni upper bound

$$\begin{aligned} \Delta_{\hat{U}} &:= \hat{P} - P_U \\ &= \sum_{k=2}^{r} (-1)^{k-1} S_k + \frac{2}{r} S_2. \end{aligned}$$

Proceeding analogously as before, cancellation occurs again, and we end up with the random variable

$$\theta_2 := \left\{ \begin{array}{cl} 0 & \text{if } \nu \le 1 \\ \frac{1}{r}(\nu - 1)(\nu - r) & \text{if } \nu \ge 2 \end{array} \right\}.$$

With $\vartheta_2 := \theta_2 + P_U$ we have $\mathbb{E}[\vartheta_2] = P_V$. Thus, we obtained a third unbiased estimator for P_V.

The final estimate is obtained as follows. Let C be the covariance matrix of $(\vartheta_0, \vartheta_1, \vartheta_2)$. The estimate will have the form $\vartheta := w_0 \vartheta_0 + w_1 \vartheta_1 + w_2 \vartheta_2$. Denoting the vector of weights by $w^{\mathrm{T}} = (w_0, w_1, w_2)$, the variance of ϑ is obviously $w^{\mathrm{T}} C w$. The weights are determined by solving the following minimum–variance problem

$$\left. \begin{array}{ll} \min & w^{\mathrm{T}} C w \\ \text{s.t.} & w_0 + w_1 + w_2 = 1 \end{array} \right\} \tag{4.28}$$

which, with Lagrange–multiplier λ, is equivalent to solving the following system of linear equations:

$$\left. \begin{array}{ll} Cw - \lambda e & = 0 \\ w_0 + w_1 + w_2 & = 1 \end{array} \right\} \tag{4.29}$$

where $e_i = 1$ for $i = 1, 2, 3$ hold. Notice that, due to the constraint in (4.28), ϑ is also an unbiased estimator of P_V.

The algorithm runs as follows:

Step 0. *Compute bounds*
Compute the Boole–Bonferroni bounds P_L and P_U according to (4.27), by numerical integration.

Step 1. *Generate a sample*
For $k = 1, \ldots, N$ proceed as follows:
- Generate a random vector according to the distribution of ξ;
- compute the realization of ν, $\hat{\nu}^k$;
- compute corresponding realizations $\hat{\vartheta}_0^k$, $\hat{\vartheta}_1^k$, and $\hat{\vartheta}_2^k$.

Step 2. *Compute first estimates*
Compute the estimates

$$\hat{\vartheta}_i := \frac{1}{N}\sum_{k=1}^{N}\hat{\vartheta}_i^k, \quad i = 1,2,3;$$

(for $\hat{\vartheta}_0$ this is implemented via a counter, of course).

Step 3. *Compute an estimate for the covariance matrix C*
Using the sample, compute an estimate $\hat{C}$ for C.

Step 4. *Compute weights which minimize variance*
Compute the weights by solving (4.29) with $C = \hat{C}$. Let the solution be $\hat{w}^{\mathrm{T}} = (\hat{w}_0, \hat{w}_1, \hat{w}_2)$.

Step 5. *Compute final estimate*
Compute $\hat{\vartheta} = \hat{w}_0\hat{\vartheta}_0 + \hat{w}_1\hat{\vartheta}_1 + \hat{w}_2\hat{\vartheta}_2$.

Step 6. *Deliver estimate for $F(x)$*
According to (4.11), deliver $\hat{F} := 1 - \hat{\vartheta}$ as an estimate for $F(x)$.

For further development of the procedure involving bounds with higher–order binomial moments and for graph–based bounds see Szántai [321]. For applying the technique to the computation of other multivariate distribution functions, including the Dirichlet distribution (cf. page 123) and the gamma distribution (cf. page 127) see Szántai [318] and [319].

4.3.6 Finite discrete distributions

If ξ has a finite discrete distribution then the SLP problem with joint probability function can be formulated as a disjunctive programming problem, see Chapter 2 Section 2.2.2. Possible solution approaches are to solve the equivalent mixed–integer linear problem (2.44) on page 94 by employing a general–purpose solver for such problems, or to apply the general techniques for disjunctive programming, see, for instance, Nemhauser and Wolsey [241].

In the case, when only the right–hand–side is stochastic, special–purpose algorithms are available. The basic idea is the following disjunctive formulation of (4.7), due to Prékopa [265] and Sen [300]:

$$\left.\begin{array}{ll} \max c^{\mathrm{T}}x & \\ \text{s.t.} & y \in \bigcup_{s=1}^{S} D_s \\ & Tx - y \geq 0 \\ & x \quad \in \mathscr{B}, \end{array}\right\} \qquad (4.30)$$

where D_s is $\{y \mid y \geq y^s\}$, with y^s, $s = 1,\ldots,S$ defined as follows. Let

$$\mathscr{D} := \{y \mid F(y) \geq \alpha \text{ and } F(y-\varepsilon) < \alpha \text{ for all } \varepsilon \in \mathbb{R}^s,\ \varepsilon > 0,\ \varepsilon \neq 0\}.$$

The set $\mathscr{D}$ is clearly a subset of all joint realizations of ξ, therefore $\mathscr{D}$ is a finite set. With S denoting the number of elements in $\mathscr{D}$, y^s is the sth element, indexed in an arbitrary order.

Clearly, $y^s \in \mathscr{D}$, if and only if $F(y^s) \geq \alpha$ holds and there exists no $y \leq y^s$, $y \neq y^s$, such that $F(y) \geq \alpha$ holds. For this reason, Prékopa[265] has coined the following terminology: the elements of $\mathscr{D}$ are called p–level efficient points (PLEP's) of the probability distribution function F. The terminology corresponds to the choice $\alpha := p$ for the probability level.

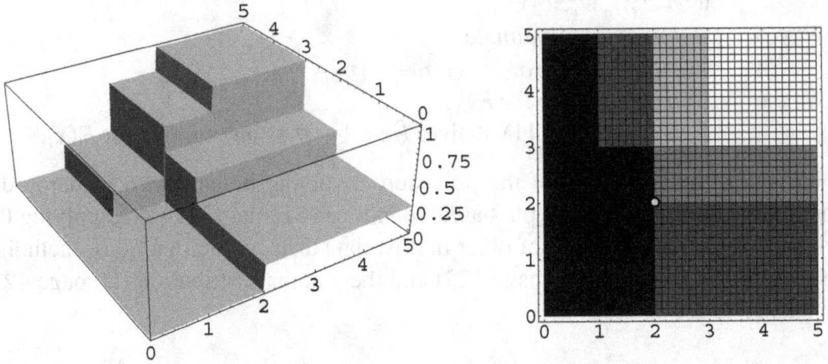

Fig. 4.4 Distribution function and level sets of a bivariate finite discrete distribution function with a PLEP at $(2,2)$, corresponding to $\alpha = 0.5$.

Figure 4.4 shows the probability distribution function for a discrete distribution in $\mathbb{R}^2$, with four equally probable realizations $(1,1)$, $(2,1)$, $(2,2)$, and $(3,3)$. For the level $\alpha = 0.5$, the realization $(2,2)$ is the single PLEP of F.

Problem (4.30) is a disjunctive programming problem with an especially simple structure. Several algorithms, based on enumeration, cuts, and Lagrangean relaxation have been proposed to its solution, see Dentcheva et al. [65], Prékopa [266], Prékopa et al. [271], Ruszczyński [294], and Sen [300], as well as the references therein. For an overview see Prékopa [268].

For a recent approach for formulating and solving equivalent mixed–integer problems see Luedtke et al. [208].

4.3.7 A guide to available software

SLP problems with logconcave distribution functions

The following solvers have been developed by Mayer, see, for instance [231]. The cutting plane method of Szántai is implemented as the solver PCSPIOR; the central cutting plane method is implemented in Fortran as the solver PROBALL, and the reduced gradient approach is implemented as the solver PROCON. These solvers are for the case when the random right–hand–side has a non–degenerate multivariate normal distribution; for computing the probability distribution function and its gradient the subroutines of Deák and Szántai have been utilized, see the next section. All implementations are in Fortran and use Minos (see Murtagh and Saunders [239], [240]) for solving the LP subproblems. The solvers are available along with the model management system SLP–IOR, as attached solvers, see Section 4.9.2 in this chapter.

For aiding the selection of an appropriate solver, comparative computational results can be found in Kall and Mayer [164], [166], and Mayer [230], [231]. According to these tests, we recommend to use PROBALL. There are no independent benchmark results available. However, as SLP–IOR is freely available for academic purposes, the reader can test the solvers herself/himself.

Szántai [320] has implemented in Fortran his cutting plane method as the solver PCSP. The authors of the methods discussed in Section 4.3.3 also report on solving some test problems, the solvers might be available on request, we suggest to contact the authors.

For the cases, when the problem is not a convex optimization problem, solvers for global optimization might prove to be useful; for an overview on solvers see Pintér [256].

Evaluating probability distribution functions

For choosing an appropriate algorithm for the evaluation of multivariate probability distribution functions, guidelines, based on numerical experimentation, have been published by Gassmann, Deák, and Szántai [116]. This paper also gives an overview on currently available software. Here we just point out the following issues:

The programs of Alan Genz, implemented in Fortran and MatLab, are available on his personal homepage
http://www.sci.wsu.edu/math/faculty/genz/homepage.

The algorithm in Section 4.3.5 has been implemented by Deák in the Fortran subroutine package NORSET, see [58], whereas the methods discussed in Section 4.3.5 have been implemented in Fortran by Szántai, see [321]. Szántai [319] has also developed a Fortran subroutine package for computing multivariate non–degenerate normal–, Dirichlet– and gamma distribution functions and their gradients. The availability of the programs is not clear from the papers, please contact the authors.

SLP problems with finite discrete distributions

According to our knowledge, there is no publicly available solver for this class of problems. The authors of the papers cited in Section 4.3.6 present illustrative numerical examples, and report on implementation of solvers. These solvers might be available on request; we suggest to contact the authors.

As we have seen in section 4.3.6, SLP problems in this class are equivalent to disjunctive programming problems, a subclass of mixed–integer programs. They belong to the class of NP–complete problems, implying that from the theoretical worst–case point of view they are difficult to solve. We are not able to provide the reader with guidance for selecting an algorithm or solver. On the one hand, we do not have personally any numerical experiences with such problems. On the other hand, as far as we see, comparative computational studies are completely missing in the literature. The authors of the papers cited in section 4.3.6 merely provide some illustrative examples, which is clearly insufficient to judge the practical value of the methods (recall that the problems dealt with are NP–complete).

Exercises

4.1. In the introductory part of Section 4.3 we have reformulated the general model with a joint probability constraint (4.5) as (4.7). Give an analogous reformulation of (4.6) involving the maximization of $F(y)$ and prove the equivalence.

4.2. On page 292 the proof of the following fact was left to the reader. Let the $(s \times s)$ symmetric matrix R be partitioned as follows:

$$R = \begin{pmatrix} 1 & \rho^{\mathrm{T}} \\ \rho & \hat{R} \end{pmatrix}$$

with $\hat{R}$ having the size $((s-1) \times (s-1))$ and $\rho \in \mathbb{R}^{s-1}$. Assume that R is positive definite. Show that this implies that $\hat{R} - \rho\rho^{\mathrm{T}}$ is positive definite too.

4.3. By utilizing SLP–IOR, perform a computational experiment involving randomly generated test problems with a joint probability constraint and several solvers. In the probability constraint only the right–hand side should be stochastic and the joint distribution of the random variables should be multivariate normal. Please proceed as follows:

(a) Generate a test problem battery consisting of 10 randomly generated SLP–problems with a joint probability constraint, having the following dimensions: $m_1 = m_2 = 5$, $n_1 = 10$, 5 random variables.
(b) Solve all problems in the battery by a selection of the available solvers and compare the computational times for the different solvers. For available solvers see Section 4.9.2.

4.4 Single–stage models based on expectation

This section is devoted to discussing algorithms for expectation–based SLP problems, presented in Section 2.4. First we review those models, for which a deterministic LP–equivalent exists, thus offering the possibility of solving them by general–purpose LP software. Subsequently we discuss the application of the dual decomposition method, presented in Section 1.2.6, for solving various single–stage expectation–based models, with a finite discrete distribution. The order of sections does not follow the order of models in Section 2.4; the discussion is governed by the logic of dual decomposition. The general idea is to solve those models via solving their equivalent representations as two–stage recourse problems.

4.4.1 Solving equivalent LP's

In this section we will summarize LP–equivalents of models based on expectation. When reporting the dimensions of these problems we will not count nonnegativity constraints, or in general, individual bounds on the variables. Let us denote the number of constraints in the definition of $\mathscr{B}$ by m.

SLP models involving integrated probability functions have been introduced and discussed in Section 2.4.1. We consider the case when ξ has a finite discrete distribution with N realizations.

For separate integrated probability functions, LP–equivalents are formulated in (2.121) and (2.122), on page 145. (2.121) corresponds to an integrated chance constraint and has $m+N+1$ rows and $n+N$ variables. (2.122) involves minimizing an integrated probability function and has $m+N$ rows and $n+N$ variables. Both problems have a dual block–angular structure, indicating, that in fact the underlying optimization models can be formulated as two–stage recourse problems. LP–equivalents for models with joint integrated probability functions are given in (2.133) and (2.134). These models grow more rapidly with N, as their counterparts with separate functions. In fact, (2.133) has $m+2Ns+1$ constraints and $n+2N$ variables, where s is the number of inequalities involved in the joint constraint. (2.134) has the same amount of variables and $m+2Ns$ constraints.

In Section 2.4.2 we have discussed a model, based on conditional expectation. Under the assumption that ξ has a logconcave density function, the stochastic constraint can be converted into a deterministic linear constraint. Consequently, the equivalent LP has essentially the same size as the original problem.

Section 2.4.3 in Chapter 2 is devoted to SLP models involving Conditional Value–at–Risk (CVaR) functions in the constraints or in the objective. Assuming that ξ has a finite discrete distribution with N realizations, the LP–equivalents are given as (2.149) on page 156 for minimizing CVaR, and as (2.156) on page 158 for a CVaR–constraint. (2.149) has $m+N$ rows and $n+N+1$ variables, whereas (2.156) has the same number of variables and a single additional constraint. Both matrices have a dual block–angular structure pointing to the fact that the underlying SLP models are essentially two–stage recourse problems.

In several cases discussed above we observed a dual block–angular structure. This suggests using a dual decomposition method, instead of the brute force application of general–purpose LP–solvers.

4.4.2 Dual decomposition revisited

The general dual decomposition algorithm has been discussed in Chapter 1, Section 1.2.6, on page 26. The basis of the method is the decomposition algorithm of Benders [14] which has been adapted to the structure of two–stage recourse problems with a finite discrete distribution by Van Slyke and Wets [336]; the latter authors named the method "L–shaped method". The method is a special case of nested decomposition, corresponding to two stages, see Section 1.2.7. In this section we will consider the case, where only the right–hand–side and the technology matrix are stochastic in the recourse subproblem. For the sake of simplicity, we will also assume that W has the complete recourse property.

Two–stage recourse problems are discussed in detail in Section 3.2. Here we will need the following, slightly modified formulation:

$$\left.\begin{array}{lll} \min\limits_{x,w} & c^{\mathrm{T}}x & +w \\ \text{s.t.} & \mathbb{E}[Q(x;T(\xi),h(\xi))] & -w \le 0 \\ & x & \in \mathscr{B}, \end{array}\right\} \quad (4.31)$$

where $\mathscr{B} = \{x \mid Ax = b, x \ge 0\}$. The recourse function $Q(x;T(\xi),h(\xi))$ is defined as

$$\left.\begin{array}{rl} Q(x;T(\xi),h(\xi)) := & \min q^{\mathrm{T}}y \\ & \text{s.t. } Wy = h(\xi) - T(\xi)x \\ & \qquad y \ge 0, \end{array}\right\} \quad (4.32)$$

which can also be expressed via the LP–dual as

$$\left.\begin{array}{rl} Q(x;T(\xi),h(\xi)) = & \max\ (h(\xi)-T(\xi)x)^{\mathrm{T}}u \\ & \text{s.t.} \qquad W^{\mathrm{T}}u \le q. \end{array}\right\} \quad (4.33)$$

We assume that $\mathscr{B} \neq \emptyset$ holds and that $\mathscr{B}$ is bounded.

Having the recourse function in the constraint, instead of the objective, recourse–constrained problems arise. These have the form

$$\left.\begin{array}{ll} \min\limits_{x} & c^{\mathrm{T}}x \\ \text{s.t.} & d^{\mathrm{T}}x + \mathbb{E}[Q(x;T(\xi),h(\xi))] \le \gamma \\ & x \qquad\qquad\qquad\qquad \in \mathscr{B} \end{array}\right\} \quad (4.34)$$

and have been introduced and first studied by Higle and Sen [138] and Yakowitz [349]. We will set up the dual decomposition method for (4.31) and (4.34) simultaneously. For this reason we consider the problem

$$\left.\begin{array}{llr} \min\limits_{x(,w)} & c^{\mathrm{T}}x & +\theta w \\ \text{s.t.} & d^{\mathrm{T}}x+\mathbb{E}[Q(x;T(\xi),h(\xi))] & -w\leq 0 \\ & x & \in \mathscr{B}. \end{array}\right\} \tag{4.35}$$

For the recourse problem (4.31) we choose $\theta = 1$, $d = 0$, and both x and w are considered as variables. For the recourse–constrained problem $\theta = 0$ is chosen and only x is considered as a variable whereas $w = \gamma$ becomes a parameter.

As we have already discussed in Section 1.2.6, the recourse function is convex and piecewise linear, and can equivalently be written in the form

$$Q(x;T(\xi),h(\xi)) = \max_{u\in\mathscr{U}}(h(\xi)-T(\xi)x)^{\mathrm{T}}u, \tag{4.36}$$

where $\mathscr{U}$ is the (finite) set of vertices of the polyhedral feasible domain of the recourse problem in the dual form (4.33). Note that the feasible domain of the dual recourse problem neither depends on x nor on ξ.

Let us assume now, that ξ has a finite discrete distribution with N realizations $\hat{\xi}^k$ and corresponding probabilities $p_k > 0$, $k = 1,\ldots,N$, $\sum\limits_{k=1}^{N} p_k = 1$. The corresponding realizations of $(h(\xi),T(\xi))$ will be denoted by (h^k,T^k), $k = 1,\ldots,N$. Thus, for any fixed $x \in \mathscr{B}$, we have to deal with N recourse subproblems, corresponding to the realizations.

The recourse constraint in (4.35) takes the form

$$d^{\mathrm{T}}x + \sum_{k=1}^{N} p_k Q(x;T^k,h^k) \; -w \;\leq\; 0,$$

which, due to the representation (4.36), can be written as

$$d^{\mathrm{T}}x + \sum_{k=1}^{N} p_k \max_{u\in\mathscr{U}}(h^k - T^k x)^{\mathrm{T}}u \; -w \;\leq\; 0.$$

Due to the nonnegativity of probabilities, this single nonlinear inequality constraint can be replaced by a system of linear inequality constraints

$$d^{\mathrm{T}}x + \sum_{k=1}^{N} p_k (h^k - T^k x)^{\mathrm{T}}u_k \; -w \;\leq\; 0, \;\; u_k \in \mathscr{U}\; \forall k, \tag{4.37}$$

where the notation is to be understood as follows. For any

$$(\hat{u}_1,\ldots,\hat{u}_N) \in \underbrace{\mathscr{U}\times\ldots\times\mathscr{U}}_{N} := \mathscr{U}_N,$$

the system (4.37) contains exactly one inequality corresponding to the choice $u_k = \hat{u}_k$, $k = 1,\dots,N$. Stating this in a different way, the inequalities in (4.37) are indexed by employing the index set $\mathscr{U}_N$. Consequently, the system of inequalities consists of M^N inequalities, where $M = |\mathscr{U}|$ is the number of elements of $\mathscr{U}$.

Hence we can rewrite (4.35) as follows:

$$\left.\begin{array}{lll} \min\limits_{x(,w)} & c^{\mathrm{T}}x & +\theta w \\ \text{s.t.} & d^{\mathrm{T}}x + \sum\limits_{k=1}^{N} p_k(u_k)^{\mathrm{T}}(h^k - T^k x) & -w \le 0,\ u_k \in \mathscr{U}, \forall k \\ & x & \in \mathscr{B}. \end{array}\right\} \tag{4.38}$$

This problem will be called the *full master problem.* It involves, in general, a tremendous amount of inequality constraints which are, as an additional difficulty, not known explicitly. From the algorithmic point of view, the basic idea of the dual decomposition method is *constraint generation.* The constraints in (4.38) are generated in a step–by–step manner in the hope that very much fewer inequalities, than in (4.38), are sufficient to reach optimality. Instead of the full master problem, *relaxed master problems* of the form

$$\left.\begin{array}{lll} \min\limits_{x(,w)} & c^{\mathrm{T}}x & +\theta w \\ \text{s.t.} & d^{\mathrm{T}}x + \sum\limits_{k=1}^{N} p_k(u_k^j)^{\mathrm{T}}(h^k - T^k x) & -w \le 0,\ j = 1,\dots,\nu \\ & \bar{u}^{\mathrm{T}}(\bar{h} - \bar{T}x) & -w \le 0 \\ & x & \in \mathscr{B} \end{array}\right\} \tag{4.39}$$

are solved, where ν is the number of constraints generated so far and $u_k^j \in \mathscr{U}$ $\forall k, j$ holds. We have added the constraint $\bar{u}^{\mathrm{T}}(\bar{h} - \bar{T}x) - w \le 0$, where $\bar{h} = \mathbb{E}[h(\xi)]$, $\bar{T} = \mathbb{E}[T(\xi)]$, and $\bar{u}$ is any feasible solution of (4.33). Due to the Jensen–inequality (see Section 3.2), the additional inequality is redundant in the full master problem (4.38). Note that, due to the assumptions concerning $\mathscr{B}$, and implied by the inclusion of the additional constraint involving expectations, the optimal solution of the relaxed master problem exists for any $\nu \ge 0$.

The decomposition method for the solution of (4.35) is an adaptation of the dual decomposition method in Section 1.2.6. The formal description follows.

Step 1. *Initialization*
Set $\nu := 0$, compute $\bar{h}$ and $\bar{T}$, determine a feasible solution of the recourse subproblem (4.32) (for instance, by the simplex method), and set up the relaxed master problem (4.39).

Step 2. *Solve the relaxed master problem*
Solve (4.39), let a solution be (x^*, w^*), where in the recourse constrained case $w^* = w(= \gamma)$ holds.

Step 3. *Evaluate the expected recourse function*

3.a. For $k = 1$ to $k = N$ do:
With $x := x^*$ and $(h(\xi), T(\xi)) := (h^k, T^k)$, solve the recourse subproblem (4.32), for instance, by the dual simplex method. Let u_k^* denote an optimal dual solution of the kth subproblem.

3.b. Compute the expected value of the recourse function as follows

$$\mathscr{Q}(x^*) = \sum_{k=1}^{N} p_k (h^k - T^k x^*)^{\mathrm{T}} u_k^*.$$

Step 4. *Check for optimality*
If $d^{\mathrm{T}} x^* + \mathscr{Q}(x^*) \le w^*$ then **Stop**, otherwise continue with the next step.

Step 5. *Add an optimality cut to the relaxed master*
Set $\nu := \nu + 1$, $u_k^\nu = u_k^*$, $k = 1, \ldots, N$ and add the corresponding cut to the set of constraints of the relaxed master (4.39). **Continue with *Step 2*.**

Using an analogous reasoning as for the dual decomposition method in Section 1.2.6, it is clear that the algorithm terminates after a finite number of iterations and that in the case of stopping, the method delivers an optimal solution.

We discuss the special case of simple recourse next, see Section 3.2.2. Following Klein Haneveld and Van der Vlerk [191], we consider the case when the technology matrix $T(\xi)$ is also stochastic. Notice that this type of problems are called generalized simple recourse (GSR) problems in our book and are discussed in Section 3.2.2.4 where a conceptual cutting–plane method is also presented.

The minimization problem (3.53) on page 226, defining the recourse function, is obviously separable in the components of $(y^+, y^-) \in \mathbb{R}^{2m_2}$ in our case, with the ith subproblem given as

$$\left.\begin{array}{rl} Q_i(x;\xi) := \min & q_i^+ y_i^+ + q_i^- y_i^- \\ & y_i^+ - y_i^- = h_i(\xi) - T_i^{\mathrm{T}}(\xi) x \\ & y_i^+, \quad y_i^- \ge 0, \end{array}\right\} \tag{4.40}$$

where T_i denotes the ith row of T, $i = 1, \ldots, n_2$. The LP dual problem is

$$\left.\begin{array}{rl} Q_i(x;\xi) = \max & (h_i(\xi) - T_i^{\mathrm{T}}(\xi) x) u_i \\ & u_i \le q_i^+ \\ & u_i \ge -q_i^- \end{array}\right\} \tag{4.41}$$

which has the optimal solution

$$u_i^* = \begin{cases} q_i^+ & \text{if } h_i(\xi) - T_i^{\mathrm{T}}(\xi) x > 0 \\ -q_i^- & \text{if } h_i(\xi) - T_i^{\mathrm{T}}(\xi) x \le 0. \end{cases} \tag{4.42}$$

The feasible domain of the dual recourse problem (3.57) is an m_2–dimensional interval, thus having 2^{m_2} vertices. The set of vertices $\mathscr{U}$ consists of vectors $u \in \mathbb{R}^{m_2}$, with either $u_i = q_i^+$ or $u_i = -q_i^-$, $i = 1, \ldots, m_2$. The explicit formula for the recourse function, corresponding to (3.58) on page 227, is the following

$$Q(x,\xi) = \sum_{i=1}^{m_2} [h_i(\xi) - T_i^{\mathrm{T}}(\xi)x]^+ q_i^+ + \sum_{i=1}^{m_2} [h_i(\xi) - T_i^{\mathrm{T}}(\xi)x]^- q_i^- . \tag{4.43}$$

Let us now turn our attention to the dual decomposition method, applied to simple recourse problems with a random technology matrix and to simple–recourse constrained problems. The full master problem (4.38) has now $2^{n_2 N} + 2$ constraints. In the dual decomposition method only *Step 3.a* changes as follows:

Step 3.a For $k = 1$ to $k = N$ do:
For $i = 1$ to $i = m_2$ do:
Compute the optimal dual solution of the recourse subproblem according to (4.42), by simply checking signs as follows

$$(u_k^*)_i = \begin{cases} q_i^+ & \text{if } h_i^k - (T_i^k)^{\mathrm{T}} x \geq 0 \\ -q_i^- & \text{if } h_i^k - (T_i^k)^{\mathrm{T}} x < 0. \end{cases}$$

The decomposition method for simple recourse problems with a random technology matrix, as outlined above, has first been proposed by Klein Haneveld and Van der Vlerk [191].

Finally let us point out that the adaptation discussed above corresponds to the dual decomposition method with aggregate cuts. The adaptation of the multi–cut version, as described in Section 1.2.6 on page 29 is left as an exercise for the reader.

4.4.3 Models with separate integrated probability functions

We consider the models (2.119) and (2.120) in Section 2.4.1, on page 144, involving separate integrated probability functions, with a finite discrete distribution. We begin with an observation, due to Higle and Sen [138]. Comparing the explicit formula (4.43) for the recourse function with the model formulations (2.119) and (2.120), we see immediately that (2.120) can be considered as a simple recourse problem and (2.119) is equivalent to a problem with a simple–recourse constraint. In both cases $m_2 = 1$, $q^+ = 0$, $q^- = 1$, and $d = 0$ hold. Thus, the dual decomposition methods, as described above can directly be applied.

We obviously have $\mathscr{U} = \{0,1\}$ and thus in *Step 3.a* of the dual decomposition method $(u_k^*)_i$ is computed as follows:

$$(u_k^*)_i = \begin{cases} 0 & \text{if } h_i^k - (T_i^k)^{\mathrm{T}} x \geq 0 \\ -1 & \text{if } h_i^k - (T_i^k)^{\mathrm{T}} x < 0. \end{cases}$$

Consequently the coefficients u_k in the formulation of the full master problem (4.38) are either 0 or 1, $k = 1,\ldots,N$. The N–dimensional binary vector $(u_1,\ldots,u_N)^{\mathrm{T}}$ can in a one–to–one manner be identified with subsets of the index set $\mathscr{N} = \{1,\ldots,N\}$, by choosing index i as an element of the subset, if and only if $u_i = 1$ holds. In this subset–language the full master problem (4.38) assumes the form

$$\left.\begin{array}{llr} \min\limits_{x(,w)} c^{\mathrm{T}}x+ & & \vartheta w \\ \text{s.t.} \sum\limits_{k\in\mathscr{K}} p_k(T^k x-h^k) & -w \le 0, & \mathscr{K}\subset\mathscr{N} \\ \quad x & \in \mathscr{B}. & \end{array}\right\} \tag{4.44}$$

Let us consider the recourse–constrained case with $w=\gamma$. Comparing (4.44) with (2.123) in Theorem 2.13. of Chapter 2, it is clear that the derivation of (4.44) includes an alternative proof of that theorem. The relaxed master problem can be formulated as

$$\left.\begin{array}{llr} \min\limits_{x(,w)} c^{\mathrm{T}}x+ & & \vartheta w \\ \text{s.t.} \sum\limits_{k\in\mathscr{K}_j} p_k(T^k x-h^k) & -w \le & 0,\ j=1,\ldots,\nu \\ \quad \bar{T}x-\bar{h} & -w \le 0 & \\ & w \ge 0 & \\ \quad x & \in & \mathscr{B}, \end{array}\right\} \tag{4.45}$$

where $\mathscr{K}_j\subset\mathscr{N}$ holds for all $j\ge 0$ and we require $\mathscr{K}_0=\emptyset$. The constraint $w\ge 0$ corresponds to the choice $\mathscr{K}=\emptyset$ in the full master whereas the constraint $\bar{T}x-\bar{h}-w\le 0$ arises from the choice $\mathscr{K}=\mathscr{N}$ and corresponds to the dual variable $\bar{u}=-1$. The relaxed master problems will be constructed in such a way that $\mathscr{K}_i\neq\mathscr{K}_j$ holds throughout, for $i\neq j$.

The final form of the decomposition algorithm is as follows.

Step 1. *Initialization*
Set $\nu:=0$, compute $\bar{h}$ and $\bar{T}$, choose $\bar{u}=1$ as the dual–feasible solution of the recourse subproblem (4.32), and set up the relaxed master problem (4.45).

Step 2. *Solve the relaxed master problem*
Solve (4.45), let a solution be (x^*,w^*), where in the recourse constrained case $w^*=w(=\gamma)$ holds.

Step 3. *Evaluate the expected recourse function*

3.a. Determine the index set $\mathscr{K}^*=\{k\mid T^k x^*-h^k>0\}$.

3.b. Compute the expected value of the recourse function as follows

$$\mathscr{Q}(x^*)=\sum_{k\in\mathscr{K}^*} p_k(T^k x^*-h^k).$$

Step 4. *Check for optimality*
If $\mathscr{Q}(x^*)\le w^*$ then **Stop**, otherwise continue with the next step.

Step 5. *Add an optimality cut to the relaxed master*
Set $\nu:=\nu+1$, $\mathscr{K}_\nu=\mathscr{K}^*$ and add the corresponding cut to the set of constraints of the relaxed master (4.45). **Continue with *Step 2*.**

For integrated chance constraints, the algorithm developed this way is identical with the cutting–plane algorithm proposed by Klein Haneveld and Van der Vlerk [191].

Let us next turn our attention to integrated probability functions of the second kind. We consider the models (2.127) and (2.128). Analogously as before, (2.127) can be interpreted as a simple–recourse constrained problem and (2.128) can be viewed as a simple recourse problem. The adaptation of the dual decomposition method can be developed along the same lines as for the previous type of integrated probability functions, and is left as an exercise for the reader (note that now $d = 0$ does not hold in general). Let us remark that similarly as before, we obtain an alternative proof for Theorem 2.14. in Chapter 2, as a by–product of constructing the method.

4.4.4 Models involving CVaR

In this section we will discuss the dual decomposition method as applied to the models (2.146) on page 155 and (2.153) on page 157, both in Section 2.4.3. As in Section 4.4.3, a comparison with (4.43) for the simple recourse function reveals the following: Model (2.146) on page 155 involving CVaR–minimization is equivalent to a generalized simple–recourse problem, whereas (2.153) on page 157 turns out to be a generalized simple–recourse constrained problem. Furthermore, in the recourse–constrained case $d \neq 0$ holds (cf. (4.34)). In the framework of two–stage recourse problems, a detailed presentation of these relationships can be found in Section 3.2.3.

Proceeding analogously, as in Section 4.4.3, we arrive at the following full master problem:

$$\left.\begin{array}{lll} \min\limits_{x,z,w} & c^{\mathrm{T}}x + \theta w & \\ \text{s.t.} & z + \frac{1}{1-\alpha} \sum\limits_{k \in \mathscr{K}} p_k \left(T^k x - h^k - z\right) - w \le 0, & \mathscr{K} \subset \mathscr{N} \\ & x & \in \mathscr{B}, \end{array}\right\} \tag{4.46}$$

where now x and the free variable z are both first–stage variables. For formulating the relaxed master problem let us introduce the notation $\kappa := \frac{1}{1-\alpha}$ and $\tau := 1 - \kappa$. Having $0 < \alpha < 1$, $\tau < 0$ obviously holds. Using our notation, the relaxed master problem can be given as

$$\left.\begin{array}{lll} \min\limits_{x,z,w} & c^{\mathrm{T}}x + \theta w & \\ \text{s.t.} & \kappa \sum\limits_{k \in \mathscr{K}_j} p_k (T^k x - h^k - z) & + z - w \le 0,\ j = 1, \ldots, \nu \\ & \kappa(\bar{T}x - \bar{h}) & + \tau z - w \le 0 \\ & & z - w \le 0 \\ & x & \in \mathscr{B}, \end{array}\right\} \tag{4.47}$$

where for all $j \geq 0$ $\mathscr{K}_j \subset \mathscr{N}$ holds with the prescription $\mathscr{K}_0 = \emptyset$. Furthermore, $\mathscr{K}_j \subset \mathscr{N}$, $\forall j$ and $\mathscr{K}_i \neq \mathscr{K}_j$, $i \neq j$, hold. The constraint $z - w \leq 0$ corresponds to $\mathscr{K} = \emptyset$ in the full master problem and the constraint $\kappa(\bar{T}x - \bar{h}) + \tau z - w \leq 0$ arises when choosing $\mathscr{K} = \mathscr{N}$. Note that the relaxed master problems (4.47) involve the free variable z. Due to the additional constraints it is easy to show, however, that the relaxed master problems have optimal solutions for any $\nu \geq 0$. Introducing the notation

$$t_{[i]} := \kappa \sum_{k \in \mathscr{K}_i} p_k T^k, \quad h_{[i]} := \kappa \sum_{k \in \mathscr{K}_i} p_k h^k, \quad p_{[i]} := \kappa \sum_{k \in \mathscr{K}_i} p_k, \tag{4.48}$$

the relaxed master problem can be written in the compact form

$$\left.\begin{array}{llll} \min\limits_{x,z,w} & c^{\mathrm{T}}x & & +\theta w \\ \text{s.t.} & t_{[i]}x & +(1-p_{[i]})z & -w \leq h_{[i]}, \; i = 1,\ldots,\nu \\ & \kappa \bar{T}x & +\tau z & -w \leq \kappa \bar{h} \\ & & z & -w \leq 0 \\ & x & & \in \mathscr{B}. \end{array}\right\} \tag{4.49}$$

Now we are prepared to formulate the dual decomposition method for the CVaR–optimization problems.

Step 1. *Initialization*
Set $\nu := 0$, compute $\bar{h}$ and $\bar{T}$, and set up the relaxed master problem (4.49).

Step 2. *Solve the relaxed master problem*
Solve (4.49), let a solution be (x^*, z^*, w^*), where in the recourse constrained case $w^* = w(= \gamma)$ holds.

Step 3. *Evaluate the expected recourse function*

3.a. Determine the index set $\mathscr{K}^* = \{k \mid T^k x - z^* - h^k > 0\}$.

3.b. Compute the expected value of the recourse function as follows

$$\mathscr{Q}(x^*, z^*) = \sum_{k \in \mathscr{K}^*} p_k (T^k x^* - z^* - h^k).$$

Step 4. *Check for optimality*
If $z^* + \kappa \mathscr{Q}(x^*, z^*) \leq w^*$ then **Stop**, otherwise continue with the next step.

Step 5. *Add an optimality cut to the relaxed master*
Set $\nu := \nu + 1$, compute $t_{[\nu]}$, $h_{[\nu]}$, and $p_{[\nu]}$ according to (4.48), and add the corresponding cut to the set of constraints of the relaxed master (4.49).
Continue with *Step 2*.

Let us finally formulate a CVaR–analogue of the polyhedral representation Theorem 2.13. in Chapter 2, given for integrated chance constraints in Section 2.4.1. Let

$$\tilde{\mathscr{D}} := \left\{ (x,z,w) \mid \sum_{k=1}^{N} p_k \left(\zeta(x,\hat{\eta}^k,\hat{\xi}^k) - z \right)^+ + z - w \leq 0 \right\}.$$

We have the following polyhedral representation:

Proposition 4.10.

$$\mathscr{D} = \bigcap_{\mathscr{K} \in \mathscr{N}} \left\{ (x,z,w) \mid \sum_{k \in \mathscr{K}} p_k \left((\hat{\eta}^k)^{\mathrm{T}} x - \hat{\xi}^k - z \right) + z - w \leq 0 \right\} \tag{4.50}$$

with the sum defined as zero for $\mathscr{K} = \emptyset$.

Proof: The proof follows directly from the method which leads to the full master problem (4.46). An alternative, direct proof can also easily be given, along the lines of the proof of Theorem 2.13. in Chapter 2; this is left as an exercise for the reader. □

For the solution of CVaR–minimization problems, the algorithm presented above has been proposed by Künzi–Bay and Mayer [198].

4.4.5 Models with joint integrated probability functions

The subject of this section is a decomposition algorithm for the SLP–problems (2.131) and (2.132), involving joint probability functions and a finite–discrete probability distribution. These problems do not fit into the general framework of the dual decomposition, as discussed Section 4.4.2. Nevertheless, we present an algorithm for the two problems simultaneously. We consider the optimization problem

$$\left.\begin{array}{lll} \min\limits_{x(,w)} c^{\mathrm{T}} x & & +\theta w \\ \text{s.t.} \sum\limits_{k=1}^{N} p_k \max\limits_{1 \leq i \leq s} (t_i^k x - h^k)^+ & & -w \leq 0 \\ \qquad x & & \in \mathscr{B}. \end{array}\right\} \tag{4.51}$$

Problem (2.131) involves a joint integrated chance constraint and can be obtained from (4.51) by choosing $\theta = 0$ and $w = \gamma$. In this case only x counts as variable. In problem (2.132) a joint integrated probability function is included into the objective function. This problem is also a special case of (4.51), corresponding to the choice $\theta = 1$; both x and w are considered as variables.

Note that (4.51) involves the expected value of a maximum of recourse functions which still fits the general framework of recourse constrained programming, as defined by Higle and Sen [138]. One way to develop an algorithm for (4.51) would be to extend the dual decomposition method to recourse constrained models of the discussed type. In this section we will chose the direct way by presenting the algorithm directly based on the polyhedral representation theorem 2.13. (Chapter 2, page 145) of Klein Haneveld and Van der Vlerk. The starting point is an equivalent representation of (4.51) in the form of a full master problem:

$$\left.\begin{array}{lllll} \min\limits_{x(,w)} & c^{\mathrm{T}}x & +\theta w & & \\ \text{s.t.} & \sum\limits_{k\in\mathcal{K}} p_k((t^k_{l_k})^{\mathrm{T}}x - h^k_{l_k}) & -w & \leq 0 & l \in \mathcal{I}^{\mathcal{K}}, \mathcal{K} \subset \mathcal{N}, \\ & x & & \in \mathcal{B}, & \end{array}\right\} \quad (4.52)$$

where $\mathcal{I} = \{1,\ldots,s\}$ is the set of row indices in the joint integrated probability function, $\mathcal{I}^{\mathcal{K}} := \{\, l := (\, l_k, k \in \mathcal{K}\,) \mid l_k \in \mathcal{I} \text{ for all } k \in \mathcal{K} \,\}$ holds, and $t^k_{l_k}$ is the l_k^{th} row of T^k.

The following type of relaxed master problems will be utilized:

$$\left.\begin{array}{lllll} \min\limits_{x(,w)} & c^{\mathrm{T}}x & +\theta w & & \\ \text{s.t.} & \sum\limits_{(k,l)\in\mathcal{M}_j} p_k((t^k_l)^{\mathrm{T}}x - h^k_l) & -w & \leq 0, & j = 1,\ldots,\nu \\ & \bar{t}_i^{\mathrm{T}}x - \bar{h}_i & -w & \leq 0, & i = 1,\ldots,s \\ & & w & \geq 0 & \\ & x & & \in \mathcal{B}, & \end{array}\right\} \quad (4.53)$$

where $\mathcal{M}_j \subset \mathcal{N} \times \mathcal{I}$ is a set of ordered pairs (k,l) with the property that each $k \in \mathcal{N}$ appears at most once. We prescribe that $\mathcal{K}_0 = \emptyset$ holds. Finally, $\bar{t}_i = \mathbb{E}[t_i(\xi)]$, $\bar{h}_i = \mathbb{E}[h_i(\xi)]$ hold for all i. The constraint $w \geq 0$ arises when choosing $\mathcal{M} = \emptyset$ in the full master problem, and $\bar{t}_i^{\mathrm{T}}x - \bar{h}_i - w \leq 0$ has its root in the choice $\mathcal{M} = \mathcal{N} \times \{i\}$. That the additional expectation–based constraints are redundant in the full master problem, can also be seen directly, by utilizing the obvious fact that $\mathbb{E}[\vartheta_i] \leq \mathbb{E}[\max\limits_{1\leq j\leq M} \vartheta_j]$ holds for any random variables ϑ_i, $i = 1,\ldots,M$ with finite expected value. It is easy to show that under our assumptions (4.53) has an optimal solution for any $\nu \geq 0$. Next we state the algorithm.

Step 1. *Initialization*
Set $\nu := 0$, compute $\bar{t}_i$ and $\bar{h}_i$, $i = 1,\ldots,s$, and set up the relaxed master problem (4.53).

Step 2. *Solve the relaxed master problem*
Solve (4.53), let a solution be (x^*, w^*), where in the case of joint integrated constraints $w^* = w(=\gamma)$ holds.

Step 3. *Evaluate the joint integrated probability function*

3.a. Determine the index set

$$\mathcal{M}^* = \{(k,l) \mid (t^k_l)^{\mathrm{T}}x^* - h^k_l := \max_{1\leq i\leq s}(\,(t^k_i)^{\mathrm{T}}x^* - h^k_i\,) > 0\}.$$

3.b. Compute the the joint integrated probability function as follows

$$K_J(x^*) = \sum_{(k,l)\in\mathscr{M}^*} p_k((t_l^k)^{\mathrm{T}}x^* - h_l^k).$$

Step 4. *Check for optimality*
If $K_J(x^*) \leq w^*$ then **Stop**, otherwise continue with the next step.

Step 5. *Add an optimality cut to the relaxed master*
Set $\nu := \nu + 1$, $\mathscr{M}_\nu = \mathscr{M}^*$ and add the corresponding cut to the set of constraints of the relaxed master (4.53).
Continue with *Step 2*.

Proposition 4.11. *The above method terminates after a finite number of iterations, with x^* being an optimal solution of our problem.*

Proof: The proof runs along the same lines as the proof of the analogous proposition for the dual decomposition. It is clear that in the case when $\mathscr{M}^* = \mathscr{M}_j$ holds for some $j \leq \nu$, then the stopping criterium in *Step 4.* will hold and the algorithm terminates. Having a finite number of different subsets in $\mathscr{N} \times \mathscr{I}$, this immediately implies finiteness. For proving the rest, let us first consider the case $\theta = 1$. In this case $c^{\mathrm{T}}x^* + w^*$ is a lower bound and $c^{\mathrm{T}}x^* + K_J(x^*)$ is an upper bound for the optimal objective value of (4.51). Thus, the stopping criterion implies optimality of x^*. In the case of an integrated chance constraint the optimal solution of the relaxed problem turns out to be feasible in the original one, thus implying optimality. □

4.4.6 A guide to available software

For several models, based on expectation, LP–equivalents exist, see the discussion in Section 4.4.1. The straightforward approach for solving these problems is to apply general–purpose LP solvers to the LP–equivalents. However, having a large number of realizations, this can become quite time consuming. Thus, if computing time matters, special–purpose algorithms are preferable.

Models with separate integrated probability functions

The recommended approach is dual decomposition. One possibility is to formulate the equivalent two–stage simple recourse problem and to employ a dual–decomposition solver for two–stage problems, see Section 4.7.5. A special–purpose solver has been developed by Klein Haneveld and Van der Vlerk [191] in MatLab, which might be available on request from the authors.

Models with joint integrated probability functions

The same comment applies as in the previous section. Again, a special–purpose solver, developed by Klein Haneveld and Van der Vlerk [191] in MatLab, might be available on request from the authors.

Models involving CVaR

For models with CVaR–minimization, a dual decomposition solver named CVaRMin has been developed in Delphi by Künzi–Bay and Mayer [198], for the LP subproblems Minos (Murtagh and Saunders [240]) has been used. It is connected to the modeling system SLP–IOR and is available along with this modeling system, see Section 4.9.2.

Exercises

4.4. Consider the dual decomposition algorithm as formulated in Section 4.4.2. A possible misinterpretation of this method is the following:
"The algorithm generates the constraints of the full master problem (4.38) in a step-by-step manner, by adding new constraints in the form of cuts to the relaxed master problem (4.39) in each of the iterations. Since the full master problem contains a tremendous amount of constraints, see page 316, this results in a hopelessly slow method for real–life problems." What is wrong with this argumentation?

4.5. Formulate the multi–cut version of the adaptation of the dual decomposition method, described in Section 4.4.2.

4.6. On the basis of the discussion in Section 4.4.3, formulate the adaptation of the dual decomposition method for models with integrated probability functions of the second kind, see (2.127) and (2.128).

4.5 Single–stage models involving VaR

Models involving quantiles have been the subject of Section 2.3 in Chapter 2. We have seen that these models can equivalently be formulated as SLP models with separate probability functions. Therefore, the considerations concerning algorithmic approaches in Section 4.2 apply also for this case.

In finance, portfolio optimization problems involving VaR are quite important. For algorithmic approaches, proposed for this particular application, see Larsen et al. [202], and the references therein.

4.6 Single–stage models with deviation measures

Models with deviation measures have been introduced in Section 2.5.

Let us discuss models involving quadratic deviation first. Having the quadratic deviation in the objective, the equivalent nonlinear programming problems (2.163) and (2.169) are convex quadratic programming problems without any special structure. Thus, the numerical approach for their solution consists of employing general–purpose algorithms of quadratic programming, see, for instance, Nocedal and Wright [243].

Regarding the models (2.162) and (2.168), these are also convex programming problems but they are much more difficult from the numerical point of view. Both of them involve a nonlinear constraint with a convex quadratic function on the left–hand–side. The straightforward approach is to apply a general–purpose solver for nonlinear programming. A better idea is the following: the problems can be reformulated as second order cone programming (SOCP) problems, see Lobo et al. [206] and primal–dual interior point methods can be employed for their solution (see also Section 4.2).

Considering models with quadratic semi–deviation, the situation is similar. We only consider the case when the underlying probability distribution is finite discrete. Having the risk measure in the objective, the convex quadratic programming problems (2.192) and (2.197) arise. With quadratic semi–deviation functions in the constraints, we get the convex programming models (2.191) and (2.196) involving a quadratic constraint. Concerning solution algorithms, the same comments apply as for the quadratic deviation case above. The models with quadratic semi–deviation have a rather special structure, which could be utilized for developing algorithms tailored to this structure.

Finally let us consider models with absolute deviation, under the assumption that the underlying probability distribution is finite discrete. The straightforward way of solving these models is via solving the corresponding equivalent LP problems (2.178), (2.179) or (2.183).

An alternative way, resulting in a much more efficient solution approach, is via equivalent simple recourse models. In Section 2.5.2 we have seen that the general model (2.173) is equivalent to the simple recourse model (2.174), provided that $\eta \equiv t$ holds. This assumption has been chosen, however, merely for the sake of simplicity of presentation. From the considerations in Section 2.5.2 it is clear that the general models (2.172) and (2.173) can equivalently be formulated as the following recourse models with a simple recourse structure:

$$\left.\begin{array}{lrl} \min & c^{\mathrm{T}}x & \\ \text{s.t.} & \mathbb{E}[y+z] & \leq \kappa \\ & \eta^{\mathrm{T}}x - \xi \; -y + z & = 0 \\ & y & \geq 0 \\ & z & \geq 0 \\ & x & \in \mathscr{B} \end{array}\right\} \qquad (4.54)$$

and

$$\left.\begin{array}{llcl} \min \mathbb{E}[y+z] & & & \\ \text{s.t. } \eta^{\mathrm{T}}x - \xi & -y & +z & = 0 \\ & y & & \geq 0 \\ & & z & \geq 0 \\ x & & & \in \mathscr{B}. \end{array}\right\} \quad (4.55)$$

The above problem (4.54) is a recourse constrained problem in the sense of Higle and Sen [138] and (4.55) is a recourse problem with a simple recourse structure and a random technology matrix.

If for (4.55) $\eta \equiv t$ holds, then the problem is a classical simple recourse problem. Consequently, the general algorithms for simple recourse problems can be applied, even without the assumption that the distribution is finite discrete.

Under the assumption that the probability distribution is finite discrete, the proposed solution approach is dual decomposition, for both problems above. In Section 4.4.2 of this chapter we have derived a general framework of dual decomposition for recourse models and for recourse constrained models, with simple recourse structure, where the technology matrix may also be stochastic. Analogously as for models involving integrated probability functions or CVaR, this approach results in specialized versions of the dual decomposition method (see Section 4.2). Working out the details is left as an exercise for the reader.

4.6.1 A guide to available software

Concerning solvers for SOCP, see Section 4.3.7 whereas for solvers for simple recourse problems consult Section 4.7.5.

4.7 Two–stage recourse models

Two–stage recourse models have been discussed in Section 3.2.

A great variety of algorithms have been proposed for the solution of this type of problems; in this book we confine ourselves to discuss some selected algorithmic approaches. For further methods see, for instance, Birge and Louveaux [26] and Ruszczyński and Shapiro [295], and the references therein.

If ξ has a finite discrete distribution, then the two–stage recourse problem can be equivalently formulated as a (typically large scale) linear programming problem. A natural idea is to apply interior–point methods for the solution of this LP problem. Interior point methods have been discussed in Section 1.2.9.

Among these methods, algorithms based on the augmented system approach (see (1.63)) turned out to be especially well–suited for the solution of the specially structured equivalent LP problem, see Maros and Mészáros [221] and Mészáros [233].

In the next section we discuss some further algorithmic issues concerning decomposition methods; the methods themselves have already been presented in Chapter 1. The subsequent section is devoted to successive discrete approximation methods. In Section 4.7.3 stochastic methods are discussed while the subsequent section 4.7.4 summarizes some algorithmic issues for the special case of simple recourse.

4.7.1 Decomposition methods

In this section we consider the two–stage problem (3.7), under the assumption that ξ has a finite discrete distribution.

The basic dual decomposition algorithm for two–stage recourse problems is essentially an application of Benders–decomposition [14], due to Van Slyke and Wets [336], and is usually called the L–shaped method in the literature. In Section 1.2.6 we have discussed the dual decomposition method, under the assumption of fixed recourse and presupposing a deterministic objective in the second stage. The algorithm for the general case is discussed as the nested decomposition algorithm for multi–stage problems in Section 1.2.7; the two–stage problem is clearly a special case corresponding to $T = 2$. A variant, also suitable for recourse–constrained problems, has been presented in Section 4.4.2. From the numerical point of view, the basic dual decomposition has some unfavorable features. On the one hand, there is no reliable way for dropping redundant cuts. On the other hand, especially at the beginning phase of iterations, the algorithm tends to make inefficient long steps.

For overcoming these difficulties, an important improvement of the basic dual decomposition algorithm is the regularized decomposition method, due to Ruszczyński [293]. This algorithm has been the subject of Section 1.2.8. For recent achievements concerning this method see Ruszczyński and Świętanowski [296].

Another way for avoiding inefficient long steps, generally known in nonlinear programming, is the trust–region method. This idea has been applied for two–stage recourse problems by Linderoth and Wright [204], by employing intervals as trust regions. The authors report quite favorable computational results concerning their method.

A common feature of all of the dual decomposition methods is the following: in each of the iterations, having the current solution x^* of the master problem, the recourse subproblem has to be solved for all realizations of ξ, in turn. Assuming fixed recourse and that q is not stochastic, the recourse subproblem (4.57) or its dual (4.58) has to be solved with the setting $\xi = \hat{\xi}^k$, for $k = 1,\ldots,N$, where N stands for the number of realizations. Now it is clear that the dual problems (4.58), corresponding to different realizations, differ only in their objective. Assume that we have solved the first recourse subproblem, corresponding to $\xi = \hat{\xi}^1$, by employing the simplex method. The optimal basis B will then be a dual feasible basis for all of the subproblems corresponding to the other realizations. Consequently, if for the kth ($k \geq 2$) subproblem

$$y^k := B^{-1}(h(\hat{\xi}^k) - T(\hat{\xi}^k)x^*) \geq 0$$

holds, then B is also primal feasible to this subproblem, therefore y^k is optimal. Consequently, for the kth subproblem we have obtained the optimal solution without starting up the simplex method at all. The idea is that, after having solved a particular subproblem, the above check is performed for the remaining subproblems, in order to identify those for which the simplex method has to be started up subsequently. This idea is called *bunching* and can reduce substantially the running time of the decomposition method. For further details, and refinements called *trickling down*, see Gassmann [112], Gassmann and Wallace [121] and Kall and Wallace [172].

Zakeri et al. [354] propose a variant of the dual decomposition method where the subproblems need to be solved only approximately thus leading to inexact cuts.

Concerning the numerical implementation of the dual decomposition method, Abaffy and Allevi [1] present a modification, based on the ABS method of numerical linear algebra, which leads to a significant decrease in the number of arithmetic operations needed to carry out the algorithm.

Another idea for decomposition is basis–decomposition. For two–stage recourse problems, an algorithm of this type has been developed by Strazicky [315].

4.7.2 Successive discrete approximation methods

In this section we will make the following assumption

Assumption 4.7.
The first and second moments exist for ξ.
The model has fixed recourse, that is, $W(\xi) \equiv W$, *i.e.* $W(\cdot)$ *is deterministic.*
The recourse matrix W *has the complete recourse property (3.12) (cf. Assumption 3.2. in Section 3.2, on page 200).*
$q(\xi) \equiv q$, *i.e.* $q(\cdot)$ *is deterministic.*
For $T(\xi)$ *and* $h(\xi)$ *the affine–linear relations (3.6) hold.*
The recourse subproblem has a finite optimum for any x *and any* ξ *(cf. Assumption 3.3. in Section 3.2, on page 201).*

Under this assumption, the recourse function $Q(x;T(\cdot),h(\cdot))$ is a convex function in ξ for any $x \in \mathbb{R}^{n_1}$, see Theorem 3.2. in Section 3.2.

Taking into account our assumptions, the two–stage recourse problem from Section 3.2 has the following form

$$\left.\begin{aligned} \min\ & c^{\mathrm{T}}x + \mathbb{E}_{\xi}[Q(x;T(\xi),h(\xi))] \\ \text{s.t.}\ & Ax = b \\ & x \geq 0, \end{aligned}\right\} \qquad (4.56)$$

where the *recourse function* $Q(x;T(\xi),h(\xi))$ is defined as

$$\left.\begin{aligned} Q(x;T(\xi),h(\xi)) := \min\ & q^{\mathrm{T}}y \\ \text{s.t.}\ & Wy = h(\xi) - T(\xi)x \\ & y \geq 0. \end{aligned}\right\} \tag{4.57}$$

Alternatively, via the duality theory of linear programming we have

$$\left.\begin{aligned} Q(x;T(\xi),h(\xi)) = \max\ & (h(\xi) - T(\xi)x)^{\mathrm{T}}u \\ \text{s.t.}\ & W^{\mathrm{T}}u \leq q. \end{aligned}\right\} \tag{4.58}$$

Let $\mathscr{B} := \{x \mid Ax = b, x \geq 0\}$ be the set of feasible solutions of (4.56). For the sake of simplicity of presentation, we will assume additionally to Assumption 4.7. that $\mathscr{B} \neq \emptyset$ holds and that $\mathscr{B}$ is bounded.

Notice, that due to our assumptions, the optimal solution for (4.56) exists. Let x^* denote an optimal solution. For later use, let us introduce the notation

$$\mathscr{Q}(x) := \mathbb{E}_\xi[Q(x;T(\xi),h(\xi))]$$

for the expected–recourse function, $f(x) := c^{\mathrm{T}}x + \mathscr{Q}(x)$ for the objective function of the recourse problem (4.56), and $f^* := c^{\mathrm{T}}x^* + \mathscr{Q}(x^*)$ for the optimal objective value of (4.56).

According to Proposition 1.18. in Chapter 1 (page 24), the recourse function $Q(x;T(\xi),h(\xi))$ is a piecewise linear convex function in x for fixed ξ. Due to the affine–linear relations (3.6), the recourse function is piecewise linear and convex also in ξ for fixed x. The proof of this fact is analogous to the proof of the above–mentioned Proposition and is left as an exercise for the reader.

Successive discrete approximation methods construct discrete approximations to the probability distribution of ξ by successively partitioning a set $\Xi \subset \mathbb{R}^r$, which is supposed to contain the support of ξ. We will proceed as follows. First we discuss algorithms, for which Ξ is supposed to be an r–dimensional interval and at each iteration the support of ξ is covered by a union of intervals. We concentrate on methods for which we have our own computational experience. The other algorithmic approaches will be summarized in the separate subsection 4.7.2.

Computing the Jensen lower bound

Our Assumption 4.7. implies that $Q(x;T(\cdot),h(\cdot))$ is a convex function in ξ for any $x \in \mathbb{R}^{n_1}$ (see Theorem 3.2. in Section 3.2). Jensen's inequality applies, see Theorem 3.4. in Section 3.2. Thus, for the expected recourse $\mathscr{Q}(x)$ we have the lower bound

$$Q(x;T(\mu),h(\mu)) \leq \mathscr{Q}(x),\ x \in \mathbb{R}^{n_1} \tag{4.59}$$

with $\mu := \mathbb{E}[\xi]$. Consequently,

$$f^L(x) := c^{\mathrm{T}}x + Q(x;T(\mu),h(\mu)) \leq c^{\mathrm{T}}x + \mathscr{Q}(x),\ \forall x \in \mathscr{B}$$

holds. A lower bound on the optimal objective value f^* of (4.56) can be obtained by solving

$$\left.\begin{array}{ll} \min & c^{\mathrm{T}}x+Q(x;T(\mu),h(\mu)) \\ \text{s.t.} & Ax=b \\ & x\geq 0, \end{array}\right\}$$

which is obviously equivalent to the following LP problem

$$\left.\begin{array}{lrl} \min & c^{\mathrm{T}}x & +q^{\mathrm{T}}y \\ \text{s.t.} & T(\mu)x & +Wy=h(\mu) \\ & Ax & \quad =b \\ & x & \quad \geq 0 \\ & & \quad y\geq 0. \end{array}\right\} \tag{4.60}$$

Problem (4.60) is called the *expected value problem*, corresponding to (4.56), cf. (3.104).

Computing the E–M upper bound for an interval

The purpose of this section is to derive a formula for computing the generalized E–M upper bound (3.38). For the sake of easy reference we begin by summarizing the derivation of this bound, as given in Section 3.2. Let $\Xi := \prod_{i=1}^{r}[\alpha_i,\beta_i]$ be an r–dimensional interval containing the support of the r–dimensional random vector ξ and let

$$|\Xi| := \prod_{i=1}^{r}(\beta_i-\alpha_i)$$

be the volume of the r–dimensional interval Ξ. Let, furthermore, $\varphi : \Xi \to \mathbb{R}$ be a convex function. Our goal is to derive an explicit formula for the generalized Edmundson–Madansky upper bound on $\mathbb{E}[\varphi(\xi)]$.

For deriving the bound, the following construction will be used. For each fixed $\xi \in \Xi$ let $\eta(\xi)$ be an r–dimensional discretely distributed random vector, with stochastically independent components having the following one–dimensional marginal distributions

$$\begin{pmatrix} \alpha_i & \beta_i \\ \dfrac{\beta_i-\xi_i}{\beta_i-\alpha_i} & \dfrac{\xi_i-\alpha_i}{\beta_i-\alpha_i} \end{pmatrix}$$

for $i=1,\dots,r$, where the first row corresponds to realizations and the second row contains the corresponding probabilities (cf. (3.33)). Thus, the set of joint realizations of η coincides with the set of vertices of Ξ. The probability of the realization corresponding to vertex v^ν $(1 \leq \nu \leq 2^r)$ is (due to the stochastic independence assumption)

$$p_\nu(\xi) := \frac{1}{|\Xi|} \prod_{i\in I_\nu} (\beta_i - \xi_i) \prod_{i\in J_\nu} (\xi_i - \alpha_i), \tag{4.61}$$

where $I_\nu = \{i \mid v_i^\nu = \alpha_i\}$ and $J_\nu = \{i \mid v_i^\nu = \beta_i\}$. Next observe that, due to the construction of η, we obviously have for each fixed $\xi \in \Xi$

$$\mathbb{E}[\eta] = \begin{pmatrix} \mathbb{E}[\eta_1] \\ \vdots \\ \mathbb{E}[\eta_r] \end{pmatrix} = \begin{pmatrix} \xi_1 \\ \vdots \\ \xi_r \end{pmatrix} = \xi.$$

Consequently, the Jensen–inequality yields

$$\varphi(\xi) = \varphi(\mathbb{E}[\xi]) \le \mathbb{E}[\varphi(\eta)] = \sum_\nu \varphi(v^\nu)\, p_\nu(\xi)$$

with $p_\nu(\xi)$ defined as (4.61). Taking expectation results in

$$\mathbb{E}[\varphi(\xi)] \le \sum_\nu \varphi(v^\nu)\, \mathbb{E}[p_\nu(\xi)], \tag{4.62}$$

which is the multivariate generalization of the Edmundson–Madansky inequality (3.38). For the independent case this inequality is due to Kall and Stoyan [171], the extension to the dependent case has been given by Frauendorfer [102].

Now we are prepared to derive a formula for $\mathbb{E}[p_\nu(\xi)]$. In the case when the components of ξ are stochastically independent, we get immediately from (4.61)

$$\mathbb{Q}(v^\nu) := \mathbb{E}[p_\nu(\xi)] = \frac{1}{|\Xi|} \prod_{i\in I_\nu} (\beta_i - \mathbb{E}[\xi_i]) \prod_{i\in J_\nu} (\mathbb{E}[\xi_i] - \alpha_i). \tag{4.63}$$

Otherwise, utilizing (4.61) a straightforward computation yields

$$p_\nu(\xi) = \frac{1}{|\Xi|} \sum_{\Lambda\subset\{1,\dots,r\}} (-1)^{|K_{\nu\Lambda}|} \prod_{i\in I_\nu\cap\Lambda^c} \beta_i \prod_{i\in J_\nu\cap\Lambda^c} \alpha_i \prod_{i\in\Lambda} \xi_i$$

with $\Lambda^c = \{1,\dots,r\}\setminus\Lambda$, $K_{\nu\Lambda} = I_\nu\cap\Lambda \cup J_\nu\cap\Lambda^c$ and with $|K_{\nu\Lambda}|$ denoting the number of elements in $K_{\nu\Lambda}$. Taking expectation leads to the formula

$$\mathbb{E}[p_\nu(\xi)] = \frac{1}{|\Xi|} \sum_{\Lambda\subset\{1,\dots,r\}} (-1)^{|K_{\nu\Lambda}|} \prod_{i\in I_\nu\cap\Lambda^c} \beta_i \prod_{i\in J_\nu\cap\Lambda^c} \alpha_i \; h_\Lambda(\xi), \tag{4.64}$$

which is an expression for $\mathbb{Q}(v^\nu) = \mathbb{E}[p_\nu(\xi)]$, where the notation $h_\Lambda(\xi) = \prod_{i\in\Lambda} \xi_i$ has been employed (see Section 3.2).

By choosing $\varphi(\xi) = Q(x; T(\xi), h(\xi))$, the above upper bound applies. In fact, due to Assumption 4.7., $Q(x; T(\cdot), h(\cdot))$ is a convex function in ξ for any $x \in \mathbb{R}^{n_1}$ (see Theorem 3.2. in Section 3.2). We get the Edmundson–Madansky inequality for two–stage recourse problems

$$\mathscr{Q}(x) \leq \sum_{\nu=1}^{2^r} Q(x;T(v^\nu),h(v^\nu))\,\mathbb{Q}(v^\nu),\ \ x \in \mathbb{R}^{n_1}, \tag{4.65}$$

see Theorem 3.5. in Section 3.2. Consequently

$$c^{\mathrm{T}}x + \mathscr{Q}(x) \leq c^{\mathrm{T}}x + \sum_{\nu=1}^{2^r} Q(x;T(v^\nu),h(v^\nu))\,\mathbb{Q}(v^\nu) := f^U(x),\ \ \forall x \in \mathscr{B}$$

holds, which immediately implies that $f^U(x)$ is an upper bound on the optimal objective value f^* of (4.56), *for any* $x \in \mathscr{B}$. The best E-M upper bound on f^* could be obtained by solving

$$\left.\begin{array}{ll} \min\ c^{\mathrm{T}}x + \displaystyle\sum_{\nu=1}^{2^r} Q(x;T(v^\nu),h(v^\nu))\,\mathbb{Q}(v^\nu) & \\ \text{s.t. }\ Ax = b & \\ \qquad\ x \geq 0, & \end{array}\right\}$$

which is equivalent to the linear programming problem

$$\left.\begin{array}{lll} \min\ c^{\mathrm{T}}x + \displaystyle\sum_{\nu=1}^{2^r} \mathbb{Q}(v^\nu)\,q^{\mathrm{T}}y^\nu & & \\ \text{s.t. }\ T(v^\nu)x + Wy^\nu & = h(v^\nu), & \nu = 1,\dots,2^r \\ \qquad Ax & = b & \\ \qquad x & \geq 0 & \\ \qquad\qquad\quad y^\nu & \geq 0, & \nu = 1,\dots,2^r. \end{array}\right\} \tag{4.66}$$

The size of this LP grows exponentially with the dimension r of the random variable ξ, which makes this approach impracticable in a successive discretization framework. In the discrete approximation method we will employ an upper bound with a fixed x.

Computing the bounds for a partition

Similarly to the previous section, let $\Xi := \prod_{i=1}^{r}[\alpha_i,\beta_i]$ be an r–dimensional interval containing the support of the r–dimensional random vector ξ. We consider a disjoint partition (see Section 3.2) $\mathscr{X} := \{\Xi_k;\ k = 1,\cdots,K\}$ of Ξ, where the Ξ_k are half-open or closed intervals, which will be called *cells*. We have $\Xi_k \cap \Xi_\ell = \emptyset$ for $k \neq \ell$ and $\bigcup_{k=1}^{K} \Xi_k = \Xi$ holds. The probability measure of the cells will be denoted by π_k, that is, $\pi_k = \mathbb{P}_\xi(\Xi_k)$, $k = 1,\dots,K$.

According to Lemma 3.8 in Section 3.2, the Jensen lower bounds corresponding to the partition will be computed as follows. We consider the conditional distribution of ξ, given $\xi \in \Xi_k$, for the cells separately, and compute the conditional moments

$$\mu_k := \mathbb{E}[\xi \mid \xi \in \Xi_k] = \frac{1}{\pi_k} \int_{\Xi_k} \xi \, \mathbb{P}_\xi(d\xi).$$

Using these, the Jensen lower bounds

$$L_k(x) := \pi_k \, Q(x; T(\mu_k), h(\mu_k)) \leq \int_{\Xi_k} Q(x; T(\xi), h(\xi)) \, \mathbb{P}_\xi(d\xi)$$

are obtained, for $k = 1, \ldots, K$, $x \in \mathbb{R}^{n_1}$, see Section 4.7.2. By summing up the inequalities

$$L_{\mathscr{X}}(x) := \sum_{k=1}^{K} L_k(x) = \sum_{k=1}^{K} \pi_k \, Q(x; T(\mu_k), h(\mu_k)) \leq \mathscr{Q}(x) \tag{4.67}$$

results and consequently

$$f^L_{\mathscr{X}}(x) := c^{\mathrm{T}} x + \sum_{k=1}^{K} L_k(x) \leq c^{\mathrm{T}} x + \mathscr{Q}(x), \ \ \forall x \in \mathscr{B}$$

holds. Finally, for obtaining a lower bound on f^*,

$$\left.\begin{array}{lllll} \min & c^{\mathrm{T}} x + \sum_{k=1}^{K} \pi_k q^{\mathrm{T}} y^k & & & \\ \text{s.t.} & T(\mu_k) x & + W y^k & = h(\mu_k), & k = 1, \ldots, K \\ & Ax & & = b & \\ & x & & \geq 0 & \\ & & y^k & \geq 0, & k = 1, \ldots, K \end{array}\right\} \tag{4.68}$$

is solved. Denoting by $x_{\mathscr{X}}$ a solution of this LP, we have

$$L_k := \pi_k Q(x_{\mathscr{X}}; T(\mu_k), h(\mu_k)) \leq \int_{\Xi_k} Q(x_{\mathscr{X}}; T(\xi), h(\xi)) \mathbb{P}_\xi(d\xi) \tag{4.69}$$

and

$$f^L_{\mathscr{X}} := c^{\mathrm{T}} x_{\mathscr{X}} + \sum_{k=1}^{K} L_k \leq c^{\mathrm{T}} x + \mathscr{Q}(x), \ \ \forall x \in \mathscr{B}. \tag{4.70}$$

In particular, $f^L_{\mathscr{X}} \leq f^*$ holds meaning that $f^L_{\mathscr{X}}$, corresponding to the current partition $\mathscr{X}$, is a lower bound on the optimal objective value of the recourse problem (4.56).

In summary, the computation of the Jensen lower bound for a partition $\mathscr{X}$ runs as follows.

<u>Computing the Jensen lower bound</u>

Step 1. *Compute moments*
Compute the conditional probabilities $\pi_k = \mathbb{P}_\xi(\Xi_k)$ and the conditional expected values $\mu_k := \mathbb{E}[\xi \mid \xi \in \Xi_k]$, for $k = 1, \ldots, K$. The

computation of these quantities is straightforward in the case when ξ has a finite discrete distribution; for continuous distributions numerical integration is needed, in general.

Step 2. *Compute the lower bounds for the cells*
Set up and solve the LP problem (4.68), let $x_{\mathscr{X}}$ be an optimal solution. Compute the lower bounds L_k for the cells according to (4.69), $k = 1,\ldots,K$.

Step 3. *Compute the lower bound for the optimal objective value*
Compute $f^L_{\mathscr{X}}$ according to (4.70).

For the E–M upper bound we proceed analogously. Again, we consider the conditional distribution of ξ, given $\xi \in \Xi_k$, for the cells separately. If the components of ξ are stochastically independent then solely the conditional probability and the conditional expected value is needed. In general, we compute

$$\mu_{\Lambda,k} := \mathbb{E}[h_\Lambda(\xi) \mid \xi \in \Xi_k] = \frac{1}{\pi_k}\int_{\Xi_k} h_\Lambda(\xi)\,\mathbb{P}_\xi(d\xi).$$

The upper bounds are computed again according to Lemma 3.8, page 218. From (3.48) it follows

$$\int_{\Xi_k} Q(x;T(\xi),h(\xi))\,\mathbb{P}_\xi(d\xi) \le \pi_k \sum_{\nu=1}^{2^r} Q(x;T(v_k^\nu),h(v_k^\nu))\mathbb{Q}_k(v_k^\nu) := U_k(x), \quad (4.71)$$

where $x \in \mathbb{R}^{n_1}$, v_k^ν is the νth vertex of cell k, $k = 1,\ldots,K$ and $\mathbb{Q}_k$ is computed according to (4.63) in the stochastically independent case and according to (4.64) in general, where in both cases the moments are replaced by the conditional moments corresponding to the cells.

Summing up the above inequalities we get

$$\mathscr{Q}(x) \le \sum_{k=1}^{K} U_k(x) := U_{\mathscr{X}}(x) \quad (4.72)$$

and consequently

$$c^{\mathrm{T}}x + \mathscr{Q}(x) \le c^{\mathrm{T}}x + \sum_{k=1}^{K} U_k(x) := f^U_{\mathscr{X}}(x), \;\; \forall x \in \mathscr{B}. \quad (4.73)$$

Notice that for any $x \in \mathscr{B}$, $f^U_{\mathscr{X}}(x)$ is an upper bound for the optimal objective value f^*. In the discrete approximation method we will choose $x = x_{\mathscr{X}}$, that is, we choose an optimal solution of the LP problem (4.68), which served for computing the Jensen lower bound. For this choice we introduce the notation $U_k := U_k(x_{\mathscr{X}})$. Thus we have

$$f^* = c^{\mathrm{T}}x^* + \mathscr{Q}(x^*) \le c^{\mathrm{T}}x_{\mathscr{X}} + \sum_{k=1}^{K} U_k = f^U_{\mathscr{X}}(x_{\mathscr{X}}) := f^U_{\mathscr{X}}. \quad (4.74)$$

Our choice also implies that the inequality

$$L_k \leq \int_{\Xi_k} Q(x_{\mathscr{X}}; T(\xi), h(\xi)) \, \mathbb{P}_\xi(d\xi) \leq U_k \tag{4.75}$$

holds for $k = 1, \ldots, K$. The interpretation of this inequality is the following. Considering the interval–wise decomposition

$$\mathscr{Q}(x) = \mathbb{E}_\xi[Q(x; T(\xi), h(\xi))] = \sum_{i=1}^{K} \int_{\Xi_k} Q(x; T(\xi), h(\xi)) \, \mathbb{P}_\xi(d\xi),$$

(4.75) provides upper and lower bounds for the kth term, corresponding to the kth cell in the partition. The overall bounds $f_{\mathscr{X}}^L$ (see (4.70)) and $f_{\mathscr{X}}^U$ (see (4.73)) are then obtained by summing up the cell–wise bounds in (4.75) and subsequently adding the term $c^{\mathrm{T}} x_{\mathscr{X}}$.

Thus, $U_k - L_k$ provides an error bound for the approximation over the kth cell. If, for example, $Q(x_{\mathscr{X}}; T(\cdot), h(\cdot))$ happens to be a linear–affine function over Ξ_k, then, as it can easily be seen, $U_k = L_k$ holds, and the error will be zero. The proof of this fact is left as an exercise for the reader.

For computing the E–M upper bound for a fixed $x \in \mathscr{B}$, we proceed as follows.

Computing the E–M upper bound

Step 1. *Compute moments*
For each of the cells in turn do:
Compute the conditional probability π_k. If the components of ξ are stochastically independent then compute the conditional expected–value vectors μ_k, otherwise compute all of the $2^r - 1$ conditional cross–moments $\mu_{\Lambda,k}$.

Step 2. *Compute distribution on the vertices and recourse function values*
For each of the different vertices v_k^ν, $\nu = 1, \ldots, 2^r$, $k = 1, \ldots, K$ do:

- Compute $\mathbb{Q}_k(v_k^\nu)$ according to (4.63) or (4.64), depending whether the components of ξ are stochastically independent or dependent, respectively. In the computations replace the moments in the formulas with the conditional moments μ_k and $\mu_{\Lambda,k}$, respectively.
- Compute $Q(x; T(v_k^\nu), h(v_k^\nu))$ by solving the linear programming problem (4.57), with $\xi := v_k^\nu$.

Step 3. *Compute the upper bounds for the cells*
Compute the upper bounds $U_k(x)$, according to (4.71), $k = 1, \ldots, K$.

Step 4. *Compute the upper bound on the optimal objective value*
The E–M upper bound $f_{\mathscr{X}}^U$ is finally computed according to (4.72).

The successive discrete approximation method

Corollary 3.3. in Section 3.2 formulates the basis for this method: Assume that $\mathscr{X}$ and $\mathscr{Y}$ are two partitions of Ξ containing the support of ξ, such that $\mathscr{Y}$ is a *refinement* of $\mathscr{X}$. This means that each of the cells in $\mathscr{X}$ is the union of one or several cells in $\mathscr{Y}$. Then for each fixed $x \in \mathscr{B}$ we have

$$L_{\mathscr{X}}(x) \le L_{\mathscr{Y}}(x) \le \mathscr{Q}(x) \le U_{\mathscr{Y}}(x) \le U_{\mathscr{X}}(x), \tag{4.76}$$

see also (4.67) and (4.72). This fact suggests the following algorithmic idea: starting with Ξ, a sequence of partitions of Ξ is generated by successive refinements of the partition. For each partition $\mathscr{X}$ an approximate solution $x_{\mathscr{X}}$ is computed by solving (4.68) along with the bounds

$$f^L_{\mathscr{X}} \le f^* \le f^U_{\mathscr{X}},$$

see (4.70) and (4.74) The algorithm is stopped when the error bound $f^U_{\mathscr{X}} - f^L_{\mathscr{X}}$ is below a prescribed stopping tolerance.

A conceptual description of the method is presented in Section 3.2.1.3, beginning at page 223. Convergence properties of this type of algorithms have been discussed in Section 3.2, see Theorem 3.7. on page 222.

In this section we concentrate on various specific algorithmic issues. An immediate implication of the monotonicity property (4.76) is that for the Jensen lower bound the inequality

$$f^L_{\mathscr{X}} \le f^L_{\mathscr{Y}} \le f^*$$

holds, see (4.70). Consequently, the lower bounds will be monotonically increasing for a sequence of successive refinements of Ξ. The same will not be true for the E–M upper bounds (4.74). The reason is that these bounds also depend on the current approximate solution $x_{\mathscr{X}}$, whereas the Jensen bounds only depend on the current partition.

Given a partition $\mathscr{X}$, the question arises, how the next, refined partition should be constructed. The key observation in this respect is that, according to (4.75), the selection of the cells to be subdivided can be performed in a cell–wise fashion.

We will proceed as follows. Next a general framework of the algorithm will be formulated and subsequently several issues related to the implementation of the method will be discussed. The details and recommendations are based on the implementation of the method, developed by the authors, and on our extensive computational experience with this solver, named DAPPROX. The current version of DAPPROX is for the case when the components of ξ are stochastically independent. Let us emphasize, however, that this assumption does not require that the random elements of the model (e.g. $(h_1(\xi), h_2(\xi), \ldots, h_{m_2}(\xi))$) should be stochastically independent, see the affine–linear relations (3.6).

For specifying the algorithm, some further notation is needed. Considering (4.69), we introduce

$$Q^L_k := Q(x_{\mathscr{X}}; T(\mu_k), h(\mu_k))$$

thus having $L_k = \pi_k Q^L_k$. Similarly for (4.71) with $x = x_{\mathscr{X}}$, let

$$Q_k^U := \sum_{\nu=1}^{2^r} Q(x_{\mathscr{X}}; T(v_k^\nu), h(v_k^\nu))\, \mathbb{Q}_k(v_k^\nu)$$

resulting in $U_k = \pi_k Q_k^U$.

Successive discrete approximation method

Step 1. *Initialization*

Let $\mathscr{X} = \{\Xi\}$ and set $K := 1$ for the number of cells in the partition. Let $\pi_1 = 1$. If the components of ξ are stochastically independent then compute the expected–value vector μ_1, otherwise compute all of the $2^r - 1$ cross–moments $\mu_{\Lambda,1}$.

Let $f^U := \infty$, this will be the best (lowest) upper bound found so far. The subdivision process will be registered by employing a rooted binary tree where the nodes correspond to cells and branching represents subdivision of the cells. Initially this tree consists of a single node, which will be the root, with Ξ associated with it. Choose a stopping tolerance ε^* and a starting tolerance ε_S for subdivision. Set the iterations counter $\iota = 1$.

Step 2. *Compute the Jensen lower bound for* $\mathscr{X}$

Apply the algorithm on page 334 for computing the Jensen lower bound. Thereby skip Step 1 of that algorithm, because the moments and probabilities are already computed. Thus we get a solution $x_{\mathscr{X}}$ of (4.68), the lower bounds L_k, $k = 1, \ldots, K$, for which (4.75) holds, and a lower bound $f_{\mathscr{X}}^L$ for the optimal objective value f^* of the recourse problem.

Step 3. *Compute the E–M upper bound for* $\mathscr{X}$

With $x = x_{\mathscr{X}}$, employ the algorithm on page 336. This delivers the upper bounds U_k, $k = 1, \ldots, K$, for which again (4.75) holds, as well as an upper bound $f_{\mathscr{X}}^U$ on the optimal objective value of the recourse problem.

Step 4. *Check the stopping criterion*

Set $f^U := \min\{f^U, f_{\mathscr{X}}^U\}$. If $\Delta_\iota := \dfrac{f^U - f_{\mathscr{X}}^L}{1 + |f_{\mathscr{X}}^L|} \leq \varepsilon^*$ then **Stop** and deliver $x_{\mathscr{X}}$ as an ε^*–optimal solution. Otherwise continue with the next step.

Step 5. *Setup a list of cells to be subdivided*

Let $S := \{k \mid \delta(Q_k^L, Q_k^U, \pi_k, \kappa_k) \geq \varepsilon_S\}$ where δ is one of the selection functions specified below. κ_k is the number of subdivisions which resulted in cell k; in the subdivision tree κ_k is the number of edges between the root and the node representing the kth cell.

If $S = \emptyset$, then set $\varepsilon_S := \frac{1}{2}\varepsilon_S$ and repeat this step, otherwise continue with the next step. With the suggested selection functions δ, this cycle is finite since the algorithm did not stop in *Step 4*.

Step 6. *Carry out the subdivision*
For each $k \in S$, with Ξ_k do

- Choose a coordinate direction. The subdivision of Ξ_k will be carried out by employing a hyperplane perpendicular to the chosen coordinate axis.
- Subdivide Ξ_k into two intervals, by applying a cutting plane across the conditional expected value μ_k and perpendicular to the chosen coordinate direction.

Step 7. *Update the partition*
Set $K := K + |S|$; renumber the cells and update $\mathscr{X}$ accordingly. For each cell which has been subdivided, do the following

- For both of the new cells compute the corresponding conditional probability and conditional moments.
- Append two edges to the corresponding node of the subdivision tree with the child–nodes corresponding to the new cells.

Set $\iota = \iota + 1$ and **continue with Step 2.**

There are several points in the algorithmic framework which need further specification.

Let us begin with the cell–selection function δ in *Step 5*. The following selection functions are used:

- $\delta_1(Q_k^L, Q_k^U, \pi_k, \kappa_k) := \dfrac{Q_k^U - Q_k^L}{1 + |Q_k^L|}$
- $\delta_2(Q_k^L, Q_k^U, \pi_k, \kappa_k) := \pi_k \, \dfrac{Q_k^U - Q_k^L}{1 + |Q_k^L|}$.
- $\delta_3(Q_k^L, Q_k^U, \pi_k, \kappa_k) := \pi_k \, 2^{\kappa_k} \dfrac{Q_k^U - Q_k^L}{1 + |Q_k^L|}$.

Each of these involves the relative approximation error. In the second and third functions the probability–multiplier enforces that, among cells with approximately the same relative error, those with a higher probability content are considered first for subdivision. The third function has been suggested by H. Gassmann. It has the effect that among cells which qualify according to the second selection function, those cells will be selected which are the result of a higher number of subdivisions. This selection function implements a depth–first selection criterion in the subdivision tree. For further selection functions and strategies see Kall and Wallace [172]. We have experimented with the above strategies and also with other, more sophisticated strategies related to the subdivision tree. Based on our experience, we recommend to use $\delta = \delta_2$.

Having selected the cells to be subdivided, the next question arises, how to choose an appropriate coordinate direction in *Step 6*. The basis for the different methods is the following observation (already discussed on page 336): If $Q(x_{\mathscr{X}}; T(\cdot), h(\cdot))$ is a linear–affine function over Ξ_k, then $U_k = L_k$ holds, that is, the approximation error is zero. Therefore that coordinate direction will be chosen, along which some measure of nonlinearity is maximal. The idea is the following.

In *Step 3* computing the E–M upper bound involved the solution of the recourse subproblem (4.57) for all of the vertices $\xi := v_k^\nu$, $\nu = 1\dots,2^r$ of Ξ_k. We assume that for all of these vertices dual optimal solutions are also available, which is always the case when the simplex method has been used. Let u_k^ν be an optimal dual solution of (4.57), corresponding to vertex v_k^ν. These optimal dual solutions are utilized to construct nonlinearity measures for pairs of adjacent vertices. Let Ψ be such a function, defined on adjacent vertices of Ξ, that is, if v_k^i and v_k^j are adjacent vertices then $\Psi(v_k^i, v_k^j)$ will be the associated nonlinearity measure.

We introduce the following notation. For $i = 1,\dots,r$ let

$$\mathscr{A}_i := \{(a,b) \mid a \text{ and } b \text{ are vertices of } \Xi \text{ and } \\ a \text{ and } b \text{ differ only in their } i^{\text{th}} \text{ coordinate}\}.$$

From a geometrical point of view, the elements of $\mathscr{A}_i$ represent the set of edges of Ξ, which are parallel to the ith coordinate direction. The coordinate–selection algorithm runs as follows.

Coordinate–selection method

Step 1. *Assign nonlinearity measures to coordinate–directions*
For each of the coordinates $i = 1,\dots,r$ compute
$\Psi_i := \min\limits_{(a,b)\in\mathscr{A}_i} \Psi(a,b).$

Step 2. *Choose coordinate*
Choose a coordinate direction for which Ψ_i is maximal.

Several nonlinearity measures have been suggested, see Kall and Wallace [172]. Here we discuss the two measures which have been proposed by Frauendorfer and Kall [105] and which are implemented in DAPPROX.

The first measure is based on the following observations. Due to our assumptions, $Q(x_{\mathscr{X}};T(\cdot),h(\cdot))$ is a convex piecewise linear function. Let us consider two adjacent vertices v_k^i and v_k^j of Ξ_k, with associated optimal dual solutions u_k^i and u_k^j. According to Theorem 3.2. in Chapter 3 (page 204), the subgradients of $Q(x;T(\xi),h(\xi))$ with respect to x are the optimal dual solutions of the recourse subproblem (4.57), multiplied by a matrix independent on x. Due to the affine–linear relations (3.6), it can be seen analogously that the subgradients of $Q(x;T(\xi),h(\xi))$ with respect to ξ have a similar form: they are again the optimal dual solutions of the recourse subproblem, multiplied by a matrix which does not depend on ξ.

Assume now that the dual solutions are equal for the two vertices, that is, we assume that $u_k^i = u_k^j$ holds. The above considerations imply that the corresponding subgradients of $Q(x_{\mathscr{X}};T(\xi),h(\xi))$ with respect to ξ are also equal. Consequently, $Q(x_{\mathscr{X}};T(\cdot),h(\cdot))$ is a linear–affine function along the edge, connecting these vertices. Thus we may expect that the difference of dual solutions for adjacent vertices indicates the degree of nonlinearity along the corresponding edge. This suggests the first nonlinearity measure for adjacent vertices v_k^i and v_k^j

$$\Psi_1(v_k^i, v_k^j) := \|u_k^i - u_k^j\|.$$

The second measure is based on Lemma 3.3 (page 203) and on the convexity of $Q(x_{\mathscr{X}};T(\cdot),h(\cdot))$. We take again two adjacent vertices v_k^i and v_k^j. With the corresponding optimal dual solutions u_k^i and u_k^j we have (see (4.58))

$$Q(x_{\mathscr{X}};T(v_k^i),h(v_k^i)) = (h(v_k^i) - T(v_k^i)x_{\mathscr{X}})^{\mathrm{T}} u_k^i$$
$$Q(x_{\mathscr{X}};T(v_k^j),h(v_k^j)) = (h(v_k^j) - T(v_k^j)x_{\mathscr{X}})^{\mathrm{T}} u_k^j.$$

Using these relations and Lemma 3.3 (page 203), we obtain by the subgradient inequality

$$(u_k^i)^{\mathrm{T}}(h(v_k^j) - T(v_k^j)x_{\mathscr{X}}) \leq Q(x_{\mathscr{X}};T(v_k^j),h(v_k^j))$$
$$(u_k^j)^{\mathrm{T}}(h(v_k^i) - T(v_k^i)x_{\mathscr{X}}) \leq Q(x_{\mathscr{X}};T(v_k^i),h(v_k^i)).$$

Let us define

$$\Delta_k^{ij} := Q(x_{\mathscr{X}};T(v_k^j),h(v_k^j)) - (u_k^i)^{\mathrm{T}}(h(v_k^j) - T(v_k^j)x_{\mathscr{X}})$$
$$\Delta_k^{ji} := Q(x_{\mathscr{X}};T(v_k^i),h(v_k^i)) - (u_k^j)^{\mathrm{T}}(h(v_k^i) - T(v_k^i)x_{\mathscr{X}}).$$

The interpretation of Δ_k^{ij} is the following: if we linearize $Q(x_{\mathscr{X}};T(\cdot),h(\cdot))$ at $\xi = v_k^i$ using the subgradient u_k^i, then Δ_k^{ij} is the linearization error at $\xi = v_k^j$. The interpretation of Δ_k^{ji} is analogous, by considering the linearization this time at $\xi = v_k^j$.

We chose

$$\Psi_2(v_k^i, v_k^j) := \min\{\Delta_k^{ij}, \Delta_k^{ji}\}$$

as our second quality measure; for the heuristics behind choosing the minimum above, see Frauendorfer and Kall [105] or Kall and Wallace [172].

According to our experience, none of the two nonlinearity measures can be considered as best. Our recommendation is the combined use of them. One possible implementation is to switch between the two strategies if the improvement in Δ_ι is small for a specified number of subsequent iterations. As a starting strategy, the use of Ψ_2 is recommended.

Implementation

The successive discrete approximation method involves the solution of several LP subproblems.

In *Step 2*, for computing the Jensen lower bound, the LP problem (4.68) has to be solved. The straightforward approach for solving (4.68) is to apply a general–purpose LP solver without any considerations concerning the special structure. This may become quite time–consuming with an increasing number of cells.

A better approach is based on the observation that (4.68) is the LP equivalent of a two–stage problem with a finite discrete distribution. The realizations of the random vector are the conditional expectations μ_k and the corresponding probabilities are the conditional probabilities π_k of the cells. Thus, the number of realizations equals the number of cells in the current partition $\mathscr{X}$. The idea is to apply solvers designed to solving two–stage recourse problems with a finite discrete distribution. With DAPPROX we have quite good experiences by employing QDECOM for solving (4.68). The solver QDECOM is an implementation of the regularized decomposition method of Ruszczyński [293], implemented by Ruszczyński. The algorithm has been discussed in Section 1.2.8.

The next idea is due to Kall and Stoyan [171]. It consists of using a general–purpose LP solver, but taking into account the specialities of the successive discrete approximation procedure. In the successive decomposition method, as discussed in the previous section, typically several cells are subdivided in a single iteration cycle. For explaining the idea, we assume that a single cell is subdivided; the extension to the general case is straightforward. The idea is that, instead of (4.68), its dual

$$\left.\begin{array}{lll} \max & \sum_{k=1}^{K} h^{\mathrm{T}}(\mu_k)u^k & + b^{\mathrm{T}}v \\ \text{s.t.} & \sum_{k=1}^{K} T^{\mathrm{T}}(\mu_k)u^k & + A^{\mathrm{T}}v \le c \\ & W^{\mathrm{T}}u^k & \le \pi_k q, \; k = 1,\ldots,K \end{array}\right\} \tag{4.77}$$

is solved. Let $\mathscr{X}$ be the partition corresponding to this LP. Assume, for the sake of simplicity, that the first cell $\Xi_1 \in \mathscr{X}$ is subdivided as $\Xi_1 = \Xi_{11} \cup \Xi_{12}$, with corresponding probabilities π_{11}, π_{12}, and conditional expected values μ_{11}, μ_{12}. Thus we have

$$\begin{aligned} \pi_1 &= \pi_{11} + \pi_{12} \\ \mu_1 &= \pi_{11}\,\mu_{11} + \pi_{12}\,\mu_{12}. \end{aligned} \tag{4.78}$$

The dual LP for the new partition will have the form

$$\left.\begin{array}{l} \max h^{\mathrm{T}}(\mu_{11})u^{11} + h^{\mathrm{T}}(\mu_{12})u^{12} + \sum_{k=2}^{K} h^{\mathrm{T}}(\mu_k)u^k + b^{\mathrm{T}}v \\ \text{s.t.}\;\; T^{\mathrm{T}}(\mu_{11})u^{11} + T^{\mathrm{T}}(\mu_{12})u^{12} + \sum_{k=2}^{K} T^{\mathrm{T}}(\mu_k)u^k + A^{\mathrm{T}}v \le c \\ \quad W^{\mathrm{T}}u^{11} \le \pi_{11} q \\ \quad W^{\mathrm{T}}u^{12} \le \pi_{12} q \\ \quad W^{\mathrm{T}}u^k \le \pi_k q, \\ \quad k = 2,\ldots,K. \end{array}\right\} \tag{4.79}$$

Let $(\bar{u}_k, k = 1,\ldots,K; \bar{v})$ be a solution of (4.77). Then with

$$\hat{v} := \bar{v};\; \hat{u}^{11} := \frac{\pi_{11}}{\pi_1}\bar{u}_1,\; \hat{u}^{12} := \frac{\pi_{12}}{\pi_1}\bar{u}_1,\; \hat{u}^k := \bar{u}_k,\; k = 2,\ldots,K \tag{4.80}$$

we have a feasible solution of (4.79), with the same objective value as the optimal objective value of (4.77). This can easily be seen by utilizing (4.78) and the affine–linear relations (3.6).

Let us discuss the solution of the LP problems involved in *Step 3* next. For computing the E–M upper bound, the recourse subproblem (4.57) has to be solved for each of the different vertices among all vertices v_k^ν, $\nu = 1,\ldots,2^r$, $k = 1,\ldots,K$, in the current partition $\mathscr{X}$. This involves solving a huge amount of LP problems, in general. The simplex method is especially well–suited for carry out this task, for the following reason. Instead of solving (4.57), its dual (4.58) is solved. Notice that the feasible domain of the dual problem does not depend on ξ. We have to solve a sequence of LP problems. Except of the first one, *hot start* can be used. This means that the optimal basis of the previous LP is taken as a starting basis for the next LP problem.

In DAPPROX we use Minos for solving the LP subproblems, see Murtagh and Saunders [239], [240].

The numerical efficiency of successive discrete approximation methods critically depends on the data–structures used. Hence we give an overview on the basic data structures used in DAPPROX.

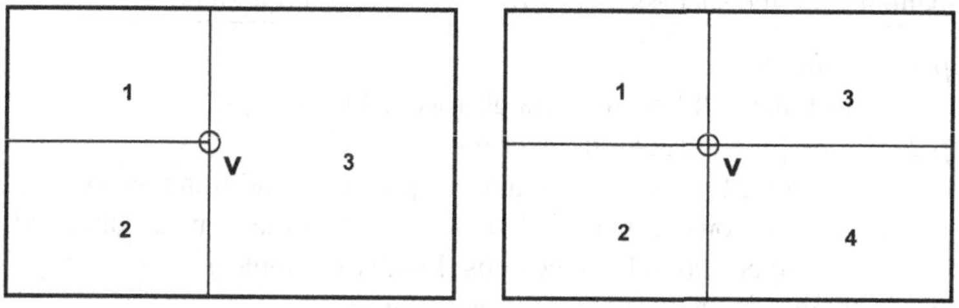

Fig. 4.5 Partitions of Ξ.

As discussed above, for obtaining the E–M upper bound, the recourse subproblem has to be solved for each of the vertices appearing in the current partition. The straightforward idea of working purely in a cell–wise fashion and solving the LP problems for the vertices of the cells in turn, is in general quite inefficient. To see this, consider a vertex of a cell which lies in the interior of Ξ, see vertex v in the partitions displayed in Figure 4.5. This vertex may have maximally 2^r neighboring cells, that is, it may belong to 2^r different cells, see the partition at the right–hand–side in Figure 4.5. Computing the E–M upper bounds cell–wise would mean that the LP belonging to that specific vertex would be solved 2^r times.

One possible remedy, implemented in DAPPROX, is the following: the different vertices are stored in a separate vertex list. For each vertex v^ν in the partition, the following quantities are stored: the coordinates of the vertex v^ν, the optimal objective value, and a pointer to an optimal dual solution. Considering the partitions in Figure 4.5, the vertex list for the partition at the left–hand–side would consist of 8 elements, and the list for the partition at the right–hand side would have 9 elements.

Notice that the feasible domain of the dual (4.58) of the recourse subproblem does not depend on ξ. According to numerical experience, the number of different optimal dual solutions which appear in the procedure is usually much smaller than the number of different vertices of the cells. Therefore, the different dual solutions are stored in a separate list, and the elements in the vertex list merely contain a pointer to the corresponding dual solution. This idea is due to Higle and Sen [139], who used it in the implementation of their stochastic decomposition method.

The information concerning the cells of the current partition is stored in a separate list, too. For each of the cells the following quantities are stored: the two diametrally opposite vertices defining the cell, the conditional probability and expectation of the cell, the upper and lower bounds corresponding to the cell, as well as a list of pointers to the vertices of the cell in the vertex list.

The subdivision procedure is recorded by employing a binary tree, the nodes of which correspond to cells and branching means subdivision. The leaves in the tree correspond to the current partition. Further information associated with the nodes includes cell probability, the bounds, and the split coordinate and split position.

The framework of an iteration of the algorithm, based on the data structures outlined above, is the following:

Implementation of the successive discrete approximation method

Step 1. *Initialization*
Initialize all lists in a straightforward fashion.

Step 2. *Compute the Jensen lower bound for $\mathscr{X}$*
Traverse the list of cells and compute the conditional probabilities and expectations. Set up and solve (4.77) and assign the obtained lower and upper bounds to the cells. Finally compute $f^L_{\mathscr{X}}$.

Step 3. *Compute the E–M upper bound for $\mathscr{X}$*
- Traverse the vertex list and solve the corresponding LP problems. For each vertex check whether a new dual solution appeared. If yes, append it to the list of dual solutions.
- Traverse the list of cells and employing the pointers to the vertices compute the E–M bound for the cells.
- Finally compute $f^U_{\mathscr{X}}$.

Step 4. *Check the stopping criterion*
This is the same as in the general method.

Step 5. *Setup a list of cells to be subdivided*
This is also the same as in the general method, too.

Step 6. *Carry out the subdivision*
The procedure is the same as for the general method, based on passing the list of cells once. For the coordinate–selection strategy Ψ_2, parallel edges are needed. This is implemented by setting up a list of the corresponding pairs of node pointers.

Step 7. *Update the partition*
This means now updating the lists. For each of the cells which is subdivided, the two new cells are added to the list of cells, one of them replacing the subdivided cell and the other appended to the list of cells. The new vertices are appended to the vertex list.

Next we discuss the case when ξ has a finite discrete distribution.

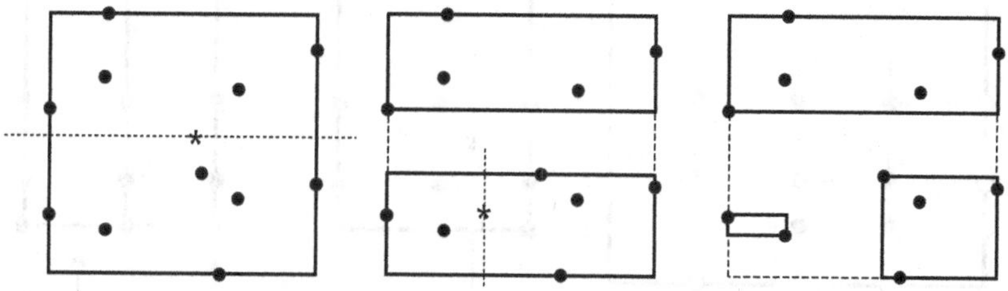

Fig. 4.6 Subdivisions of Ξ for a finite discrete distribution.

For explaining the idea let us consider Figure 4.6 first. In the figure, realizations of a two dimensional random variable are indicated by black bullets. The left–hand–side of the figure displays Ξ, which is in this case the smallest interval containing all realizations. The assumed first cut is indicated by the horizontal dotted line. The resulting first partition is shown in the middle part of the figure. Notice that for both cells the smallest interval, containing all realizations, has been taken. Assume, that the subsequent cut is performed according to the vertical dotted line. The resulting partition is displayed in the right–hand–side part of the figure. Again, the intervals for the new cells have been shrank. This means a change in the interpretation of a partition. This is no more a partition of the original interval, but a partition of the realizations, covered by the smallest possible intervals cell–wise.

This is also the general idea: after carrying out a subdivision, for each of the cells in the new partition we take the smallest interval which contains all realizations belonging to the cell. This has the obvious disadvantage, that now typically there are no common vertices of the cells. Thus, in the general case, all of the 2^r vertices for each of the cells have to be dealt with separately. According to numerical experience, however, the smaller cells result in much better E–M upper bounds, and the overall numerical efficiency becomes significantly better (the overall number of cells needed to achieve the required accuracy is much smaller). Edirisinghe and Ziemba [82] call this kind of partitioning a cell redefining strategy.

From the point of view of implementation, an additional feature appears. To see this, compare the partitions in the middle part and in the right–hand side part of the figure. The point is, that some vertices in the middle partition vanish when carrying out the next cut. These dummy vertices have to be removed from the vertex list, which can either be done by appropriately modifying the update algorithm after subdivision, or by periodically running a "garbage collection" procedure.

Finally we consider the case, when ξ has a finite discrete distribution and the components of ξ are stochastically independent. Such a situation is displayed in Figure 4.7. In the left–hand–side of the figure the interval Ξ contains all realizations. On the boundary of the interval, circles indicate the one dimensional marginal distributions and as before, the black bullets represent the joint realizations. Unlike in the general case (see Figure 4.6), the joint realizations are located now in a lattice.

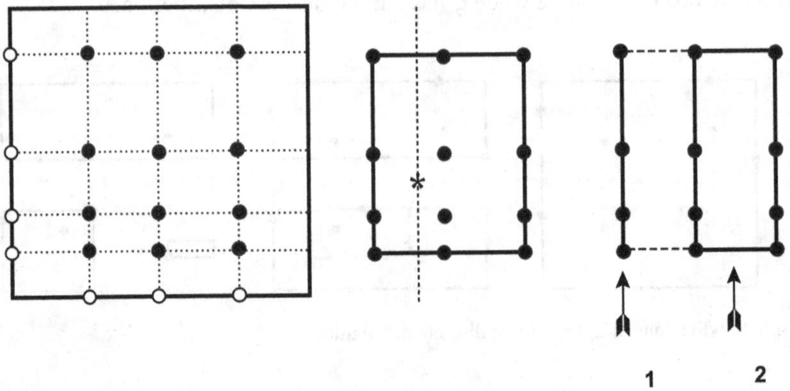

Fig. 4.7 Subdivisions of Ξ for independent finite discrete distributions.

This regular pattern has important implications concerning the efficient implementation. For explaining the idea, we consider again Figure 4.7. Similarly as in the general case with finite discrete distributions, the smallest interval Ξ containing all realizations is taken as the starting point of the method, see the middle part of the figure. Assume that the first cut is performed according to the vertical dotted line. Performing the subdivision, the partition shown in the right–hand–side of the figure results, where again the smallest intervals containing all realizations have been taken. Observe, that the cell Ξ_1 is now one–dimensional, thus having just 2 vertices.

This is an important special feature also in the general case. According to numerical experience, the dimensions of the cells collapse rapidly as the subdivision process proceeds. Thus, instead of 2^r, for a significant number of cells a much smaller amount of LP problems need to be solved for the E–M upper bound. This presupposes, of course, that the implementation is tailored to account for this possibility. Note, that the "collapsing dimensions" phenomenon has two roots: on the one hand, the components of Ξ should be stochastically independent, and on the other hand, the cells should be intervals.

Simple recourse

Simple recourse models have been the subject of Section 3.2.2. In this section we discuss how the successive discrete approximation method specializes in this case. The resulting algorithm is due to Kall and Stoyan [171]; a conceptual presentation of the method is given in Section 3.2.2.2, see page 235.

The main special feature of simple recourse models is separability, see (3.60), (3.63), and (3.62). For the sake of easy reference we reproduce some of the key relations, with slightly changed notation: let $z = Tx$ then with

$$
\begin{aligned}
\tilde{Q}_i(z_i,\xi_i) &:= q_i^+ \cdot [\xi_i - z_i]^+ + q_i^- \cdot [\xi_i - z_i]^- \\
\tilde{\mathscr{Q}}_i(z_i) &:= \mathbb{E}_{\xi_i}[\tilde{Q}_i(z_i,\xi_i)]
\end{aligned}
\tag{4.81}
$$

we have

$$
\begin{aligned}
\tilde{Q}(z,\xi) &= \sum_{i=1}^{m_2} \tilde{Q}_i(z_i,\xi_i) \\
\tilde{\mathscr{Q}}(z) &:= \sum_{i=1}^{m_2} \tilde{\mathscr{Q}}_i(z_i).
\end{aligned}
\tag{4.82}
$$

The separability property implies that the discrete approximation can be built in a coordinate–wise fashion. Instead of working with the r–dimensional interval $\Xi := \prod_{i=1}^{r} [\alpha_i, \beta_i]$ containing the support, the approximation is built for the one–dimensional intervals $[\alpha_i, \beta_i]$, $i = 1, \ldots, r$ separately, by considering the corresponding one–dimensional marginal distributions of ξ_i.

In the general complete recourse case we have constructed an upper bound for the expected recourse function $\mathscr{Q}(x)$, at the point $x = x_{\mathscr{X}}$ (see *Step 3* of the algorithm on page 338). In the simple recourse case the expected recourse function can be computed by an explicit formula, hence we use the function value itself as an upper bound on the optimal objective value. Next we discuss the formula for computing the expected recourse.

Due to the separability property, for deriving the formula it is sufficient to consider the case $r = 1$. Dropping the subscript 1, the recourse function has the form

$$
\tilde{Q}(z,\xi) = q^+ \cdot [\xi - z]^+ + q^- \cdot [\xi - z]^-,
$$

see (4.81).

Let $[\alpha, \beta]$ be an interval containing the support of the random variable ξ, subdivided as $\alpha = a_0 < a_1 < \ldots < a_K = \beta$. Let $I_1 := [a_0, a_1]$ and $I_k := (a_{k-1}, a_k]$ for $k \ge 2$, $\pi_k = \mathbb{P}_\xi(I_k)$, and $\mu_k := \mathbb{E}[\xi \mid \xi \in I_k]$, $k = 1, \ldots, K$. Let, furthermore, $\bar{k}$ be the index of the interval in the partition which contains z, that is, $z \in I_{\bar{k}}$ holds.

According to Lemma 3.14 on page 232, we have the formula

$$
\begin{aligned}
\tilde{\mathscr{Q}}(z) &= \mathbb{E}_\xi[\tilde{Q}(z,\xi)] = \hat{\pi}_1\, \tilde{Q}(z,\hat{\mu}_1) + \hat{\pi}_2\, \tilde{Q}(z,\hat{\mu}_2) \\
&= \hat{\pi}_1\, q^- \, [\hat{\mu}_1 - z]^- + \hat{\pi}_2\, q^+ \, [\hat{\mu}_2 - z]^+
\end{aligned}
\tag{4.83}
$$

with $\hat{\pi}_1 = \mathbb{P}_\xi([\alpha, z])$, $\hat{\pi}_2 = \mathbb{P}_\xi((z, \beta])$, $\hat{\mu}_1 = \mathbb{E}[\xi \mid \xi \in [\alpha, z]]$, and $\hat{\mu}_2 = \mathbb{E}[\xi \mid \xi \in (z, \beta]]$. This approach has the following drawback: it does not depend on the current partition; except of $K = 2$, the quantities $\hat{\pi}_1$, $\hat{\pi}_2$, $\hat{\mu}_1$, and $\hat{\mu}_2$ serve solely for computing $\tilde{\mathscr{Q}}(z)$, the rest of the discrete approximation method makes no use of them.

To see, how a better formula should look like, observe that $\tilde{Q}(z,\xi)$ as a function of ξ has a single kink at $\xi = z$ (see Figure 3.2 on page 229). Thus it is linear over all subintervals I_k in the partition, except of interval $I_{\bar{k}}$ which contains z. Therefore the approximation error is zero for all intervals I_k, $k \neq \bar{k}$, see the discussion in Section 4.7.2. Consequently, in the approximation scheme it makes only sense to consider $I_{\bar{k}}$ for further subdivision. It is also clear that the subdivision point should be $\xi = z$, because after the subdivision the recourse function $\tilde{Q}(z,\xi)$, as a function of ξ, will be linear on all of the subintervals for the current fixed z.

It is easy to see, that the following extension of (4.83) to several subintervals holds:

$$\mathbb{E}_\xi[\tilde{Q}(z,\xi)] = \sum_{k=1}^{K} \tilde{Q}_k(z) \tag{4.84}$$

with

$$\tilde{Q}_k(z) := \begin{cases} q^-(z-\mu_k)\,\pi_k & \text{if } k < \bar{k} \\ q^-(z-\mu_{\bar{k}}^1)\,\pi_{\bar{k}}^1 + q^+(\mu_{\bar{k}}^2 - z)\,\pi_{\bar{k}}^2 & \text{if } k = \bar{k} \\ q^+(\mu_k - z)\,\pi_k & \text{if } k > \bar{k}, \end{cases}$$

where for $k = \bar{k}$, $\pi_{\bar{k}}^1 := \mathbb{P}_\xi((a_{k-1}, z])$, $\pi_{\bar{k}}^2 := \mathbb{P}_\xi((z, a_k])$, $\mu_{\bar{k}}^1 := \mathbb{E}[\xi \mid \xi \in (a_{k-1}, z]]$, and $\mu_{\bar{k}}^2 := \mathbb{E}[\xi \mid \xi \in (z, a_k]]$ hold. If the interval $I_{\bar{k}}$ happens to be subdivided in the current iteration then these newly computed quantities can directly be used in the next iteration.

For specifying the discrete approximation method some further notation is needed. For $j = 1,\ldots,r$, let $[\alpha_j, \beta_j]$ be an interval containing the support of ξ_j, $\mathscr{X}_j$ be the current partition of $[\alpha_j, \beta_j]$ into K_j subintervals I_{jk}, let $\pi_{jk} = \mathbb{P}_{\xi_j}(I_{jk})$, $\mu_{jk} = \mathbb{E}[\xi_j \mid \xi_j \in I_{jk}]$, for $k = 1,\ldots,K_j$.

In the subsequent description of the method we will just stress those parts which are different with respect to the general method; for a detailed description we refer to the general algorithm on page 338.

<u>Successive discrete approximation for simple recourse</u>

Step 1. *Initialization*
This is basically the same as in the general method, except that now the initialization is carried out separately for $j = 1,\ldots,r$.

Step 2. *Compute the Jensen lower bound*
This is the same as in the general method, too. With $z := x_{\mathscr{X}}$ we also get, due to separability, the separate Jensen lower bounds $Q_j^L := \sum_{k=1}^{K_j} \pi_{jk} Q_j(z, \mu_{jk})$, for $j = 1,\ldots,r$.

Step 3. *Compute the recourse objective value*
With $z = Tx_{\mathscr{X}}$ apply formula (4.84) for computing the marginal expected recourse function values $Q_j^U := \mathscr{Q}_j(z_j)$, for $j = 1,\ldots,r$. According to (4.82), compute $f_{\mathscr{X}}^U := c^{\mathrm{T}} x_{\mathscr{X}} + \mathscr{Q}(z)$, which will be an upper bound on f^*.

Step 4. *Check the stopping criterion*
This step is the same as in the general method.

Step 5. *Setup a list of coordinates for subdivision*
Let $\mathscr{J} = \{ j \mid \delta(Q_j^L, Q_j^U, \pi_{j\bar{k}_j}, \kappa_{j\bar{k}_j}) \geq \varepsilon_S \}$, where $\bar{k}_j$ is the index of

the interval containing z_j, that is, $z_j \in I_{j\bar{k}_j}$ holds. If $\mathscr{J} = \emptyset$, then set $\varepsilon_S := \frac{1}{2}\varepsilon_S$ and repeat this step, otherwise continue with the next step.

Step 6. *Carry out the subdivision*
For each $j \in \mathscr{J}$, split $I_{j\bar{k}_j}$ at the point z_j into two intervals.

Step 7. *Update the partition*
For each $j \in \mathscr{J}$ set $K_j := K_j + 1$; renumber the cells and update $\mathscr{X}$ accordingly. Notice that for the new cells in the partition the probabilities and the conditional expectations have already been computed in *Step 3*, see (4.84). Update the subdivision trees; set $\iota = \iota + 1$ and **continue with Step 2.**

In *Step 2* of the algorithm, for computing the Jensen lower bound and the next iteration point, the LP problem (4.68) must be solved. According to the discussion in Section 4.7.2, an efficient way for solving this is solving the dual problem, which assumes in the simple recourse case the following form.

$$\left.\begin{array}{lrcl}
\max & \sum_{i=1}^{r}\sum_{k=1}^{K_j} h^{\mathrm{T}}(\mu_{jk})u^{jk} & + b^{\mathrm{T}}v & \\
\text{s.t.} & \sum_{j=1}^{r}\sum_{k=1}^{K_j} t_j^{\mathrm{T}}u^{jk} & + A^{\mathrm{T}}v \le c & \\
& -\pi_{jk}q_j^- \le \quad u^{jk} & \le \pi_{jk}q_j^+, & k = 1,\dots,K, \\
& & & j = 1,\dots,r,
\end{array}\right\} \qquad (4.85)$$

where the components of the n_1–dimensional vector t_j are the elements of the jth row of T, for all j.

For a sequence of such problems, the Kall–Stoyan method (see page 342) can be utilized to provide feasible starting points. Comparing (4.85) and the general counterpart (4.79), we notice that in (4.85) we simply have individual lower and upper bounds on the variables u^{jk}, instead of the corresponding parts in (4.79), where the recourse matrix W is involved. Due to this special structure, the Kall–Stoyan idea can be improved to provide a feasible basic solution for the next iteration, with the same or a better objective function value. This can be used for a hot start, which, according to numerical experience, reduces dramatically the solution time for solving (4.85). We assume for the sake of simplicity of presentation, that I_{11} has been split as $I_{11} = I_{11}^1 \cup I_{11}^2$, with corresponding probabilities π_{11}^1, π_{11}^2, and conditional expected values μ_{11}^1, μ_{11}^2. For these quantities relation (4.78) can be formulated analogously. The LP problem has the following form after the split:

$$
\left.\begin{array}{rl}
\max & h^{\mathrm{T}}(\mu_{11}^1)u_1^{11} + h^{\mathrm{T}}(\mu_{11}^2)u_2^{11} + \sum_{(j,k)\neq(1,1)} h^{\mathrm{T}}(\mu_{jk})u^{jk} + b^{\mathrm{T}}v \\
\text{s.t.} & t_1^{\mathrm{T}}u_1^{11} + t_1^{\mathrm{T}}u_2^{11} + \sum_{(j,k)\neq(1,1)} t_j^{\mathrm{T}}u^{jk} + A^{\mathrm{T}}v \leq c \\
& -\pi_{11}^1 q_1^- \leq \quad u_1^{11} \quad \leq \pi_{11}^1 q_1^+ \\
& -\pi_{11}^2 q_1^- \leq \quad u_2^{11} \quad \leq \pi_{11}^2 q_1^+ \\
& -\pi_{jk} q_j^- \leq \quad u^{jk} \quad \leq \pi_{jk} q_j^+, \\
& \qquad k = 1,\ldots,K, \\
& \qquad j = 1,\ldots,r, \\
& \qquad (j,k) \neq (1,1).
\end{array}\right\} \quad (4.86)
$$

The Kall–Stoyan feasible starting point will be, analogously as in (4.80),

$$
\hat{u}_1^{11} := \frac{\pi_{11}^1}{\pi_{11}} \bar{u}^{11}
$$

$$
\hat{u}_2^{11} := \frac{\pi_{11}^2}{\pi_{11}} \bar{u}^{11}
$$

and $\hat{u}^{jk} := \bar{u}_1^{jk}$ for $(j,k) \neq (1,1)$ as well as $\hat{v} := \bar{v}$, where $(\bar{u},\bar{v})$ is a solution of (4.85). If $\bar{u}^{11}$ was a non–basic variable, then its value is either the corresponding lower bound or the corresponding upper bound in (4.85). Then both $\hat{u}^{11}$ and $\hat{u}^{12}$ can be declared as non–basic variables, both of them being on the analogous lower or upper bound in (4.86). If $\bar{u}^{11}$ was a basic variable, then the following can be done: one of the variables $\hat{u}_1^{11}$ or $\hat{u}_2^{11}$ is shifted to the corresponding lower or upper bound in (4.86) and the other one is shifted by the same amount in the opposite direction. The variable shifted to a bound will be declared as non–basic and the other one as basic. This can be done in such a way, that the objective function does not decrease. The details are left as an exercise for the reader.

The authors have implemented the method as the solver SRAPPROX. To illustrate the efficiency, we refer to our paper Kall and Mayer [168] where we report on computational results with test problem batteries consisting of simple recourse problems with $r = 300$, which have been solved using SRAPPROX on a 660 MHz PC in approximately half a minute.

Other successive discrete approximation algorithms

As discussed in Section 3.2.1, there are basically two different algorithmic approaches, depending on the geometry of Ξ. The approach which has been discussed so far in this section, employs intervals. Thus, Ξ was an r–dimensional interval, and at each iteration the support of ξ was covered by a union of intervals.

Employing also intervals, Edirisinghe and Ziemba [82] report on the implementation of their variant of the successive discrete approximation method, with an extension to the case when also the recourse objective q is allowed to be stochastic.

A different approach, also based on interval–partitions, is due to Fábián and Szőke [90]. The authors combine a bundle–type convex programming method with a successive discrete approximation scheme. At each iteration a linear and a quadratic programming problem is to be solved. For the underlying NLP method see the references in the cited paper.

In the second approach Ξ is a regular simplex, which is partitioned into sub–simplices as the procedure progresses. For this approach and its implementation see Frauendorfer [103], who has also extended the algorithm for the case, when in (4.57) the second stage objective vector q may also contain stochastic entries. The approach has the advantageous property that for computing the E–M upper bound, the recourse subproblem (4.57) has to be solved only on the $r+1$ vertices of the simplex representing a cell, whereas when employing intervals, 2^r LP problems have to be solved for a cell. The price for this algorithmic advantage is that the simplex–based upper bounds may be much higher than the interval–based bounds; for an example see Kall [160]. According to our knowledge, there is no comparative computational study available in the literature for comparing the two approaches.

4.7.3 Stochastic algorithms

Sample average approximation (SAA)

In this section we will make the same assumptions and consider the same problem formulation as in Section 4.7.2.

Employing the notation from Section 4.7.2, we consider the two–stage problem

$$\left.\begin{array}{ll} \min & c^{\mathrm{T}}x + \mathscr{Q}(x) \\ \text{s.t.} & x \in \mathscr{B} \end{array}\right\} \tag{4.87}$$

with $\mathscr{Q}(x) := \mathbb{E}_{\xi}[Q(x;T(\xi),h(\xi))]$ and with the recourse function Q defined by the recourse–subproblem (second–stage problem) (4.57). Let f^* be the optimal objective value of (4.87) and let x^* be an optimal solution. Finally, Ξ denotes in this section the support of ξ.

We also introduce the notation

$$f(x;\xi) := c^{\mathrm{T}}x + Q(x;T(\xi),h(\xi)) \tag{4.88}$$

which results in the reformulation of (4.87)

$$\left.\begin{array}{ll} \min & \mathbb{E}[f(x;\xi)] \\ \text{s.t.} & x \in \mathscr{B}. \end{array}\right\} \tag{4.89}$$

Let $\xi^1,\ldots,\xi^N$ be a sample according to the distribution of ξ. This means that $\xi^1,\ldots,\xi^N$ are independent and identically distributed (i.i.d.) random variables, having the same distribution as ξ. Let us consider the following random variable

$$\mathscr{Q}_N(x;\xi^1,\ldots,\xi^N) := \frac{1}{N}\sum_{k=1}^{N} Q(x;T(\xi^k),h(\xi^k)), \tag{4.90}$$

which is the sample–mean estimator for the expected value $\mathscr{Q}(x)$, for each fixed x. From the viewpoint of simulation, $\mathscr{Q}_N(x;\xi^1,\ldots,\xi^N)$ is the crude Monte–Carlo approximation to $\mathscr{Q}(x)$, see, for instance, Ripley [278]. For each fixed x, this is clearly an unbiased estimator of $\mathscr{Q}(x)$:

$$\mathbb{E}[\mathscr{Q}_N(x;\xi^1,\ldots,\xi^N)] = \frac{1}{N}\sum_{k=1}^{N} \mathbb{E}[Q(x;T(\xi^k),h(\xi^k))] = \mathscr{Q}(x), \tag{4.91}$$

due to the fact that $\mathbb{E}[Q(x;T(\xi^k),h(\xi^k))] = \mathbb{E}[Q(x;T(\xi),h(\xi))]$ holds for all k.

In particular, choosing an arbitrary $\hat{x} \in \mathscr{B}$,

$$\vartheta_N^U(\hat{x};\xi^1,\ldots,\xi^N) := c^{\mathrm{T}}\hat{x} + \frac{1}{N}\sum_{k=1}^{N} Q(\hat{x};T(\xi^k),h(\xi^k)) = \frac{1}{N}\sum_{k=1}^{N} f(\hat{x};\xi^k) \tag{4.92}$$

is an unbiased estimator of $\mathbb{E}[f(\hat{x};\xi)]$ and due to

$$\mathbb{E}[\vartheta_N^U(\hat{x};\xi^1,\ldots,\xi^N)] = \mathbb{E}[f(\hat{x};\xi)] \geq f^* \tag{4.93}$$

we have an upper bound on f^*.

Based on the Monte–Carlo approximation (4.90), let us formulate the proplem

$$\left.\begin{array}{rl} \vartheta_N^L(\xi^1,\ldots,\xi^N) := \min & c^{\mathrm{T}}x + \mathscr{Q}_N(x;\xi^1,\ldots,\xi^N) \\ \text{s.t.} & x \in \mathscr{B}, \end{array}\right\} \tag{4.94}$$

which, under our assumptions, defines the random variable on the left–hand–side. Let $x_N(\xi^1,\ldots,\xi^N)$ be a solution of this problem. Problem (4.94) will be called a sample average approximation (SAA) problem for the original two–stage problem (4.87). Notice that the problem on the right–hand–side of (4.94) is not a single nonlinear optimization problem but a family of such problems, corresponding to the different realizations of $\xi^1,\ldots,\xi^N$. Considering a realization $\hat{\xi}^1,\ldots,\hat{\xi}^N$ of $\xi^1,\ldots,\xi^N$, and substituting the random variables with their realization in the minimization problem above, results in a deterministic optimization problem. In accordance with the literature, besides (4.94), this deterministic optimization problem will also be called a SAA–problem. Viewing (4.94) as a random optimization problem, the deterministic optimization problem resulting from the substitution of a realization, can be viewed as a realization of the SAA problem (4.94).

Based on the SAA problem, Mak, Morton, and Wood [214] proposed a lower bound for f^*:

Proposition 4.12. *The following inequality holds:*

$$\mathbb{E}[\,\vartheta_N^L(\xi^1,\ldots,\xi^N)\,] \le f^*. \tag{4.95}$$

Proof: We obviously have that $\vartheta_N^L(\xi^1,\ldots,\xi^N) \le c^T x + \mathscr{Q}_N(x;\xi^1,\ldots,\xi^N)$ holds for all $x \in \mathscr{B}$ and all realizations of $(\xi^1,\ldots,\xi^N)$, $\xi^k \in \Xi$, for all k. Taking expectation and utilizing (4.91) leads to

$$\mathbb{E}[\,\vartheta_N^L(\xi^1,\ldots,\xi^N)\,] \le c^T x + \mathscr{Q}(x).$$

Finally, taking the minimum over $x \in \mathscr{B}$ on the right–hand–side, yields the desired inequality. □

Notice that for $N = 1$ the above lower bound reduces to the wait–and–see lower bound WS, see Proposition 3.5. in Section 3.2.4.

The following monotonicity property has been discovered by Mak et al. [214] and, independently, by Norkin, Pflug, and Ruszczyński [244].

Proposition 4.13. *Let $\xi^1,\ldots,\xi^N,\xi^{N+1}$ be (i.i.d.) random variables, having the same distribution as ξ. Then*

$$\mathbb{E}[\,\vartheta_N^L(\xi^1,\ldots,\xi^N)\,] \le \mathbb{E}[\,\vartheta_{N+1}^L(\xi^1,\ldots,\xi^N,\xi^{N+1})\,]$$

holds.

Proof: Let $\mathscr{J} := \{1,\ldots,N,N+1\}$. We utilize the following obvious reformulation for sums of real numbers $\gamma_1,\ldots,\gamma_N,\gamma_{N+1}$

$$\sum_{k=1}^{N+1} \gamma_k = \sum_{k=1}^{N+1} \frac{1}{N} \sum_{j\in\mathscr{J},\, j\neq k} \gamma_j.$$

Thus we get

$$\begin{aligned}
\mathbb{E}[\,\vartheta_{N+1}^L(\xi^1,\ldots,\xi^N,\xi^{N+1})\,] &= \frac{1}{N+1}\,\mathbb{E}\left[\min_{x\in\mathscr{B}} \sum_{k=1}^{N+1} f(x;\xi^k)\right] \\
&= \frac{1}{N+1}\,\mathbb{E}\left[\min_{x\in\mathscr{B}} \sum_{k=1}^{N+1} \frac{1}{N} \sum_{j\in\mathscr{J},\, j\neq k} f(x;\xi^j)\right] \\
&\ge \frac{1}{N+1}\,\mathbb{E}\left[\sum_{k=1}^{N+1} \frac{1}{N} \min_{x\in\mathscr{B}} \sum_{j\in\mathscr{J},\, j\neq k} f(x;\xi^j)\right] \\
&= \frac{1}{N+1} \sum_{k=1}^{N+1} \frac{1}{N}\,\mathbb{E}\left[\min_{x\in\mathscr{B}} \sum_{j\in\mathscr{J},\, j\neq k} f(x;\xi^j)\right] \\
&= \mathbb{E}[\,\vartheta_N^L(\xi^1,\ldots,\xi^N)\,].
\end{aligned}$$

□

This is an attractive property, implying that increased sample–size leads in average to the same, or to an improved lower bound.

A second look on the facts and their proofs, discussed so far, reveals that only the following properties of ξ and f have been used: $f(x,\xi)$ should be finite for all $\xi \in \Xi$ and for all $x \in \mathscr{B}$, $\mathbb{E}[f(x,\xi)]$ should exist for all $x \in \mathscr{B}$, and the solutions of the minimization problems involved should exist. In particular, the convexity of $f(\cdot,\xi)$ and of $f(x,\cdot)$ did not play any role. In fact, the generality of results of the above type allows for designing algorithms for stochastic global optimization, see Norkin et al. [244].

Notice that the stochastic independence assumption concerning $\xi^1,\ldots,\xi^N$ has not been used in the argumentations and proofs above; they remain valid by merely assuming that the random variables are identically distributed and that they have the same probability distribution as ξ.

Let us now consider a sample (observations) of sample–size N, $\hat{\xi}^1,\ldots,\hat{\xi}^N$, that is, we take a realization of the (i.i.d.) random variables $\xi^1,\ldots,\xi^N$.

For computing the corresponding realization $\vartheta_N^U(\hat{x};\hat{\xi}^1,\ldots,\hat{\xi}^N)$ of the statistic ϑ_N^U, the recourse subproblem (4.57) has to be solved with fixed $x=\hat{x}$ for the realizations $\hat{\xi}^k$, for $k=1,\ldots,N$.

Concerning the computation of the realization of the statistic ϑ_N^L, we observe that the corresponding realization of he random program (4.94) is the two–stage recourse problem

$$\left.\begin{array}{ll} \min c^{\mathrm{T}}x+\dfrac{1}{N}\sum\limits_{k=1}^{N} Q(x;T(\hat{\xi}^k),h(\hat{\xi}^k)) & \\ \text{s.t.} & x\in\mathscr{B} \end{array}\right\} \tag{4.96}$$

with a finite discrete distribution having the equally probable realizations $\hat{\xi}^k$, for $k=1,\ldots,N$. This can be solved with any one of the methods designed for two–stage recourse problems with a finite discrete distribution. For instance, under our assumptions the successive discrete approximation method discussed in Section 4.7.2 can be used.

The question arises, how good the approximate solution obtained this way is. Since both ϑ_N^U and ϑ_N^L are random variables, adequate answers to this question have a probabilistic nature.

Mak et al. [214] propose to use confidence intervals for judging the quality of a candidate solution $\hat{x}\in\mathscr{B}$. The idea is to construct confidence intervals on the optimality gap $\mathbb{E}[f(\hat{x};\xi)]-f^*$ by utilizing (4.93) and (4.95), which imply the following upper bound on the optimality gap

$$\mathbb{E}\left[\vartheta_N^U(\hat{x};\xi^1,\ldots,\xi^N)-\vartheta_N^L(\xi^1,\ldots,\xi^N)\right] \ge \mathbb{E}[f(\hat{x};\xi)]-f^*.$$

The point is that, instead of estimating the upper and lower bounds from separate samples, the same sample is used for both of them according to the above formula. This corresponds to the variance–reduction technique *common random numbers* in

Monte Carlo simulation, see, for instance, Ross [291]. The confidence intervals are computed by utilizing the central limit theorem of probability theory; for the details see the above–cited paper [214]. In summary, the method works as follows. Let $M > 0$ be fixed and choose a sample–size N. For $\nu = 1,\ldots,M$ carry out the following procedure:

Testing the quality of $\hat{x} \in \mathscr{B}$

Step 1. *Generate a sample*
Generate a sample of size N, $\hat{\xi}^1,\ldots,\hat{\xi}^N$, according to the probability distribution of ξ, and independently of previously generated samples.

Step 2. *Solve a realization of SAA*
Solve the corresponding realization of (4.94), thus obtaining $\vartheta_N^L(\hat{\xi}^1,\ldots,\hat{\xi}^N)$.

Step 3. *Solve recourse subproblems*
Solve the recourse subproblems (4.57) for $\xi = \hat{\xi}^k$, $k = 1,\ldots,N$ and compute $\vartheta_N^U(\hat{x};\hat{\xi}^1,\ldots,\hat{\xi}^N)$ according to (4.92).

Step 4. *Compute the νth term for the estimator of the optimality gap*
Compute $\Delta_\nu := \vartheta_N^U(\hat{x};\hat{\xi}^1,\ldots,\hat{\xi}^N) - \vartheta_N^L(\hat{\xi}^1,\ldots,\hat{\xi}^N)$.

Having executed the above procedure M times, construct the estimator $\frac{1}{M}\sum\limits_{\nu=1}^{M}\Delta_\nu$ for the duality gap and compute a confidence interval as described in [214].

So far we have discussed, how the quality of a given approximate solution $\hat{x} \in \mathscr{B}$ can be judged. For obtaining an approximate solution of the two–stage recourse problem (4.87), the SAA–based approach relies on solving realizations of the approximate SAA problem (4.94). Before specifying how the algorithm works, let us summarize some theoretical results.

Let $\hat{x} \in \mathscr{B}$ be fixed. As discussed above, $\mathscr{Q}_N(\hat{x};\xi^1,\ldots,\xi^N)$ is an unbiased estimator of $\mathscr{Q}(\hat{x})$, for all N. Moreover, due to Kolmogorov's strong law of large numbers, $\mathscr{Q}_N(\hat{x};\xi^1,\ldots,\xi^N)$ converges to $\mathscr{Q}(\hat{x})$ almost surely. The question arises, whether we also have almost sure convergence of the optimal objective values $\vartheta_N^L(\xi^1,\ldots,\xi^N)$ of the SAA problems, to the true optimal objective value f^*. This question can be investigated by employing the theory of epi–convergence. For the case of deterministic approximations, the main results based on this theory are summarized in Theorem 3.7. on page 222. In the stochastic case we have epi–convergence in an almost sure sense, see King and Rockafellar [183] and King and Wets [184], and the references therein.

Results are also available concerning the speed of convergence of the solutions of (4.94). Assuming, for instance, that the original problem (4.87) has a unique solution x^*, under appropriate assumptions we have that

$$\mathbb{P}(\|x_N(\xi^1,\ldots,\xi^N) - x^*\| \geq \varepsilon) \to 0 \quad \text{for} \quad N \to \infty$$

holds for any $\varepsilon > 0$, and the rate of convergence is exponential, see Kaniovski, King, and Wets [175]. Under specific assumptions regarding convexity properties of f or considering the case when ξ has a finite discrete distribution, improved results of this type have been found by Shapiro and Homem–de–Mello [308], see also Linderoth et al. [205], and the references in these papers.

The SAA–algorithm relies on "external sampling", meaning that sampling is performed prior to solving the (approximate) problem. In contrast to this, "internal sampling" means that sampling is performed as the algorithm proceeds; for an example see stochastic decomposition in the next section.

Sample average approximation algorithm

Step 1. *Initialization*
Choose $N > 0$, $M > 0$.

Step 2. *Generate samples*
Generate M independent samples (batches) $\hat{\xi}^{1,\nu},\ldots,\hat{\xi}^{N,\nu}$, according to the probability distribution of ξ, $\nu = 1,\ldots,M$, each of which has the sample–size N.

Step 3. *Solve realizations of SAA*
For each of these samples solve the corresponding realization of (4.94), let $\vartheta_{N,\nu}$ be the optimal objective value, $\nu = 1,\ldots,M$.

Step 4. *Estimate f^**
Use $\frac{1}{M}\sum\limits_{\nu=1}^{M}\vartheta_{N,\nu}$ as an estimator of f^*.

Step 5. *Test the quality of solution*
This step involves statistical techniques for judging solution quality. For instance, the method for estimating the optimality gap can be used, as discussed on page 355.

For implementing this method, several important points have to be specified in a much more detailed fashion.

In general, the crude Monte–Carlo method is notoriously slow, therefore variance–reduction techniques have to be included, see, for instance Ross [291]. One such method, relying on common random numbers, has been mentioned above, regarding the optimality gap.

Assuming that the two–stage problem has a unique solution x^*, the solutions of the realizations of the SAA–problems converge rapidly to x^* for $N \to \infty$, in the sense as discussed in this section. Consequently, for N large enough, we may expect that the solution of SAA will be a good approximation to x^*. The question, how large N should be for getting a good solution, remains open.

Consequently, testing the quality of an obtained approximate solution is of vital importance. Two kinds of statistical approaches have been proposed for this. In the first class of methods the optimality gap is estimated; we have discussed an example for this technique above. For further methods belonging to this class see Bayraksan and Morton [10]. The second class of methods tests the Kuhn–Tucker optimality conditions, see Shapiro and Homem–de–Mello [307]. The practical procedure runs as follows: the above algorithm is carried out for a starting selection of M and N.

Subsequently the solution obtained this way is tested and if it turns out that it is not yet satisfactory, the algorithm is repeated with increased N and/or M.

If the solution of (4.87) is not unique, then recognizing an optimal solution may involve quite large samples. For further discussions of these problems and for other variance–reduction techniques see Shapiro and Homem–de–Mello [307], [308] and Linderoth, Shapiro, and Wright [205].

For statistical tests of optimality, based on duality theory, see Higle and Sen [140].

Stochastic decomposition

The stochastic decomposition (SD) method is a stochastic analog of the dual decomposition method, developed by Higle and Sen, see [136], [137] and [139]. The dual decomposition method has been presented in Section 1.2.6 (page 23) and has been further discussed in Section 4.4.2 of this Chapter.

The monograph [139] by Higle and Sen presents a detailed discussion of the method, along with the statistical tests involved, and including issues related to the implementation. For recent developments see Sen et al. [302]. Therefore, in this book we confine ourselves to pointing out some of the main ideas of the algorithm. Regarding (deterministic) dual decomposition, we will use the notation introduced in Section 4.4.2.

Similarly as in Section 4.7.2, we consider the two–stage recourse problem (4.56) under Assumption 4.7. on page 329. Additionally, for the sake of simplicity of presentation, we will suppose that $Q(x;T(\xi),h(\xi)) \geq 0$ holds for all $x \in \mathscr{B}$, almost surely. For a weaker assumption see Higle and Sen [139]. Our assumption is fulfilled, for instance, if $q \geq 0$ holds, which will be presupposed for the sake of simplicity.

Let Ξ denote the support of ξ in this section.

The SD algorithm relies on "internal sampling"; at iteration k we will have a sample of sample-size k. Let $\xi^1,\ldots,\xi^k$ be (i.i.d.) random variables having the same distribution as ξ. The idea is to construct a lower bounding approximation to the sample–average approximation $\mathscr{Q}_k(x;\xi^1,\ldots,\xi^k)$ of $\mathscr{Q}(x)$ (cf. (4.90)), and to update this approximation as iterations proceed.

Let us recall that due to weak duality and due to the fact that the feasible domain of the dual (4.58) of the recourse subproblem does not depend on x nor on ξ, we have the inequality

$$\frac{1}{k}\sum_{t=1}^{k}(h(\xi^t)-T(\xi^t)x)^{\mathrm{T}}u^t \leq \frac{1}{k}\sum_{t=1}^{k}Q(x;T(\xi^t),h(\xi^t)), \tag{4.97}$$

which holds for any $\xi^t \in \Xi$, any $x \in \mathscr{B}$, and any $u^t \in \mathscr{D}$, $t = 1\ldots,k$, where $\mathscr{D}$ denotes the feasible domain of the dual (4.58) of the recourse subproblem. The lower–bounding function on the left–hand–side of (4.97) will be utilized to generate a cut in the algorithm, and u^t will be an optimal dual solution, $t = 1,\ldots,k$.

In the subsequent iteration we deal with a sample $\xi^1,\ldots,\xi^k,\xi^{k+1}$. For ensuring that the previously generated cut has the lower bounding property also for the new sample–average approximation $\mathscr{Q}_{k+1}(x;\xi^1,\ldots,\xi^k,\xi^{k+1})$, the previous cut must be updated. The most natural update relies on the following obvious inequality

$$\begin{aligned}\frac{1}{k+1}\sum_{t=1}^{k}(h(\xi^t)-T(\xi^t)x)^{\mathrm{T}}u^t &\leq \frac{1}{k+1}\sum_{t=1}^{k}Q(x;T(\xi^t),h(\xi^t))\\ &\leq \frac{1}{k+1}\sum_{t=1}^{k+1}Q(x;T(\xi^t),h(\xi^t)),\end{aligned} \tag{4.98}$$

which holds for any $\xi^t \in \Xi$, any $x \in \mathscr{B}$, and any $u^t \in \mathscr{D}$, $t = 1\ldots,k$. The relaxed master problem (cf. (4.39)) will have the form

$$\left.\begin{array}{rll}\min & c^{\mathrm{T}}x+w & \\ \text{s.t} & (\beta_t^k)^{\mathrm{T}}x-w \leq -\alpha_t^k, & t=1,\ldots,k\\ & x \in \mathscr{B}, & \end{array}\right\} \tag{4.99}$$

where the coefficient vectors and constant terms concerning cuts have double indices, due to the above–mentioned updating. The basic (conceptual) SD algorithm can be specified as follows.

Basic stochastic decomposition method

Step 1. *Initialization*
Let $k := 0$, $\xi^0 := \mathbb{E}[\,\xi\,]$, and solve the corresponding expected–value (EV) problem (3.104). Let x^1 be a solution of the EV–problem. Set $\mathscr{V}_0 := \emptyset$. $\mathscr{V}_k$ will be the set of the different optimal dual solutions of the recourse subproblem (4.57) (vertices of $\mathscr{D}$), encountered up to iteration k.

Step 2. *Generate the next sample point*
Set $k := k+1$ and generate the next sample point ξ^k of ξ.

Step 3. *Solve a recourse subproblem*
With $\xi = \xi^k$ solve the dual recourse subproblem (4.58) by using the simplex method, let $u_k^k \in \mathscr{D}$ be an optimal basic solution. If $u_k^k \notin \mathscr{V}_{k-1}$ then let $\mathscr{V}_k := \mathscr{V}_{k-1} \cup \{u_k^k\}$, otherwise let $\mathscr{V}_k := \mathscr{V}_{k-1}$.

Step 4. *Generate a new cut*

- Taking the current feasible solution $x^k \in \mathscr{B}$, for each of the previous realizations choose the best vertex from $\mathscr{V}_k$, that is, compute

$$u_t^k \in \operatorname{argmax}\{\,(h(\xi^t)-T(\xi^t)x^k)^{\mathrm{T}}u \mid u \in \mathscr{V}_k\,\},$$

$t = 1,\ldots,k-1$.
- Compute the kth cut

$$\alpha_k^k + (\beta_k^k)^{\mathrm{T}} x := \frac{1}{k} \sum_{t=1}^{k} (h(\xi^t) - T(\xi^t)x)^{\mathrm{T}} u_t^k.$$

Step 5. *Update previous cuts*
For $t = 1, \ldots, k-1$ compute

$$\alpha_t^k := \frac{k-1}{k} \alpha_t^{k-1} \text{ and } \beta_t^k := \frac{k-1}{k} \beta_t^{k-1}.$$

Step 6. *Solve the relaxed master problem*
Solve (4.99); let x^{k+1} be an optimal solution.
Continue with *Step 2*.

Notice that due to the fact that (4.97) holds for *any* $u_t \in \mathscr{D}$, the newly generated cut in *Step 4* has the lower bounding property. Due to the "argmax" procedure, the best such cut is generated taking into account the dual–vertex information available so far. The update formulas of the previous cuts in *Step 5* imply that the lower bounding property is preserved, see (4.99).

The algorithm above employs aggregate cuts. A version of the SD algorithm with disaggregate cuts has been developed by Higle, Lowe, and Odio [135].

From the theoretical point of view, all that could be proved for the basic algorithm, was the existence of a subsequence of the sequence of generated points x^k, $k = 1, 2, \ldots$, such that every accumulation point of this subsequence is an optimal solution of the recourse problem (4.56), almost surely (see Higle and Sen [139]).

Therefore, the full version of the SD method of Higle and Sen employs *incumbent solutions*. Initially, the first incumbent solution is just the solution of the expected value problem, obtained in *Step 1* of the basic algorithm. The current solution of the relaxed master problem becomes the new incumbent, if the actual objective value of the relaxed master problem is sufficiently lower than the approximate objective value at the incumbent. The cut corresponding to the current incumbent is updated in each iteration, using the analogous "argmax" procedure as for constructing the new cut in *Step 4*. Considering an appropriate subsequence of iterations, where the incumbent changes, a numerically implementable procedure results for identifying approximate solutions of (4.56), see Higle and Sen [139].

The idea of working with incumbent solutions is also the basis of the regularized dual decomposition method of Ruszczyński [293], see Section 1.2.8. One of the attractive features of regularized decomposition is that it provides a safe way of removing redundant cuts. The accumulation of redundant cuts can become in fact a numerical problem for the version of the SD algorithm discussed so far. Consequently, Higle and Sen [137], [139] and Yakowitz [350] developed the regularized SD algorithm, which can be viewed as a stochastic version of the regularized dual decomposition method. Let us denote the incumbent solution at iteration k by $\bar{x}^k$. In the regularized SD method, the objective function of the relaxed master problem (4.99) includes a regularizing term, thus becoming

$$c^{\mathrm{T}} x + w + \frac{1}{2} \|x - \bar{x}^k\|^2,$$

otherwise the method is basically the same as SD. This regularized version of the SD method can currently be considered as the best version of SD, see Higle and Sen [139] for the details.

The reader might wonder that the basic SD method, as specified above, does not contain a stopping rule. This is merely for the purpose of simplicity of presentation. For any stochastic method, the most important questions are how to stop the algorithm and how to identify an approximate optimal solution of the two–stage recourse problem, on the basis of results delivered by the method. We have discussed this problem in the previous section, in connection with the SAA method. In fact, most of the stopping rules proposed for the SAA method are essentially generalizations of stopping rules proposed by Higle and Sen for the SD method, see [139]. Three classes of stopping rules have been proposed. The first class contains rules which are based on asymptotic properties regarding the sequence of incumbents. The second type of rules utilizes estimates on the optimality gap, including also bootstrap schemes. Finally, the third group is based on optimality conditions. For the details see [139].

The authors have implemented stochastic decomposition as the solver SDECOM, following [139] and some additional guidelines of Higle and Sen, which were highly appreciated by the authors. The present version implements the SD method with incumbents (not yet the regularized version). The stopping rule is a rule based on asymptotic properties. The solver is connected to SLP-IOR, see Section 4.9.2.

Other stochastic algorithms

The stochastic methods not yet discussed belong to the class of methods with "internal sampling".

The stochastic quasi–gradient methods are stochastic versions of subgradient methods. The basic idea is to work with stochastic quasi–gradients. At iteration ν, a random variable v^ν is a stochastic quasi–gradient at x^ν, if

$$\mathbb{E}[v^\nu \mid x^1, \ldots, x^\nu] \in \partial_x (c^{\mathrm{T}} x + \mathcal{Q}(x^\nu))$$

holds. With step–size ρ_ν, the next iteration point is computed by the projection onto the feasible domain $\mathcal{B}$: $x^{\nu+1} := \prod_{\mathcal{B}}(x^\nu - \rho_\nu v^\nu)$. Under appropriate assumptions, in particular, by choosing suitable sequences of step–sizes ρ_ν, the algorithm converges to a solution of the two–stage problem, almost surely. For details concerning these methods see Ermoliev [89] and Gaivoronski [109], and the references therein. For an introduction see Kall and Wallace [172].

For stabilizing the sequence of points in stochastic quasi–gradient methods, Marti [223] and Marti and Fuchs [226], [227] propose algorithms where at certain iterations deterministic descent directions are used, instead of stochastic quasi–gradients. The authors call the methods in this class semi–stochastic approximation methods.

Under appropriate assumptions concerning the probability distribution, these methods also converge to a solution, almost surely.

Besides stochastic decomposition, another stochastic version of the dual decomposition method has also been developed, relying on importance sampling. For this method see Dantzig and Glynn [52] and Infanger [146], [147].

Deák [57], [59] proposes an algorithm for the special case where the random entries in the right–hand–side and in the technology matrix have a joint multivariate normal distribution. The expected recourse function is computed via Monte–Carlo integration, in the framework of a solution method based on successive regression approximation technique.

4.7.4 Simple recourse models

Simple recourse models have been the subject of Section 3.2.2. From the point of view of applications, simple recourse problems are an important subclass of two–stage recourse problems; they can be solved numerically for a large amount of random variables. Several authors have proposed algorithms for simple recourse problems; below we just mention some of the approaches.

One of the algorithms, based on successive discrete approximation, has been the subject of Section 4.7.2. For the case, when ξ has a finite discrete distribution, methods, utilizing the special basis–structure of the equivalent LP–problem have been developed by Prékopa [265] and Wets [344]. Further methods include the algorithms of Cleef [43] which employs a sequence of linear substitute problems, and the method of Qi [273], who proposes an algorithm which involves solving linear and nonlinear convex programming subproblems, in an alternating fashion. For the other methods see the references in the above–cited papers.

Let us point out, that for several classes of probability distributions, simple recourse problems can equivalently be formulated as nonlinear programming problems in algebraic terms, see, for instance, Kall [154].

Finally we consider models with multiple simple recourse, discussed in Section 3.2.2. In the case when ξ has a finite discrete distribution, such models can be transformed into a simple recourse problem, see Theorem 3.8. in Section 3.2.2. Consequently, such problems can be efficiently solved by solving the equivalent simple recourse problem.

4.7.5 A guide to available software

In the listing of solvers below, we include also solvers for multistage recourse problems; two–stage problems are clearly a special case for them.

Let us begin with SLP solvers for recourse problems, available at the NEOS Server for optimization, http://www-neos.mcs.anl.gov/. The general idea of the

NEOS server is, that users select a solver available at the server, send their problems, and obtain the solution, via the Internet. The SLP–problem must be sent to the server in the SMPS format; for this see Gassmann [115], and the references therein.

- Bnbs (Bouncing nested Benders solver), is an implementation of the nested decomposition method, for multistage recourse problems with a finite discrete distribution. It has been developed by Fredrik Altenstedt, Department of Mathematics, Chalmers University of Technology, Sweden. The source code of the solver can also be downloaded from the author's homepage http://www-neos.mcs.anl.gov/.
- FortSP (the Stochastic Programming extensions to FortMP). The current version is for two–stage recourse problems with a finite discrete distribution. It is the SLP–solver in the stochastic programming integrated environment (SPinE), see Valente et al. [332].
- MSLiP is an implementation of the nested decomposition algorithm, for multistage recourse problems with a finite discrete distribution, developed by Gassmann [112]. The code is available to universities and academic institutions for academic purposes, please contact the author.

The IBM stochastic programming system, OSLSE, designed for multistage recourse problems with finite discrete distributions, is available for academic purposes, in executable form. For OSLSE see King et al. [185]. Recently, IBM initiated the project "COmputational INfrastructure for Operations Research" (COIN–OR). As far as we know, a version of OSLSE is now available with an added facility, which enables for the user to connect her/his LP solver to OSLSE, instead of the LP solver OSL of IBM. For the details see http://www-124.ibm.com/developerworks/opensource/coin/.

The solver SQG is an implementation of stochastic quasi–gradient methods, see Gaivoronski [110]; the author of this paper encourages interested readers to contact him.

An interior point method based on the augmented system approach has been implemented by Csaba Mészáros [233] as the solver BPMPD. We do not know the present status of this solver, interested readers might contact the author of BPMPD.

Almost all authors of algorithms, discussed in this section, report on computational experience. Concerning the availability of solvers, we suggest to contact the authors.

For commercially available solvers we refer to the solvers OSLSE and DECIS, both available with the algebraic modeling system GAMS, Brooke et al. [35]. OSLSE has already been mentioned above, DECIS is an implementation of the importance sampling algorithm, implemented by G. Infanger.

Finally we give a short list of solvers which are connected to our model management system SLP–IOR and have not been discussed so far in this section. They are available for academic purposes along with SLP–IOR, in executable form. For further details see Section 4.9.2. The following solvers, all of them developed for the case of a finite discrete distribution, have been provided to us by their authors:

- QDECOM, regularized decomposition method, implemented by A. Ruszczyński, for two–stage fixed recourse problems.
- SHOR2, decomposition scheme of Shor, implemented by N. Shor and A. Likhovid, for complete recourse problems.
- SHOR1, the same method and authors as for SHOR2, for simple recourse.
- SIRD2SCR, for simple integer recourse, implemented by J. Mayer and M.H. van der Vlerk.
- MScr2Scr, for multiple simple recourse, implemented by J. Mayer and M.H. van der Vlerk.

Finally we list our own solvers, which have already been mentioned in the preceding sections. The solvers have been implemented by the authors of this book.

- DAPPROX implements the successive discrete approximation method, for complete recourse problems, with a deterministic objective in the second stage, and assuming the stochastic independence of the components of ξ. Probability distributions: finite discrete, uniform, exponential, and normal distributions.
- SRAPPROX is an implementation of the successive discrete approximation algorithm for simple recourse problems. Stochastic independence is not required; the marginal distributions should belong to one of the classes of distributions listed with DAPPROX.
- SDECOM is an implementation of of the stochastic decomposition method.

The question, which of the available solvers should be chosen for solving a specific instance of a two–stage recourse problem, is a difficult one. There exists no general answer to this question, the performance of algorithms and solvers may depend substantially on the specific characteristics of the problem instance. The main factors influencing solver performance are the type of the probability distribution, the stochastic dependence properties of the components of ξ, which parts of the model are stochastic, the number of random variables (dimension of ξ), the number of joint realizations in the discretely distributed case. For instance, having a complete recourse problem with a 10–dimensional random vector ξ with stochastically independent components, and each of the components having 10 realizations, results in 10^{10} joint realizations. This rules out all solvers, based on solving the equivalent LP problem, including solvers based on dual decomposition or the regularized version of it.

Selecting an appropriate solver is clearly supported by comparative computational results; this seems to be a scarce resource in the SLP literature, though. Concerning comparative computational results we refer to Kall and Mayer [164], [166], [168] and to Mayer [230].

Exercises

4.7. Assume that in the successive discrete approximation algorithm, as presented in Section 4.7.2, the recourse function $Q(x; T(\xi), h(\xi))$ is a linear function of ξ on an interval Ξ_k in the subdivision of Ξ. Show that in this case the Jensen lower bound $L_k(x)$ and the Edmundson–Madansky upper bound $U_k(x)$ are equal.

4.8. Let us consider the stochastic decomposition method, discussed in Section 4.7.3.

(a) On page 357 we assumed that $Q(x;T(\xi),h(\xi)) \geq 0$ holds for all $x \in \mathscr{B}$, almost surely. Explain where this assumption has been utilized in the design of the method.

(b) Discuss the main differences between the SAA method (Section 4.7.3) and the SD algorithm. Why is the SD method called an internal sampling method?

4.8 Multistage recourse models

Multi–stage recourse models have been discussed in Section 3.3. Analogously to the two–stage case (see Section 4.7), many algorithmic proposals have been published for multistage recourse problems; we will discuss some of the main approaches. For further algorithms see Birge and Louveaux [26] and Ruszczyński, and Shapiro [295], and the references therein.

Most of the available algorithms are for the case, when ξ has a finite discrete distribution specified in the form of a scenario tree. If the distribution of ξ is continuous, then the usual approach consists of generating a discrete approximation to the distribution, in the form of a scenario tree, and subsequently solving the resulting multistage problem with the original distribution replaced by the approximate discrete distribution. Constructing approximate scenario trees is called *scenario generation* and will be the subject of Section 4.8.2. Another class of methods consists of algorithms, which combine the building of the scenario tree with the optimization process. One of the algorithmic approaches relies of successive discrete approximation, employing a simplicial cover of the support of the random vectors, see Frauendorfer [104], Frauendorfer and Schürle [106], [107], and the references therein. These algorithms allow that also the objective function is stochastic, and have been successfully applied in financial engineering.

4.8.1 Finite discrete distribution

The multistage recourse problem with a finite discrete distribution, the distribution being specified in the form of a scenario tree, has been the subject of Section 3.3.1.

A great majority of solution methodologies for this type of problems has its roots in the nested decomposition method, presented in Section 1.2.7. In that section we have pointed out, that in the framework of the nested decomposition method, several different variants of the algorithm can be built. The difference is in the sequence, in which nodes of the tree are processed in the algorithm. Different *sequencing protocols* are possible, the description in Section 1.2.7 corresponds to the *FFFB* (fast–

forward–fast–backward) protocol. For other sequencing protocols see, for instance, Gassmann [112] and Dempster and Thompson [62], [63].

The above–mentioned (restricted) freedom of choice is due to Propositions 1.20. and 1.21., both in Section 1.2.7. These propositions may also serve as guidelines for building valid variants of nested decomposition.

A further remark concerns the presentation of the nested decomposition method. For the sake of simplicity of presentation, we have assumed a form, where $A_{t\tau} = 0$ holds for $\tau < t - 1$ (for the general form see (3.114)). In Section 3.3.1 we have shown that the general formulation can always be transformed into the special form. Note, however, that this conceptual transformation is not needed when implementing the algorithm; the method can be reformulated for the general case in a straightforward way, see, for instance, Dempster and Thompson [63].

For recovering dual variables from the solution delivered by the nested decomposition algorithm, see Gassmann [113].

Instead of employing a fixed sequencing protocol, the above–mentioned freedom in choosing the next node to be processed allows also for dynamic sequencing algorithms. Methods of this type have been developed by Dempster and Thompson [62], [63]. The basic idea is using the expected value of perfect information (EVPI), attached in this case to the nodes, to choose the next node to be processed among the nodes having the highest EVPI–value. EVPI has been discussed in Section 3.2.4. The multistage extension is due to Dempster [60], see also [63]. Another useful idea, due to Dempster and Thompson [63], concerns stage–aggregation. According to this, in the equivalent deterministic LP, stages can be aggregated, leading to equivalent formulations of the MSLP problem involving fewer stages. The price for this is an increase in the dimension of matrices $A_{t\tau}$, involved in the problem formulation. This idea has also been utilized by Edirisinghe [80] for constructing bound–based approximation for MSLP problems. Another idea in this paper concerns bounds based on nonanticipativity aggregation. This leads us to our next subject.

The equivalent LP problem (1.28) of the MSLP problem is also called the *compact form* or *implicit form*. The reason is that the nonanticipativity requirement is ensured implicitly, by assigning the decision variables to the nodes of the scenario tree. The compact form has the disadvantage that in the case, when the underlying LP problem has some special structure (for instance, it is a transportation problem), this structure will be partially lost in the equivalent LP.

Another idea for formulating an equivalent LP preserves the problem structure. In this approach the decision variables are assigned to scenarios and nonanticipativity is enforced by explicit constraints. The resulting LP problems are called *explicit forms* or *split–variable forms*. "split–variable" has the following interpretation: the variables in the implicit form become split into several variables, according to scenarios. Below we present one variant of this type of problem formulation, for other variants see, for instance, Kall and Mayer [169].

$$
\left.\begin{array}{ll}
\min \sum_{s=1}^{S} p_s \left(c_1 x_1^s + \sum_{t=2}^{T} c_t^s x_t^s \right) & \\
\text{s.t. } A_{11} x_1^s \qquad\qquad = \quad b_1 & \\
A_{t1}^s x_1^s + \sum_{\tau=2}^{t} A_{t\tau}^s x_\tau^s = \quad b_t^s,\ t = 2,\ldots,T;\ \forall s \in \mathscr{S} & \\
\qquad\qquad x_t^s \geq \quad 0,\ \forall t,\ \forall s \in \mathscr{S} & \\
x_t^{\rho(n)} = x_t^s,\ \forall s \in \mathscr{S}(n) \forall n \in \mathscr{D}(t), t = 1,\ldots,T, &
\end{array}\right\} \tag{4.100}
$$

where we assume that the scenarios, belonging to the same bundle, have been (arbitrarily) ordered, and $\rho(n)$ is the index of the first scenario in the scenario bundle corresponding to node n, according to the ordering. $A_{t\tau}^s$, b_t^s, and c_t^s denote the realization of the corresponding random arrays, according to scenario $s \in \mathscr{S}$.

The last group of constraints obviously enforces the nonanticipativity requirement; we will call these constraints *nonanticipativity constraints.*

This form is ideally suited for *Lagrangean relaxation.* In fact, formulating the Lagrange function with respect to the nonanticipativity constraints, the following Lagrange–relaxation results:

$$
\left.\begin{array}{ll}
\min \sum_{s=1}^{S} p_s \left(c_1 x_1^s + \sum_{t=2}^{T} c_t^s x_t^s \right) + \sum_{t,s} \lambda_{ts} (x_t^s - x_t^{1_{\mathscr{S}(n)}}) & \\
\text{s.t. } A_{11} x_1^s \qquad\qquad = b_1 & \\
A_{t1}^s x_1^s + \sum_{\tau=2}^{t} A_{t\tau}^s x_\tau^s = b_t^s,\ t = 2,\ldots,T;\ \forall s \in \mathscr{S} & \\
\qquad\qquad x_t^s \geq 0,\ \forall t,\ \forall s \in \mathscr{S}, &
\end{array}\right\} \tag{4.101}
$$

which is separable with respect to the scenarios $s \in \mathscr{S}$, and decomposes into $S = |\mathscr{S}|$ separate subproblems.

Based on Lagrangean relaxation, several algorithms have been proposed for solving multistage recourse problems with finite discrete distributions. As the most well–known example, let us mention the progressive hedging algorithm of Rockafellar and Wets [287], where augmented Lagrangians are utilized. For further methods based on Lagrangean relaxation see, for instance, Birge and Louveaux [26] and Ruszczyński and Shapiro (editors) [295].

4.8.2 Scenario generation

In stochastic programming, scenario generation means generating a discrete approximation to the probability distribution of ξ, in the form of a scenario tree. In the multistage recourse problem, the original probability distribution is then replaced by this discrete approximation. The resulting multistage recourse problem is considered as an approximation of the original problem and can be solved, for instance, by the nested decomposition method.

The asymptotic properties of this discrete approximation are well–understood, see, for instance, Pennanen [251], and the references therein. Considering for $T \geq 3$ the present state of the art in scenario generation, there does not exist, at least according to our knowledge, any practically implementable scenario generation method, which would deliver for any (reasonable) error bound $\varepsilon > 0$ a scenario tree, such that the deviation between the true objective value of the multistage problem and the optimal objective value of the approximating problem is less than ε. By "practically implementable" we mean that all constants in the method are computable with a reasonable numerical effort, and that the resulting scenario trees (and consequently the equivalent LP problems) have a manageable size, for most problem instances. The difficulty has its roots in computing upper bounds on the optimal objective value of the original problem, see, for instance, Shapiro [305].

Therefore, according to our view, the presently available scenario generation techniques are essentially heuristic algorithms. For overviews on scenario generation see Dupačová, Consigli, and Wallace [76] and the references therein. The book Dupačová, Hurt, and Štěpán [79] contains a summary on scenario generation, along with applications in economics and finance. A comparison of the different techniques can be found in Kaut and Wallace [179]. In this book we confine ourselves to discuss some of the main approaches and present two techniques in a more detailed form.

For continuous distributions, a possible way for arriving at a discrete distribution leads via sampling, followed by scenario reduction. The scenario reduction phase can also be used in cases when the original probability distribution is already discrete but involves an unmanageable amount of scenarios.

In a first step a sample $\hat{\xi}^k = (\hat{\xi}_2^k, \ldots, \hat{\xi}_T^k)$, $k = 1, \ldots, N$, is generated, according to the joint probability distribution of ξ. This can either be done directly, by generating random vectors, or by simulating sample paths of the underlying stochastic process. For appropriate techniques see the literature on simulation, for instance Deák [56], Devroye [69], Ripley [278] or Ross [291].

The sample can be considered as a scenario tree, where each realization defines a root–to–leaf path, each scenario has the same probability $\frac{1}{N}$, and the single branching point is the root. In the second step, this tree is reduced by employing distances defined between probability distributions. For methods belonging to this class see Dupačová, Gröwe–Kuska, and Römisch [78], Heitsch and Römisch [133], and the references therein. Pflug [255] presents a related algorithm based on optimal discretization, in a financial application framework.

Another algorithmic approach proceeds in the reverse direction. The starting point is a scenario tree with a single scenario, corresponding to the expected value of ξ. This tree is then grown in an iterative fashion, by employing a cut–and–paste operation, based on successive partitioning of the supports of the random variables ξ_t $(t \geq 2)$. This method has been discussed in Section 3.3.2.

Next we discuss two of the main approaches in a more detailed form.

Bundle–based sampling

The idea is to partition the support of ξ into a finite number of subsets which are utilized for generating a scenario tree via sampling. We discuss the method under some simplifying assumptions, the extension to the general case is straightforward. Let $\Xi_t \subset \mathbb{R}^{r_t}$ be an interval containing the support of the random variable ξ_t, $t = 2,\ldots,T$, thus $\Xi := \prod_{t=2}^{T} \Xi_t$ contains the support of ξ. For the sake of simplicity let us assume that $r_t = r$ holds for all t.

Let us partition Ξ along each coordinate into d subintervals, resulting altogether in $d^{r(T-1)}$ cells. This implies a partition of $\hat{\Xi}_t := \prod_{\tau=2}^{t} \Xi_\tau$ into $d^{r(t-1)}$ cells, for $t = 2,\ldots,T$. With the partition we associate a rooted tree as follows. The root corresponds to $t = 1$. The child–nodes of the root correspond to the cells in the partition of $\hat{\Xi}_2 = \Xi_2$. In general, assume that the tree has been built up to stage $t-1$ such that the nodes in stage $t-1$ are associated with the cells in the partition of $\hat{\Xi}_{t-1}$. For each of the nodes in stage $t-1$, define d^r children, corresponding to the partition of Ξ_t. Consequently, the nodes in stage t will correspond to the partition of $\hat{\Xi}_t$.

Taking a realization of ξ, we associate with it the cell of the partition of Ξ, which contains it. In the tree, this implies an assignment to a scenario, that is, to the set of nodes along a root–to–leaf path. The algorithm runs as follows:

Bundle–based sampling

Step 1. *Initialize*
Choose a sample–size $N > 0$ and choose the parameter d, defining the number of coordinate–wise subintervals in the partition. Set up the tree corresponding to the partition, as described above. With each node of the tree associate a counter and initialize it with 0.

Step 2. *Generate a sample*
Choose $N > 0$ and randomly generate a sample $\hat{\xi}^k = (\hat{\xi}_2^k,\ldots,\hat{\xi}_T^k)$, $k = 1,\ldots,N$, according to the joint probability distribution of ξ.

Step 3. *Assign probabilities and realizations to nodes*
For each k, $k = 1,\ldots,N$, in turn, increase the counter by 1 for all nodes along the path corresponding to realization $\hat{\xi}^k$ in the tree, and store the corresponding realizations $\hat{\xi}_t^k$ node–wise. Subsequently, for each of the nodes $n \in \mathcal{N}$ do:
- Assign the probability $p_n := \frac{N_n}{N}$, where N_n is the value of the counter associated with node n.
- Compute a realization as the conditional sample mean of the realizations associated with the node. Assign this realization to node n.

Step 4. *Drop superfluous nodes*
Drop all nodes with associated counter values zero. Obviously, after this the graph remains still a tree.

The algorithm, as it stands above, is a conceptual framework. For instance, there is no need to store realizations at the nodes, the conditional sample means can be updated at the same pass over the realizations, which serves for assigning counter values.

For consistency properties of the above scheme see King and Wets [184]. The approach clearly has its limitations, due to the combinatorial explosion. The number of scenarios is $d^{r(T-1)}$, which grows exponentially with the dimension r of the random vector, and with the number of stages T.

A moment–matching heuristics

The subject of this section is a heuristic algorithm of Høyland, Kaut, and Wallace [142]. According to this method, the scenario tree is being built in a node–wise fashion, according to the following scheme:

Sequential framework for scenario generation

Step 1. *Initialize*
Set $t = 1$, assign probability 1 to the root node.

Step 2. *Generate nodes in the next stage*
For each of the nodes in stage t ($t \geq 1$) proceed as follows: specify conditional distributional properties (for instance, moments and correlations), given the outcome corresponding to the specific node. Generate outcomes (realizations of a random variable with a finite discrete distribution), which are consistent with the specification. Define the corresponding child–nodes and assign to them the realizations and associated probabilities. If $t = T-1$ then **stop**, otherwise set $t := t+1$ and **repeat *Step 2***.

In the rest of this section we will discuss the subproblem arising at the nodes: given some distributional properties of a random vector, generate a finite discrete distribution having the prescribed distributional properties. More closely, we consider the following problem: Let ζ be an r–dimensional random vector with a finite discrete distribution. We prescribe the number of realizations N, the probabilities $p_1, \ldots, p_N$ of the realizations, the expected values, standard deviations, skewness, and kurtosis for the 1–dimensional marginal distributions, as well as the correlation matrix of ζ. Given these quantities, we wish to compute the realizations in such a way that the resulting discrete distribution has the prescribed properties. The data concerning the marginal distributions and the realizations z_{ij} which we wish to compute, are summarized in Table 4.2, where $\mu_i = \mathbb{E}[\zeta_i]$, $\sigma_i = \left(\mathbb{E}[(\zeta_i - \mu_i)^2]\right)^{\frac{1}{2}}$, and

- $s_i = \dfrac{\mathbb{E}[(\zeta_i - \mu_i)^3]}{\sigma_i^3}$ is the skewness, and
- $k_i = \dfrac{\mathbb{E}[(\zeta_i - \mu_i)^4]}{\sigma_i^4}$ is the kurtosis

of ζ_i, $i = 1, \ldots, r$.

	p_1	p_2	$\cdots$	p_N				
ζ_1	z_{11}	z_{12}	$\cdots$	z_{1N}	μ_1	σ_1	s_1	k_1
ζ_2	z_{21}	z_{22}	$\cdots$	z_{2N}	μ_2	σ_2	s_2	k_2
$\vdots$	$\vdots$	$\vdots$		$\vdots$	$\vdots$	$\vdots$	$\vdots$	$\vdots$
ζ_r	z_{r1}	z_{r2}	$\cdots$	z_{rN}	μ_r	σ_r	s_r	k_r

Table 4.2 Marginal distribution data and realizations.

Additionally to the data summarized in the table, the correlation matrix of ζ is also prescribed. Let R be the correlation matrix of ζ defined as

$$R_{ij} = \frac{\mathbb{E}[(\zeta_i - \mu_i)(\zeta_j - \mu_j)]}{\sigma_i \sigma_j}, \quad i,j = 1,\ldots,r.$$

We assume throughout that R is nonsingular, consequently it is positive definite. The Cholesky–factorization of R is $R = LL^{\mathrm{T}}$, where L is a lower triangular matrix.

It is clearly sufficient to solve the problem for standardized random variables. In fact, let ξ be the standardized of ζ, that is,

$$\xi_i = \frac{\zeta_i - \mu_i}{\sigma_i}, \quad \forall i.$$

Then we have

$$\mathbb{E}[\xi_i] = 0, \quad \mathbb{E}[\xi_i^2] = 1, \quad \mathbb{E}[\xi_i^3] = s_i, \quad \mathbb{E}[\xi_i^4] = k_i,$$

and the correlation matrices of ζ and ξ are the same. Therefore, it is sufficient to solve the above problem for standardized random variables. Having generated the realizations x_{ij} $\forall i,j$ for ξ, then, according to $\zeta_i = \sigma_i \xi_i + \mu_i$, we get the solution $z_{ij} = \sigma_i x_{ij} + \mu_i$ for the original problem. Consequently, $\mu_i = 0$ and $\sigma_i = 1$ will be assumed in the sequel, for all i.

We will utilize some transformations of random variables and random vectors.

The first one will be called *moment–matching* transformation and is defined for random variables. Let ξ be a standardized random variable ($r = 1$) and assume that the first 12 moments of this random variable exist and that these moments are known. We consider a nonlinear transformation of the following form

$$\eta = \Gamma_{s,k}^{mom}(\xi) := a + b\xi + c\xi^2 + d\xi^3$$

and wish to determine the coefficients a, b, c, and d of this cubic polynomial in such a way, that $\mathbb{E}[\eta] = 0$, $\mathbb{E}[\eta^2] = 1$, $\mathbb{E}[\eta^3] = s$, and $\mathbb{E}[\eta^4] = k$ hold, with s and $k > 0$ prescribed. This requirement can be formulated as the following system of nonlinear equations

$$
\begin{array}{rlll}
0 = \mathbb{E}[\eta] & = \mathbb{E}[a+b\xi+c\xi^2+d\xi^3] & = \mathscr{P}_1(a,b,c,d), \\
1 = \mathbb{E}[\eta^2] & = \mathbb{E}[(a+b\xi_i+c\xi_i^2+d\xi_i^3)^2] & = \mathscr{P}_2(a,b,c,d), \\
s = \mathbb{E}[\eta^3] & = \mathbb{E}[(a+b\xi+c\xi^2+d\xi^3)^3] & = \mathscr{P}_3(a,b,c,d), \\
k = \mathbb{E}[\eta^4] & = \mathbb{E}[(a+b\xi+c\xi^2+d\xi^3)^4] & = \mathscr{P}_4(a,b,c,d),
\end{array} \tag{4.102}
$$

where $\mathscr{P}_i(a,b,c,d)$, denotes a polynomial function of order i, in the variables a, b, c, and d. The coefficients of these polynomials involve the moments of ξ, with the highest moment having order 12 and appearing in $\mathscr{P}_4(a,b,c,d)$. The analytical form of these polynomials can be obtained by straightforward calculation, see [142].

If the system of equations (4.102) has a real solution, then we have the desired transformation. It may happen, however, that there exists no real solution of it. For accounting also for this case, the suggested way of numerical solution relies on minimizing the sum of quadratic deviations, for instance, by employing the Levenberg–Marquardt method. Thus, if there does not exist a solution, the method will deliver a, b, c and d, for which the deviation is minimal. For the sake of simplicity of presentation, we will assume that whenever this transformation is applied, the corresponding system (4.102) has a real solution.

The second transformation will be called *correlation–matching* transformation, or alternatively *forward transformation*, and is defined for random vectors. Let now ξ be an r–dimensional standardized random vector and assume that the components of ξ are uncorrelated (the correlation matrix of ξ is the identity matrix I). The transformation is defined as the following nonsingular linear transformation

$$
\eta = \Gamma_R^{cr}(\xi) := L\xi,
$$

where L is a lower–triangular matrix with $R = LL^{\mathrm{T}}$ (Cholesky–factorization). Clearly we have $\mathbb{E}[\eta] = \mathbb{E}[L\xi] = L\mathbb{E}[\xi] = 0$. Furthermore

$$
\mathbb{E}[\eta\,\eta^{\mathrm{T}}] = L\mathbb{E}[\xi\xi^{\mathrm{T}}]L^{\mathrm{T}} = LIL^{\mathrm{T}} = R,
$$

consequently the covariance matrix of η is R. In particular, we get that the variance of η_i is 1, for all i and thus the correlation matrix of η is also R.

Next we take a look, how this transformation changes the third and fourth moments.

Proposition 4.14. *Let us assume that the components of ξ are stochastically independent. Then*

$$
\begin{aligned}
\mathbb{E}[\eta_i^3] &= \sum_{j=1}^{i} L_{ij}^3\mathbb{E}[\xi_j^3] \\
\mathbb{E}[\eta_i^4] &= 3 + \sum_{j=1}^{i} L_{ij}^4(\mathbb{E}[\xi_j^4] - 3)
\end{aligned} \tag{4.103}
$$

holds.

Proof: The first equality follows from

$$\mathbb{E}[\eta_i^3] = \mathbb{E}[(L_i\xi)^3] = \sum_{j,k,l=1}^{r} L_{ij}L_{ik}L_{il}\mathbb{E}[\xi_j\xi_k\xi_l] = \sum_{j=1}^{i} L_{ij}^3\mathbb{E}[\xi_j^3],$$

where L_i is the i'th row of L and we have used the fact that, due to the stochastic independence assumption, we have $\mathbb{E}[\xi_j\xi_k\xi_l] = \mathbb{E}[\xi_j]\mathbb{E}[\xi_k]\mathbb{E}[\xi_l] = 0$ for three different indices j, k, l, and $\mathbb{E}[\xi_j\xi_k^2] = \mathbb{E}[\xi_j]\mathbb{E}[\xi_k^2] = 0$ for $k = l$, $j \neq k$.

For the second equality in (4.103) observe:

$$\mathbb{E}[\eta_i^4] = \mathbb{E}[(L_i\xi)^4] = \sum_{j,k,l,m=1}^{r} L_{ij}L_{ik}L_{il}L_{im}\mathbb{E}[\xi_j\xi_k\xi_l\xi_m],$$

where, again implied by the stochastic independence assumption, all terms are zero, except those where either all four indices are equal, or there exist two pairs of equal indices. The number of possibilities for selecting the latter is 6, therefore, observing $\mathbb{E}[\xi_j^2] = 1\ \forall j$, we have

$$\mathbb{E}[\eta_i^4] = \sum_{j=1}^{i} L_{ij}^4\mathbb{E}[\xi_j^4] + 3\sum_{j=1}^{i} L_{ij}^2\left(\sum_{k=1}^{j-1} L_{ik}^2 + \sum_{k=j+1}^{i} L_{ik}^2\right).$$

We utilize that $\sum_{k=1}^{i} L_{ik}^2 = 1$ holds for the Cholesky–factor, for all i, thus getting

$$\sum_{k=1}^{j-1} L_{ik}^2 + \sum_{k=j+1}^{i} L_{ik}^2 = 1 - L_{ij}^2,$$

which proves the second equality in (4.103). □

Notice that in (4.103) we have two nonsingular triangular systems of linear equations which, given $\mathbb{E}[\eta_i^3]$ and $\mathbb{E}[\eta_i^4]$ for all i, can be solved for $\mathbb{E}[\xi_j^3]$ and $\mathbb{E}[\xi_j^4]\ \forall j$, respectively:

$$\begin{aligned}\mathbb{E}[\xi_i^3] &= \tfrac{1}{L_{ii}^3}\left(\mathbb{E}[\eta_i^3] - \sum_{j=1}^{i-1} L_{ij}^3\mathbb{E}[\eta_j^3]\right)\\ \mathbb{E}[\xi_i^4] &= 3 + \tfrac{1}{L_{ii}^4}\left(\mathbb{E}[\eta_i^4] - 3 - \sum_{j=1}^{i-1} L_{ij}^4\left(\mathbb{E}[\eta_j^4] - 3\right)\right).\end{aligned} \tag{4.104}$$

Now we are prepared to presenting a *perfect matching* algorithm for the solution of our problem. Assume that the standardized random vector ξ has independent components and that $\hat{s}_i := \mathbb{E}[\xi^3]$ and $\hat{k}_i := \mathbb{E}[\xi^3]$ are computed according to (4.104), with the setting $\mathbb{E}[\eta_i^3] = s_i$ and $\mathbb{E}[\eta_i^4] = k_i$, $\forall i$. The quantities $\hat{s}_i$ and $\hat{k}_i$ will be called *transformed target moments*. Applying the forward transformation Γ_R^{cr} to ξ, $\eta := \Gamma_R^{cr}(\xi) = L\xi$ will be a solution of our problem. Thus we have the following conceptual algorithm:

Perfect matching conceptual algorithm

Step 1. *Initialization*
Compute the Cholesky–factorization $R = LL^{\mathrm{T}}$. According to (4.104), compute $\hat{s}_i := \mathbb{E}[\xi^3]$ and $\hat{k}_i := \mathbb{E}[\xi^3]$, using the target moments and the Cholesky–factor of the target correlation matrix.

Step 2. *Choose a starting distribution*
Take any discretely distributed standardized r–dimensional random vector $\tilde{\xi}$, which has stochastically independent components, N joint realizations and the prescribed probabilities $p_1, \ldots, p_N$ for the joint realizations.

Step 3. *Match the transformed target moments*
Component–wise apply the moment matching transformation $\tilde{\xi}_i := \Gamma^{mom}_{\hat{s}_i, \hat{k}_i}(\tilde{\xi}_i)$, $i = 1, \ldots, r$. This results in a random vector $\tilde{\xi}$, which has independent components and moments $0, 1, \hat{s}_i, \hat{k}_i$.

Step 4. *Match the correlations*
Apply the forward transformation $\eta := \Gamma^{cr}_R(\tilde{\xi}) = L\tilde{\xi}$, then η will be a solution of our problem, with a perfect matching both for the first four moments and for the correlation matrix.

The difficulty with the perfect matching method is that, using simulation, it is not possible to generate a random vector with theoretically independent components. Therefore, we will also need a transformation which decreases the "degree" of dependence. Instead of ensuring independence, we are merely able to remove correlations. Let ξ be a standardized random vector with correlation matrix R and

$$\eta = \Gamma^{0cr}_R(\xi) := L^{-1}\xi,$$

then the correlation matrix of η will be I. In fact

$$\mathbb{E}[\eta\eta^{\mathrm{T}}] = L^{-1}\mathbb{E}[\xi\xi^{\mathrm{T}}](L^{-1})^{\mathrm{T}} = I.$$

This transformation will also be called *backward transformation.*

The heuristic scenario–generation method of Høyland, Kaut, and Wallace [142] (HKW–method) is designed along the lines of the perfect matching algorithm. *Step 1* is carried out without changes. The implementation of *Step 2* consists of randomly and independently generating N random vectors with independent components, where the components are taken from a standard normal or from a uniform distribution. Let us denote this discretely distributed random vector by $\tilde{\xi}$ and its correlation matrix matrix by $\tilde{R}$. The problem is, that the components of $\tilde{\xi}$ will not be independent in a theoretical sense.

After having carried out *Steps 1* and *2* of the conceptual algorithm, the HKW–method proceeds in two phases.

Phase I corresponds to *Step 3* of the conceptual method. The goal is to construct a discrete distribution, such that the components of a corresponding random vector

$\tilde{\xi}$ are stochastically independent and have the first four prescribed marginal moments: $(0, 1, \hat{s}_i, \hat{k}_i)$. Instead of the theoretically required independence, we only try to achieve approximately zero correlations, and a hopefully good–enough approximation to the moments. The algorithm runs as follows.

Phase I: removing correlations and matching moments

Step 1. *Initialization*
Choose $\varepsilon^I_{cr} > 0$ for the stopping tolerance concerning correlations. Apply $\tilde{\xi}_i := \Gamma^{mom}_{\hat{s}_i,\hat{k}_i}(\tilde{\xi}_i)$ componentwise $\forall i$ (target moments are 0, 1 $\hat{s}_i$, $\hat{k}_i$ $\forall i$).
$\{\rightarrow$ right transformed target moments $\leftarrow\}$

Step 2. *Compute the correlation matrix and factorize it*
Compute $\tilde{R}$. If $\|\tilde{R}-I\| < \varepsilon^I_{cr}$ then **Stop**, otherwise continue. Perform the Cholesky–factorization of $\tilde{R}$, resulting in $\tilde{R} = \tilde{L}\tilde{L}^{\mathrm{T}}$, with $\tilde{L}$ being lower triangular.

Step 3. *Remove correlations*
Perform backward transformation:
$\tilde{\xi} := \Gamma^{0cr}_{\tilde{R}}(\tilde{\xi}) = \tilde{L}^{-1}\tilde{\xi}$. Store $\tilde{\xi}^I := \tilde{\xi}$, which has zero correlations.
$\{\rightarrow$ zero correlations $\leftarrow\}$

Step 4. *Achieve transformed target moments*
Apply $\tilde{\xi}_i := \Gamma^{mom}_{\hat{s}_i,\hat{k}_i}(\tilde{\xi}_i)$ componentwise $\forall i$; $\tilde{\xi}_i$ has the desired moments.
$\{\rightarrow$ right transformed target moments $\leftarrow\}$
Continue with *Step 2*.

In the subsequent Phase II, *Step 4* of the conceptual algorithm is implemented. Similarly to Phase I, this is also carried out in an iterative manner. The method is the following.

Phase II: simultaneous moment and correlation matching

Step 1. *Initialization*
Choose ε^{II}_{cr} for the stopping tolerance regarding correlations. Set $\tilde{\xi} := \tilde{\xi}^I$, where $\tilde{\xi}^I$ is the distribution, saved in *Step 3* of Phase I. Apply $\tilde{\xi}_i := \Gamma^{cr}_{\tilde{R}}(\tilde{\xi}) = L\tilde{\xi}$.

Step 2. *Compute and factorize the correlation matrix*
Compute $\tilde{R}$. If $\|\tilde{R}-R\| < \varepsilon^{II}_{cr}$ then **Stop**, otherwise continue. Compute the Cholesky–factorization $\tilde{R} = \tilde{L}\tilde{L}^{\mathrm{T}}$.

Step 3. *Remove correlations*
Perform backward transformation: $\tilde{\xi} := \Gamma^{0cr}_{\tilde{R}}(\tilde{\xi}) = \tilde{L}^{-1}\tilde{\xi}$.
$\{\rightarrow$ zero correlations $\leftarrow\}$

Step 4. *Forward transformation*
Compute $\tilde{\xi}_i := \Gamma^{cr}_{\tilde{R}}(\tilde{\xi}_i) = L\tilde{\xi}$. Store $\tilde{\xi}^{II} := \tilde{\xi}$, which has the right correlations.
$\{\rightarrow$ right target correlations $\leftarrow\}$

Step 5. *Achieve target moments*
Apply $\tilde{\xi}_i := \Gamma^{mom}_{s_i,k_i}(\tilde{\xi}_i)$ componentwise $\forall i$.
$\{\rightarrow$ right target moments $\leftarrow\}$
Continue with *Step 2*.

The HKW–algorithm is a heuristic scenario-tree generation procedure; there exist no proofs of finite termination or of convergence for the iterative cycles involved, neither for Phase I nor for Phase II. Høyland, Kaut, and Wallace [142] report on successful practical applications and present some quite favorable computational results. The authors of this book have also implemented the method; it is one of the scenario–tree generation methods, available with SLP–IOR, see Section 4.9.2. Our computational experience is also in favor of the algorithm.

4.8.3 A guide to available software

The solvers, available for multistage problems, have been discussed in Section 4.7.5.

The scenario generation algorithm in Section 4.8.2 is part of the modeling system OSLSE of IBM, see King et al. [185]. We have also implemented a version, which is available with SLP-IOR, see Section 4.9.2.

An experimental implementation of the HKW–method for scenario generation, presented in Section 4.8.2, has been developed by the authors of the algorithm, and can be downloaded in executable form from the homepage of Michal Kaut, http://work.michalkaut.net/.

A commercial version of a scenario reduction method, presented in Dupačová et al. [78], has been implemented by Nicole Gröwe as the solver SCENRED. It is available with the algebraic modeling system GAMS, Brooke et al. [34], [35].

A comparison of the different scenario–generation methods can be found in Kaut and Wallace [179], which may serve as a guide to choose an appropriate method.

4.9 Modeling systems for SLP

Modeling systems are aimed to provide support to the various stages in a model's life cycle including building, solving, and analyzing problem instances, and their solution. They have a specified scope concerning model types and differ in their scope, in the extent of support provided to the different stages in the model's life–cycle, and in the range of modeling tools offered by them. Modeling systems can also be integrated systems, including links to modeling languages and solvers. Some of the modeling systems are based on modeling languages.

4.9.1 Modeling systems for SLP

Considering SLP, presently several modeling systems and tools are available; for an introduction to a selection of these systems see Kopa [195]; see also the Stochastic Programming Community home page at http://www.stoprog.org/.

Below we provide a short list of the most well–known systems. We just list some of the major characteristic features of the systems, for the details see the cited papers.

- **OSLSE** is the stochastic programming system of IBM, see King, Wright, Parija, and Entriken [185], for multistage recourse models with scenario trees. It is an optimization system and a library of tools, supporting model building including scenario generation, and the solution phase. The MSLP solver with the same name OSLSE, is also separately available.
- **SETSTOCH,** developed by C. Condevaux–Lanloy and E. Fragnière [46]. This is a modeling tool, with the main goal of supporting the linking of SLP solvers to algebraic modeling systems. The authors report on the application of this tool for linking the solver OSLSE to the algebraic modeling system GAMS. A generalized version, called SET (Structure–Exploiting Tool), has been developed by E. Fragnière, J. Gondzio, R. Sarkissian and J.–P. Vial [101].
- **SLP–IOR** is our model management system for SLP and will be discussed in a detailed fashion in Section 4.9.2.
- **SMI**, Stochastic Modeling Interface from the open–source COIN–OR project, developed by Alan King. The system supports the building of multistage recourse models with scenarios, it includes a scenario generation method and the generation of deterministic equivalents, see https://projects.coin-or.org/Smi.
- **SPInE,** a Stochastic Programming Integrated Environment, developed by P. Valente, G. Mitra, and C.A. Poojari, see [332]. The scope of the system consists of multistage recourse models with scenario trees and of chance constrained models. It serves for supporting the entire modeling life–cycle and has integrated facilities for accessing databases. A unique feature of this system is that it includes an extension of the algebraic modeling language MPL, adding SLP–specific language constructs to it.
- **StAMPL** A modeling tool for multistage recourse problems, with an emphasis on the language constructs concerning the filtration process, developed by Fourer and Lopes [99].
- **Stochastics**^{TM} is a modeling system for generating large–scale MSLP problems with scenario trees, developed by Dempster et al. It has a link to the algebraic modeling system AMPL and to XPRESS-MP. Its component for stochastic modeling is called **stochgen**. For an overview see Dempster, Scott, and Thompson [61]. The main emphasis in this system is on modeling.

A modeling system for supporting different LP–equivalent formulations, according to the needs of decomposition solvers, and including stage–aggregation, has been developed by Fourer and Lopes [98]. The targeted model class consists of MSLP models with scenario trees.

An integrated modeling environment has been developed by Gassmann and Gay [117], for MSLP models with scenario trees. The integration involves the algebraic modeling language AMPL (Fourer, Gay, and Kernighan [97]) and Microsoft's MS Access and MS Excel.

A further modeling tool, also based on AMPL, is the open source tool of Thénié, Van Delft and Vial [325] which supports the automatic formulation of scenario tree based MSLP models.

Shapiro, Powell, and Bernstein [309] developed a Java–representation for stochastic online operations research models.

General problems related to formulating SLP models in algebraic modeling systems are discussed by Entriken [87], Fragniére and Gondzio [100] and Gassmann and Ireland [118]; for modeling languages and systems see Kallrath [174]. Specific issues related to modeling support for SLP are the subject of the papers Gassmann [114] and Kall and Mayer [165], [162]. M. Bielser [19] has developed in his Thesis a new algebraic modeling language called SEAL, specifically designed for SLP. Colombo, Grothey, Hogg, Wooksend and Gondzio [45] present a structure–conveying algebraic modeling language as an extension of AMPL, which contains features for modeling SLP problems with recourse.

In the development process of modeling systems and solvers it is important to have a generally accepted standard input format for test problem instances. In the field of SLP this is the SMPS input format. Originally it was developed for multistage recourse problems, see Birge et al. [24]. Gassmann and Schweitzer [120] proposed an extension to other classes of SLP problems involving e.g., continuous distributions and SLP–problems with probabilistic constraints. For detailed explanations and examples concerning the SMPS format see also Gassmann [115] and Gassmann and Kristjánsson [119]. A Fortran 90 toolkit for supporting the reading of model instances in SMPS format has been developed by Gassmann and is freely available at http://myweb.dal.ca/gassmann/. Recently an XML–based representation of SLP model–instances has been proposed by Fourer et al. [96].

4.9.2 SLP–IOR

Our model management system SLP–IOR was one of the first modeling systems for stochastic linear programming. The system design was published in Kall and Mayer [161] in 1992, the first version of the system was available in the same year. The scope of this version consisted of two–stage recourse models and models with joint probability constraints. Since then, the system has been continually further–developed, by extending the scope with new model types, by adding new modeling tools, and by developing and connecting new solvers. For an overview see Kall and Mayer [163] and Mayer [230], for the present state of development see Kall and Mayer [167] and [169], as well as the user's guide to SLP–IOR, available via the

Internet at http://www.ior.uzh.ch/research/stochOpt.html.

For using the former versions, the user had to have her/his own version of the algebraic modeling system GAMS (Brooke, Kendrick and Meeraus [34], Brooke et al.[35], and Bussieck and Meeraus [40]). The reason is that the solver interface of that versions was based on GAMS. Since 2001, this is no more a requirement, SLP–IOR can be used in a stand–alone mode.

In the rest of this section we give a short overview on SLP–IOR, for the details see our papers, cited above, and the user's guide of SLP–IOR.

General issues

The scope of the present version of SLP–IOR consists of the following model types:

- Single stage models.
  - Deterministic LP.
  - Probability constraints (Section 2.2); separate (Section 2.2.3) and joint (Section 2.2.5).
  - Integrated probability functions as constraint or in the objective (Section 2.4.1); separate and joint.
  - CVaR as constraint or in the objective (Section 2.4.3).
- Multistage models.
  - Deterministic LP.
  - Two–stage recourse models (Section 3.2).
    - Random recourse (Section 3.2).
    - Fixed recourse (Section 3.2).
    - Complete fixed recourse (Section 3.2.1).
    - Simple recourse (Section 3.2.2); continuous recourse and integer recourse.
    - Multiple simple recourse (Section 3.2.2).
  - Multistage recourse.

The random entries of the arrays in the model are represented via affine linear relations, see (3.6). The random variables may be independent, may form a single group of dependent variables, or in the general case, mutually independent groups of random variables can be specified.

Concerning probability distributions, the choice for univariate distributions consists of 16 continuous and 7 discrete distributions, containing most of the well–known statistical distributions. In the multivariate case, normal and uniform distributions are available in the continuous case, empirical and uniform distributions in the discrete case.

Deterministic LP's, two–stage and multistage recourse models can be exported and imported according to the SMPS data–format, see Gassmann [115]. The present version uses the original specification by Birge, Dempster, Gassmann, Gunn, King, and Wallace [24], that is, the extensions proposed in [115] are not yet implemented.

Deterministic linear programs, formulated in the algebraic modeling language GAMS (Brooke et al. [34], [35]), can be imported with the aim of formulating stochastic versions of them.

The system includes an interface to the algebraic modeling system GAMS. Consequently, if the user has a copy of GAMS, all solvers available with that particular GAMS distribution can also be used for solving SLP problems formulated in SLP–IOR, provided, that an algebraic equivalent exists and the formulation of it is supported by SLP–IOR. For instance, multistage recourse problems with a scenario tree can also be solved this way via GAMS.

The user communicates with the system via an interactive, menu–driven interface.

Analyze tools and workbench facilities

The analyze tools provide support for analyzing a model instance or its solution. The tools are presently available for recourse models and include

- for two–stage models computing the solutions for the following associated problems: the expected value problem (EV), the wait–and–see solution (WS), the expected result (EEV), as well as computing the derived quantities expected value of perfect information (EVPI), and value of stochastic solution (VSS). For the definition of these characteristic values see Section 3.2.4. Computations are done in the discretely distributed case directly, for continuous distributions sampling is available.
- For two–stage models checking whether the model instance has the complete recourse property and analyzing the model for finding out whether it has a hidden simple recourse structure.
- For two–stage recourse models, computing the solution of the SAA problem, see Section 4.7.3.
- Computing the recourse objective for a fixed first–stage vector x^*. The implementation of the procedures for testing the quality of a solution, discussed in Section 4.7.3, is in progress.

The primary aim of the workbench facilities is to support the testing of solvers. They include

- our test–problem generator GENSLP. This serves for randomly generating test problem batteries consisting of model instances of recourse problems or problems with joint probability constraints. Several parameters can be chosen to control, for instance the nonzero density of the arrays, the type of recourse matrix (for instance, complete fixed recourse can be prescribed), or the number of random entries in the stochastic arrays (via the affine–linear relations (Section 3.6)).
- Generating test problem batteries by randomly perturbing the array–elements of a single model instance.
- Running a selected collection of solvers on a battery of test problems, with the aim of supporting comparative computational studies.

Transformations

Two types of model transformations are supported.

On the one hand, a model instance can be transformed into an algebraic equivalent provided that such an algebraic equivalent exists. As an example, let us consider multistage recourse models with scenario trees. Such models can either be transformed to an equivalent LP having the compact form, or into explicit forms (presently 4 different such forms are supported), see Section 4.8.1 for a discussion concerning the different equivalent LP forms. These LP problems can be subsequently exported in MPS form, for the sake of testing LP solvers, for instance.

On the other hand, a model instance can be transformed into an instance of another model type, e.g., a two–stage recourse problem can be transformed into a chance constrained model. Missing data are replaced by default values. The aim of this facility is to support the formulation of different types of SLP models, on the basis of the same underlying data–set.

Scenario generation

Scenario generation has been discussed in Section 4.8.2. In SLP–IOR two algorithms are implemented: the bundle–based sampling method, discussed in Section 4.8.2, and the moment matching heuristics of Høyland, Kaut, and Wallace [142], see Section 4.8.2.

Besides these, the user can also build manually a scenario tree, via a graphical interface. Several tools are available for supporting this, for instance a cut–and–paste procedure, discussed in Section 3.3.2 in connection with discretization methods for MSLP problems.

For the bundle–based simulation method several probability distributions are available, see Section 4.9.2.

The solver interface

The solver interface of SLP–IOR is an open interface, in the sense that the user can connect her/his own solver to the executable of SLP–IOR. For the details see the user's guide.

Several solvers are connected to SLP–IOR. Some of them are commercial solvers, others have been developed by ourselves. Some have been obtained from the authors of the solver, which we would like to gratefully acknowledge also in this place. If the user has a copy of GAMS, then the general–purpose GAMS solvers can also be called within SLP–IOR. A student version of GAMS with dimensional limitations is freely available at http://www.gams.com/download/.

Here we confine ourselves to listing some of the solvers, for a full list see Kall and Mayer [167], or the user's guide.

- *General–purpose LP solvers*

- **HiPlex**, variant of the simplex method by Maros [218], [219], Maros and Mitra [220], implemented by I. Maros.
- **HOPDM**, an interior–point method of Gondzio [129], implemented by J. Gondzio.
- **Minos**, a commercial solver for NLP, for LP problems it implements the simplex method. See Murtagh and Saunders [240].

- *Solvers for two–stage recourse problems*
  - **BPMPD** general–purpose LP solver, interior point method, implemented by Cs. Mészáros [233], see also Section 4.7.5. Although a general–purpose solver, it is especially well–suited for recourse problems with a finite discrete distribution.
  - **DAPPROX**, successive discrete approximation method, see Section 4.7.2, implemented by P. Kall and J. Mayer.
  - **MSLiP**, nested decomposition, implemented by H. Gassmann [112]. For the nested decomposition method see Section 1.2.7.
  - **QDECOM**, regularized decomposition method of Ruszczyński [293], implemented by A. Ruszczyński. For the algorithm see Section 1.2.8.
  - **SDECOM**, stochastic decomposition method of Higle and Sen [136], [139], implemented by P. Kall and J. Mayer. For the method see Section 4.7.3.
  - **SHOR2**, decomposition scheme of Shor [310], Shor, Bardadym, Zhurbenko, Likhovid, Stetsyuk [311], implemented by N. Shor and A. Likhovid.

- *Specialized solvers for simple recourse*
  - **SHOR1**, the same method and authors, as for SHOR2, the algorithm has been adapted to the special structure.
  - **SRAPPROX**, successive discrete approximation method, see Section 4.7.2, implemented by P. Kall and J. Mayer.

- *Simple integer recourse*
  - **SIRD2SCR**, implements the convex hull algorithm of Klein Haneveld, Stougie, and Van der Vlerk [189], implemented by J. Mayer and M.H. van der Vlerk.

- *Multiple simple recourse*
  - **MScr2Scr**, transformation of Van der Vlerk [334], see Section 3.2.2, implemented by J. Mayer and M.H. van der Vlerk.

- *Models with integrated probability functions*
  - **ICCMIN**, dual decomposition method, see Sections 4.4.3 and 4.4.5, implemented by A. Künzi–Bay and J. Mayer.

- *Models involving CVaR*
  - **CVaRMin**, dual decomposition method, see Section 4.4.4, implemented by A. Künzi–Bay and J. Mayer.

- *Joint probability constraints*
  - **PCSPIOR**, supporting hyperplane method of Szántai [320], implemented by J. Mayer. For the method see Section 1.3.2 and Section 4.3.2.
  - **PROBALL**, central cutting plane method, Mayer [230], implemented by J. Mayer, see Section 1.3.2 and Section 4.3.2.
  - **PROCON**, reduced gradient method, see Mayer [230], implemented by J. Mayer. See Section 1.3.2 and Section 4.3.2.

System requirements and availability

The system runs under the Microsoft Windows32 operating system family; it has been tested under Windows 95, Windows NT, Windows 2000, and Windows XP.

If the user has a copy of GAMS, then the GAMS–solvers can also be used from SLP–IOR, see Section 4.9.2. Having GAMS is, however, not a prerequisite for using SLP–IOR.

SLP–IOR, in executable form, is available free of charge for academic purposes. For obtaining a copy please visit http://www.ior.uzh.ch/research/stochOpt.html, and follow the instructions. From this site you can also download the User's Guide. Furthermore, tutorials concerning the installation and usage of the system, developed by Humberto Bortolossi are also available.

References

[1] J. Abaffy and E. Allevi. A modified L–shaped method. *J. Opt. Theory Appl.*, 123:255–270, 2004.

[2] P.G. Abrahamson. A nested decomposition approach for solving staircase structured linear programs. In Y. M. Ermoliev and R. J.-B. Wets, editors, *Numerical Techniques for Stochastic Optimization*, pages 367–381. Springer, Berlin, 1988.

[3] C. Acerbi. Spectral measures of risk: a coherent representation of subjective risk aversion. *J. Banking & Finance*, 26:1505–1518, 2002.

[4] C. Acerbi and D. Tasche. On the coherence of expected shortfall. *J. Banking & Finance*, 26:1487–1503, 2002.

[5] R. K. Ahuja, T. L. Magnanti, and Orlin J. B. *Network Flows: Theory, Algorithms, and Applications*. Prentice-Hall Publ. Co., 1993.

[6] E.D. Andersen, C. Roos, and T. Terlaki. On implementing a primal–dual interior–point method for conic quadratic optimization. *Math. Prog.*, 95:249–277, 2003.

[7] P. Artzner, F. Delbaen, J.-M. Eber, and D. Heath. Coherent measures of risk. *Math. Finance*, 9:203–228, 1999.

[8] M. Avriel, W. E. Diewert, S. Schaible, and I. Zang. *Generalized Concavity*. Plenum Press, New York and London, 1988.

[9] M. Avriel and A. Williams. The value of information and stochastic programming. *Oper. Res.*, 18:947–954, 1970.

[10] G. Bayraksan and D.P Morton. Assessing solution quality in stochastic programs. *Math. Prog.*, 108:495–514, 2006.

[11] M.S. Bazaraa and C.M. Shetty. *Nonlinear Programming—Theory and Algorithms*. John Wiley & Sons, New York, 1979.

[12] E.M.L. Beale. The use of quadratic programming in stochastic linear programming. Rand Report P-2404, The RAND Corporation, 1961.

[13] E.F. Beckenbach and R. Bellman. *Inequalities*. Springer, 1971.

[14] J.F. Benders. Partitioning procedures for solving mixed-variables programming problems. *Numer. Math.*, 4:238–252, 1962.

[15] A. Ben-Tal, L. El Ghaoui, and A. Nemirovski. Robustness. In H. Wolkowicz, R. Saigal, and L. Vandenberghe, editors, *Handbook of Semidefinite Programming*, pages 139–162. Kluwer Academic Publ., 2000.

[16] A. Ben-Tal, L. El Ghaoui, and A. Nemirovski. *Robust Optimization*. Princeton Univ. Press, 2009.

[17] D.P. Bertsekas. *Constrained Optimization and Lagrange Multiplier Methods*. Academic Press, New York, 1982.

[18] D.P. Bertsekas. *Nonlinear Programming*. Athena Scientific, Belmont, Mass., 1995.

[19] M.T. Bielser. *SEAL: Stochastic Extensions for Algebraic Languages*. PhD thesis, Oekonomische Abteilung, Universität Zürich, 2009.

[20] P. Billingsley. *Convergence of Probability Measures*. John Wiley & Sons, New York, second edition, 1999.

[21] P. Billingsley. *Probability and Measure*. John Wiley & Sons, 1986.

[22] J.R. Birge. The value of the stochastic solution in stochastic linear programs with fixed recourse. *Math. Prog.*, 24:314–325, 1982.

[23] J.R. Birge. Decomposition and partitioning methods for multi–stage stochastic linear programs. *Oper. Res.*, 33:989–1007, 1985.

[24] J.R. Birge, M.A.H. Dempster, H.I. Gassmann, E. Gunn, A.J. King, and S.W. Wallace. A standard input format for multiperiod stochastic linear programs. *COAL Newsletter*, 17:1–19, 1987.

[25] J.R. Birge and F.V. Louveaux. A multicut algorithm for two stage linear programs. *Eur. J. Oper. Res.*, 34:384–392, 1988.

[26] J.R. Birge and F. Louveaux. *Introduction to Stochastic Programming*. Springer Series in Operations Research. Springer, Berlin/Heidelberg, 1997.

[27] J.R. Birge and R.J.-B. Wets. Designing approximation schemes for stochastic optimization problems, in particular for stochastic programs with recourse. *Math. Prog. Study*, 27:54–102, 1986.

[28] J.R. Birge and R.J.-B. Wets. Computing bounds for stochastic programming problems by means of a generalized moment problem. *Math. Oper. Res.*, 12:149–162, 1987.

[29] M.P. Biswal, N.P. Biswal, and D. Li. Probabilistic linear programming problems with exponential random variables: A technical note. *Eur. J. Oper. Res.*, 111:589–597, 1998.

[30] E. Blum and W. Oettli. *Mathematische Optimierung*, volume XX of *Ökonometrie und Unternehmensforschung*. Springer, 1975.

[31] C. Borell. Convex set–functions in d–space. *Periodica Mathematica Hungarica*, 6:111–136, 1975.

[32] E. Boros and A. Prékopa. Closed-form two-sided bounds for probabilities that at least r and exactly r out of n events occur. *Math. Oper. Res.*, 14:317–342, 1989.

[33] H.J. Brascamp and E.H. Lieb. On extensions of the Brunn–Minkowski and Prékopa–Leindler theorems, including inequalities for log-concave functions, and with an application to the diffusion equation. *J. Funct. Anal.*, 22:366–389, 1976.

[34] A. Brooke, D. Kendrick, and A. Meeraus. *GAMS. A User's Guide, Release 2.25*. Boyd and Fraser/The Scientific Press, Danvers, MA, 1992.

[35] A. Brooke, D. Kendrick, A. Meeraus, and R. Raman. GAMS. a user's guide. Technical report, GAMS Development Corporation, Washington DC, USA, 1998. It can be downloaded from http://www.gams.com/docs.

[36] H. Brunn. *Über Ovale und Eiflächen*. PhD thesis, München, 1887. Inauguraldissertation.

[37] J. Bukszár and A. Prékopa. Probability bounds with cherry trees. *Math. Oper. Res.*, 26:174–192, 2001.

[38] V.P. Bulatov. *Approximation Methods for Solving some Extremal Problems*. PhD thesis, University of Tomsk, SSSR, 1967. In Russian.

[39] A. Burkauskas. On the convexity problem of probabilistic constrained stochastic programming problems. *Alkalmazott Matematikai Lapok (Papers in Applied Mathematics)*, 12:77–90, 1986. In Hungarian.

[40] M.R. Bussieck and A. Meeraus. General algebraic modeling system (GAMS). In Kallrath [174], chapter 8, pages 137–157.

[41] A. Charnes and W.W. Cooper. Chance–constrained programming. *Management Sci.*, 6:73–79, 1959.

[42] E.W. Cheney and A.A. Goldstein. Newton's method for convex programming and Tchebycheff approximation. *Numer. Math.*, 1:253–268, 1959.

[43] H.J. Cleef. A solution procedure for the two–stage stochatic program with simple recourse. *Z. Operations Research*, 25:1–13, 1981.

[44] P. Cogneau and G. Hubner. The 101 ways to measure portfolio performance. Technical report, SSRN Technical Report, January 2009. Available at SSRN: http://ssrn.com/abstract=1326076.

[45] M. Colombo, A. Grothey, J. Hogg, K. Wooksend, and J. Gondzio. A structure-conveying modelling language for mathematical and stochastic programming. *Math. Prog. Comp.*, 1:223–247, 2009.

[46] C. Condevaux-Lanloy and Fragnière E. SETSTOCH: A tool for multistage stochastic programming with recourse. Logilab Technical Report, Dept. of Management Studies, University of Geneva, Switzerland, August 1998.

[47] H. Cramér. *Mathematical Methods of Statistics*. Princeton Landmarks in Mathematics and Physics. Princeton Univ. Press, 1999. Nineteenth Printing.

[48] S. Dancs and B. Uhrin. On a class of integral inequalities and their measure–theoretic consequences. *J. Math. Analysis Appl.*, 74:388–400, 1980.

[49] G.B. Dantzig. Linear programming under uncertainty. *Management Sci.*, 1:197–206, 1955.

[50] G.B. Dantzig. *Linear Programming and Extensions*. Princeton University Press, Princeton, New Jersey, 1963.

[51] G.B. Dantzig. Time–staged linear programs. Technical Report SOL 80–28, Systems Optimization Laboratory, Stanford University, 1980.

[52] G. B. Dantzig and P. W. Glynn. Parallel processors for planning under uncertainty. *Ann. Oper. Res.*, 22:1–22, 1990.

[53] G.B. Dantzig and A. Madansky. On the solution of two-stage linear programs under uncertainty. In I.J. Neyman, editor, *Proc. 4th Berkeley Symp. Math. Stat. Prob.*, pages 165–176, Berkeley, 1961.

[54] S. Das Gupta. Brunn–Minkowsi inequality and its aftermath. *J. Multivariate Analysis*, 10:296–318, 1980.

[55] I. Deák. Three digit accurate multiple normal probabilities. *Numer. Math.*, 35:369–380, 1980.

[56] I. Deák. *Random Number Generators and Simulation.* Akadémiai Kiadó, Budapest, 1990.

[57] I. Deák. Linear regression estimators for multinormal distributions in optimization of stochastic progamming problems. *Eur. J. Oper. Res.*, 111:555–568, 1998.

[58] I. Deák. Subroutines for computing normal probabilities of sets–computer experiences. *Ann. Oper. Res.*, 100:103–122, 2002.

[59] I. Deák. Two-stage stochastic problems with correlated normal variables: computational experiences. *Ann. Oper. Res.*, 142:79–97, 2006.

[60] M.A.H. Dempster. On stochastic programming II: Dynamic problems under risk. *Stochastics*, 25:15–42, 1988.

[61] M.A.H. Dempster, J.E. Scott, and G.W.P. Thompson. Stochastic modelling and optimization using Stochastics^{TM}. In S.W. Wallace and W.T. Ziemba, editors, *Applications of Stochastic Programming*, Series in Optimization, chapter 9, pages 137–157. MPS and SIAM, Philadelphia, 2005.

[62] M.A.H. Dempster and R.T. Thompson. Parallelization and aggregation of nested Benders decomposition. *Ann. Oper. Res.*, 81:163–187, 1998.

[63] M.A.H. Dempster and R.T. Thompson. EVPI–based importance sampling solution procedures for multistage stochastic linear programmes on parallel MIMD architectures. *Ann. Oper. Res.*, 90:161–184, 1999.

[64] M. Densing. *Hydro-Electric Power Plant Dispatch-Planning— Multi-Stage Stochastic Programming with Time-Consistent Constraints on Risk.* PhD thesis, ETHZ, 2007. Diss. ETHZ Nr. 17244.

[65] D. Dentcheva, A. Prékopa, and A. Ruszczyński. Concavity and efficient points of discrete distributions in probabilistic programming. *Math. Prog.*, 89:55–77, 2000.

[66] D. Dentcheva and A. Ruszczyński. Optimization with stochastic dominance constraints. *SIAM J. Opt.*, 14:548–566, 2003.

[67] D. Dentcheva and A. Ruszczyński. Portfolio optimization with stochastic dominance constraints. *J. Banking & Finance*, 30:433–451, 2006.

[68] C.L. Dert. A dynamic model for asset liability management for defined benefit pension funds. In W.T Ziemba and J.M. Mulvey, editors, *Worldwide Asset and Liability Modeling*, pages 501–536. Cambridge University Press, 1998.

[69] L. Devroye. *Non–Uniform Random Variate Generation.* Springer, 1986. Out of print; available on the Internet at Luc Devroye's homepage http://www.nrbook.com/devroye/.

[70] I.I. Dikin. Iterative solutions of problems of linear and quadratic programming. *Dokl. Akad. Nauk SSSR*, 174:747–748, 1967.

[71] I.I. Dikin and C. Roos. Convergence of the dual variables for the primal affine scaling method with unit steps in the homogeneous case. *J. Opt. Theory Appl.*, 95:305–321, 1997.

[72] A. Dinghas. Über eine Klasse superadditiver Mengenfunktionale von Brunn–Minkowski–Lusternikschem Typus. *Math. Zeitschrift*, 68:111–125, 1957.

[73] J.H. Dulá. An upper bound on the expectation of simplicial functions of multivariate random variables. *Math. Prog.*, 55:69–80, 1992.

[74] J. Dupačová. Minimax stochastic programs with nonconvex nonseparable penalty functions. In A. Prékopa, editor, *Progress in Operations Research*, pages 303–316. North-Holland Publ. Co., 1976.

[75] J. Dupačová. Minimax stochastic programs with nonseparable penalties. In K. Iracki, K. Malanowski, and S. Walukiewicz, editors, *Optimization Techniques, Part I*, volume 22 of *Lecture Notes in Contr. Inf. Sci.*, pages 157–163, Berlin, 1980. Springer.

[76] J. Dupačová, G. Consigli, and S.W. Wallace. Scenarios for multistage recourse programs. *Ann. Oper. Res.*, 100:25–53, 2000.

[77] J. Dupačová, A. Gaivoronski, Z. Kos, and T. Szántai. Stochastic programming in water management: A case study and a comparison of solution techniques. *Eur. J. Oper. Res.*, 52:28–44, 1991.

[78] J. Dupačová, N. Gröwe-Kuska, and W. Römisch. Scenario reduction in stochastic programming: An approach using probability metrics. *Math. Prog.*, 95:493–511, 2003.

[79] J. Dupačová, J. Hurt, and J. Štěpán. *Stochastic Modeling in Economics and Finance*. Kluwer Academic Publ., 2002.

[80] N.C.P. Edirisinghe. Bound–based approximations in multistage stochastic programming: Nonanticipativity aggregation. *Ann. Oper. Res.*, 85:103–127, 1999.

[81] N.C.P. Edirisinghe and W.T. Ziemba. Bounds for two-stage stochastic programs with fixed recourse. *Math. Oper. Res.*, 19:292–313, 1994.

[82] N.C.P. Edirisinghe and W.T. Ziemba. Implementing bounds–based approximations in convex–concave two–stage stochastic programming. *Math. Prog.*, pages 295–325, 1996.

[83] H.P. Edmundson. Bounds on the expectation of a convex function of a random variable. Technical Report Paper 982, The RAND Corporation, 1956.

[84] A. Eichhorn and W. Römisch. Polyhedral risk measures in stochastic programming. *SIAM J. Opt.*, 16:69–95, 2005.

[85] E.J. Elton, M.J. Gruber, S.J. Brown, and W.N. Goetzmann. *Modern Portfolio Theory and Investment Analysis*. John Wiley & Sons, sixth edition, 2003.

[86] J. Elzinga and T.G. Moore. A central cutting plane method for the convex programming problem. *Math. Prog.*, 8:134–145, 1975.

[87] R. Entriken. Language constructs for modeling stochastic linear programs. *Ann. Oper. Res.*, 104:49–66, 2001.

[88] P. Erdős and A.H. Stone. On the sum of two Borel sets. *Proc. Amer. Math. Soc.*, 25:304–306, 1970.

[89] Y. Ermoliev. Stochastic quasigradient methods. In Y. M. Ermoliev and R. J.-B. Wets, editors, *Numerical Techniques for Stochastic Optimization*, pages 143–185. Springer, Berlin, 1988.

[90] Cs.I. Fábián and Z. Szőke. Solving two–stage stochastic programming problems with level decomposition. *Comp. Management Sci.*, 4:313–353, 2007.

[91] W. Feller. *An Introduction to Probability Theory and its Applications*, volume 1. John Wiley & Sons, 1968.

[92] W. Feller. *An Introduction to Probability Theory and its Applications*, volume 2. John Wiley & Sons, 1991.

[93] A.V. Fiacco and G.P. McCormick. *Nonlinear Programming: Sequential Unconstrained Minimization Techniques*. John Wiley & Sons, New York, 1968.

[94] H. Föllmer and A. Schied. Convex measures of risk and trading constraints. *Finance and Stochastics*, 6:429–447, 2002.

[95] H. Föllmer and A. Schied. *Stochastic Finance. An Introduction in Discrete Time*. Walter de Gruyter, Berlin, New York, 2002.

[96] R. Fourer, H.I. Gassmann, J. Ma, and R.K. Martin. An XML–based schema for stochastic programs. *Ann. Oper. Res.*, 166:313–337, 2009.

[97] R. Fourer, D.M. Gay, and B.W. Kernighan. *AMPL: A Modeling Language for Mathematical Programming*. Duxbury Press, Belmont CA, 2nd edition, 2003.

[98] R. Fourer and L. Lopes. A management system for decompositions in stochastic programming. *Ann. Oper. Res.*, 142:99–118, 2006.

[99] R. Fourer and L. Lopes. StAMPL: A filtration-oriented modeling tool for multistage stochastic recourse problems. *INFORMS J. Computing*, 21:242–256, 2009.

[100] E. Fragnière and J. Gondzio. Stochastic programming from modeling languages. In S.W. Wallace and W.T. Ziemba, editors, *Applications of Stochastic Programming*, Series in Optimization, chapter 7, pages 95–113. MPS and SIAM, Philadelphia, 2005.

[101] E. Fragnière, J. Gondzio, R. Sarkissian, and J.-P. Vial. Structure exploiting tool in algebraic modeling languages. *Management Sci.*, 46:1145–1158, 2000.

[102] K. Frauendorfer. Solving SLP recourse problems with arbitrary multivariate distributions—the dependent case. *Math. Oper. Res.*, 13:377–394, 1988.

[103] K. Frauendorfer. *Stochastic Two-Stage Programming*, volume 392 of *Lecture Notes in Econ. Math. Syst.* Springer, Berlin, 1992.

[104] K. Frauendorfer. Barycentric scenario trees in convex multistage stochastic programming. *Math. Prog.*, 75:277–293, 1996.

[105] K. Frauendorfer and P. Kall. A solution method for SLP recourse problems with arbitrary multivariate distributions – the independent case. *Probl. Contr. Inf. Theory*, 17:177–205, 1988.

[106] K. Frauendorfer and M. Schürle. Multistage stochastic programming: Barycentric approximation. In P.M. Pardalos and C.A. Floudas, editors, *Encyclopedia of Optimization*, volume 3, pages 577–580. Kluwer Academic Publ., 2001.

[107] K. Frauendorfer and M. Schürle. Stochastic linear programs with recourse and arbitrary multivariate distributions. In P.M. Pardalos and C.A. Floudas, editors, *Encyclopedia of Optimization*, volume 5, pages 314–319. Kluwer Academic Publ., 2001.

[108] P. Fúsek, P. Kall, J. Mayer, S. Sen, and S. Siegrist. Multistage stochastic linear programming: Aggregation, approximation, and some open problems. Technical report, Inst. Oper. Res., University of Zurich, 2000. Presented at the conference on Stochastic Optimization, Gainesville, FL (Feb. 2000).

[109] A. Gaivoronski. Stochastic quasigradient methods and their implementation. In Y. M. Ermoliev and R. J.-B. Wets, editors, *Numerical Techniques for Stochastic Optimization*, pages 313–351. Springer, Berlin, 1988.

[110] A.A. Gaivoronski. SQG: A software for solving stochastic programming problems with stochastic quasi–gradient methods. In S.W. Wallace and W.T. Ziemba, editors, *Applications of Stochastic Programming*, Series in Optimization, chapter 4, pages 37–60. MPS and SIAM, Philadelphia, 2005.

[111] J. Galambos and I. Simonelli. *Bonferroni–Type Inequalities with Applications*. Springer, 1996.

[112] H.I. Gassmann. MSLiP: A computer code for the multistage stochastic linear programming problem. *Math. Prog.*, 47:407–423, 1990.

[113] H.I. Gassmann. Decomposition methods in stochastic linear programming: Dual variables. Technical Report WP–91–17, Dalhousie School of Business Administration, 1991.

[114] H. I. Gassmann. Modelling support for stochastic programs. *Ann. Oper. Res.*, 82:107–137, 1998.

[115] H.I. Gassmann. The SMPS format for stochastic linear programs. In S.W. Wallace and W.T. Ziemba, editors, *Applications of Stochastic Programming*, Series in Optimization, chapter 2, pages 9–19. MPS and SIAM, Philadelphia, 2005.

[116] H.I. Gassmann, I. Deák, and T. Szántai. Computing multivariate normal probabilities: A new look. *J. Comp. Graph. Stat.*, 11:920–949, 2002.

[117] H.I. Gassmann and D.M. Gay. An integrated modeling environment for stochastic programming. In S.W. Wallace and W.T. Ziemba, editors, *Applications of Stochastic Programming*, Series in Optimization, chapter 10, pages 159–175. MPS and SIAM, Philadelphia, 2005.

[118] H. I. Gassmann and A. M. Ireland. On the formulation of stochastic linear programs using algebraic modeling languages. *Ann. Oper. Res.*, 64:83–112, 1996.

[119] H.I. Gassmann and B. Kristjánsson. The SMPS format explained. *IMA J. Management Math.*, 19:347–377, 2008.

[120] H.I. Gassmann and E. Schweitzer. A comprehensive input format for stochastic linear programs. *Ann. Oper. Res.*, 104:89–125, 2001.

[121] H.I. Gassmann and S.W. Wallace. Solving linear programs with multiple right–hand–sides: Pricing and ordering schemes. *Ann. Oper. Res.*, 64:237–259, 1996.

[122] H. Gassmann and W.T. Ziemba. A tight upper bound for the expectation of a convex function of a multivariate random variable. *Math. Prog. Study*, 27:39–53, 1986.

[123] C. Geiger and C. Kanzow. *Theorie und Numerik restringierter Optimierungsaufgaben*. Springer, 2002.

[124] A. Genz. Numerical computation of rectangular bivariate and trivariate normal and t probabilities. *Statistics and Computing*, 14:151–160, 2004.

[125] A. Genz and F. Bretz. Comparison of methods for the computation of multivariate t–probabilities. *J. Comp. Graph. Stat.*, 11:950–971, 2002.

[126] K. Glashoff and S.A. Gustafson. *Einführung in die Lineare Optimierung*. Wissenschaftliche Buchgesellschaft, Darmstadt, 1978.

[127] G.H. Golub and C.F. Van Loan. *Matrix Computations*. The John Hopkins Univ. Press, second edition, 1990.

[128] M.A. Goberna and M.A. López. *Linear Semi–Infinite Optimization*. John Wiley & Sons, 1998.

[129] J. Gondzio. HOPDM (version 2.12) – A fast LP solver based on a primal–dual interior point method. *Eur. J. Oper. Res.*, 85:221–225, 1995.

[130] N. Gröwe. Estimated stochastic programs with chance constraints. *Eur. J. Oper. Res.*, 101:285–305, 1997.

[131] P.R. Halmos. *Measure Theory*. D. van Nostrand, Princeton, New Jersey, 1950.

[132] G.H. Hardy, J.E. Littlewood, and G. Pólya. *Inequalities*. Cambridge University Press, 1988. 2nd ed., reprinted 1988.

[133] H. Heitsch and W. Römisch. Scenario reduction algorithms in stochastic programming. *Comp. Opt. Appl.*, 24:187–206, 2003.

[134] D. den Hertog. *Interior-Point Approach to Linear, Quadratic and Convex Programming: Algorithms and Complexity*. Kluwer Academic Publishers, 1994.

[135] J.L. Higle, W.W. Lowe, and R. Odio. Conditional stochastic dcomposition: An algorithmic interface for optimization and simulation. *Oper. Res.*, 42:311–322,1994.

[136] J.L. Higle and S. Sen. Stochastic decomposition: An algorithm for two stage linear programs with recourse. *Math. Oper. Res.*, 16:650–669, 1991.

[137] J.L. Higle and S. Sen. Finite master programs in regularized stochastic decomposition. *Math. Prog.*, 67:143–168, 1994.

[138] J.L. Higle and S. Sen. Statistical approximations for recourse constrained stochastic programs. *Ann. Oper. Res.*, 56:157–175, 1995.

[139] J.L. Higle and S. Sen. *Stochastic Decomposition. A Statistical Method for Large Scale Stochastic Linear Programming*. Kluwer Academic Publ., 1996.

[140] J.L. Higle and S. Sen. Duality and statistical tests of optimality for two stage stochastic programs. *Math. Prog.*, 75:257–275, 1996.

[141] R.A. Horn and C.R. Johnson. *Matrix Analysis*. Cambridge University Press, 1996.

[142] K. Høyland, M. Kaut, and S.W. Wallace. A heuristic for moment–matching scenario generation. *Comp. Opt. Appl.*, 24:169–185, 2003.

[143] C.C. Huang, I. Vertinsky, and W.T. Ziemba. Sharp bounds on the value of perfect information. *Oper. Res.*, 25:128–139, 1977.

[144] C.C. Huang, W.T. Ziemba, and A. Ben-Tal. Bounds on the expectation of a convex function of a random variable: With applications to stochastic programming. *Oper. Res.*, 25:315–325, 1977.

[145] D. Hunter. An upper bound for the probability of a union. *J. Appl. Prob.*, 13:597–603, 1976.

[146] G. Infanger. Monte Carlo (importance) sampling within a Benders decomposition algorithm for stochastic linear programs. *Ann. Oper. Res.*, 39:69–95, 1992.

[147] G. Infanger. *Planning under Uncertainty: Solving Large–Scale Stochastic Linear Programs*. Boyd & Fraser Publ. Co., Danvers, MA, 1994.

[148] J.L. Jensen. Sur les fonctions convexes et les inégalités entre les valeurs moyennes. *Acta Math.*, 30:173–177, 1906.

[149] N.L. Johnson and S. Kotz. *Distributions in Statistics: Continuous Multivariate Distributions*. John Wiley & Sons, 1972.

[150] N.L. Johnson, S. Kotz, and N. Balakrishnan. *Continuous Univariate Distributions, Volume 1*. John Wiley & Sons, second edition, 1992.

[151] P.H. Jorion. *Value at Risk: A New Benchmark for Measuring Derivatives Risk*. Irwin Professional Publ., 1996.

[152] P. Kall. Qualitative Aussagen zu einigen Problemen der stochastischen Programmierung. *Z. Wahrscheinlichkeitstheorie u. verwandte Gebiete*, 6:246–272, 1966.

[153] P. Kall. Approximations to stochastic programs with complete fixed recourse. *Numer. Math.*, 22:333–339, 1974.

[154] P. Kall. *Stochastic Linear Programming*. Springer, 1976.

[155] P. Kall. Approximation to optimization problems: An elementary review. *Math. Oper. Res.*, 11:9–18, 1986.

[156] P. Kall. On approximations and stability in stochastic programming. In J. Guddat, H. Th. Jongen, B. Kummer, and F. Nožička, editors, *Parametric Optimization and Related Topics*, pages 387–407. Akademie-Verlag, Berlin, 1987.

[157] P. Kall. Stochastic programs with recourse: An upper bound and the related moment problem. *Z. Operations Research*, 31:A119–A141, 1987.

[158] P. Kall. Stochastic programming with recourse: Upper bounds and moment problems—A review. In J. Guddat, B. Bank, H. Hollatz, P. Kall, D. Klatte, B. Kummer, K. Lommatzsch, K. Tammer, M. Vlach, and K. Zimmermann, editors, *Advances in Mathematical Optimization (Dedicated to Prof. Dr. Dr. hc. F. Nožička)*, pages 86–103. Akademie-Verlag, Berlin, 1988.

[159] P. Kall. An upper bound for SLP using first and total second moments. *Ann. Oper. Res.*, 30:267–276, 1991.

[160] P. Kall. Bounds for and approximations to stochastic linear programs–Tutorial. In K. Marti and P. Kall, editors, *Stochastic programming methods and technical applications*, pages 1–21. Springer, 1998.

[161] P. Kall and J. Mayer. SLP-IOR: A model management system for stochastic linear programming, system design. In A.J.M. Beulens and H.-J. Sebastian, editors, *Optimization–Based Computer–Aided Modelling and Design*, Lecture Notes in Control and Information Sciences 174, pages 139–157. Springer, 1992.

[162] P. Kall and J. Mayer. Computer support for modeling in stochastic linear programming. In K. Marti and P. Kall, editors, *Stochastic Programming: Numerical Techniques and Engineering Applications*, Lecture Notes in Economics and Math. Systems 423, pages 54–70. Springer, 1995.

[163] P. Kall and J. Mayer. SLP-IOR: An interactive model management system for stochastic linear programs. In A. King, editor, *Approximation and Computation in Stochastic Programming*, volume 75, pages 221–240. Math. Prog. B, 1996.

[164] P. Kall and J. Mayer. On testing SLP codes with SLP–IOR. In F. Giannessi, T. Rapcsák, and S. Komlósi, editors, *New Trends in Mathematical Programming*, pages 115–135. Kluwer Academic Publ., 1998.

[165] P. Kall and J. Mayer. On solving stochastic linear programming problems. In K. Marti and P. Kall, editors, *Stochastic Programming Methods and Technical Applications*, Lecture Notes in Economics and Math. Systems 458, pages 329–344. Springer, 1998.

[166] P. Kall and J. Mayer. On the role of bounds in stochastic linear programming. *Optimization*, 47:287–301, 2000.

[167] P. Kall and J. Mayer. Building and solving stochastic linear programming models with SLP-IOR. In S.W. Wallace and W.T. Ziemba, editors, *Applications of Stochastic Programming*, Series in Optimization, chapter 6, pages 79–93. MPS and SIAM, Philadelphia, 2005.

[168] P. Kall and J. Mayer. Some insights into the solution algorithms for SLP problems. *Ann. Oper. Res.*, 142:147–164, 2006.

[169] P. Kall and J. Mayer. Modeling support for multistage recourse problems. In K. Marti, Y. Ermoliev, and G. Pflug, editors, *Dynamic stochastic optimization*, Lecture Notes in Economics and Math. Systems 532, pages 21–41. Springer, 2004.

[170] P. Kall and W. Oettli. Measurability theorems for stochastic extremals. *SIAM J. Contr. Opt.*, 13:994–998, 1975.

[171] P. Kall and D. Stoyan. Solving stochastic programming problems with recourse including error bounds. *Math. Operationsforsch. Statist., Ser. Opt.*, 13:431–447, 1982.

[172] P. Kall and S.W. Wallace. *Stochastic Programming*. John Wiley & Sons, Chichester, 1994. Out of stock; see http://www.stoprog.org.

[173] J.G. Kallberg and W.T. Ziemba. Generalized concave functions in stochastic programming and portfolio theory. In S. Schaible and W.T. Ziemba, editors, *Generalized Concavity in Optimization and Economics*, pages 719–767. Academic Press, 1981.

[174] J. Kallrath, editor. *Modeling Languages in Mathematical Optimization*, volume 88 of *Applied Optimization*. Kluwer Academic Publ., 2004.

[175] Y.M. Kaniovski, A.J. King, and R.J-B Wets. Probabilistic bounds (via large deviations) for the solutions of stochastic programming problems. *Ann. Oper. Res.*, pages 189–208, 1995.

[176] S. Karlin and W.J. Studden. *Tschebycheff Systems: With Applications in Analysis and Statistics*. Interscience Publ., New York, 1966.

[177] N. Karmarkar. A new polynomial-time algorithm for linear programming. *Combinatorica*, 4:373–395, 1984.

[178] S. Kataoka. A stochastic programming model. *Econometrica*, 31:181–196, 1963.

[179] M. Kaut and S.W. Wallace. Evaluation of scenario–generation methods for stochastic programming. *Pacific J. of Optimization*, 3:257–271, 2007.

[180] J.E. Kelley. The cutting plane method for solving convex programs. *SIAM J. Appl. Math.*, 8:703–712, 1960.

[181] J.M.B. Kemperman. The general moment problem, a geometric approach. *Ann. Math. Statist.*, 39:93–122, 1968.

[182] A.I. Kibzun and Y.S. Kan. *Stochastic Programming Problems with Probability and Quantile Functions*. John Wiley & Sons, 1996.

[183] A.J. King and R.T. Rockafellar. Asymptotic theory for solutions in statistical estimation and stochastic programming. *Math. Oper. Res.*, 18:148–162, 1993.

[184] A.J. King and R.J.-B. Wets. Epi–consistency of convex stochastic programs. *Stochastics*, 34:83–92, 1991.

[185] A.J. King, S.E. Wright, G.R. Parija, and R. Entriken. The IBM stochatic programming system. In S.W. Wallace and W.T. Ziemba, editors, *Applications of Stochastic Programming*, Series in Optimization, chapter 3, pages 21–36. MPS and SIAM, Philadelphia, 2005.

[186] M. Kijima, , and M. Ohnisi. Mean–risk analysis of risk aversion and wealth effects on optimal portfolios with multiple investment opprtunities. *Ann. Oper. Res.*, 1993.

[187] K. Kleibohm. *Ein Verfahren zur approximativen Lösung von konvexen Programmen.* PhD thesis, Universität Zürich, 1966. Short description in C.R. Acad. Sci. Paris 261:306–307 (1965).

[188] W.K. Klein Haneveld. *Duality in Stochastic Linear and Dynamic Programming*. Springer, 1986.

[189] W.K. Klein Haneveld, L. Stougie, and M.H. van der Vlerk. On the convex hull of the simple integer recourse objective function. *Ann. Oper. Res.*, 56:209–224, 1995.

[190] W.K. Klein Haneveld, M.H. Streutker, and M.H. van der Vlerk. An ALM model for pension funds using integrated chance constraints. *Ann. Oper. Res.*, 177:47–62, 2010.

[191] W.K. Klein Haneveld and M.H. van der Vlerk. Integrated chance constraints: reduced forms and an algorithm. *Comp. Management Sci.*, 3:245–269, 2006.

[192] D.E. Knuth. *The Art of Computer Programming; Volumes 1–3*. Addison-Wesley Publ. Co., third edition, 1997.

[193] É. Komáromi. A dual method for probabilistic constrained problems. *Math. Prog. Study*, 28:94–112, 1986.

[194] H. Konno and H. Yamazaki. Mean–absolute deviation portfolio optimization model and its application to tokyo stock market. *Management Sci.*, 37:519–531, 1991.

[195] M. Kopa, editor. *On selected software for stochastic programming*. MATFYZPRESS, Publishing House of the Faculty of Mathematics and Physics, Charles University in Prague, 2008.

[196] M.G. Krein and A.A. Nudel'man. *The Markov Moment Problem and Extremal Problems*, volume 50 of *Transactions of Math. Mon.* AMS, 1977.

[197] Alexandra Künzi-Bay. *Mehrperiodige ALM-Modelle mit CVaR-Minimierung für Schweizer Pensionskassen.* PhD thesis, Oekonomische Abteilung, Universität Zürich, 2007.

[198] A. Künzi-Bay and J. Mayer. Computational aspects of minimizing conditional value-at-risk. *Comp. Management Sci.*, 3:3–27, 2006.

[199] Y.-J. Kuo and H.D. Mittelmann. Interior point methods for second–order cone programming and OR applications. *Comp. Opt. Appl.*, 28:255–285, 2004.

[200] T. Kuosmanen. Performance measurement and best-practice benchmarking of mutual funds: combining stochastic dominance criteria with data envelopment analysis. *J. Productivity Analysis*, 28:71–86, 2007.

[201] S. M. Kwerel. Most stringent bounds on aggregated probabilities of partially specified dependent probability systems. *J. American Stat. Ass.*, 70:472–479, 1975.

[202] N. Larsen, H. Mausser, and S. Uryasev. Algorithms for optimization of value–at–risk. In P. Pardalos and V.K. Tsitsiringos, editors, *Financial Engineering, e-Commerce and Supply Chain*, pages 129–157. Kluwer Academic Publ., 2002.

[203] L. Leindler. On a certain converse of Hölder's inequality ii. *Acta. Sci. Math. (Szeged)*, 33:217–223, 1972.

[204] J.T. Linderoth and S. J. Wright. Decomposition algorithms for stochastic programming on a computational grid. *Comp. Opt. Appl.*, 24:207–250, 2003.

[205] J. Linderoth, A. Shapiro, and S. Wright. The empirical behavior of sampling methods for stochastic programming. *Ann. Oper. Res.*, 142:219–245, 2006.

[206] M.S. Lobo, L. Vandenberghe, S. Boyd, and H. Lebret. Applications of second–order cone programming. *Linear Algebra and its Appl.*, 284:193–228, 1998.

[207] J. Luedtke and S. Ahmed. A sample approximation approach for optimization with probabilistic constraints. *SIAM J. Opt.*, 19:674–699, 2008.

[208] J. Luedtke, S. Ahmed, and G.L. Nemhauser. An integer programming approach for linear programs with probabilistic constraints. *Math. Prog.*, 122:247–272, 2010.

[209] D.G. Luenberger. *Introduction to Linear and Nonlinear Programming*. Addison-Wesley, Reading, Massachusetts, 1973.

[210] A. Madansky. Bounds on the expectation of a convex function of a multivariate random variable. *Ann. Math. Statist.*, 30:743–746, 1959.

[211] A. Madansky. Inequalities for stochastic linear programming problems. *Management Sci.*, 6:197–204, 1960.

[212] A. Madansky. Methods of solutions of linear programs under uncertainty. *Oper. Res.*, 10:463–470, 1962.

[213] A. Madansky. Linear programming under uncertainty. In R.L. Graves and P. Wolfe, editors, *Recent Advances in Mathematical Programming*, pages 103–110. McGraw–Hill, 1963.

[214] W-K. Mak, D.P. Morton, and R.K. Wood. Monte Carlo bounding techniques for determining solution quality in stochastic programs. *Oper. Res. Lett.*, 24:47–56, 1999.

[215] R. Mansini, W. Ogryczak, and M.G. Speranza. Conditional value at risk and related linear programming models for portfolio optimization. In H. Vladimirou, editor, *Financial Optimization*, volume 152 of *Ann. Oper. Res.*, pages 227–256. Springer, 2007.

[216] K.V. Mardia, J.T. Kent, and J.M. Bibby. *Multivariate Analysis*. Academic Press, 1979.

[217] H. Markowitz. *Portfolio Selection. Efficient Diversification of Investments*. John Wiley & Sons, 1959.

[218] I. Maros. A general Phase–I method in linear programming. *Eur. J. Oper. Res.*, 23:64–77, 1986.

[219] I. Maros. *Computational Techniques of the Simplex Method*. Kluwer Academic Publ., Boston/Dordrecht/London, 2003.

[220] I. Maros and G. Mitra. Strategies for creating advanced bases for large-scale linear programming problems. *INFORMS J. Computing*, 10(2):248–260, Spring 1998.

[221] I. Maros and Cs. Mészáros. The role of the augmented system in interior point methods. *Eur. J. Oper. Res.*, 107(3):720–736, 1998.

[222] K. Marti. Konvexitätsaussagen zum linearen stochastischen Optimierungsproblem. *Z. Wahrscheinlichkeitstheorie u. verwandte Gebiete*, 18:159–166, 1971.

[223] K. Marti. *Descent Directions and Efficient Solutions in Discretely Distributed Stochastic Programs*. Springer, 1988.

[224] K. Marti. Optimal engineering design by means of stochastic optimization methods. In J. Blachut and H.A. Eschenauer, editors, *Emerging Methods for Multidisciplinary Optimization*, pages 107–158. Springer, 2001.

[225] K. Marti. *Stochastic Optimization Methods*. Springer, 2nd edition, 2008.

[226] K. Marti and E. Fuchs. Computation of descent directions and efficient points in stochastic optimization problems without using derivatives. *Math. Prog. Study*, 28:132–156, 1986.

[227] K. Marti and E. Fuchs. Rates of convergence of semi-stochastic approximation procedures for solving stochastic optimization problems. *Optimization*, 17:243–265, 1986.

[228] J. Mayer. Probabilistic constrained programming: A reduced gradient algorithm implemented on PC. Working Paper WP-88-39, IIASA, 1988.

[229] J. Mayer. Computational techniques for probabilistic constrained optimization problems. In K. Marti, editor, *Stochastic Optimization: Numerical Methods and Technical Applications*, pages 141–164. Springer, 1992.

[230] J. Mayer. *Stochastic Linear Programming Algorithms: A Comparison Based on a Model Management System.* Gordon and Breach Science Publishers, 1998. (Habilitationsschrift, Wirtschaftswiss. Fakultät, Universität Zürich, 1996).

[231] J. Mayer. On the numerical solution of jointly chance constrained problems. In S. Uryasev, editor, *Probabilistic Constrained Optimization: Methodology and Applications*, pages 220–233. Kluwer Academic Publ., 2000.

[232] G.P. McCormick. *Nonlinear Programming. Theory, Algorithms, and Applications.* John Wiley & Sons, 1983.

[233] Cs. Mészáros. The augmented system variant of IPMs in two–stage stochastic linear programming computation. *Eur. J. Oper. Res.*, 101:317–327, 1997.

[234] B. L. Miller and H. M. Wagner. Chance constrained programming with joint constraints. *Oper. Res.*, 13:930–945, 1965.

[235] H. Minkowski. *Geometrie der Zahlen.* Teubner, Leipzig und Berlin, 1896.

[236] H.D. Mittelmann. An independent benchmarking of SDP and SOCP solvers. *Math. Prog.*, 95:407–430, 2003.

[237] A. Müller and D. Stoyan. *Comparison Methods for Stochastic Models and Risks.* John Wiley & Sons, 2002.

[238] G.L. Nemhauser and W.B. Widhelm. A modified linear program for columnar methods in mathematical programming. *Oper. Res.*, 19:1051–1060, 1971.

[239] B. A. Murtagh and M. A. Saunders. Large scale linearly constrained optimization. *Math. Prog.*, 14:41–72, 1978.

[240] B. A. Murtagh and M. A. Saunders. MINOS 5.4. User's Guide. Technical report sol 83-20r, Department of Operations Research, Stanford University, 1995.

[241] G.L. Nemhauser and L.A. Wolsey. *Integer and Combinatorial Optimization.* John Wiley & Sons, 1990.

[242] Yu. Nesterov and A. Nemirovsky. *Interior Point Polynomial Methods in Convex Programming.* SIAM, Philadelphia, 1994.

[243] J. Nocedal and S.J. Wright. *Numerical Optimization.* Springer, 1999.

[244] V.I. Norkin, G.Ch. Pflug, and A. Ruszczyński. A branch and bound method for stochastic global optimization. *Math. Prog.*, 83:425–450, 1998.

[245] V.I. Norkin and N.V. Roenko. α–convave functions and measures and their applications. *Kibernet. Sistem. Analiz.*, 189:77–88, 1991. In Russian, English translation in Cybernetics and Sytems Analysis 27(1991) 860–869.

[246] N. Noyan. Two-stage stochastic programming involving CVaR with an application to disaster management. *Optimization-Online*, pages 1–28, March 2010.

[247] P. Olsen. When is a multistage stochastic programming problem well defined? *SIAM J. Contr. Opt.*, 14:518–527, 1976.

[248] B.K. Pagnoncelli, S. Ahmed, and A. Shapiro. Sample Average Approximation method for chance constrained programming: theory and applications. *J. Opt. Theory Appl.*, 142:399–416, 2009.
[249] J. Palmquist, S. Uryasev, and P. Krokhmal. Portfolio optimization with conditional Value–at–Risk objective and constraints. *The Journal of Risk*, 4(2), 2002.
[250] J. Peng, C. Roos, and T. Terlaky. *Self–Regularity. A new Paradigm for Primal–Dual Interior–Point Methods*. Princeton Univ. Press, 2002.
[251] T. Pennanen. Epi-convergent discretizations of multistage stochastic programs. *Math. Oper. Res.*, 30:245–256, 2005.
[252] T. Pennanen. Epi-convergent discretizations of multistage stochastic programs via integration quadratures. *Math. Prog.*, 116:461–479, 2009.
[253] J. Pfanzagl. Convexity and conditional expectations. *Ann. Probability*, 2:490–494, 1974.
[254] G. Ch. Pflug. Some remarks on the Value–at–Risk and the Conditional Value–at–Risk. In S.P. Uryasev, editor, *Probabilistic Constrained Optimization, Methodology and Applications*, pages 272–281. Kluwer Academic Publ., 2000.
[255] G. Ch. Pflug. Scenario tree generation for multiperiod financial optimization by optimal discretization. *Math. Prog.*, 89:251–271, 2001.
[256] J. Pintér. Global optimization: Software, test problems, and applications. In P. M. Pardalos and H. E. Romeijn, editors, *Handbook of Global Optimization, Volume 2*, pages 515–569. Kluwer Academic Publ., 2002.
[257] A. Prékopa. On probabilistic constrained programming. In H. W. Kuhn, editor, *Proc. Princeton Symp. on Math. Programming*, pages 113–138. Princeton Univ. Press, 1970.
[258] A. Prékopa. Logarithmic concave measures with applications to stochastic programming. *Acta. Sci. Math. (Szeged)*, 32:301–316, 1971.
[259] A. Prékopa. On logarithmic concave measures and functions. *Acta. Sci. Math. (Szeged)*, 34:335–343, 1973.
[260] A. Prékopa. Contributions to stochastic programming. *Math. Prog.*, 4:202–221, 1973.
[261] A. Prékopa. Programming under probabilistic constraints with a random technology matrix. *Math. Oper. Res.*, 5:109–116, 1974.
[262] A. Prékopa. Boole-bonferroni inequalities and linear programming. *Oper. Res.*, 36:145–162, 1988.
[263] A. Prékopa. Numerical solution of probabilistic constrained programming problems. In Y. Ermoliev and R.J-B. Wets, editors, *Numerical Techniques for Stochastic Optimization*, pages 123–139. Springer, 1988.
[264] A. Prékopa. Sharp bounds on probabilities using linear programming. *Oper. Res.*, 38:227–239, 1990.
[265] A. Prékopa. Dual method for the solution of a one–stage stochastic programming problem with random RHS obeying a discrete probability distribution. *ZOR–Methods and Models of Operations Research*, 34:441–461, 1990.
[266] A. Prékopa. *Stochastic Programming*. Kluwer Academic Publ., 1995.

[267] A. Prékopa. The use of discrete moment bounds in probabilistic constrained stochastic programming models. *Ann. Oper. Res.*, 85:21–38, 1999.

[268] A. Prékopa. Probabilistic programming. In A. Ruszczyński and A. Shapiro, editors, *Stochastic Programming*, volume 10 of *Handbooks in Operations Research and Management Science*, pages 267–351. Elsevier, Amsterdam, 2003.

[269] A. Prékopa, S. Ganczer, I. Deák, and K. Patyi. The STABIL stochastic programming model and its experimental application to the electricity production in Hungary. In M. A. H. Dempster, editor, *Stochastic Programming*, pages 369–385. Academic Press, 1980.

[270] A. Prékopa and T. Szántai. Flood control reservoir system design using stochastic programming. *Math. Prog. Study*, 9:138–151, 1978.

[271] A. Prékopa, B. Vizvári, and T Badics. Programming under probabilistic constraint with discrete random variable. In F. Giannessi, T. Rapcsák, and S. Komlósi, editors, *New Trends in Mathematical Programming*, pages 235–255. Kluwer Academic Publ., 1998.

[272] G.B. Price. *Multivariable Analysis*. Springer, 1984.

[273] L. Qi. A alternating method for stochastic linear programming with simple recourse. *Math. Prog. Study*, 27:183–190, 1986.

[274] W.M. Raike. Dissection methods for solution in chance constrained programming problems under discrete distributions. *Management Sci.*, 16:708–715, 1970.

[275] T. Rapcsák. *On the numerical solution of a reservoir model.* PhD thesis, University of Debrecen, 1974. In Hungarian.

[276] H. Richter. Parameterfreie Abschätzung und Realisierung von Erwartungswerten. *Blätter Dt. Ges. Versicherungsmath.*, 3:147–161, 1957.

[277] Y. Rinott. On convexity of measures. *Ann. Probability*, 4:1020–1026, 1976.

[278] B.D. Ripley. *Stochastic Simulation.* John Wiley & Sons, 1987.

[279] S.M. Robinson and R.J.-B. Wets. Stability in two stage programming. *SIAM J. Contr. Opt.*, 25:1409–1416, 1987.

[280] R.T. Rockafellar. Measurable dependence of convex sets and functions on parameters. *J. Math. Anal. Appl.*, 28:4–25, 1969.

[281] T.R. Rockafellar. *Convex Analysis.* Princeton University Press, 1970.

[282] T.R. Rockafellar and S.P. Uryasev. Optimization of Conditional Value–at–Risk. *Journal of Risk*, 2:21–41, 2000.

[283] T.R. Rockafellar and S.P. Uryasev. Conditional Value–at–Risk for general loss distributions. *J. Banking & Finance*, 26:1443–1471, 2002.

[284] T.R. Rockafellar, S.P. Uryasev, and M. Zabarankin. Deviation measures in risk analysis and optimization. Research Report 2002–7, ISE Dept., University of Florida, 2002.

[285] R.T. Rockafellar and R.J.-B. Wets. Stochastic convex programming: Relatively complete recourse and induced feasibility. *SIAM J. Contr. Opt.*, 14:574–589, 1976.

[286] R.T. Rockafellar and R.J.-B. Wets. Nonanticipativity and $\mathscr{L}^1$-martingales in stochastic optimization problems. *Math. Prog. Study*, 6:170–186, 1976.

[287] R.T. Rockafellar and R.J.-B. Wets. Scenarios and policy aggregation in optimization under uncertainty. *Math. Oper. Res.*, 16:119–147, 1991.

[288] W. Römisch and R. Schultz. Multistage stochastic integer programs: An introduction. In M. Grötschel, S.O. Krumke, and J. Rambau, editors, *Online Optimization of Large Scale Systems*, pages 579–598. Springer, 2001.

[289] W.W. Rogosinski. Moments of nonnegative mass. *Proc. Roy. Soc. London*, A 245:1–27, 1958.

[290] T. Roos, T. Terlaky, and J.-Ph. Vial. *Theory and Algorithms for Linear Optimization—An Interior Point Approach—*. John Wiley & Sons, 1997.

[291] S.M. Ross. *Simulation.* Academic Press, third edition, 2002.

[292] A.D. Roy. Safety–first and the holding of assets. *Econometrica*, 20:434–449, 1952.

[293] A. Ruszczyński. A regularized decomposition method for minimizing a sum of polyhedral functions. *Math. Prog.*, 35:309–333, 1986.

[294] A. Ruszczyński. Probabilistic programming with discrete distributions and precedence constrained knapsack polyhedra. *Math. Prog.*, 93:195–215, 2002.

[295] A. Ruszczyński and A. Shapiro, editors. *Stochastic Programming*, volume 10 of *Handbooks in Operations Research and Management Science*. Elsevier, Amsterdam, 2003.

[296] A. Ruszczyński and A. Świętanowski. Accelerating the regularized decomposition method for two stage stochastic linear programming problems. *Eur. J. Oper. Res.*, 101:328–342, 1997.

[297] S. Schaible. Fractional programming. In R. Horst and P.M. Pardalos, editors, *Handbook of Global Optimization*, pages 495–608. Kluwer Academic Publ., 1995.

[298] S. Schaible. Fractional programming. I. Duality. *Management Sci.*, 22:858–867, 1976.

[299] H. Schramm and J. Zowe. A version of the bundle idea for minimizing a nonsmooth function: Conceptual idea, convergence analysis, numerical results. *SIAM J. Opt.*, pages 121–152, 1992.

[300] S. Sen. Relaxations for probabilistically constrained programs with discrete random variables. *Oper. Res. Lett.*, 11:81–86, 1992.

[301] S. Sen. Algorithms for stochastic mixed-integer programming models. In K. Aardal, G.L. Nemhauser, and R. Weismantel, editors, *Handbook on Discrete Optimization*, pages 515–558. North-Holland Publ. Co., 2005.

[302] S. Sen, Z. Zhou, and K. Huang. Enhancements of two–stage stochastic decomposition. *Computers and Op. Res.*, 36:2434–2439, 2009.

[303] D. Sengupta and A.K. Nanda. Log–concave and concave distributions in reliability. *Naval. Res. Logist. Quart.*, 46:419–433, 1999.

[304] Y. Seppälä and T. Orpana. Experimental study on the efficiency and accuracy of a chance–constrained programming algorithm. *Eur. J. Oper. Res.*, 16:345–357, 1984.

[305] A. Shapiro. Inference of statistical bounds for multistage stochastic programming problems. *Math. Meth. Oper. Res.*, 58:57–68, 2003.

[306] A. Shapiro, D. Dentcheva, and A. Ruszczyński. *Lectures on Stochastic Programming—Modeling and Theory*. MPS-SIAM Series on Optimization. SIAM and MPS, Philadelphia, 2009.

[307] A. Shapiro and T. Homem–de–Mello. A simulation–based approach to two–stage stochastic programming with recourse. *Math. Prog.*, 81:301–325, 1998.

[308] A. Shapiro and T. Homem–de–Mello. On the rate of convergence of optimal solutions of Monte Carlo approximations of stochastic programs. *SIAM J. Opt.*, 11:70–86, 2000.

[309] J.A. Shapiro, W.B. Powell, and D. Bernstein. A flexible Java representation for uncertainty in online operations–research models. *INFORMS J. Computing*, 13:29–55, 2001.

[310] N. Z. Shor. *Nondifferentiable Optimization and Polynomial Problems*. Kluwer Academic Publ., 1998.

[311] N. Shor, T. Bardadym, N. Zhurbenko, A. Likhovid, and P. Stetsyuk. The use of nonsmooth optimization methods in stochastic programming problems. *Kibernetika i Sistemniy Analiz*, 5:33–47, 1999. (In Russian); English translation: Cybernetics and System Analysis, vol. 35, No. 5, pp. 708–720, 1999.

[312] Gy. Sonnevend. An analytical centre for polyhedrons and new classes for linear (smooth, convex) programming. In A. Prékopa et al., editors, *System Modelling and Optimization*, pages 866–878. Springer, 1986.

[313] D. Stoyan. *Comparison Methods for Queues and other Stochastic Models*. John Wiley & Sons, New York, 1983. Revised English Edition by D.J. Daley of *Qualitative Eigenschaften und Abschätzungen stochastischer Modelle*, Oldenbourg-Verlag, München, 1977.

[314] S.V. Stoyanov, S.T. Rachev, and F.J Fabozzi. Optimal financial portfolios. *Applied Math. Finance*, 14:401–436, 2007.

[315] B. Strazicky. Some results concerning an algorithm for the discrete recourse problem. In M.A.H. Dempster, editor, *Stochastic Programming*, pages 263–271. Academic Press, 1980.

[316] J. F. Sturm. Implementation of interior point methods for mixed semidefinite and second order cone optimization problems. *Optimization Methods and Software*, 17:1105–1154, 2002. Special 10th anniversary issue.

[317] T. Szántai. An algorithm for calculating values and gradient vectors for the multivariate normal distribution function. *Alkalmazott Matematikai Lapok*, 2:27–39, 1976. In Hungarian.

[318] T. Szántai. Evaluation of a special multivariate gamma distribution function. *Math. Prog. Study*, 27:1–16, 1986.

[319] T. Szántai. Calculation of the multivariate probability distribution function values and their gradient vectors. Working Paper WP-87-82, IIASA, 1987.

[320] T. Szántai. A computer code for solution of probabilistic-constrained stochastic programming problems. In Y. M. Ermoliev and R. J.-B. Wets, editors, *Numerical Techniques for Stochastic Optimization*, pages 229–235. Springer, Berlin, 1988.

[321] T. Szántai. Improved bounds and simulation procedures on the value of the multivariate normal probability distribution function. *Ann. Oper. Res.*, 100:85–101, 2000.

[322] T. Szántai and J. Bukszár. Probability bounds given by hypercherry trees. *Opt. Meth. Softw.*, 17:409–422, 2002.

[323] S. Talluri, R. Narasimhan, and A. Nair. Vendor performance with supply risk: A chance-constrained DEA approach. *Int. J. Production Economics*, 100:212–222, 2006.

[324] E. Tamm. On g–concave functions and probability measures. *Eesti NSV Teaduste Akademia Toimetised, Füüsika– Matemaatika (News of the Estonian Academy of Sciences, Math.–Phys.)*, 26:376–379, 1977. In Russian.

[325] J. Thénié, Ch. van Delft, and J.-Ph. Vial. Automatic formulation of stochastic programs via an algebraic modeling language. *Comp. Management Sci.*, 4:17–40, 2007.

[326] G. Tintner. Stochastic linear programming with applications to agricultural economics. In H.A. Antosiewicz, editor, *Proc. 2nd Symp. Linear Programming*, volume 2, pages 197–228, Washington D.C., 1955. National Bureau of Standards.

[327] K.C. Toh, M.J. Todd, and R.H. Tütüncü. SDPT3 — a Matlab software package for semidefinite programming. *Optimization Methods and Software*, 11:545–581, 1999.

[328] Y.L. Tong. *The Multivariate Normal Distribution*. Springer, 1990.

[329] D.M. Topkis and A.F. Veinott. On the convergence of some feasible direction algorithms for nonlinear programming. *SIAM J. Contr. Opt.*, 5:268–279, 1967.

[330] V.V. Uchaikin and V.M. Zolotarev. *Chance and Stability. Stable Distributions and their Applications*. VSP, Utrecht, The Netherlands, 1999.

[331] S.P. Uryasev. Introduction to the theory of probabilistic functions and percentiles (Value–at–Risk). In S.P. Uryasev, editor, *Probabilistic Constrained Optimization: Methodology and Applications*, pages 1–25. Kluwer Academic Publ., 2000.

[332] P. Valente, G. Mitra, and C.A. Poojari. A stochastic programming integrated environment (SPInE). In S.W. Wallace and W.T. Ziemba, editors, *Applications of Stochastic Programming*, Series in Optimization, chapter 8, pages 115–136. MPS and SIAM, Philadelphia, 2005.

[333] C. van de Panne and W. Popp. Minimum cost cattle feed under probabilistic problem constraint. *Management Sci.*, 9:405–430, 1963.

[334] M.H. van der Vlerk. On multiple simple recourse models. *Math. Meth. Oper. Res.*, 62:225–242, 2005.

[335] M.H. van der Vlerk. Convex approximations for complete integer recourse models. *Math. Prog.*, A 99:297–310, 2004.

[336] R. Van Slyke and R. J-B. Wets. *L*-shaped linear programs with applications to optimal control and stochastic linear programs. *SIAM J. Appl. Math.*, 17:638–663, 1969.

[337] R.J. Vanderbei. *Linear Programming: Foundations and Extensions.* Kluwer Academic Publ., Boston, 1996.

[338] A.F. Veinott. The supporting hyperplane method for unimodal programming. *Oper. Res.*, 15:147–152, 1967.

[339] S. von Bergen. *Stochastische "Data Envelopment Analysis"-Modelle — Anwendung verschiedener Risikofunktionen in DEA-Modellen.* PhD thesis, Oekonomische Abteilung, Universität Zürich, 2009.

[340] D.W. Walkup and R.J.B. Wets. Stochastic programs with recourse. *SIAM J. Appl. Math.*, 15:1299–1314, 1967.

[341] A. Weintraub and J. Vera. A cutting plane approach for chance constrained linear programs. *Oper. Res.*, 39:776–785, 1991.

[342] R. Wets. Programming under uncertainty: The complete problem. *Z. Wahrscheinlichkeitstheorie u. verwandte Gebiete*, 4:316–339, 1966.

[343] R. Wets. Stochastic programming: Solution techniques and approximation schemes. In A. Bachem, M. Grötschel, and B. Korte, editors, *Mathematical Programming: The State-of-the-Art, Bonn 1982*, pages 566–603. Springer-Verlag, Berlin, 1983.

[344] R.J-B. Wets. Solving stochastic programs with simple recourse. *Stochastics*, 10:219–242, 1983.

[345] J.R. Wittrock. Advances in nested decomposition algorithm for solving staircase linear programs. Technical Report SOL 83–2, Systems Optimization Laboratory, Stanford University, 1983.

[346] J.R. Wittrock. Dual nested decomposition of staircase linear programs. *Math. Prog. Study*, 24:65–86, 1985.

[347] S.E. Wright. Primal-dual aggregation and disaggregation for stochastic linear programs. *Math. Oper. Res.*, 19:893–908, 1994.

[348] Stephen J. Wright. *Primal-Dual Interior-Point Methods.* SIAM, Philadelphia, 1997.

[349] D.S. Yakowitz. An exact penalty algorithm for recourse–constrained stochastic linear programs. *Appl. Math. Comp.*, 49:39–62, 1992.

[350] D.S. Yakowitz. A regularized stochastic decomposition algorithm for two–stage stochastic linear programs. *Comp. Opt. Appl.*, 3:59–81, 1994.

[351] Y. Ye. *Interior Point Algorithms—Theory and Analysis—.* John Wiley & Sons, 1997.

[352] M.R. Young. A minimax portfolio selection rule with linear programming solution. *Management Sci.*, 44:673–683, 1998.

[353] A.C. Zaanen. *Linear Analysis—Measure and Integral, Banach and Hilbert Space, Linear Integral Equations*, volume II of *Bibliotheca Mathematica*. North-Holland Publ. Co., Amsterdam, 1964.

[354] G. Zakeri, A.B. Philpott, and D.M. Ryan. Inexact cuts in Benders decomposition. *SIAM J. Opt.*, 10:643–657, 2000.

[355] G. Zoutendijk. *Methods of Feasible Directions.* Elsevier, Amsterdam/D. Van Nostrand, Princeton, New Jersey, 1960.

[356] G. Zoutendijk. Nonlinear programming: A numerical survey. *SIAM J. Contr. Opt.*, 4:194–210, 1966.

[357] A.I. Zukhovitskii, P.A. Poljak, and M.E. Primak. Two methods for determining equilibrium points for concave n– person games. *Doklady Akademii Nauk SSSR*, 185:24–27, 1969. In Russian.

[358] A.I. Zukhovitskii and M.E. Primak. On the convergence of the method of Tchebychev–centers and the central cutting plane method for solving convex programming problems. *Doklady Akademii Nauk SSSR*, 222:273–276, 1975. In Russian.

Exercises: Hints for answers

Section 1.1 (page 6)

1.1 In the standard formulation (1.3) (page 2) the problems read as

(a)
$$\begin{aligned} &\min d^{\mathrm{T}}y^+ - d^{\mathrm{T}}y^- \\ &Ax - y^+ + y^- - z = b \\ &x, y^+, y^-, z \geq 0 \end{aligned}$$

(b)
$$\begin{aligned} &\max -f^{\mathrm{T}}z - g^{\mathrm{T}}y^+ + g^{\mathrm{T}}y^- \\ &-Az + By^+ - By^- + s = d \\ &-Cz = e \\ &z, y^+, y^-, s \geq 0 \end{aligned}$$

1.2 The results to be derived are:

(a) The solution is $(\hat{x}, \hat{y}) = (1, 2)$ with the optimal value $\gamma = 7$.
The expected supply shortage of the second product amounts to
$\mathbb{E}[(\xi - \hat{x} - \hat{y})^+] = \frac{1}{2}\int_3^4 (\xi - 3)d\xi = \frac{1}{4}.$

(b) Here the second contraint—with the distribution function F_ξ of ξ—reads as $\mathbb{P}(x + y \geq \xi) = F_\xi(x + y) \geq 0.95$ and hence as $x + y \geq 3.9$. Together with the first constraint the solution can be found (e.g. graphically) as $(\tilde{x}, \tilde{y}) = (0.1, 3.8)$ with the minimal value $\gamma = 7.9$.

(c) Now for the feasible supply shortage it is required that
$\mathbb{E}[(\xi - x - y)^+] \leq 0.05 \cdot \overline{\xi} = 0.15 < 0.25$, i.e. it has to be less than that resulting in case (a). It follows that $x + y > \hat{x} + \hat{y}$ has to hold. Assuming that $x + y < 4$, for the expected shortage follows

$$\mathbb{E}[(\xi - x - y)^+] = \frac{1}{2}\int_{x+y}^4 (\xi - x - y)d\xi = \frac{(x+y)^2}{4} - 2(x+y) + 4,$$

which is to be bounded above by $0.05 \cdot \overline{\xi} = 0.15$. Hence this model reads as

$$\begin{aligned} \min\ & 3x+2y \\ \text{s.t.}\ & 2x+y \ge 4 \\ & (x+y)^2-8(x+y)+15.4 \le 0 \\ & x,y \ge 0, \end{aligned}$$

which—as to be seen later—is solved at $(\hat{x},\hat{y}) = (\sqrt{0.6},\ 4-2\cdot\sqrt{0.6})$ with the optimal value $\gamma = 8-\sqrt{0.6}$.

Section 1.2, part 1 (page 22)

1.3 Let $\hat{x}$ be a feasible solution of $\{Ax=b, x\ge 0\}$ and $I(\hat{x}) = \{j \mid \hat{x}_j > 0\}$. If the columns $\{A_j,\ j\in I(\hat{x})\}$ are linearly independent, then $\hat{x}$ is basic.
Otherwise, the homogeneous system $\sum\limits_{j\in I(\hat{x})} \xi_j A_j = 0$ has a nontrivial solution ξ with at least one $\xi_j > 0$ (and $\xi_j = 0\ \forall j \notin I(\hat{x})$). Then, for $\bar{\mu} := \max\{\mu \mid \hat{x}-\mu\xi \ge 0\}$ follows that $\bar{x} := \hat{x}-\bar{\mu}\xi$ is feasible and $|I(\bar{x})| \le |I(\hat{x})|-1$ (with $|I(x)|$ the cardinality of $I(x)$). Now either the reduced feasible solution $\bar{x}$ is basic, or this reduction can be repeated. Obviously, this process has to end after finitely many reductions with a basic solution.

1.4 Let $\tilde{x}$ be a solution of $\tilde{\gamma} := \min\{c^{\mathrm{T}}x \mid Ax=b, x\ge 0\}$. If $\tilde{x}$ is not basic, then the homogeneous system $\sum\limits_{j\in I(\tilde{x})} \xi_j A_j = 0$ has a nontrivial solution ξ with at least one $\xi_j > 0$ (and $\xi_j = 0\ \forall j \notin I(\tilde{x})$). Furthermore, $c^{\mathrm{T}}\xi = 0$ has to hold due to the optimality of $\tilde{x}$. Hence, $z := \tilde{x}-\tilde{\mu}\xi$ for $\tilde{\mu} := \max\{\mu \mid \tilde{x}-\mu\xi \ge 0\}$ is optimal again, i.e. $c^{\mathrm{T}}z = \tilde{\gamma}$, and $|I(z)| \le |I(\tilde{x})|-1$. If z is not basic, this reduction may be repeated finitely often, at most.

1.5 For the dual pair (PP) and (DP) follows immediately:

(a) $(x_1 = 4, x_2 = 0)$ is feasible in (PP), whereas the dual constraints

$$\begin{aligned} -u_1 - 3u_2 &\le -10 \\ 2u_1 + 9u_2 &\le -2 \end{aligned}$$

obviously cannot be satisfied with $u_i \ge 0, i=1,2$;

(b) (DP) is unsolvable due to its infeasibility, whereas (PP) is feasible but unsolvable as well (with $\gamma = -\infty$); see Prop. 1.13. and Prop. 1.7..

1.6 The correct answers are

(DP) / (PP)	*FS*	*FU*	*NF*
FS	**YES**	**NO**	**NO**
FU	**NO**	**NO**	**YES**
NF	**NO**	**YES**	**YES**

Here the **"YES"** follows for the case FS/FS from Prop. 1.11., for the case FU/NF from Prop. 1.9., and for the case NF/FU from Prop. 1.9. as well, whereas for the case NF/NF examples may be constructed, using the Farkas lemma Prop. 1.13. for instance.

1.7 With the primal solution $(\hat{x}_1, \hat{x}_2) = (0.5, 1.5)$, from the complementarity conditions follows the dual solution $(\hat{u}_1, \hat{u}_2, \hat{u}_3) = (0.5, 0.5, 0.0)$.

1.8 To verify your solution, use SLP-IOR and any LP solver to confirm the correct result $(\hat{x}_1, \hat{x}_2) = (\frac{5}{3}, \frac{2}{3})$.

Section 1.2, part 2 (page 53)

1.9 The claimed facts follow immediately since

(a) the feasible sets of the successive master programs are monotonically decreasing, implying that the minima are monotonically increasing;

(b) by the general assumption it holds that $\{u \mid W^{\mathrm{T}}u \leq q\} \neq \emptyset$, and from the additional assumption $\{y \mid Wy = \zeta, y \geq 0\} \neq \emptyset\; \forall \zeta$ then follows the solvability of $\min\{q^{\mathrm{T}}y \mid Wy = \zeta, y \geq 0\}\; \forall \zeta$ due to Prop. 1.10..

1.10 The following answers are straight forward:

(a) The interior-point condition is satisfied for instance with $x_j = \frac{1}{3}\; \forall j$ and $s_1 = s_2 = \frac{1}{3}$, $s_3 = \frac{4}{3}$.

(b) From the dual constraints follows $s_1(\lambda) = s_2(\lambda) = -1 - u(\lambda)$ and $s_3(\lambda) = -u(\lambda)$, where $u(\lambda) < -1$ has to hold due to the requirement $s_j(\lambda) > 0\; \forall j$. From the above equations for $s_j(\lambda)$, the conditions $(x_j(\lambda) \cdot s_j(\lambda) = \lambda)$, and the primal constraint $\sum_{j=1}^{3} x_j(\lambda) = 1$ then follows that $\tilde{u} := u(\lambda)$ has to satisfy the quadratic equation $\tilde{u}^2 + (1 + 3\lambda)\tilde{u} + \lambda = 0$; due to $u(\lambda) < -1$ the only possible solution is $\tilde{u} = -\frac{1+3\lambda}{2} - \sqrt{(\frac{1+3\lambda}{2})^2 - \lambda}$, from which $s_j(\lambda)$ follow immediatly observing the above equations, which then imply the $x_j(\lambda)$ due to the central path conditions.

1.11 Hint: Show that the Jacobian $\mathscr{J}(x(\mu), u(\mu), s(\mu))$ is a regular matrix for all $\mu > 0$; then the assertion follows from the implicit function theorem.

1.12 Due to the lower triangularity of L the Cholesky factorization of D, and hence the solution of the system $L \cdot L^{\mathrm{T}} = D$ for L, can be performed immediately.
With $d = (7, 18, 10)^{\mathrm{T}}$, solving the system $Dx = d$ by solving successively $Ly = d$ for y and then $L^{\mathrm{T}}x = y$ for x yields $y = (7, 2, 3)^{\mathrm{T}}$ and $x = (2, 1, 3)^{\mathrm{T}}$.

Section 1.3 (page 69)

1.13 For a given $\zeta_0 \in \mathbb{R}^m$ holds

$$\varphi(\zeta_0) = \min\{c^{\mathrm{T}}x \mid Ax = \zeta_0, x \geq 0\} = \max\{\zeta_0^{\mathrm{T}} u \mid Au \leq c\} = \zeta_0^{\mathrm{T}} u_0$$

for any optimal dual feasible u_0.
For any $\zeta \in \mathbb{R}^m$ then follows $\varphi(\zeta) = \max\{\zeta^{\mathrm{T}} u \mid A^{\mathrm{T}} u \leq c\} \geq \zeta^{\mathrm{T}} u_0$ and therefore

$$\varphi(\zeta) - \varphi(\zeta_0) \geq u_0^{\mathrm{T}}(\zeta - \zeta_0) \text{ and hence } u_0 \in \partial\varphi(\zeta_0).$$

1.14 The pair $(\hat{x}, \hat{u})$ satisfying the KKT conditions (1.70) for this LP have the following properties:

(*i*) feasibility of dual constraints,
(*ii*) complementarity (primal variables vs. dual constraints),
(*iii*) feasibility of primal constraints,
(*iv*) complementarity (dual variables vs. primal constraints),
(*v*) nonnegativity (primal),
(*vi*) nonnegativity (dual).

Hence, $\hat{x}$ and $\hat{u}$ solve the primal LP and its dual, respectively.

1.15 The Lagrange function of $\min\{c^{\mathrm{T}}x \mid Ax \geq 0\}$ is $L(x,u) = c^{\mathrm{T}}x + u^{\mathrm{T}}(b - Ax)$. It follows that

$$\max_{u \geq 0} L(x,u) = \begin{cases} c^{\mathrm{T}}x \text{ if } b - Ax \leq 0 \text{ (implying } u^{\mathrm{T}}(b - Ax) = 0) \\ +\infty \text{ if } b - Ax \not\leq 0 \end{cases}$$

and

$$\begin{aligned} \min_{x \in \mathbb{R}^n} L(x,u) &= \min_x\{c^{\mathrm{T}}x + u^{\mathrm{T}}(b - Ax)\} = \min_x\{(c^{\mathrm{T}} - u^{\mathrm{T}}A)x + u^{\mathrm{T}}b\} \\ &= \begin{cases} -\infty \text{ if } A^{\mathrm{T}}u - c \neq 0 \\ b^{\mathrm{T}}u \text{ if } A^{\mathrm{T}}u - c = 0 \end{cases} \end{aligned}$$

It follows that

$$\begin{aligned} L(\hat{x}, \hat{u}) &= \min_{x \in \mathbb{R}^n} \max_{u \geq 0} L(x,u) = \min_x\{c^{\mathrm{T}}x \mid b - Ax \leq 0\} \\ &= \max_{u \geq 0} \min_{x \in \mathbb{R}^n} L(x,u) = \max\{b^{\mathrm{T}}u \mid A^{\mathrm{T}}u - c = 0\} \end{aligned}$$

such that the saddle point $(\hat{x}, \hat{u})$ is a pair of primal-dual optimal solutions.

1.16 For the NLP $\min\{f(x,y) \mid g(x,y) \leq 0\}$ with $f(x,y) = x^2 + 4xy + y^2$ and $g(x,y) = x^2 + y^2 - 1$ the KKT conditions $\nabla f + u\nabla g = 0,\ u \cdot g(x,y) = 0$ yield either, if $x = 0$, that $y = u = 0$ and hence $f(x,y) = 0$, or else, if $x \neq 0$, that $u = 1$ and hence $y = -x$, $(x^2 + y^2 - 1) = 0$, yielding either $x = \sqrt{\frac{1}{2}}, y = -\sqrt{\frac{1}{2}}$ or $x = -\sqrt{\frac{1}{2}}, y = \sqrt{\frac{1}{2}}$ with the optimal value $f(x,y) = -1$. The nonoptimal KKT point $(0, 0)$ appears due to the nonconvexity of f (show: relation (1.67) (page 55) does not hold in general).

1.17 It is seen immediately, that $(\hat{x} = 0, \hat{y} = 1, \hat{z} = 0)$ is the unique solution of the NLP; however, the KKT conditions cannot be satisfied with $(\hat{x}, \hat{y}, \hat{z})$. (Hint: check the regularity condition $\underline{\mathscr{RC}\ 0}$ on page 56.)

1.18 That a solution $\hat{x}$ of (P) solves (A), should be obvious.
For the particular problem, to find the minimum of a linear function $c^{\mathrm{T}}x$ over the unit ball is also geometrically immediate and yields as the unique solution $\hat{x} = \frac{-c}{\sqrt{c^{\mathrm{T}}c}}$.
From the KKT conditions follows immediately an $\hat{u}$ such that $(\hat{x}, \hat{u})$ is a saddle point of the Lagrangian, implying strong duality by Prop. 1.28..

1.19 With $w = (x,y)$ the iteration begins with the optimal (infeasible) vertex $\hat{w}(1) = (0,0)$ and the corresponding boundary point $wb(1) \approx (3.307, 2.480)$ of the feasible set (between $\hat{w}$ and $\tilde{w}$) with the objective value $zb(1) \approx 15.71$), and hence an error estimate of $\Delta(1) \approx 15.71$, yielding after the third cycle $wb(3) \approx (3.85, 2.02)$ with $zb(3) \approx 13.96$ and an error of $\Delta(3) \approx 1.28$. The exact solution is easily seen to be $\hat{w} = (3.9, 2.01)$ with the optimal value $\hat{z} = 13.95$.

Section 2.1 (page 87)

2.1

(a) This follows immediately from the fact that the upper level sets of a concave function are convex sets.

(b) For proving this assertion employ the following inequality for concave functions:

$$f(y) - f(x) \leq \nabla^{\mathrm{T}} f(x)(y - x), \quad \forall x, y \in C.$$

Having $\nabla^{\mathrm{T}} f(x)(y-x) \leq 0$, the above inequality directly implies that $f(y) \leq f(x)$ holds.

(c) $g(x) := log[f(x)]$ is a transformation of the concave function $f(x)$ via the monotonically increasing concave function log, consequently g is concave.

2.2

$f_1(x) = e^x$: Notice that this function is strictly convex. Nevertheless, since $log f_1(x) = x$ holds, f_1 is both logconcave and logconvex. Consequently, it belongs to all of the classes of generalized concave functions, listed in the exercise.
$f_2(x) = x^3$: Since both the upper and the lower level sets are convex sets, the function is both quasi–convex and quasi-concave, that is, it is quasi–linear. It is neither

pseudo–concave nor pseudo convex, since with $\bar{x} = 0$ we have $\nabla f_2(\bar{x}) = f_2'(\bar{x}) = 0$ and the implication in Definition 2.12. on page 83 does not hold. Because of missing nonnegativity, f_2 is neither logconcave nor logconvex.
$f_3(x_1,x_2) = e^{-x_1^2-x_2^2}$: Taking logarithm shows that f_3 is logconcave, consequently it is also pseudo–concave and quasi–concave. Since the lower level sets $\{(x_1,x_2)^{\mathrm{T}} \mid f_3(x_1,x_2) \le \gamma\}$ are non–convex for $0 < \gamma < 1$, the function is not quasi–convex. Consequently, it is not pseudo–convex or log–convex, either.
$f_4(x) = \begin{cases} 1, \text{ if } x \in \mathscr{B}, \\ 0, \text{ if } x \notin \mathscr{B}. \end{cases}$: This function is logconcave. In fact, considering the defining inequality on page 84, this inequality holds trivially for f_4 if the right–hand side is zero. For $0 < \lambda < 1$, the right–hand side can only be 1 if $x, y \in \mathscr{B}$ holds. Due to the convexity of $\mathscr{B}$, $\lambda x + (1-\lambda)y \in \mathscr{B}$ follows, implying that for f_4 the left–hand side in the inequality is also 1. f_4 is clearly quasi–concave since the upper level sets are convex. The lower level set corresponding e.g., to level $\gamma = \frac{1}{2}$ is clearly nonconvex, thus f_4 is not quasi–convex and consequently it is not logconvex either. Since f_4 is not differentiable, it is neither pseudo–convex nor pseudo–concave, according to our definition.

2.3
Apply Hölder's inequality (see, e.g., Hardy et al. [132]):

$$a_1 b_1 + a_2 b_2 \le \left(a_1^p + a_2^p\right)^{\frac{1}{p}} \left(b_1^q + b_2^q\right)^{\frac{1}{q}},$$

with $a_1, a_2, b_1, b_2 > 0$, $\frac{1}{p} + \frac{1}{q} = 1$, $p, q > 1$.
For $0 < \lambda < 1$ choose $p = \frac{1}{\lambda}$, $q = \frac{1}{1-\lambda}$, $a_1 = f(x)^\lambda$, $a_2 = g(x)^\lambda$, $b_1 = f(y)^{1-\lambda}$, $b_2 = g(y)^{1-\lambda}$ to arrive at

$$f(x)^\lambda f(y)^{1-\lambda} + g(x)^\lambda g(y)^{1-\lambda} \le [f(x) + g(x)]^\lambda \, [f(y) + g(y)]^{1-\lambda}.$$

Utilizing at the left–hand–side the logconvexity of f and of g completes the proof.

Section 2.2 (page 136)

2.4 With D^c denoting the complement of event D, we have

$$\mathbb{P}((A \cap B)^c) = \mathbb{P}(A^c \cup B^c) \le \mathbb{P}(A^c) + \mathbb{P}(B^c) = 0,$$

from which the proposition immediately follows.

2.5
(a) The equivalent LP is (cf. Section 2.2.3):

$$\left.\begin{array}{lrl} \min & 2x_1 + x_2 & \\ \text{s.t.} & x_1 & \geq 1 \\ & x_1 + x_2 & \geq 1.9, \end{array}\right\}$$

where 1.9 on the right–hand side is the 0.9–quantile of the distribution of ξ, obtained by solving $F(x) = 0.9$ with F being the distribution function of ξ:

$$F(x) = \begin{cases} 0, & \text{if } x < 1 \\ x-1, & \text{if } 1 \leq x \leq 2 \\ 1, & \text{if } x > 2. \end{cases}$$

(b) The optimal solution is $x_1^* = 1$, $x_2^* = 0.9$.

2.6
Let $x, y \in \mathbb{R}^n$, $0 \leq \lambda \leq 1$. Utilizing the triangle inequality and the positive homogeneity of the norm we get:

$$\begin{aligned} h(\lambda x + (1-\lambda)y) &= \|D^{\mathrm{T}}(\lambda x + (1-\lambda)y) - d\| = \|\lambda(D^{\mathrm{T}}x - d) + (1-\lambda)(D^{\mathrm{T}}y - d)\| \\ &\leq \|\lambda(D^{\mathrm{T}}x - d)\| + \|(1-\lambda)(D^{\mathrm{T}}y - d)\| \\ &= \lambda\|D^{\mathrm{T}}x - d\| + (1-\lambda)\|D^{\mathrm{T}}y - d\| = \lambda h(x) + (1-\lambda)h(y). \end{aligned}$$

2.7

(a) The solution of (S) is $x_S^* = (1.82, 1.00)$ with optimal objective value $z_S^* = 4.65$, whereas the solution of (J) is $x_J^* = (1.97, 1.00)$ with optimal objective value $z_S^* = 4.95$.

(b) The following inequality

$$\mathbb{P}\left(t_i^{\mathrm{T}} x \geq \xi_i,\ i = 1, \ldots, r\right) \ \leq\ \mathbb{P}(t_k^{\mathrm{T}} x \geq \xi_k)$$

holds for $k = 1, \ldots, r$, since the event that all random inequalities hold simultaneously implies the event that a particular selected inequality holds. Consequently, the feasible domain of (J) is a subset of the feasible domain of (S). This implies that the inequality $z_S^* \leq z_J^*$ must hold.

Section 2.4 (page 158)

2.8 The equivalent LP formulation for (2.127) on page 147 is

$$\left.\begin{array}{rrl} \min c^{\mathrm{T}}x & & \\ \text{s.t.}\;\; \alpha\bar{t}x + (1-2\alpha)\sum_{k=1}^{N} p_k y^k & \leq & \alpha\bar{h} \\ t^k x \qquad\qquad\qquad - y^k & \leq & h^k,\; k=1,\ldots,N \\ y^k & \geq & 0,\; k=1,\ldots,N \\ x & \in & \mathscr{B}; \end{array}\right\}$$

the LP–reformulation of (2.128) can be done analogously.

2.9

(a) The solution of (ICC) is $x^*_{ICC} = (2.6, 0.4)$ with optimal objective value $z^*_{ICC} = 3.4$; the solution of $(CVaR)$ is $x^*_{CVaR} = (1.8, 1.2)$ with $z^*_{CVaR} = 4.2$. Thus we have $z^*_{ICC} < z^*_{CVaR}$.

(b) The phenomenon is best to understand by considering the case when the distribution function of (ξ_1, ξ_2) is continuous. With positive values of ϑ representing losses and for α high enough we have:

$$\rho^{+}_{\text{sic}}(\vartheta) = \mathbb{E}[\vartheta^{+}] \;\leq\; \mathbb{E}[\vartheta \mid \vartheta \geq \nu(\vartheta,\alpha)] = \rho^{\alpha}_{\text{CVaR}}(\vartheta),$$

where $\nu(\vartheta,\alpha)$ is the Value–at–Risk corresponding to ϑ and α (cf. Sections 2.4.1 and 2.4.3). This implies that the feasible domain of $(CVaR)$ is contained in the feasible domain of (ICC) thus implying $z^*_{ICC} < z^*_{CVaR}$ for a minimization problem.
For our case with a discrete distribution this is a heuristic reasoning, of course, which can be made precise by utilizing the upper–tail distribution function, see Pflug [254] and Rockafellar and Uryasev [283].

Section 2.5 (page 171)

2.10 With $0 \leq \lambda \leq 1$ we have

$$\begin{aligned} A(\lambda x+(1-\lambda)y) &= \mathbb{E}[|\eta^{\mathrm{T}}(\lambda x+(1-\lambda)y)-\xi|] \\ &= \mathbb{E}[|\lambda(\eta^{\mathrm{T}}x-\xi)+(1-\lambda)(\eta^{\mathrm{T}}y-\xi)|] \\ &\leq \mathbb{E}[\lambda|\eta^{\mathrm{T}}x-\xi|+(1-\lambda)|\eta^{\mathrm{T}}y-\xi|] \\ &= \lambda\mathbb{E}[|\eta^{\mathrm{T}}x-\xi|]+(1-\lambda)\mathbb{E}[|\eta^{\mathrm{T}}y-\xi|] \\ &= \lambda A(x)+(1-\lambda)A(y). \end{aligned}$$

2.11 We sketch a proof of the equivalence of (2.189) and (2.191); for the other pair of problems the proof is analogous.
Let x be a feasible solution of (2.189) and take $y_k := (t^k x - h^k)^-$, $\forall k$. Then

$(x, y_1, \ldots, y_N)$ is a feasible solution of (2.191) and the corresponding objective function values are equal.
Conversely, let $(x, y_1, \ldots, y_N)$ be feasible in (2.191). Implied by the observation outlined in the paragraph next to (2.192), x is a feasible solution of (2.189) and the objective function values in both problems are again equal.
Thus we were able to associate to any feasible solution of (2.189) a feasible solution of (2.192) having the same objective function value, and vice versa. Consequently the two problems are equivalent.

Section 2.7 (page 188)

2.12

(a) For any set $\mathscr{B} \subset \mathbb{R}$ we have

$$\max_{x \in \mathscr{B}} \min_{1 \le i \le N} (\hat{\eta}^i)^{\mathrm{T}} x = -\min_{x \in \mathscr{B}} \max_{1 \le i \le N} -(\hat{\eta}^i)^{\mathrm{T}} x.$$

Since the negative portfolio return $-(\hat{\eta}^i)^{\mathrm{T}} x$ is the loss, this explains the model formulation.

(b) An equivalent LP formulation is:

$$\left.\begin{array}{lrl} \max\limits_{x,z} & & z \\ & (\hat{\eta}^i)^{\mathrm{T}} x - z & \ge 0,\ i = 1, \ldots, N \\ & r^{\mathrm{T}} x & \ge \mu_p \\ & \mathbb{1}^{\mathrm{T}} x & = 1 \\ & x & \ge 0. \end{array}\right\}$$

For showing the equivalence of the two formulations, the basic observation is the following: since z is maximized, it is sufficient to consider feasible solutions (x,z) of the above problem for which $z = \min\limits_{1 \le i \le N} (\hat{\eta}^i)^{\mathrm{T}} x$ holds.

2.13 Compute the expected asset returns r from the realizations tableau; you should get $r^{\mathrm{T}} = (r_1, r_2) = (0.0111, 0.019)$. With $\alpha = 0.99$ now set up the portfolio selection problem (cf. (2.206)) to get

$$\left.\begin{array}{ll} \min \rho_{\mathrm{CVaR}}^{0.99}(-(\eta_1 x_1 + \eta_2 x_2)) & \\ \text{s.t. } r_1 x_1 + r_2 x_2 \ge 0.018 & \\ \quad\ x_1 + x_2 = 1 & \\ \quad\ x_1, \quad x_2 \ge 0. & \end{array}\right\}$$

Solving this by employing SLP–IOR yields the optimal asset allocation $x^* = (0.13, 0.87)$ with the optimal (minimal) CVaR–value $z^* = 0.0027$.

2.14 We prove (a); the proof of (b) runs analogously.
Assume that x^* is not optimal in (2.206). Then there exists a feasible solution $\hat{x}$ of (2.206) for which

$$\rho(-\eta^{\mathrm{T}}x^*) > \rho(-\eta^{\mathrm{T}}\hat{x}) \quad \text{and} \quad r^{\mathrm{T}}x^* = \mu_p \leq r^{\mathrm{T}}\hat{x}$$

holds. Multiplying the first inequality with $-\nu$ and adding the two inequalities results in:

$$r^{\mathrm{T}}x^* - \nu\rho(-\eta^{\mathrm{T}}x^*) < r^{\mathrm{T}}\hat{x} - \nu\rho(-\eta^{\mathrm{T}}\hat{x}),$$

which contradicts the optimality of x^* in (2.208) since $\hat{x}$ is obviously feasible in (2.208).

Section 3.2, part 1 (page 224)

3.1 The conditions of Lemma 3.1 are not satisfied, and the induced constraints (resulting as $x_1 + x_2 \leq 2$) show that the problem is not even of relatively complete recourse.
The first stage solution follows (with the induced constraint) as $\hat{x} = (2, 0)^{\mathrm{T}}$.

3.3 With $\bar{\xi} = 0$ and (approximately) $\bar{\eta} = 0.686965$ the first Jensen lower bound of $\mathbb{E}[\psi(\xi, \eta)]$ amounts to $\varphi(\bar{\xi}) + \theta(\bar{\eta}) = 0.5 + 1.373929 = 1.873929$. Furthermore, it turns out that

(b) The first E–M bound amounts approximately to $ub(\bar{\psi}) = 2.6239$ yielding as first error estimate $\Delta = 2.6239 - 1.873929 \approx 0.75$.

(c) with $I_1^{(\xi)} = [-1, 0]$ and $I_2^{(\xi)} = [0, 1]$ follows
$lb_{|\bar{\xi}} = \frac{1}{2} \cdot \frac{1}{2} + \frac{1}{2} \cdot 1 + 1.373929 = 2.123929$ and hence an increasing lower bound, whereas with $I_1^{(\eta)} = [0, 0.686965]$ and $I_2^{(\eta)} = [0.686965, 2]$ we get
$lb_{|\bar{\eta}} = \varphi(\mathbb{E}[\xi]) + \theta(\mathbb{E}[\eta]) = 1.873929$;

(d) partitioning the η-interval yields no increase of the lower bound as compared to $\psi(\bar{\xi}, \bar{\eta})$, which is due to the linearity of $\theta(\cdot)$; consequently, also in a further partioning step an increase of the lower bound can only be expected by dividing the ξ-interval $I_1^{(\xi)}$ at $\bar{\xi}_1 = \mathbb{E}[\xi \mid I_1^{(\xi)}] = -\frac{1}{2}$ since $\varphi(\cdot)$ and $\theta(\cdot)$ are linear on $I_2^{(\xi)}$ and on the support of η, respectively. So far our best error estimate is $\Delta = 2.6239 - 2.123929 \approx 0.5$.

3.4 For (a) just use the definition of convexity of $\psi(\cdot)$ and the linearity of the function to be a majorant of $\psi(\cdot)$. Then (b) is an immediate consequence of (a), where the uniqueness follows from the unique solvability of the constraints (due to $\alpha < \beta$).

3.5 As is well known, that a quadratic function is convex if an only if it is positive definite, which is true iff the eigenvalues are nonnegative.

(a) From linear algebra it is well known that there exists a transformation T (here a 2×2-matrix T) such that $\Lambda = T^{\mathrm{T}}MT$, where Λ is the diagonal matrix with the eigenvalues of M and the columns of T are the corresponding eigenvectors. In our case it follows that $\Lambda = \begin{pmatrix} 2.1716 & 0 \\ 0 & 7.8284 \end{pmatrix}$ and for the transformation $T = \begin{pmatrix} -0.9239 & 0.3827 \\ 0.3827 & 0.9239 \end{pmatrix}$, and hence the convexity of F.

(b) For $\mu = (2;2)^{\mathrm{T}}$ the Jensen bound is $F(\mu) = -72$.

(c) The vertices of $\mathscr{B}$ are obviously given as $\{x^{(1)} = (0;0)^{\mathrm{T}},\, x^{(2)} = (0;5)^{\mathrm{T}},\, x^{(3)} = (2;4)^{\mathrm{T}},\, x^{(4)} = (8;0)^{\mathrm{T}}\}$ with $F(x^{(1)}) = 0,\ F(x^{(2)}) = -55,\ F(x^{(3)}) = -64,\ F(x^{(4)}) = 48$. Hence the LP to solve is

$$\max\left\{ \sum_{i=1}^{4} p_i \cdot F(x^{(i)}) \;\middle|\; \sum_{i=1}^{4} p_i \cdot x^{(i)} = \mu,\ \sum_{i=1}^{4} p_i = 1,\ p_i \geq 0\ \forall i \right\}.$$

Results: The LP optimal value (the upper bound) is -10 with the optimal probabilities of the vertices $\{p_1, \cdots, p_4\} = \{0.35, 0.40, 0.00, 0.25\}$, and Jensen's lower bound amounts to -72.

3.6 The answers and results, respectively, are:

(a) The conditions for complete fixed recourse are satisfied.

(b) The first bounds are $lb = 23.0590$ and $ub = 28.2166$.

(c) For 4 subintervals (after crosswise partition) the bounds approximately are $lb = 23.7$ and $ub = 24.5$; with the default stopping tolerance of $\Delta = 10^{-5}$ (relative error), DAPPROX yields after 33 Iterations and with 383 subintervals the bounds $lb = 24.0795$; $ub = 24.0760$.

Section 3.2, part 2 (page 249)

3.7 The following consideration may serve as a heuristic argument for the different performance of the two solvers on the given model instances:

For a prescribed accuracy of the solution and 4 random variables, for a simple recourse problem the required accuracy Δ is certainly achieved if for each of the 4 components of the separable objective an accuracy of $\Delta/4$ is obtained. If this is the case for 5 subintervals per component, say, then for the overall accuracy at most 20 splits are needed. The separability and the fact, that for any component it is clear in advance whether partitioning a (sub-)interval of it improves the accuracy, is explicitely made use of in SRAPPROX.

In contrast, DAPPROX is designed to operate—in our case—on the 4-dimensional support of the given random vector and on (again 4-dimensional) subintervals of it. Although, from the (conditional) Jensen and E–M inequalities on any subinterval, it is also clear in this case if its subpartition can (and will) improve the accuracy, the difficulty is to find out the (most) "profitable" of the 4 coordinates to be chosen for subpartition; in the worst case this may need 4 subdivisions (one per coordinate) to be successful. Hence, the most unfavourable incidence could require $4^5 = 1024$ subdivisions.

The results for the given problems (3.93) are somewhat better than our worst case consideration. Also, you'll observe at the SLP-IOR output, that for both problems the number of splits/subdivisions grows much faster than the number of iterations; the reason is that due to step III. of DAPPROX (page 223)—and similar for SRAPPROX—at one iteration step it may be decided to subdivide various intervals simultaneously.

3.8 To prove (a), show first that for all (z_1,z_2,z_3) feasible in (3.94) holds $\Psi(z) \leq \Theta(z_1,z_2,z_3)$, and then determine for any of the three possibilities, $z \leq a$, $a < z \leq b$, and $z > b$, a feasible solution $(\hat{z}_1,\hat{z}_2,\hat{z}_3)$ of (3.94) such that $\Psi(z) = \Theta(\hat{z}_1,\hat{z}_2,\hat{z}_3)$.

As to (b), the uniform distribution of ξ on $[a,b]$ leads for feasible (z_1,z_2,z_3) to $\Theta(z_1,z_2,z_3) = z_3 + \frac{1}{b-a}\int_a^{a+z_2}(z_2+a-\xi)d\xi = z_3 + \frac{1}{2(b-a)}z_2^2$ and hence to convex quadratic functions $\Psi(z)$ and $\Phi(z)$. The fact, that simple recourse problems with uniformly distributed right-hand sides can be reformulated as quadratic progams, was first discovered by Beale [12].

3.9 Using for the simple recourse formulation of (3.93) the solvers SRAPPROX and DAPPROX in SLP-IOR, and for the quadratic programming formulation, according to Exercise 3.8, the solver MINOS (e.g. in GAMS), you should get for the first stage variables $\{x_1,\cdots,x_4\}$, the first stage objective value $Zf1$, and the overall objective ZF

	x_1	x_2	x_3	x_4	$Zf1$	ZF
SRAPPROX	4.25	0.14	3.53	2.56	30.16	62.64
DAPPROX	4.26	0.14	3.53	2.55	30.16	62.64
MINOS	4.25	0.14	3.53	2.55	30.17	62.66

Section 4.3 (page 312)

4.1 The equivalent formulation of (4.6) is:

$$\left.\begin{array}{lrl} \max & F(y) & \\ \text{s.t.} & Tx\ -y & \geq 0 \\ & x & \in \mathscr{B}. \end{array}\right\} \quad (E)$$

Showing the equivalence:
Let x be a feasible solution of (4.6) and choose $y = Tx$. Then (x,y) is a feasible solution of (E) and the corresponding objective function values are equal.
Conversely, let (x,y) feasible in (E). Since F is monotonically increasing in all

of its arguments and we have a maximization problem, it is sufficient to consider feasible solutions of (E) where $y = Tx$ holds. Then x is feasible in (4.6) with the same objective value.

4.2 Let $z^{\mathrm{T}} = (x, y^{\mathrm{T}})$ with $x \in \mathbb{R}$ and $y \in \mathbb{R}^{s-1}$ The positive definiteness of R implies that

$$z^{\mathrm{T}} R z = x^2 + 2x\rho^{\mathrm{T}} y + y^{\mathrm{T}} \hat{R} y > 0$$

holds for any $z \neq 0$. Choosing $x := -\rho^{\mathrm{T}} y$, we get that $y^{\mathrm{T}}(\hat{R} - \rho\rho^{\mathrm{T}})y > 0$ must hold for any $y \neq 0$. Consequently this matrix is positive definite.

4.3

(a) For generating the test problem battery choose the main-menu item *Workbench* and subsequently on the pull–down menu *GENSLP: joint chance constraint.* A pop–up menu appears where the parameters of the battery to be generated can be specified. Clicking the <*Generate*> button results in generating the battery.

(b) For performing the run with the different solvers choose again *Workbench* first and subsequently *Test Problem Batteries.* On the pop-up menu you can choose the battery, the solvers and subsequently you can start up the run by clicking the button <*run test battery*>. During this run all of the test problems in the battery will be solved in turn by all of the selected solvers. After termination the computational results can be viewed/saved by clicking <*view results*>.

See also Section 4.9.2 and for a detailed description see the User's Guide of SLP–IOR.

Section 4.4 (page 325)

4.4 The goal of the algorithm is by no means to generate *all* of the constraints of the full master problem (4.39). Since the new constraints (cuts) are generated on the basis of an optimal solution of the previous relaxed master problem (4.39), the general aim is to generate in a step–by–step manner constraints of the full master problem which are binding in the vicinity of the optimal solutions of (4.39), including finally those which are binding at an optimal solution of the full master problem.

4.5 The full master problem takes the following form:

$$\left.\begin{array}{llllll}
\min\limits_{x, w_1, \ldots, w_N (,w)} & c^{\mathrm{T}} x & & +\theta w & & \\
\text{s.t.} & d^{\mathrm{T}} x & + \sum\limits_{k=1}^{N} p_k w_k & -w & \leq 0 & \\
 & (u_k)^{\mathrm{T}}(h^k - T^k x) & -w_k & & \leq 0, & u_k \in \mathscr{U}, \ \forall k \\
 & x & & & \in \mathscr{B} &
\end{array}\right\}$$

and the relaxed master problems have to be changed accordingly. *Step 5.* of the algorithm changes also: $\nu := \nu + N$, in the multi–cut version N cuts are added in each of the iterations.

4.6 In the equivalent simple recourse model we have now $m_2 = 1$, $q^+ = 0$, $q^- = 1 - 2\alpha$ and $d = \alpha \bar{t} x$. The full master problem will be the following:

$$\left.\begin{array}{lll} \min\limits_{x(,w)} c^{\mathrm{T}}x + & & \theta w \\ \text{s.t. } \alpha \bar{t} x + (1-2\alpha) \sum\limits_{k\in\mathscr{K}} p_k(T^k x - h^k) & -w \le \alpha \bar{h}, & \mathscr{K} \subset \mathscr{N} \\ x & \in \mathscr{B}; & \end{array}\right\}$$

the relaxed master problems are to be formulated in accordance with this.

Section 4.7 (page 363)

4.7 Assume that $Q(x;T(\xi),h(\xi)) = a^{\mathrm{T}}\xi + b$ holds for $\xi \in \Xi_k$, for a fixed x. Regarding the Jensen lower bound we get (cf. page 334):

$$\begin{aligned} \int_{\Xi_k} Q(x;T(\xi),h(\xi))\,\mathbb{P}_\xi(d\xi) &= \int_{\Xi_k} (a^{\mathrm{T}}\xi + b)\,\mathbb{P}_\xi(d\xi) = \pi_k\,(a^{\mathrm{T}}\mu_k + b) \\ &= \pi_k\, Q(x;T(\mu_k),h(\mu_k)) = L_k(x). \end{aligned}$$

Concerning the Edmundson–Madansky upper bound we refer to the inequality (4.62) on page 332, formulated for a convex function φ. If this function is linear then the inequality holds obviously as equality. Taking $\varphi(\xi) = Q(x;T(\xi),h(\xi))$ and the conditional probability and expected value, given $\xi \in \Xi_k$, we get (see (4.71) on page 335):

$$\int_{\Xi_k} Q(x;T(\xi),h(\xi))\,\mathbb{P}_\xi(d\xi) = \pi_k \sum_{\nu=1}^{2^r} Q(x;T(v_k^\nu),h(v_k^\nu))\mathbb{Q}_k(v_k^\nu) = U_k(x).$$

Thus $L_k(x) = U_k(x)$ follows.

4.8

(a) The assumption has been utilized in the second inequality in (4.98) on page 358, i.e., in the updating of the previous cuts.

(b) In the SAA method a full sample is drawn first and the resulting two–stage recourse problem with a discrete distribution is solved, e.g., by utilizing regularized decomposition or a general–purpose LP solver. Based on the quality tests, this procedure might be repeated with a larger sample.
In the SD method the sampling occurs within a dual decomposition type method, at each of the iterations of the SD method a single new sample point

is drawn. The sample is steadily growing as the iterations of the algorithm proceed.

Index

U

V

W

Z